Lecture Notes in Mathematics

Volume 2295

This series reports on new developments in all areas of mathematics and their applications - quickly, informally and at a high level. Mathematical texts analysing new developments in modelling and numerical simulation are welcome. The type of material considered for publication includes:

1. Research monographs
2. Lectures on a new field or presentations of a new angle in a classical field
3. Summer schools and intensive courses on topics of current research.

Texts which are out of print but still in demand may also be considered if they fall within these categories. The timeliness of a manuscript is sometimes more important than its form, which may be preliminary or tentative.

Titles from this series are indexed by Scopus, Web of Science, Mathematical Reviews, and zbMATH.

More information about this series at http://www.springer.com/series/304

Robert Lockhart

The Theory of Near-Rings

Robert Lockhart
Abingdon-on-Thames
UK

ISSN 0075-8434 ISSN 1617-9692 (electronic)
Lecture Notes in Mathematics
ISBN 978-3-030-81754-1 ISBN 978-3-030-81755-8 (eBook)
https://doi.org/10.1007/978-3-030-81755-8

Mathematics Subject Classification: 16Y30

This Springer imprint is published by the registered company Springer Nature Switzerland AG.
The registered company address is: Gewerbestrasse 11, 6330 Cham, Switzerland

Foreword by Günter Pilz

As the name suggests, a near-ring is a generalized ring. Two axioms for rings are missing: the commutativity of addition, and one of the distributive laws.

There are plenty of natural examples for (proper) near-rings: take, for instance, a group $(M, +)$ (not necessarily abelian), and form the collection $T(M)$ of all self-maps (transformations) on M with pointwise addition $+$ of functions and their compositions $\circ$. $(T(M), +, \circ)$ becomes a near-ring, as do the collections of all continuous maps on a topological group M or the sets of all polynomials or polynomial maps on a commutative ring R with identity.

It is well-known that every ring can be viewed as a subring of the endomorphism ring $End(M)$ on an abelian group M. The author shows (with a completely different proof) that every near-ring can be embedded in some $T(M)$, so rings "describe linear maps", while near-rings treat arbitrary mappings and so near-rings can be seen as the "non-linear generalizations of rings". Near-ring theory started in 1905, when L.E. Dickson demonstrated that the second distributive law was independent of the remaining axioms for a skew-field, by constructing the first proper near-ring (actually: a "near-field"), [33]. A little later, Veblen and Wedderburn, [140], used these new structures to coordinatize geometric planes.

Any successful new theory should not simply be an exercise in "playing with the axioms" but should, after a suitable time, be sound, have interesting examples, and have connections (or even applications) to other mathematical theories. The reader will see that near-rings fulfil all these wishes.

Even the examples mentioned above indicate that near-ring theory spreads out to group theory, topology, and the theory of polynomials (and hence equations). Concerning groups: the author proves in Sect. 11.4.5 that precisely for finite simple non-abelian groups M (the "famous groups"), all maps $M \to M$ are sums of inner automorphisms.

Topological aspects are discussed in several places in this book such as Sect. 8.9.2. Applications to polynomials are touched on in Sect. 8.10 and you can find out much more on them and the solvability of algebraic equations over universal algebras in [1].

Other important applications have their origin in planar near-rings (Sect. 2.7) which deliver the so-called Balanced Incomplete Block Designs, in turn giving rise to excellent experimental designs. By their use, one can not only determine the effect of single factors (which would be easy), but also all double factors, as occur in cross sensitivities, in unbelievably few tests. From my side, I can safely say that this application saved my life [129].

Of course, it would be impossible to cover all these things in detail without expanding the book to a thousand pages, but the reader should realize the variety and depth of this subject.

The book begins with gentle introductions to topics which are needed in the sequel: foundations of group theory, homological algebra, categories, and topology.

The author's approach to near-rings is via more general structures ("stems", "pre-near-rings"). This is good, because there do seem to be connections between "generalized near-rings" and quantum mechanics (Bernd Henschenmacher, private communications—see [131]), based on old ideas of Pascual Jordan, regarding space quantization involving the rejection of the standard associative matrix algebras of more conventional mathematics.

After that, the "elementary theory" of near-rings (ideals, (sub)direct products, (semi)prime ideals, F-near-rings, distributively generated near-rings, etc.) follows. Several new concepts like "phomomorphisms" (zero-fixing maps from a group G to a group H which are homomorphisms modulo the commutator group $[H, H]$) enter the scene.

Chapter 3 deserves a special "Thanks!" to the author. He describes near-fields in detail, including the famous classification of all finite near-fields by H. Zassenhaus (without proof). The usual reference to near-fields is Wähling's book, [141], but, unfortunately, this book is written in German and quite hard to obtain. Bob Lockhart's book gives the first detailed English account on this fascinating subject. It also contains a lot of new information on near-fields, for example on the product of all non-zero elements.

Chapters 4–6 investigate the additive group G of a near-ring (one says that G "hosts" a near-ring), including questions like "which groups host near-rings with identity?" and the like. These chapters again contain a lot of new material and are really detailed (the largest finite groups considered have order 720). I know that the author has done substantial research in this field.

Transformation near-rings $T(M)$, as mentioned above, and their subnear-rings represent essentially *all* near-rings (up to isomorphism). So, it is natural to consider them in detail. The author does this in Chaps. 7 and 8 (including endomorphism and centralizer near-rings, the structures of their multiplicative semigroups and of their minimal left ideals, dense subnear-rings, and so on). These chapters contain much interesting new material such as "Stemhome near-rings" and the tribe of near-rings defined in Notation 227.

The short Chap. 9 picks up on phomomorphisms again and studies new sorts of cohomology groups. The connection to this and their utility with respect to near-rings remain unclear to me, however; but there are specific concrete examples in Chap. 10.

Chapter 11 is on near-ring modules M over R. The key definition offered is unusual, involving homomorphisms between near-rings and near-rings of phomomorphisms; but, low and behold—one can find the famous density theorem for primitive near-rings there.

Chapter 12 covers the radical theory of near-rings, and Chap. 13 gives a very short account of matrix near-rings.

The material in Chap. 14 is entirely new and concerns "F-near-rings". These are defined by the property that all elements are "distributive modulo the commutator subgroup". Every d.g. near-ring is an F-near-ring, and, remarkably, the class of F-near-rings is closed with respect to homomorphic images, and also closed with respect to direct sums, direct products, and ultraproducts. This is an interesting new concept!

The remaining Chaps. 15 and 16 investigate questions relating to extensions, liftings, and representations. If one has two near-rings N and M, one might consider group extensions of $(N, +)$ by $(M, +)$. The insights the author has provided in Chaps. 4–6 on groups hosting near-rings now really pay off. In Chap. 16, these ideas are specialized to the case when $(M, +)$ is finite and elementary abelian (hence a finite vector space) and this actually permits the use of matrix representations of the homomorphisms involved. Finite near-fields fall into this scheme, giving new insight into their structure (especially into the "exceptional" near-fields). As a side result, this even demonstrates "Wedderburn's Little Theorem" that a finite skew field is actually a field. The enormous amount of new material in these chapters is due entirely to the author.

Isotopies and other generalizations of the concept of isomorphisms are treated in the Appendix, as well as generalized associativity rules. Also, all near-rings hosted by the dihedral group D_4 are determined. Related structures like seminear-rings, near-domains, and quasi-fields are presented and their use in geometry is outlined.

How does this book differ from existing texts on near-rings? Well, the first edition of my book is 45 years old, [128]. John Meldrum's book [119] runs along similar lines but contains less than one third of the topics of this text. And the books by Clay [23] and the Ferreros [40] are more specialized, concentrating on planar near-rings and their applications.

In summary, the book records many results that were previously known only to Bob Lockhart. His excellent background in group theory has obviously helped substantially with his ideas. Some readers might be surprised by his idiosyncratic and personal style (for example, instead of the usual "we will now show..." he writes "I will now show...") but I rather like that, because that academic "we" usually means "I" anyway, reminding one of the language of kings and popes!

Linz, Austria Günter Pilz
May 2021

Preface

This is a personal account of some aspects of the modern theory of near-rings, which includes some new material. There are perhaps four standard texts on near-rings, the books by Pilz [128], Meldrum [119], Clay [23], and the Ferreros [40], and there are a number of less well-known books covering conference proceedings or specialised areas of the subject.

The classic treatises [128] and [119] were published over forty years ago and perform the important function of introducing the key ideas to a wider mathematical audience and are necessarily expository rather than thematic. The books [23] and [40] also a have an expository role but are a little more recent and devote a lot of space to research ideas pioneered by their authors, particularly involving planar near-rings.

This book also has an expository function (almost inevitable in a subject such as near-rings) but additionally treats subjects either not mentioned by the standard texts (non-abelian group cohomology, phomomorphisms, F-near-rings, product theory) or augments existing accounts (near-fields, transformation near-rings, host problems and representation theory). Some of these ideas have only previously been available as little known conference proceedings or, in the case of near-fields, in German [141].

I do not cover planar near-rings here, feeling that [23] and [40] have covered that topic extensively. About sixty percent of this book is entirely new to book form. I shall now say something about the really new material and why it is of interest.

The subject of near-rings evolved from several considerations. One of these was the early twentieth century passion for axiomatisation plus the work of Dickson, Zassenhaus and Wielandt relating the such things as Frobenius groups (giving birth to near-fields).

Another was research by Albrecht Fröhlich attempting to formulate a non-abelian group cohomology [44–48]. This work is mentioned approvingly by each of the texts I have listed above and most of them say something to the effect that it would be worthwhile for someone to a have another look at. It spawned important ideas such as d.g. near-rings and generated sufficient interest for a number of mathematicians in the United Kingdom and elsewhere to achieve interesting results mirroring the Artin-Wedderburn results in ring theory (cf. [99]). Interestingly, the German tradition associated with Helmut Wielandt did similar work (cf. [11]) and our present account of this is a happy fusion of these ideas, which I describe in my chapter on modules.

I began my own researches with a review of Fröhlich's work and that involved reading the early papers on group cohomology. I came to realise that it was possible to generalise the delta function to non-abelian situations. It became, not a homomorphism, but a phomomorphism, and using that idea I was able to obtain a simple collection of cohomology groups, as well as a set of new concepts—including a generalisation of d.g. near-rings called F-near-rings, which seem rather more simple to handle than the d.g. near-rings they generalise. These ideas are fully described in what follows.

The last two chapters of the book describe product theory. Although the papers describing this appeared in the 1990s (cf. [105, 106]) there has never been an account in book form. A simplistic view of product theory is that it is rather like Schreier's theory of group extensions. Starting with a group one obtains a group-exact sequence involving a near-ring and a left ideal. Products hosted by the group correspond to liftings of the exact sequence. Such liftings may split in two distinct ways. When they split additively the product is right distributive, when they split multiplicatively, it is associative. Near-rings correspond to the lifting, splitting both ways. If the original group was finite elementary abelian there is a very simple matrix representation of this (which of course applies to finite near-fields) and I use that to fix the unital near-rings hosted by the Klein group in just three A4 pages (try doing it with pencil and paper!).

I devote several chapters of this book to standard near-ring topics, but they don't merely regurgitate known results. For example, the largest chapter in the book is on transformation near-rings and I develop that ab initio. At least half of its material is original.

Working on this book, I continually faced non-isomorphic symmetries and structures with more general properties than my subject matter. I used one of the appendices to give a formal account of these symmetries which I believe to be original and hope to be of interest outside the domain of near-rings. I believe that this book presents a number of approachable problems which would merit further research.

Near-ring theory is a generalisation of the theory of rings. The fact that one works with groups which might not be abelian and multiplications which might not satisfy both distributivity laws augments the kinds of structures that one wishes to discuss and considerably complicates the notation one needs. I have made a conscious effort to cross-reference unfamiliar definitions and notation intensively although this may try the patience of some readers. In particular, I have opted for a numbering system that is strictly sequential, throughout.

More, I have attempted to specify any restrictions that are to apply within a section or subsection at its start, and I consciously repeat these strictures when that seems appropriate, on the basis of my personal belief that very few people read mathematics in a linear and sequential way; and, in any case, the book is not intended to be an entirely developmental story (the *Leitfaden* is more of a Midgard serpent!).

The reader may find an outline account of the history of near-rings as a subject in itself in Meldrum [119], although that account was written some forty years ago. Gerhard Betsch wrote another short history which was published in the papers for a 1993 near-ring conference in Canada [13].

I shall sometimes need to appeal to particular facts about groups or families of groups, which while not always obvious (to me!), can be regarded as well-known. I do not always give references for such facts but they can usually be found either in the standard texts that I mention—such as: [62, 92, 93, 132]—or perhaps via Internet searches—Wikipedia is often an excellent first stop for something you are unsure of.

My topic is near-rings, therefore ring theory provides a sort of ground-level motivation for at least some of the material. Anything I use can be found in Herstein, [66], Beachy [7], Cohn [27], or any other introductory book on ring theory. One certainly needs to know about Schur's lemma and primitive[1] rings in Chap. 11, which is partly about generalising those ideas.

In Chap. 3 I shall assume some basic information on the theory of finite fields. Any modern textbook should confirm these facts, [96], for instance. The same chapter contains Sect. 3.6.2, which requires some knowledge of permutation groups. It's largely standard stuff—such as *primitivity* and the various flavours of transitivity, and I list a number of readily obtained sources. I also make considerable use of rather less well-known results from the celebrated textbook by Joseph Wolf [144].

Chapter 15 probably requires more background knowledge than the other chapters. Appeals are made to: linear algebra, field theory, and group representations. I think that all that is required is available in [27] or perhaps [28].

Late in the construction of this book, I acquired a copy of [71]. Readers finding themselves generally at sea (something most mathematicians have experienced[2]) might obtain this book. It is beautifully clear and gives a wonderful background account.

I myself have little access to academic libraries and I regret inconveniencing readers in the same situation, with the over-large bibliography constituting a further offence. My families have always tolerated my incessant book purchasing and, with the exception of some of the material cited by Professor Pilz, I do own most of the listed references.

[1] The term 'primitive' is used in at least three distinct senses in this book—and that is simply the result of following standard terminology. This is regrettable but seems unavoidable.

[2] Come to think of it, I actually *was* at sea when I first got this, my wife and I did a Covid—19 bedevilled trip from Tilbury to Sydney and I bought it for the voyage.

I am simply not as conversant with recent research in near-rings as I would wish and it's probably true that many of the things that most concern me here have either not been investigated for some years or are completely new. A reader motivated to find out more about near-rings is referred to the excellent online bibliography of near-ring papers, maintained by the University of Linz, [101], or the bibliographies in the books by Pilz, Meldrum, Clay and the Ferreros, [23, 40, 119, 128].

Much of the early research in near-rings involved attempts to generalise ideas and constructions from ring theory. To an extent, this is still the case, and the apparent closeness of near-rings to rings gives some grounds for optimism. There have been real successes in some areas with this approach, notably in the radical theory of near-rings [64].

However, the seminal nature of some of these ideas is already the subject of what one might describe as *abstraction generalisations*, involving subjects such as category theory, model theory and universal algebra. So, if one is going to consider the idea of a polynomial in near-rings, say, one should at least be aware of how it is handled in universal algebra and in these other subjects.

I wanted to write this book without forever glancing at the driving mirror showing ring theory in pursuit. If I were to start the whole project again, I would hope to find space for a bigger mirror, and maybe even learn how to start the car.

Most of my near-rings will be *right* near-rings: $(a+b)\cdot c = (a\cdot c+b\cdot c)$ but there are chapters in which this is not reliably the case (Chaps. 15 and 16, for instance). I ask the reader to believe that I have reasons for this vacillation.

Much of the material in this book is original and many of my definitions are appearing in book form for the first time. I explain the background and history of some of these ideas on page 521.

Abingdon-on-Thames, UK Robert Lockhart

Notation

This is an attempt to get some notational complications out of the way. The first set of notations involves material so ubiquitous that subsequently it is often used without referring back; but first I list the Gothic symbols used.

Gothic Symbols

I find deciphering Gothic notation difficult, and thought the reader might appreciate a list of what I use. I try not to be too confusing with this (for example, I don't use $\mathfrak{A}$ and $\mathfrak{U}$ in the same context, nor $\mathfrak{I}$ and $\mathfrak{J}$—which, frankly, I can hardly tell apart myself).

I usually employ Gothic notation either when the corresponding standard letter might be confusing, or when I wish to use a special symbol for an important idea. Table 1 gives an example of each letter used, but the same letter may well be used elsewhere in this book, and with a different meaning (page xiv). I try to indicate when the symbol has special significance. 'Incidental usage' means it does not.

I do also use a fair amount of calligraphic notation, availing myself of the LaTeX mathcal facility. I find these symbols rather more intelligible so I don't list them here, but they are usually indexed unless their use is very local indeed. Some of these symbols are reserved for particular ideas.

Notation 1

(i) A^n means $A \times A \times \ldots \times A$ (n times) (where A is an additive group). That is, the usual **Cartesian** or **Direct** product. It is to be understood that I impose no finiteness conditions on the associated coordinates. However, when I refer to **direct sums** it is supposed that 'tuples' have all but finitely many coordinate values set to zero and the $\sum$ notation reliably indicates that.

In some systems I refer to *sums* of sub-structures, and use $\sum$ merely to indicate elements which are sums of a finite number of elements. With direct

Table 1 Gothic notation used

Gothic	English	Page	Special significance
$\mathfrak{A}$	'A'	Page 289	Injective endomorphism group
$\mathfrak{a}$	'a'	Page 350	Inverse image of generalised coboundaries
$\mathfrak{B}$	'B'	Page 239	Additive group of bijective mappings
$\mathfrak{C}$	'C'	Page 402	Incidental usage
$\mathfrak{c}$	'c'	Page 336	Closure operation on tuples
$\mathfrak{D}$	'D'	Page 101	Incidental usage
$\mathfrak{E}$	'E'	Page 474	Incidental usage
$\mathfrak{F}$	'F'	Page 461	Incidental usage
$\mathfrak{G}$	'G'	Page 336	Smallest sub-near-ring containing
$\mathfrak{g}$	'g'	Page 213	$\mathfrak{g}_r(G)$ indicates a direct sum of r copies of G
$\mathfrak{H}$	'H'	Page 262	Additive holomorph
$\mathfrak{I}$	'I'	Page 101	Incidental usage
$\mathfrak{i}$	'i'	Page 255	Incidental usage
$\mathfrak{J}$	'J'	Page 274	Incidental usage
$\mathfrak{K}$	'K'	Page 276	Incidental usage
$\mathfrak{k}$	'k'	Page 22	Incidental usage
$\mathfrak{L}$	'L'	Page 335	Special sub-near-ring of transformations
$\mathfrak{l}$	'l'	Page 60	Modular left ideal
$\mathfrak{M}$	'M'	Page 232	Synonym for T_0
$\mathfrak{m}$	'm'	Page 405	Maximal, left ideals, which are also modular
$\mathfrak{N}$	'N'	Page 120	The normaliser of the distributive elements
$\mathfrak{P}$	'P'	Page 6	Power set
$\mathfrak{p}$	'p'	Page 412	Prime radical
$\mathfrak{PC}$	'PC'	Page 235	Pseudo-finite topology
$\mathfrak{Ph}_S$	'Ph'	Page 296	Centraliser for phomomorphism transforms
$\mathfrak{Q}$	'Q'	Page 464	Incidental usage
$\mathfrak{R}$	'R'	Page 323	Neumann near-ring
$\mathfrak{r}$	'r'	Page 402	The radical
$\mathfrak{S}$	'S'	Page 321	Incidental usage
$\mathfrak{T}$	'T'	Page 264	Additive group of additive transforms
$\mathfrak{t}$	't'	Page 337	Topological space defined on $T(N)$
$\mathfrak{U}$	'U'	Page 101	Incidental usage
$\mathfrak{V}$	'V'	Page 286	Incidental usage
$\mathfrak{W}$	'W'	Page 286	Incidental usage
$\mathfrak{X}$	'X'	Page 335	Semi-group closure in $T_0(N)$
$\mathfrak{Y}$	'Y'	Page 335	Subgroup closure in $T_0(N)$
$\mathfrak{Z}$	'Z'	Page 122	Groups with Sylow subgroups cyclic

sums, this still applies but the elements involved in the sum are unique—that is, there is only one way to do this (see page 390).

Direct sums of finite algebraic systems are often represented using $\oplus$. Thus, $(A \oplus B)$.

(ii) Angled brackets always denote subgroup closure, usually *additive*. So $\langle A \rangle$ is the smallest subgroup containing the set A. In near-fields, (Definition 44), it can be useful to use the same notation for *multiplicative* subgroup closure. Where confusion may occur, I use suffixes: $\langle\rangle_{(add)}$ or $\langle\rangle_{(mult)}$ (Chap. 3).

(iii) $\langle A \rangle_{\mathcal{N}}^{H}$ is the normal closure of $\langle A \rangle$ in the group H (smallest normal subgroup of H containing it). I also have use for the **normaliser** of a subset (largest subgroup in which the subset is normal). Recall that the index of the normaliser gives the number of distinct conjugates, in finite cases.

I also have use for the related idea of the **normal core** of a subgroup—which is the largest normal subgroup contained in it (Page 117). This concerns me primarily in the chapters on near-fields.

(iv) Aut$(A, +)$, (sometimes Aut(A)), represents the multiplicative group of automorphisms of the additive group $(A, +)$, (see page xviii).

Characteristic subgroups of a group are simply subgroups mapping to themselves under the action of any automorphisms of the group. **Characteristic subsets** are defined similarly; and, correspondingly, one defines **fully invariant subgroups** by requiring them to be mapped to themselves under the action of group endomorphisms. Commutator subgroups are always fully invariant, and, therefore, always characteristic subgroups. Groups having no proper, non-trivial, fully invariant subgroups are called **fully invariantly simple groups**. Groups having no proper, non-trivial, characteristic subgroups are called **characteristically simple groups**.

The term 'transversal' is used to mean 'set of (distinct) coset representatives'.

(v) [A,A] represents the commutator subgroup of the additive group $(A, +)$ but the same notation is employed for multiplicative commutators, particularly in Chap. 3. There are even occasions when more than one multiplicative product is at issue—where there may some opaqueness I sometimes include the product involved in the notation, [,]^, for instance; and I also write $[\ ,\ ]_{(mult)/(add)}$.

(vi) When H and K are non-empty subsets of the group A,

$$[H, K] = \langle [h, k] \ : \ h \in H, \ k \in K \rangle \tag{1}$$

—'the commutator subgroup of H and K'.

(vii) $\delta_{i,j}$ is the standard Kronecker-delta, throughout.

(viii) I shall write $H < G$, to mean that H is an (additive) subgroup of G.

(ix) When G is some structure upon which is defined a binary addition (+), and H and K are subsets of G, the set $H + K$ is defined as

$$H + K = \{h + k \ : \ h \in H, \ k \in K\} \tag{2}$$

(x) The symbol $\mathbb{N}$ is to mean the set of positive integers, (not including zero).

(xi) Similarly, $\mathbb{Z}$ represents the set of all integers (sometimes, the ring of integers), and the symbol $\mathbb{Q}$ denotes the set of rational numbers.

(xii) If $a, b \in \mathbb{N}$, then (a, b) is the highest common factor of a and b, so $(a, b) = 1 \Leftrightarrow$ a and b are coprime.

(xiii) If $a, b \in \mathbb{N}$, then $a \mid b$ means $b = ma$ for some $m \in \mathbb{N}$ ('divides').

(xiv) If $S \subseteq \mathbb{N}$ then $\chi_S : \mathbb{N} \to \{0, 1\}$ is the characteristic function defined as

$$\chi_S(x) = \begin{cases} 1 & \text{if } x \in S \\ 0 & x \notin S \end{cases}$$

(xv) When B, and C are additive groups, I sometimes use the notation

$$\prod_{b \neq 0} C$$

to represent a direct product of copies of $(C, +)$, indexed by the non-zero elements of the group $(B, +)$. This occurs mainly in Chap. 7 or in Chap. 8. I would similarly write

$$\bigoplus_{b \neq 0} C$$

for the corresponding direct sum, distinguishing the two, as always, on the basis of whether the 'tuples' involve infinitely many non-zero elements, or not. Most of my groups will be finite and I am probably rather lax about distinguishing direct products and direct sums.

(xvi) The abbreviation *p.p.*—'presque partout' is intended to mean 'all but finitely many', it occurs only in my chapter on transformation near-rings (page 241, (7.30)). I also talk about the 'support' of functions in that chapter, meaning the elements for which they take non-zero values.

(xvii) The symbol c is used for set complements when the context is clear: S^c. I cheerfully use the usual set-theoretic notation because it is so standard. ϕ sometimes represents the empty set and where it does the meaning should be clear from the context. For example, if one is suddenly told $A \neq \phi$, where A is some set (page 438).

(xviii) A **subdirect product** is a subset of a direct product with the property that each associated projection function is surjective (see Sect. 1.7).

(xix) I occasionally use the symbol **Id** to represent an identity mapping, but more often I shall use **I**, or perhaps 1. I sometimes use $\mathbf{ID}_G$ to represent the identity of a group G.

(xx) Special arrows: epimorphic mappings are often represented by $\twoheadrightarrow$, and injections by $\rightarrowtail$.

(xxi) If I define a symbol or symbol string, such as $\mathcal{B}$, to represent an algebraic structure related to some group G, I may remove ambiguity, if there are several groups in consideration, by writing $\mathcal{B}(G)$ or $\mathcal{B}_G$ without further explanation (example, page 276).

(xxii) The **exponent** of a group is the least common multiple of the orders of its elements. I write that **exp**$(A, +)$.

Commutation Identities

I list the standard commutation identities for additive groups.

(i) $[x, y] = -[y, x]$
(ii) $[x + y, z] = [x, z]^y + [y, z]$
iii) $[x, y + z] = [x, z] + [x, y]^z$

and I use the following standard group theory notation.

Notation 2

(i) G_n often represents the nth term of the **lower central series** for the additive group $(G, +)$. Thus, $\mathbf{G_1} = \mathbf{G}$, $\mathbf{G_2} = [\mathbf{G}, \mathbf{G}], \ldots, \mathbf{G_{n+1}} = [\mathbf{G_n}, \mathbf{G}]$ etc. However, the reader is warned that I sometimes find it convenient to adopt this sort of suffix notation to represent other things (see, for example, Notation 63).
(ii) $\mathcal{Z}(\mathbf{G})$ is the additive centre of $(G, +)$. That is

$$\mathcal{Z}(\mathbf{G}) = \{g \in G : [a, g] = 0, \forall a \in G\}$$

Warning!—do not confuse this with the zeroiser ideal (Definition 55), or with the multiplicative centre (page 53). I augment the confusion in Chap. 3 because it is convenient to use $\mathcal{Z}$ for the multiplicative centre in just that chapter (page 117).
(iii) When G is a semi-direct product of the normal subgroup N by the group H, (so N and H generate the group G and have trivial intersection) I may write: $\boldsymbol{G = N \rtimes H}$. There will be the corresponding *split* exact sequence

$$N \rightarrowtail G \twoheadrightarrow H$$

This is the same idea as that of H having a normal complement. I would sometimes refer to this as a 'split extension'. My group theory notation is additive, so it would perhaps be more appropriate to refer to **semi-direct sums**, rather than semi-direct products, and I may do this, occasionally.
(iv) If g is an element of an additive group $(G, +)$ then $|g|$ nearly always refers to additive order and this will be apparent from the context. **Torsion groups** are groups in which all elements have finite order. **Torsion free** groups, such as the real numbers under addition, are groups in which no non-trivial element has finite order.
(v) When $(G, +)$ is some additive group the set End$(G, +)$ represents group homomorphisms from G to itself. This may alternatively be written as **End**(G) or as **Hom**(G, G). In the same way, **Hom**(G, H) represents the group homomorphisms from the additive group $(G, +)$ to the additive group $(H, +)$.

Class Equations and Cayley Tables
Recall that the **class equation** partitions a group into its disjoint conjugacy classes and that the number of such classes is called the **class number**. I often need to talk about the **Cayley table** of a group, or indeed of any groupoid (see Appendix A, page 478). This is just the usual multiplication table of the group, though in our case, often expressed additively. **Automorphisms** are just bijective homomorphisms from a structure to itself. In groups, these divide into **inner automorphism**—maps: $x \mapsto x^y = (-y + x + y)$ in additive notation; and **outer automorphisms**—automorphisms which are not inner! The inner automorphisms of a group themselves form a group under the usual composition, as do the collection of all automorphisms; and the inner automorphisms is a normal subgroup of that group. The corresponding quotient group is unfortunately called the **outer automorphism group**—unfortunate, because its elements are not outer automorphisms.

There is at least one situation in which I discuss an injective homomorphism from a group to itself which is not surjective, (page 289). Inner automorphisms are also defined for rings and near-rings—page 46.

The following notation and terms are used for particular kinds and classes of groups.

Notation 3

(i) The symbol $\mathbf{C_n}$ denotes the cyclic group of order n. $\mathbf{D_n}$ represents the dihedral group of degree n and order $2n$ (see page 83). It might be more modern for me to write this group as $\mathbf{D_{2n}}$, and it would certainly fit better with similar notation given below, so I suppose this is an indulgence, which I hope the reader will forgive. In Chaps. 9 and 10, I find it more appropriate to use the symbol $\mathbb{Z}_\mathbf{n}$ for C_n.

(ii) I shall sometimes overload the symbol C_n, thinking of it as the ring of integers modulo n.

(iii) $\mathbf{A_n}$ and $\mathbf{S_n}$ usually represent the familiar alternating and symmetric groups; however, I sometimes write $\mathbf{S_p}$ to mean a Sylow p-subgroup of a particular group (described page 4), particularly in the case of Sylow 2-subgroups, in Chap. 3; and I even violate this convention in (5), below.

(**a**, **b**) usually signifies the **transposition** which switches a with b—the reader will be familiar with this standard notation. When I feel I absolutely have to, I write the symmetric group on the set X as **Symm**(**X**). I think this only happens in Chap. 3.

(iv) I write $\mathbf{Q_8}$ for the quaternion group of order eight throughout. This group is recurrently important to me and I found it convenient to represents its elements in a variety of ways depending on context (see page 191). Q_8 is the smallest member of the family of **generalised quaternion groups**, which are finite non-abelian 2-groups with order 2^n $(n \geq 3)$ and presentations, (written additively)

$$\mathbf{Q}_{2^n} = \left\langle x, y \, : \, 2^{(n-1)}x = 0, \;\; 2y = 2^{(n-2)}x, \;\; (-y + x + y) = -x \right\rangle \quad (3)$$

In the literature, Q_8 is often called 'the quaternion group' to distinguish it from its larger cousins and it does have distinctive properties which they do not share—for example, that all subgroups of it are normal.

My references either to 'generalised quaternion groups' or to 'quaternion groups' are intended as encompassing all members of the family, including Q_8, which is a little non-standard. I use results from the famous book by Joseph Wolf in Chap. 3, [144] and the same difficulties occur there, being resolved with his use of the adjective 'quaternionic' in this broader sense, and the fact that he too permits Q_8 to be a generalised quaternion group.

Notice, above, that $(4y) = 0$, a relation that will fix $2y$ as -1 in the context of near-fields, in Chap. 3. Also, note that the commutator subgroup of a generalised quaternion group is cyclic, equal to $\langle 2x \rangle$, and has order $2^{(n-2)}$; whereas the centre of the group is $\langle 2y \rangle$ and has order two. The group exponent is 2^{n-1}. Elements of Q_{2^n} have the (additive) form $(ax + by)$, $0 \leq a < 2^{(n-1)}$, $0 \leq b < 1$, and all such elements, when $b = 1$, have order four.

Reference [121] is a good reference for such properties of these groups as I need, for instance; that subgroups are either cyclic or themselves quaternion; or that all such groups contain copies of Q_8; or that any two elements with order four in Q_8 are linked by a group automorphism. Reference [126] is another good reference for these groups. A theorem given there states that in quaternion groups with order larger than eight, proper, normal, non-abelian subgroups have index two. It also says the group has a characteristic cyclic subgroup with index two; the automorphism group is a two-group; and that there are exactly three distinct conjugacy classes containing elements with order four.

(v) The dicyclic group with order $4n$ ($n > 1$) is a group with presentation

$$\mathbf{Dic_n} = \langle a, x : 2na = 0, 2x = na, (-x + a + x) = -a \rangle \tag{4}$$

These groups have exponents $4n$, when n is odd, $2n$, otherwise. When $n = 2^\alpha$, $\text{Dic}_n \cong Q_{2^{(\alpha+2)}}$, so this family forms a superclass of the quaternion groups. Dicyclic groups are called 'binary dihedral groups' in some texts. Dicyclic groups have unique involutions (elements with order two). Their commutator subgroups are cyclic and have order n. There is more about them on page 208.

(vi) The alternating group A_5 occurs repeatedly here (page 210, for instance). This is a simple group with order sixty. Each of its non-trivial elements has prime order, twenty-four have order five, twenty have order three, and fifteen have order two. The group is intimately related to the structure of finite non-Dickson near-fields (Chap. 3).

(vii) Groups with presentations of the form

$$\mathbf{S_m} = \left\langle x, y : 2^{m-1}x = 2y = 0, (-y + x + y + x) = 2^{m-2}x \right\rangle \tag{5}$$

($m > 3$), are called semi-dihedral groups [55]. They have order 2^m and are, of course, nilpotent. Their commutator subgroups are cyclic with order 2^{m-2}; and their centres are cyclic with order 2 and quotients isomorphic to $D_{2^{m-2}}$.

(viii) Gorenstein [55] describes a further non-abelian p-group

$$\mathbf{M_m(p)} = \left\langle x, y : p^{m-1}x = py = 0, (-y + x + y - x) = p^{m-2}x \right\rangle \quad (6)$$

In this presentation, p is prime, and $p = 2 \Rightarrow m > 3$, while $p > 2 \Rightarrow m > 2$. This group has order p^m. Its commutator subgroup is cyclic and had order p. This family of groups is sometimes called the family of **modular p-groups** (although that family is often taken to include the p^2 (abelian) case).

(ix) **GL(m, n)** represents the **general linear group** of $m \times m$ matrices with non-zero determinant over the Galois field with order n. If K is a field I might also write this as GL(m, K); and may talk of the 'm-dimensional general linear group'. GL(m, n) has order

$$\left(\left(n^m - 1\right) . \left(n^m - n\right) . \left(n^m - n^2\right) \ldots \left(n^m - n^{(m-1)}\right) . \right) \quad (7)$$

(x) **SL(m, n)** represents the **special linear group** of $m \times m$ matrices with determinant 1 over the Galois field of order n. I am particularly interested in SL(2,3) and SL(2,5) here. The former has order 24 and is solvable but not supersolvable; the latter has order 120. It is not solvable, and it is a perfect group (defined below). **SL(2, 5)** has a single, non-trivial, proper, normal subgroup. This is its centre, which is cyclic and has quotient isomorphic to the alternating group A_5. The group has a unique element with order two, twenty elements with order three, thirty with order four, twenty-four with order five, twenty with order six, and twenty-four with order ten. SL(2, 5) has exponent sixty : SL(2, 3) is a subgroup of SL(2, 5).

In general, the order of the two-dimensional special linear group over the Galois field with order q is $q.(q^2 - 1)$, with q equal to 3 and five, respectively, in the above special cases. For odd primes, p, the centre of the group SL$(2, p)$ is cyclic with order two [132]. SL$(2, p)$ is non-solvable for all primes $p \geq 5$, [55].

(xi) A **perfect** group is one which is identical with its own commutator subgroup.

(xii) One sometimes needs to consider expressions in indeterminates describing properties of groups—**words** which may represent relations—**laws**—applying in a collection of groups. For example, $[x, y] = 0$ describes abelian groups—any substitution of group elements for x and y results in a true relation for such groups. I sometimes misuse my definition, referring to 'words' which, strictly, would be objects produced from words by substituting group elements.

If W represents a collection of words in a free group, then $\mathbf{V}(W)$ represents the **variety of groups** defined by W (see [124] or [132] for more details). When $(G, +)$ is an additive group, $\mathbf{V}(G)$ means the verbal subgroup of $(G, +)$ generated by W. This is, of course, fully invariant in $(G, +)$.

Hanna Neumann reports a wonderful theorem due to her husband, Bernard, to the effect that any word is *equivalent* to a pair of words, one of the form x^m (multiplicative notation), the other a commutator, [124]. Here, equivalence means that they are laws in exactly the same class of groups.[3]

(xiii) The **interior** of a square matrix is the square matrix obtained by deleting the first row and column. The **rank** of a matrix is the dimension of its space of rows, or its space of columns (for they are the same thing). I only apply this term to matrices over commutative rings.

(xiv) **Diagonal matrices** are (square) matrices in which only the leading diagonal contains non-zero elements. **Anti-diagonal matrices** are (square) matrices in which only the bottom left to top right diagonal contains non-zero elements. **Symmetric matrices** are square matrices which are identical to their own transposes. I denote the transpose of a square matrix **A** by $\mathbf{A^t}$. In general, $\mathbf{A_{i,j}}$ is supposed to represent the element of the matrix A lying on then $i^{th.}$ row and $j^{th.}$ column. In Chap. 7 I use $\boldsymbol{\theta}_{\mathbf{i,j}}$ to represent the matrix form of a transformation θ (page 250). Two square matrices, A and B are held to be **equivalent** provided there are invertible matrices, S and T such that $B = (S.A.T)$, [95].

(xv) Any additive group $(G, +)$ is isomorphic to a subgroup of the group S_G of permutations of the set G, in which elements of G are mapped to the translation of the set G that they induce (this is Cayley's theorem, of course). The subgroup of S_G generated by this, together with all of the additive automorphisms of $(G, +)$ is called the **Holomorph** of G (see [132], or [62]). The holomorph, **Hol(G)**, is a semi-direct product of the automorphisms by one or other of the multiplicative groups of left or right translations—either will do (see Definition 2). As is well-known, the holomorph is the normaliser of *either* of these translation groups, within S_G; and those mappings in the holomorph which fix the additive identity of the original group form the group of all possible automorphisms of that group. Thus, multiplicatively, (and using $\mathcal{T}(G)$ for the group of left translations), we have the split exact sequence

$$\mathcal{T}(G) \rightarrowtail \mathbf{Hol(G)} \twoheadrightarrow \mathrm{Aut}(G) \tag{8}$$

Although most of the groups occurring here are written additively, I shall be referring to the holomorph in contexts in which multiplicative notation is appropriate (Sect. 7.4).

[3] The Neumann's and their son, Peter, were formidable mathematicians. Peter, died in December, 2020—another victim of Covid 19.

(xvi) A **permutation group** G is simply a group for which there is a homomorphism: $G \mapsto S_\Omega$, where S_Ω is the symmetric group on the set Ω. G is usually regarded as acting directly on Ω, identifying elements with their images under the mapping. The action is **k-transitive**, provided that every pair of k-tuples of *distinct* elements of Ω, $(\alpha_1, \ldots, \alpha_k)$ and $(\beta_1, \ldots, \beta_k)$ can be connected by some element $g \in G$, as: $(\beta_j) = (\alpha_j)g$ for $j = 1, 2, \ldots, k$. When k=2, (or 3), I may speak of **doubly transitive** (or of **triply transitive**) groups. I am only concerned with finite permutation groups here. **Sharp k-transitivity** means that the element g is unique. Sharply k-transitive groups are only really interesting for the first small values of k. I am particularly concerned with sharp 1—and 2—transitivity. When k is 1 we have **fixed-point-free**, or **regular** action. So, the action is fixed-point-free, or regular provided that $(\alpha)g = \alpha \Leftrightarrow g = ID_G$. Sharp 2-transitivity occurs in my discussion of near-fields and Frobenius groups (Chap. 3, Definition 197).

If G is a permutation group then there will be a partition of Ω into subsets of the form

$$G.\gamma = \{g.\gamma \; : \; g \in G\} \tag{9}$$

which are called the **orbits** of G. Transitive permutation groups have but a single orbit. Orbits are fixed by any single element that they contain. Associated with any element ω of Ω, there is a subgroup **Fixed**(ω) of G, defined as

$$\mathbf{Fixed}(\omega) = \{g \in G \; : \; g(\omega) = \omega\} \tag{10}$$

and known as the **fixed set of** ω, (see (11.14)). The **orbit-stabiliser theorem** says that when the orbit of ω is finite, its size is the same as the index of its fixed set in G, [35]. If $\alpha \in \Omega$, the subgroup $Fixed(\alpha)$ is alternatively denoted by

$$\boldsymbol{G_\alpha} = \{g \in G \; : \; g(\alpha) = \alpha\} \tag{11}$$

We can extend this, so that

$$\mathbf{G}_{\boldsymbol{\alpha},\boldsymbol{\beta}} = \left(G_\alpha \cap G_\beta\right) = \{g \in G \; : \; g(\alpha) = \alpha \; and \; g(\beta) = \beta\} \tag{12}$$

and so on. I am a bit lax with this notation. Here, the emphasis is on the permutation group G, and Fixed(ω) is the subset of G that fixes it. In Chap. 11 the considerations can be the other way round, looking at, effectively, the subset of Ω fixed by a single element of G. I use the same notation and hope it is not too confusing.

(xvii) **Permutation matrices** are square matrices obtained by interchanging some of the rows or columns of a square identity matrix. As such, they are $(0 - 1)$ matrices—all entries are either 0's or 1's; and they are both row and column

stochastic—the sum of entries for any row or column is 1. These invertible matrices represent finite permutations.

(xviii) (Finite) elementary abelian groups are simply directs sums of cyclic groups with prime-power order (the 'Fundamental Theorem of Abelian Groups'). In particular, $V = (C_2 \oplus C_2)$ will often be referred to as the **Klein group** or as the **Klein four-group** (German—'Viergruppe').

(xix) The **centraliser** of an element of a group is the subgroup of elements that commute with it. In finite groups, the group order is the product of the order of the centraliser with the order of the class of elements conjugate to it (an application of the orbit-stabiliser theorem, [35, 56]). The centraliser of a set of elements is, or course, the set of those elements which commute with all elements in the original set. Centralisers based on particular near-ring products crop up in Chap. 7, page 298.

(xx) I do mention free groups and reduced free groups (in fact, I already have!)

Vectors

When I need vectors, I usually represent them as $1 \times n$ matrices of the form

$$\underline{v} = \begin{bmatrix} v_1 \\ v_2 \\ \dots \\ v_n \end{bmatrix}$$

As such, they would normally be operated on the *left* by square matrices. I do, sometimes, represent vectors as $1 \times n$ matrices although in Chap. 7, where this most often occurs, I try to be consistent by using transpose notation, with row vectors of the form $\underline{v}^t$, operated on the *right*. When this notation is in use I try to refer to **'tuples'** rather than 'vectors'. Tuples are just row vectors, except they could have an infinite number of coordinates.

Contents

Part I Structure Theory

Part I
Structure Theory

Chapter 1
Stems, Mappings and Near-Rings

1.1 Basic Group Theory

A Useful Result from Group Theory

This lemma is diluted version of the one used by Betsch in his work on primitive near-rings. It is essentially due to Wielandt [11].

Lemma 4 *Suppose* $(G, +)$ *is an additive group and* A, B, C *are all normal subgroups of it. Then the group*

$$\frac{((A+C)\cap(B+C))}{((A\cap B)+C)} \tag{1.1}$$

is abelian.

Proof I use the commutation identities on page xvii.

Elements of numerator, above, have the form $(a+\mathrm{C}_1)$ and the form $(b+\mathrm{C}_2)$ for suitable $a \in A$, $b \in B$, $\mathrm{C}_1, \mathrm{C}_2 \in C$.

$$[a_1+\mathrm{C}_1, a_2+\mathrm{C}_2] = \left(([a_1, \mathrm{C}_3]+[a_1, a_2])^{\mathrm{C}_1}+[\mathrm{C}_1, \mathrm{C}_3]+[\mathrm{C}_1, a_2]^{\mathrm{C}_3}\right)$$

This is congruent to something in A (modulo C). A similar argument, applying to $[b_1+\mathrm{C}_1, b_2+\mathrm{C}_2]$, gives something congruent to something in B (modulo C). □

1.1.1 Sylow Theory

Sylow p-subgroups are subgroups with maximal prime power order. In a finite group with order $(p^m.n)$, with $(p, n) = 1$, there will always be Sylow p-subgroups (with order p^m). Any two are conjugate, and the number of distinct Sylow p-subgroups

R. Lockhart, *The Theory of Near-Rings*, Lecture Notes in Mathematics 2295,
https://doi.org/10.1007/978-3-030-81755-8_1

is a divisor of n and has the form $(1 + kp)$. There is an important body of work generalising and extending these ideas, and all basic group theory texts cover them [92, 93, 132].

Primes are important in ring theory too. For example, McDonald defines a ring to be **decomposable** if it may be expressed as a direct sum of two other non-trivial rings [118]. He then gives the pleasing, if obvious, result that a finite ring with identity is a direct sum of rings each having prime power order—mimicking the fundamental theorem of abelian groups. He goes on to say that a finite **indecomposable ring** has prime power order.

1.1.2 The Jordan-Hölder Theorem

Another basic result from group theory which I assume is the Jordan-Hölder theorem. Again, any group theory text covers this. The theorem covers series of subgroups of a group. The basic set-up involves series of subgroups of a group G in which each subgroup is normal in the next

$$G = G_0 \rhd G_1 \rhd G_2 \rhd \ldots \rhd G_n \tag{1.2}$$

There is the associated sequence of quotient groups $\left\{\frac{G_j}{G_{j+1}}\right\}_{j=0}^{n-1}$. This sort of series is called a **subnormal** or **subinvariant** series. If we have $G_j \lhd G$ for all j, the series is called a **normal series** or an **invariant series**. Normal series in which each $\left\{\frac{G_j}{G_{j+1}}\right\}$ is a simple group are called **principal series** or, sometimes, **chief series**. Finally, a subnormal series in which each factor group $\left\{\frac{G_j}{G_{j+1}}\right\}$ is a simple group is called a **composition series**.

There are a number of theorems about these series, mostly to the effect that given two series one may extend each of them, producing series with isomorphic quotient groups. The most famous of these theorems is usually called the **Jordan-Hölder** theorem. It says that if you have two composition series linking a group to the identity element, then the series have the same length, and the quotient groups are isomorphic in pairs (although not necessarily occurring in the same order). Not all groups have composition series, of course.

1.1.3 Solvable, Supersolvable and Nilpotent Groups

I do discuss **solvable**, **nilpotent** and **supersolvable** groups and, indeed, have already mentioned them.

Solvable groups are simply groups formed as iterated extensions of abelian groups. Finite nilpotent groups are the groups having all Sylow subgroups normal (a more general definition is needed for infinite cases). Supersolvable groups are groups having a finite sequence of normal subgroups, each successively containing the previous, such that the corresponding quotients are all cyclic (one can define solvable groups by insisting the quotients are all merely abelian). Finite supersolvable groups are precisely the groups whose maximal subgroups always have prime index.

Interestingly, finite supersolvable groups are **Lagrangian**, which means that they contain subgroups with orders equal to any integral divisor of their own order.

I do appeal to the famous **Feit and Thompson** theorem on at least three occasions. What I need from it is the observations that finite groups with odd order are inescapably solvable and that finite simple non-abelian groups have even composite order.

Group-theoretic **complements** (otherwise—*normal complements*) occur in Chap. 3 and are defined there (page 127). The, more specialised, concept of a **$\mathfrak{Z}$-group** arises in Chap. 3 and I reserve the Gothic "Z", $\mathfrak{Z}$, for such groups. This is simply a (in this case, finite) group for which all Sylow subgroups are cyclic (see page 122). There is a theorem to the effect that in such finite groups both the commutator subgroup and its factor group are cyclic [144]. I also refer to **subnormal subgroups** of groups—these being subgroups which may be linked to the main group by a sequence of subgroups, each containing the last and each normal in the next—as happens in subnormal series (see (1.2)).

These definitions are too sketchy to constitute one's only resource for these ideas, and they, and any other terms not specifically defined here, can be resolved in any of the standard group-theoretic texts that I cite, [132], for instance. My experience of group theory texts is that they usually stand out among mathematics texts for the quality of their writing even when quite advanced [55, 62, 92, 93, 121, 139, 143].

1.2 Homological Algebra and Category Theory

I mention a few ideas from homological algebra and category theory, particularly in the chapters in Section III. These ideas are more than covered in the two books by Saunders Maclane, [108] and [109], but the reader without much experience of "abstract nonsense" would be better referring to [27] or [96], or, indeed, any basic algebra text of their choice, and my use for these ideas is hardly profound—one really just needs to know what a category is and what a functor is. Incidentally, the adjective "categorial", here used, I think, only a handful of times, is supposed to mean "pertaining to category theory" and so distinguishes from the inappropriate "categorical" [53].

I was very influenced by the papers in [37] when originally constructing the material in Chap. 9, partly because my work started by considering Fröhlich's work on non-abelian homological algebra; and those papers, plus the classic textbook [18], probably provided the main impetus to his work. These days slicker, but perhaps less intelligible, presentations are available [67].

1.3 Topology

I don't use as much topology in this book as I had intended or expected and rather hope the reader can already handle whatever I throw at them. Basic ideas about bases, neighbourhood bases, closure, compactness and separation axioms are needed in Chaps. 7 and 8; and there is a small amount of stuff about function spaces there too.

I do need the idea of **density**. Recall that a subset of a topological space is **dense** in that space provided that every neighbourhood of every point in the topological space contains points in the subspace—the rational numbers are dense in the real numbers, with the usual topology, for instance; but the integers are not [68].

In Chap. 8 **disconnected** spaces are mentioned, and **clopen sets** come up too. Clopen sets are simply sets which are both closed and open—interesting, because such sets have no boundary points and consequently important distinction properties. **Disconnected spaces** are spaces which have multiple components—they may be written as a union of two disjoint non-empty open sets. **Totally disconnected spaces** have the points of the space as the components. The totally disconnected spaces that concern me occur in compact Hausdorff spaces, and the definition is equivalent to insisting that the set of clopen sets constitute a basis [52]. In any topological space, the whole space and the empty set are clopen sets. If these are the only clopen sets of the space, the space is said to be **connected** (a concept normally met *before* one encounters disconnected spaces). Connected spaces cannot be the union of two non-empty, disjoint open sets.

Other, rather different, notions of connectivity, relating to the presence of *paths* between points, occur [68]. This is a topologically vital concept, but not one I use here. I do need the following well-known idea, which is covered in [87] and in [68].

1.3.1 The Kuratowski Closure Axioms

Suppose N is a set and $\mathfrak{P}(N)$ is the **power set** of N (i.e. the set of all subsets).

Suppose there is a mapping

$$\Theta : \mathfrak{P}(N) \longmapsto \mathfrak{P}(N)$$

such that

(i) $\Theta(\phi) = \phi$ (empty set).
(ii) For each $P \in \mathfrak{P}(N)$, we have $P \subseteq \Theta(P)$.
(iii) For each $P \in \mathfrak{P}(N)$, we have $\Theta(\Theta(P)) = \Theta(P)$.
(iv) For each $P_1, P_2 \in \mathfrak{P}(N)$, we have $\Theta(P_1 \cup P_2) = (\Theta(P_1) \cup \Theta(P_2))$.

Then the collection of all sets

$$\{P \in \mathfrak{P}(N) : \Theta(P) = P\}$$

are the closed sets of a topology defined on N. The corresponding open sets are, of course, the complements of these sets; and for each $P \in \mathfrak{P}(N)$, the closure of P is the set $\Theta(P)$.

1.4 Stems and Near-Rings

Definition 5 A **stem** is an additive group $(A, +)$ (not necessarily abelian) plus a binary product $(\cdot)$.

The stem $(A, +, \cdot)$ is **bi-zero-symmetric** provided that $0 \cdot x = x \cdot 0 = 0$, for all $x \in A$.

Definition 6 If, for all $x, y \in N$ we have that $x \cdot y = 0$, the stem is called **trivial**.

Elements a of stems $(N, +, \cdot)$ for which finite powers give zero ($a^n = 0$) are termed **nilpotent**.

1.4.1 Star Notation

Notation 7 When $(A, +, \cdot)$ is a stem, I sometimes write A^* to mean the non-zero elements of A. Quite generally B^* means "non-zero elements of the set B". There are occasions (e.g. in Chap. 3) where I first define a set S^*, expecting the reader to recognise S as that same set, with zero adjoined.

1.4.2 Pre-Near-Rings

Definition 8 Stems $(N, +, \cdot)$ in which $(\cdot)$ is right-distributive—by which I mean

$$((x + y) \cdot z) = ((x \cdot z) + (y \cdot z)) \tag{1.3}$$

are called (right) **pre-near-rings**.

Definition 9 A (right, or *right-distributive*) **near-ring** $(N, +, \cdot)$ is a (right-distributive) pre-near-ring in which the product operation is associative, that is, $(N, \cdot)$ is a semi-group.

One may make the obvious definitions of left-distributive stems and near-rings. Stems which are both left- and right-distributive are described as **bi-distributive** (see Sect. 1.26).

Generally, my structures will be right-distributive, and if that is not stated, it will be assumed, but there are places in which it becomes convenient to discuss left-distributive near-rings—in Sect. 15.6, for instance, or in Chap. 16.

Pre-near-rings $(N, +, \cdot)$ automatically have the property $0 \cdot x = 0 \; \forall \, x \in N$, but the reader may usually assume that all stems and near-rings discussed here will be bi-zero-symmetric stems.

Pre-near-rings, and near-rings, with the property $x \cdot 0 = 0 \; \forall \, x \; \in N$ are said to be **zero-symmetric**.

Some classes of near-rings, near-fields and d.g. near-rings, for instance, are automatically zero-symmetric.

I often use the abbreviation **p.n.r.** for "pre-near-ring". This abbreviation occurs so often in this text that I only reference it on the first few occasions of its use. It is intended to be read as singular (pre-near-ring) or plural (pre-near-rings) as the context demands.

I occasionally describe p.n.r. with multiplicative identities as **unital p.n.r** and use the same terminology for near-rings (see page 65 for more on this terminology).

1.4.3 Conventions and Notation

I shall sometimes speak of "the near-ring N" (leaving the operations implicit). In the main, I shall avoid expressions such as xy for products $x \cdot y$ because, in keeping with right distributivity, the expression xy should most obviously be interpreted as the group addition

$$\overbrace{(y + y + y \ldots + y)}^{\text{(x times)}} \tag{1.4}$$

However, in left-distributive cases (e.g. in Sect. 15.6 and in Sect. 16), it should mean

$$\overbrace{(x + x + x \ldots + x)}^{\text{(y times)}} \tag{1.5}$$

There are occasions in which I shall use other symbols for near-ring products and I shall do that without additional comment. The sort of symbols used will include elements from the set

$$\{\wedge, \; *, \; \overset{*}{*}, \; :, \; \overset{+}{+}\} \tag{1.6}$$

I often work with additive generators of groups. If the symbol a represents a generator, or even a general group element, I might sometimes write na to mean $a + a + \ldots + a$ (n times) without specifically identifying n as an integer, when this seems immediate. The alternative is to add in intrusive and pointless explanations which my policy of specifically identifying products makes otiose.

Finally, right-distributive stems do have the property $(-x) \cdot y = -(x \cdot y)$, but the opposite condition, that $x \cdot (-y) = -(x \cdot y)$, which one might describe as **negative symmetry**, is not usually true. There are important classes of near-rings for which this holds, near-fields, for instance; and there are important classes of near-rings for which it does not, transformation near-rings, for instance.

Definition 10 If $(A, +, \cdot)$ is a stem, and H, K are two subsets of A, then the **product of the sets H, K** is the set

$$H \cdot K = \{h \cdot k : h \in H, k \in K\} \tag{1.7}$$

This idea returns in Sect. 1.17.

1.4.4 Examples of p.n.r. and of Near-Rings

(i) The standard vector cross product in Euclidean three-dimensional space gives an example of a bi-distributive p.n.r. which is not a near-ring (the cross product is not associative). This p.n.r. is also not unital.

(ii) The classic example of a near-ring is the transformation near ring, $\mathrm{T}_0(M)$ (Definition 222). Transformation near-rings occur all through this book and are of major importance to it. A reader unused to near-rings would be advised to get the hang of them at this point.

(iii) Of course, any ring is necessarily both a bi-distributive stem and a near-ring.

(iv) I often identify small near-rings by giving tables explicitly showing the additions and multiplications involved. There are several examples of this for the near-rings hosted by the dihedral group D_4, in Appendix B.

(v) Near-fields form an important proper subclass of the class of unital near-rings. These are the near-rings in which the semi-group of non-zero elements forms a multiplicative group. They are closely related to the standard fields of algebra. I give an account of the smallest near-field which is not actually a field in Sect. 3.3.1. The multiplicative elements form a group isomorphic to Q_8; the additive group involved is $(C_3 \oplus C_3)$. In this account I take the multiplication (of the standard quaternions) as known, and I specify the *addition* table in terms of i, j, k—the **basic quaternions.**

(vi) I represent near-ring products in several ways. The multiplication table for the example just noted—which is of course essentially that of Q_8—is shown in matrix form in Sect. 16.4.

(vii) A class of near-rings occurring in computer science is given on page 90.

(viii) The unital near-rings that can be defined on the Klein group are listed in Sect. 16.1.6. Again, I represent the products involved using matrices, but the matrices occurring encapsulate both product and sum information involving an idea which I refer to as "product theory". This idea, which is original to this book, is described fully in Chap. 15.

(ix) There are a number of structures that are related to near-rings—some of them well-known, others more obscure. Appendix C gives several examples which might motivate the study of these generalisations.

(x) Endomorphisms of abelian groups may be added using pointwise addition of function values. In this way one obtains an **endomorphism ring**. For non-abelian group this addition lacks closure although one can still use it to generate a sub-near-ring of the transformation near-ring of the group, which is commonly called an **endomorphism near-ring** (see page 240). Endomorphism near-rings are examples of d.g. near-rings—near-rings additively generated by some collection of bi-distributive elements; and these have been intensively studied for many years. Phomomorphisms (Definition 47) generalise even this idea and are covered in several chapters of this book.

1.5 Hosting

I shall speak of additive groups $(N, +)$ **hosting** stems, or near-rings $(N, +, \cdot)$. Some of my near-rings will have multiplicative identities and so are, multiplicatively speaking, **monoids**—semi-groups with 1.
I shall also occasionally speak of semi-groups $(A, \cdot)$ hosting near-rings or near-ring-like structures $(A, +, \cdot)$ (see Definition 333).

Notation 11 Suppose $(B, +)$ is any additive group.

(i) The set of distinct, left-distributive, bi-zero-symmetric, stem products hosted by $(B, +)$ (i.e. left, bi-zero-symmetric p.n.r.) will be denoted by

$$\mathcal{D}_l(B, +)$$

(ii) The set of distinct, right-distributive, bi-zero-symmetric, stem products hosted by $(B, +)$ (i.e. right, bi-zero-symmetric, p.n.r.) will be denoted by

$$\mathcal{D}_r(B, +)$$

(iii) The set of distinct, associative, bi-zero-symmetric, stem products hosted by $(B, +)$ will be denoted by

$$A(B, +)$$

Multiplicative Identities
Since my near-rings are not generally assumed to possess a multiplicative identity element, my rings will not necessarily have them either.

Zassenhaus invented the term whose English translation is "near-ring", in 1936 [128], Wielandt calling these structures "Stamm(s)" a couple of years later.[1]

1.6 Ideals

Suppose that $(N, +, \cdot)$ is a pre-near-ring and that $I \lhd (N, +)$.

Definition 12

(i) I is a **left ideal** provided that $\forall\, r, s, \in N$, and $\forall\, j \in I$ we have $r \cdot (j+s) - r \cdot s \in I$
(ii) I is a **right ideal** provided that $I \cdot N \sqsubseteq I$
I is an **ideal** provided that it is both a left and a right ideal.
I sometimes refer to ideals as "two-sided ideals", for reasons of clarity.

Notation 13 I may use the notation $J \lhd_x N$ where

$$x = \begin{cases} r & \text{if J is a right ideal} \\ l & \text{if J is a left ideal} \end{cases}$$

Remark 14

(i) When N is zero-symmetric, any left ideal I has the property $N \cdot I \subseteq I$.
(ii) Ideals are precisely the kernels of pre-near-ring homomorphisms.
(iii) Pre-near-rings devoid of proper, non-trivial, ideals are termed **simple** pre-near-rings. The same definition gives **simple near-rings**.
(iv) Left ideals are completely characterised by the property that for all $r, s, \in N$ and for all $j \in I$, $r \cdot (j+s) \equiv r \cdot s$ (modulo I).
(v) Associativity has no bearing on the definitions given here. Ideals of near-rings are defined exactly as above, but I shall naturally be more concerned with simple near-rings than simple pre-near-rings.
(vi) All flavours of ideal are preserved under intersections and the usual additions

$$I + J = \{(i + j) \,:\, i \in I, j \in J\}$$

[1] In case the story should be lost. A distinguished student of Wielandt told me that Zassenhaus habitually wrote to him as "Wieland". Losing patience, the great algebraist started writing back to "Zassenhau". My own term "stem" was chosen to directly reference Wieland's work. Incidentally, "stamms" is *not* the German plural!

It is important, later, to distinguish between sums of ideals and direct sums of ideals. One clarifying distinction is that sums usually suggest that elements can always be written as sums of elements from the constituent ideals, perhaps in many ways, whereas with direct sums, this representation should be unique (see page 390).

Very occasionally, I talk about subsets S, $H, \subseteq N$, where N is a near-ring, and form a set

$$(S+T) = \{(s+t) : s \in S, t \in T\} \tag{1.8}$$

This is just that—a set of sums. Algebraically, it probably had little structure, but in general I do not overload the notation by requiring, for example, that the $(S+T)$ should refer to a larger set, such as the smallest ideal containing the set of sums.

(vii) The left ideal condition reduces to the standard ring definition, $R \cdot J \subseteq J$, in the case of distributively generated pre-near-rings (see Definition 75).

(viii) Suppose H is a fully invariant subgroup of the right ideal J. Then H is itself a right ideal. Consequently, minimal right ideals are fully invariantly simple.

(ix) If follows from the previous point that the commutator subgroup, and all other fully invariant subgroups, are right ideals in any near-ring in which they occur.

(x) In ring theory, if I and J are ideals, then

$$I \cdot J = \{(i \cdot j) : i \in I, j \in J\} \tag{1.9}$$

is not usually itself an ideal. The smallest ideal containing this set consists of all finite sums of its elements because one has distributivity on each side. Ring theorists sometimes define $I \cdot J$ to mean not (1.9) but this set of finite sums—the ideal they generate; and this is sometimes also done in near-rings, but not in most of this book (see Definition 66).

In zero-symmetric near-rings, (1.9) will be both a left and a right sideal (defined below), and the set of its finite sums will be both a subgroup and a right sideal, contained in $(I \cap J)$. The corresponding set of finite sums of these product elements is a subgroup and a right sideal but seems to have diminished importance.

(xi) Ideals which are simple when regarded as near-rings are called **simple ideals**. Terminology defined for rings is frequently extended to ideals in ring theory, and so it is in near-rings. One sees **decomposable ideals**, **completely reducible ideals** and so on. I usually make an effort to define the term explicitly in its new context when this is done here.

Alternatively, sometimes a concept is first defined for ideals and extended to near-rings in an alternative way. So, for example, a **prime ideal**, P, is often defined by insisting that if I and J are two ideals and $I \cdot J \subset P$, then one of I or J must themselves be contained in P. Extending this to a **prime near-ring**, one simply insists that the trivial ideal $\{0\}$ should be a prime ideal.

Another way one moves to ideals from rings is via quotients. So, [84] defines a **primitive ideal** in a ring to be an ideal for which the corresponding quotient is a primitive ring (see page 407).

(xii) If $(N, +, \cdot)$ is a near-ring and there is a group epimorphism $\theta : (N, +) \twoheadrightarrow (G, +)$ whose kernel happens to be an ideal of N, then, of course, the prescription $(g_1 * g_2) = \theta(\theta^{-1}(g_1) \cdot \theta^{-1}(g_2))$ defines a near-ring product on $(G, +, \cdot)$ which is just the quotient near-ring (see page 480). Well-definedness is consequential on the kernel being an ideal.

Definition 15 In any near-ring, a **maximal ideal** is a proper ideal not contained in any other proper ideal. **Minimal ideals** are defined by insisting they should not contain other proper ideals.

These terms can be extended to other similar structures, left and right ideals, sideals, etc. Of course, there are near-rings which do not possess maximal ideals and near-rings which do not possess minimal ones.

Lemma 16 *Suppose I and J are left ideals, $I \subseteq J$, and that, given any other left ideal C*

$$(I \cap C) = (J \cap C)$$
$$(I + C) = (J + C)$$

Then it must be the case that $I = J$.

Proof If $j \in J$, then we may write $j = (i + c)$ for appropriate $i \in I, c \in C$.
Now, $(j - i) \in (J \cap C)$. This means $(j - i) \in I$, and the result follows. □

The same result holds true for right ideals.

Lemma 17 *Suppose I and J are left ideals of the zero-symmetric near-ring $(N, +, \cdot)$ and $(I \cap J) = \{0\}$.*
If $i \in I$, $j \in J$, $n \in N$, then

$$n \cdot (i + j) = ((n \cdot i) + (n \cdot j))$$

Proof We know

$$n \cdot (i + j) = (n \cdot i) + j' \textit{ for some } j' \in J$$

and

$$n \cdot (i + j) = (i') + (n \cdot j) \textit{ for some } i' \in I$$

So, $(n \cdot i) = (i')$ and $(n \cdot j) = j'$. □

Ideal Generation
The intersection of all the ideals containing a given subset of a near-ring will itself be an ideal, the ideal generated by the subset. If S, then, is a subset, the ideal it generates will normally be written $\langle S\rangle_{(ideal)}$, with obvious modifications for left and right ideals. Of course, there is no simple general formula for the form of the elements of this structure.

1.7 Subdirect Products of Near-Rings

Given any family $\{N_a\}_{a\in\alpha}$ of near-rings, the **direct product** is the near-ring of all functions

$$f\ :\ \alpha \mapsto \bigcup_{a\in\alpha}(N_a) \tag{1.10}$$

for which $f(a) \in N_a$ for all $a \in \alpha$.

This is a standard definition for algebraic systems.

Sums are defined coordinate-wise

$$(f+g)(a)\ =\ (f(a)+g(a)) \tag{1.11}$$

and products coordinate-wise

$$(f\cdot g)(a)\ =\ (f(a)\cdot g(a)) \tag{1.12}$$

This near-ring is habitually written $\prod_{a\in\alpha}(N_a)$.

Associated with this near-ring are surjective mappings

$$\pi_a\ :\ \prod_{a\in\alpha}(N_a) \longmapsto N_a \tag{1.13}$$

defined by $\pi_a(f) = f(a) \in N_a$ and called the **projection functions**.

The sub-near-ring of functions with finite support is then called the **direct sum** $\oplus_{a\in\alpha}(N_a)$. Elements of this near-ring can be thought of as the set of "finite tuples" of elements $(a_1, a_2, \ldots a_n)$ where coordinates are from distinct members of the original family of near-rings.

Definition 18 Suppose $\{N_a\}_{a\in\alpha}$ is some family of near-rings and suppose S is a sub-near-ring of the direct product $\prod_{a\in\alpha}(N_a)$ which has the property that the projection functions, applied to S, are still surjective. Then S is said to be a **subdirect product** of the near-rings $\{N_a\}_{a\in\alpha}$.

If we have a subdirect product S of the family $\{N_a\}_{a\in\alpha}$, the projection functions $\pi_a : S \mapsto N_a$ are actually homomorphisms of near-rings. Each has a kernel, which would be an ideal $K_a \subset S$, and we then have that

$$\bigcap_{a\in\alpha} (K_a) = \{0\} \tag{1.14}$$

It is interesting to note that if we are given a near-ring N and some family of ideals of it $\{K_a\}_{a\in\alpha}$, we obtain quotient near-rings $N_a = \left(\frac{N}{K_a}\right)$. By working through $\prod_{a\in\alpha} (N_a)$ in the obvious way, we can obtain a subdirect product of these near-rings which is isomorphic to the original near-ring N.

Subdirect products are relatively loose structures compared to the tightness of direct sums and products. In general there are many non-isomorphic subdirect products of a given family of near-rings. Robinson relates subdirect products to *residual properties* in [132].

1.8 Sideals and Near-Ring Groups

Near-ring multiplication is associative, so zero-symmetric near-rings are semi-groups with two-sided zeros. In semi-group theory, an **ideal** is defined as a subset S of $(N, \cdot)$ such that both $S \cdot N \subseteq N$ and $N \cdot S \subseteq N$ (two-sided ideal) and one defines one-sided ideals in the obvious way (see [70] or [49]). I cannot use the term "ideal" for this concept here, so I call subsets of the near-ring semi-group $(N, \cdot)$ **sideals** if they are ideals in the semi-group sense with respect to the near-ring multiplication. There is no presumption of additive normality here.

I believe this result is due to S.D. Scott.

Lemma 19 *When S is a right sideal of the right-distributive near-ring, $(N, +, \cdot)$ (i.e. $S \cdot N \subset S$), there will be a smallest left ideal containing it. This left ideal is actually an ideal.*

Proof The intersection of all the left ideals containing S must itself be a left ideal, which I call J. This is the smallest left ideal containing it.

Set

$$K = \{n \in N : n \cdot N \subseteq J\} = \binom{N}{J}_{\mathcal{L}}$$

[Here I introduce notation defined in Sect. 1.15, not to facilitate the proof, but for future reference.]

Suppose $k_1, k_2 \in K$. Clearly $(k_1 - k_2) \in K$, since J is a subgroup, and because J is a normal subgroup, so is K. K is a right sideal, and therefore a right ideal. Taking $m, n, r \in N$ and $k \in K$, there is some $j \in J$ such that

$$\begin{aligned}(m \cdot (n + k) - (m \cdot n)) \cdot r &= (m \cdot (n \cdot r + k \cdot r) - (m \cdot n \cdot r)) \\ &= (m \cdot (n \cdot r + j) - (m \cdot n \cdot r)) \in J\end{aligned}$$

so, K is a left ideal. The conclusion is that K is a two-sided ideal of N. K certainly contains S. Interestingly, J must be contained in K, because K is a left ideal and J is the smallest left ideal containing S, but that means that $J \cdot N \subseteq J$ and that J is an ideal. □

Definition 20 Suppose N is a near-ring. Subgroups $G < (N, +)$ which are (left/right/two-sided) sideals are called **near-ring groups** (left/right/two-sided).

I tend to use the unqualified term "near-ring group" to mean *left* near-ring group.

1.8.1 Generalisations

Most published accounts of near-ring groups refer to "N-groups", terminology which changes depending on the near-ring being considered (see later); and I sometimes have situations in which a near-ring N acts on the left on a group $(\Gamma, +)$ not immediately related to $(N, +)$, in the sense

- $N \times \Gamma \mapsto \Gamma$, $(x, \gamma) \mapsto (x \cdot \gamma) \in \Gamma$.
- For all $m, n \in N, \gamma \in \Gamma$, $((m \cdot n) \cdot \gamma) = (m \cdot (n \cdot \gamma))$.
- For all $m \in N$, $(m \cdot 0_\Gamma) = 0_\Gamma$ (where 0_Γ here represents the additive identity of $(\Gamma, +)$).
- For all $m, n \in N, \gamma \in \Gamma$, $((m + n) \cdot \gamma) = ((m \cdot \gamma) + (n \cdot \gamma))$.

I shall use the term (left) "near-ring group" here, too. This mostly occurs in the chapters on transformation near-rings—in Chaps. 7 and 8 and in Chap. 11.

When the near-ring has an identity element e (i.e. is *unital*), I also usually assume

$$(e \cdot \gamma) = \gamma, \quad \forall \gamma \in \Gamma \tag{1.15}$$

that is, the near-ring group is **unitary**.

I am often concerned with monogenic near-ring groups (page 20) over unital near-rings. These are automatically unitary (see Lemma 30).

Definition 21 If Γ_1 and Γ_2 are both (left) near-ring groups, a **near-ring group homomorphism** is a mapping, $\theta : \Gamma_1 \longrightarrow \Gamma_2$, such that, for all $\gamma, \gamma_1, \gamma_2 \in \Gamma_1$, and all $n \in N$:

(i) $\theta(\gamma_1 + \gamma_2) = (\theta(\gamma_1) + \theta(\gamma_2))$.
(ii) $\theta(n \cdot \gamma) = n \cdot \theta(\gamma)$.

I also call this a **NR-linear mapping** and might write $\mathrm{Hom}_N(\Gamma_1, \Gamma_2)$ for the collection of such mappings. Later on, near-ring group homomorphisms from a near-ring group to itself become important—**near-ring group endomorphisms**—I would write $\mathrm{Hom}_N(\Gamma, \Gamma)$ as $\mathrm{Hom}_N(\Gamma)$ here.

I also discuss **near-ring group automorphisms**. In these, I tend to act the mappings on the *right* (see page 299).

1.8.2 Right Near-Ring Groups

On a few occasions, I discuss *right* near-ring groups (page 307, page 392). These are modelled on right sideals of near-rings, and the salient properties, in zero-symmetric cases, are

(i) $(\Gamma \times N) \longmapsto \Gamma,\ (\gamma, n) \mapsto (\gamma \cdot n)$.
(ii) If 0_N is the additive zero of $(N, +)$ and $\gamma \in \Gamma$, then $(\gamma \cdot 0_N) = 0_\Gamma$.
(iii) For all $m, n, \in N, \gamma \in \Gamma$, $((\gamma \cdot m) \cdot n) = ((\gamma) \cdot (m \cdot n))$.
(iv) For all $\gamma_1, \gamma_2 \in \Gamma$, and all $n \in N$,

$$((\gamma_1 + \gamma_2) \cdot n) = ((\gamma_1 \cdot n) + (\gamma_2 \cdot n)) \tag{1.16}$$

If the near-ring has an identity, e, I would also assume $\gamma = (\gamma \cdot e)$, for all $\gamma \in \Gamma$. The obvious modification is made if one is discussing NR linearity in the context of right near-ring groups.

1.8.3 Highly Non-standard Terminology

Previous authors have almost universally referred to "N-groups", instead of near-ring groups. It is one of the few notational constants in near-rings to do so and I do find myself slipping easily into it myself. The problem, of course, is its inappropriate elevation of the particular near-ring involved. It is a category error. One finds oneself discussing "R-groups", "F-groups" and so on, and I think that that constant confusion of terminology is worth this change, although it is utterly non-standard.

The change brings its own difficulties in that one does not specifically identify the near-ring involved. I hope, wherever I have done it, there is no uncertainty about this, but that, of course, is a judgement for the reader. If one wishes to stress that, say, left near-ring groups are under discussion, one could talk of $NR_{(l)}$**-groups**, or, conversely, of $NR_{(r)}$**-groups**, for right ones.

$_N\Gamma$ Notation

When discussing near-ring groups and other related structures, it can be important to spell out both the near-ring involved and whether a left or right action is involved.

I use notation such as ${}_N\Gamma$ and Γ_N to indicate the near-ring N, the group acted upon, Γ, and, respectively, left and right action.

1.8.4 Sub-Structures and Mideals

Definition 22 Suppose we have a left near-ring group ${}_N\Gamma$.

(i) A subgroup $\Omega \subseteq \Gamma$ which is also a left near-ring group ${}_N\Omega$ is a **sub-near-ring group**.
(ii) A normal subgroup $\Omega \lhd \Gamma$ with the property

$$m \cdot (\gamma + \omega) - (m \cdot \gamma) \in \Omega, \ \forall m \in N \ \forall \omega \in \Omega, \ \forall \gamma \in \Gamma \quad (1.17)$$

is called a **mideal**. I sometimes refer to mideals as "left mideals" since the important action occurs on the left.
(iii) These definitions are applied unchanged to left modules in Chap. 11.

Notice that mideals are specific to the near-ring group being discussed. If we have two near-ring groups $A \subset B$ and a mideal $J \subset A$, it may well be that J is not a mideal of B. Of course, if J is a mideal of B, it will be a mideal of A. This is not a problem in the case of sub-near-ring groups. A sub-near-ring group of A will be one of B too. I have only found time to discuss mideals of the sort defined above in this book. The reader will be able to guess at the formulation of right and two-sided mideals, but I have not space for that here. In F-near-rings (page 47), the additive commutator subgroup is a (left) mideal.

Lemma 23 *Suppose ${}_N\Gamma_1$ and ${}_N\Gamma_2$ are left near-ring groups and*

$$\theta : \Gamma_1 \longmapsto \Gamma_2$$

is a near-ring group homomorphism.
Then,

$$Ker(\theta) = \{\gamma \in \Gamma_1 : \theta(\gamma) = 0\} \quad (1.18)$$

is a mideal.

Lemma 24 *Suppose ${}_N\Gamma$ is a left near-ring group and $M \subset \Gamma$ is a mideal.*
Then the quotient group $\frac{\Gamma}{M}$ is a left near-ring group ${}_N(\frac{\Gamma}{M})$ under the action on cosets

$$n \cdot \{\gamma + M\} = \{(n \cdot \gamma) + M\}$$

Lemma 25 *Suppose* $_N\Gamma$ *is a left near-ring group and* Ω, Λ *are mideals.*

(i) $(\Omega + \Lambda)$ *is a mideal contained in* Γ.
(ii) $(\Omega \cap \Lambda)$ *is a mideal contained in* Γ.

1.8.5 Faithfulness

Definition 26 A near-ring N acts **faithfully** upon a left near-ring group Γ provided that, if for any $n \in N^*$ we have that

$$(n \cdot \Gamma) = \{(n \cdot \gamma) : \gamma \in \Gamma\} = \{0\}$$

(i.e. n *annihilates* Γ), then it must be the case that $n = 0$.
In the notation of Sect. 1.15

$$\binom{\Gamma}{0}^N_{\mathcal{L}} = \{0\}$$

The near-ring group is then referred to as a **faithful near-ring group**.

Faithful actions are ubiquitous in modern algebra and I apply the idea in other situations without formally spelling out any slight modifications that might be needed. When N acts faithfully on M as a near-ring group, we may assume $M \subseteq T_0(M)$ (Definition 222).
The following lemma appears in [11].

Lemma 27 *Suppose* Γ *is a left near-ring group for the near-ring* $(N, +, \cdot)$ *which is faithful and abelian and with the property that the mapping*

$$\Gamma \mapsto \Gamma, \gamma \mapsto (n \cdot \gamma)$$

is an additive homomorphism for all elements $n \in N$.
Then $(N, +, \cdot)$ *is actually a ring.*

Proof Select $a, b, c \in N, \gamma \in \Gamma$.
Then,

$$(a + b) \cdot \gamma = (a \cdot \gamma) + (b \cdot \gamma) = (b \cdot \gamma) + (a \cdot \gamma) = (b + a) \cdot \gamma$$

so $(a + b) = (b + a)$ by faithfulness.
Similarly,

$$a \cdot (b + c) \cdot \gamma = a \cdot ((b \cdot \gamma) + (c \cdot \gamma)) = ((a \cdot b \cdot \gamma) + (a \cdot c \cdot \gamma)) =$$
$$((a \cdot c \cdot \gamma) + (a \cdot b \cdot \gamma)) = ((a \cdot c) + (a \cdot b)) \cdot \gamma$$

so $a \cdot (b + c) = (a \cdot b + a \cdot c)$. □

This corollary involves definitions that are made later: phomomorphisms on page 31, modules on page 371 and $\mathfrak{Ph}_S(\Gamma)$ on page 296. The semi-group S mentioned is defined on page 295. The reader working sequentially can safely ignore it, for now.

Corollary 28 *Suppose Γ is a left near-ring module for the near-ring $(N, +, \cdot)$ which is faithful and has the property that the mapping*

$$\Gamma \mapsto \Gamma, \gamma \mapsto (n \cdot \gamma)$$

is an additive phomomorphism for all elements $n \in N$.

Then $(N, +, \cdot)$ is a sub-near-ring of $\mathfrak{Ph}_S(\Gamma)$.

1.8.6 Monogenicity

Irreducible modules are central to ring theory. They automatically have a property which usually has to be specifically demanded in their near-ring analogues: **monogenicity**.

Definition 29

(i) A left near-ring group ${}_R\Gamma$ is **monogenic** provided there is some $\gamma \in \Gamma$ such that

$$\Gamma = \{R \cdot \gamma\} = \{(r \cdot \gamma) : r \in R\} \tag{1.19}$$

(ii) The obvious definition $\Gamma = \{\gamma \cdot R\}$ applies to right near-ring groups.
(iii) And the same obvious definition applies to right modules (defined page 371).
(iv) I may refer to γ as a **monogenic generator**.

Jacobson considered monogenic ring modules in [76], though he calls them **strictly cyclic modules**.

Lemma 30 *Suppose $(N, +, \cdot)$ is a near-ring and ${}_N\Gamma$ is a monogenic left near-ring group with monogenic generator γ.*

(i) If N has a left identity e, then, for all $w \in \Gamma$, we have $e \cdot w = w$. That is, the near-ring group is unitary.
(ii) If J is a left ideal of N, then $J \cdot \gamma = \{(j \cdot \gamma) : j \in J\}$ is a mideal.
(iii) If Ω is a mideal in Γ, then $\binom{\gamma}{\Omega}^N_{\mathcal{L}} = \{n \in N : (n \cdot \gamma) \in \Omega\}$ is a left ideal of N.
(iv) If N acts faithfully on Γ and e is a left identity, then it is also a right identity.

Proof

(i) We know $\Gamma = \{N \cdot \gamma\}$ for some $\gamma \in \Gamma$. If $\omega \in \Gamma$, then $\omega = (n \cdot \gamma)$; consequently

$$(e \cdot \omega) = (e \cdot (n \cdot \gamma)) = ((e \cdot n) \cdot \gamma) = (n \cdot \gamma) = \omega \tag{1.20}$$

(ii) We only need to establish that if $(j \cdot \gamma) \in \Omega$, then $(m \cdot (n+j) - (m \cdot n)) \cdot \gamma \in \Omega$. This is obvious.
(iv) Taking any $n \in N$, and any $w \in \Gamma$,

$$0 = (n \cdot \gamma - (n \cdot e).\gamma) = (n - (n \cdot e)) \cdot \gamma$$

□

N is a left near-ring group ${}_N N$ and the possible monogenicity of that near-ring group is of interest. If it is monogenic, as it will be if the near-ring contains either a left or a right identity, the collection of elements that are monogenic generators seems important, and indeed that collection of monogenic generators is important in more general near-ring groups (see page 381).

Generally, I tend to refer to sideals rather than near-ring groups, particularly when stressing various sorts of multiplicative closure. I also talk about left and right sideals sometimes and I shall sometimes extend these ideas to p.n.r. (Sect. 1.4.2). Observe that, in p.n.r., left sideals automatically contain the additive zero element.

Right ideals of zero-symmetric p.n.r. also contain it.[2] When the p.n.r. has a left identity element ($e \cdot x = x$), left sideals contain all "additive powers" of their members (i.e. all expressions mx, where $m \in \mathbb{Z}$ and $x \in S$). Two-sided sideals are themselves semi-groups.

1.8.7 Some Two-Sided Sideals

This subsection requires some of the basic notation for transformation near-rings. This can be found in Chap. 7 and on or around page 232.

First Example

The transformation near-ring $\mathrm{T}_0(M)$ is defined on page 232. $(M, +)$ is an additive group, and the near-ring is simply the zero-fixing mappings of M to itself, with addition based on the group addition and multiplication based on the composition of functions.

We may consider subsets Σ_N of $\mathrm{T}_0(M)$ involving functions whose image sets have cardinality less than some fixed positive integer, N. For each N, Σ_N is a multiplicative semi-group and a two-sided sideal of $\mathrm{T}_0(M)$. Extending the idea, we can consider the subset Σ of $\mathrm{T}_0(M)$ consisting of mappings which lie in some Σ_N—that is, mappings with finite image sets—mappings having finite *support*.

$$\Sigma = \bigcup_N (\Sigma_N) \tag{1.21}$$

[2] My first chapter on transformation near-rings points out that in $\mathrm{T}(N)$, $\{0\}$ is an ideal but not a left sideal.

This too is a semi-group and a two-sided sideal (see (7.28)). One obtains a strictly ascending sequence of two-sided sideals (so long as N remains smaller than the group size). The idea can be extended to infinite cardinalities.

Second Example
Taking an additive group, $(M, +)$, we can define constant mappings C_n in $T(M)$ as is given in Sect. 7.2.1; and then, these properties are apparent for all $n, m \in N$, and all $\phi \in T(N)$.

(i) $(C_n + C_m) = C_{(n+m)}$.
(ii) $(\phi \circ C_n) = C_{\phi(n)}$.
(iii) $(C_n \circ \phi) = C_n$.

Thus, the set

$$\mathfrak{k} = \{C_n \ : n \in N\} \tag{1.22}$$

is both a subgroup of $T(N)$ and a two-sided sideal of it.

For this to be a normal subgroup of $T(N)$, we should require, among other things, that (considering the identity mapping)

$$(-m + C_n + m) \in \mathfrak{k}, \quad \textit{for all}\ m, n \in N$$

This can only be true for elements n which are in the additive centre of the group $(N, +)$. So the possibility of an ideal of $T(N)$ contained in $\mathfrak{k}$ only looms for mappings C_n in which $n \in \mathcal{Z}(N, +)$. But notice that property (ii), above, ensures that any left sideal of $T(M)$ containing *any* mapping C_n must also contain all of $\mathfrak{k}$. If we have normality, for the additive group $\mathfrak{k}$, it must be that the group $(N, +)$ is actually abelian.

1.8.8 The Weak Left Ideal Property

On at least one occasion, I become interested in additive subgroups, H, of a near-ring N, which exhibit the property

$$a \cdot (b + h) - (a \cdot b) \in H, \ \forall\, a, b, \in N, \ \text{ and } \forall\, h \in H \tag{1.23}$$

I shall call this property, which is similar to the group-theoretic idea of marginality [132], the **weak left ideal** property, and note that normality is not required here (see Definition 22). If we do consider normal subgroups with the weak left ideal property, then, clearly, there will be a unique maximal such subgroup, and that will be a left ideal.

1.8.9 Rings and Modules

A clear majority of texts on the theory of rings require rings to be unital, and some of the most striking results of the subject are restricted to unital rings. Respected texts argue that non-unital rings are of lesser importance [56]. This may be due to the existence of the Dorroh extension (page 102); but whatever validity that viewpoint may have is diluted for near-rings, although I find unital near-rings a useful subclass when looking at the host problem (Chaps. 4, 5, 6).

Modules are fundamental constructs in ring theory. These are simply abelian groups upon which the ring acts in a "vector space-like" way. In my own terminology, they are near-ring groups (for the abelian and bi-distributive near-ring in question) for which $r \cdot (m_1 + m_2) = (r \cdot m_1 + r \cdot m_2)$, for all $r \in R, m_1, m_2 \in M$. Modules are said to be **simple** provided they have no proper non-trivial submodules.

I purloin the term "module" for a slightly different purpose in Chap. 11, but there are occasions in this account where I must report classic ring theory results that mention modules and simple modules, and for those I mean the structures just described.

1.9 Semi-simplicity

The **socle** is a useful construct in the theory of groups. It is the subgroup generated by its collection of minimal normal subgroups, or the trivial subgroup, if the group has none. It is standard group theory that this subgroup is actually a direct sum of some of these minimal normal subgroups. For finite groups, the socle is a direct sum of simple groups.

William Scott actually refers to it as the "sockel" and says the word is German and usually untranslated [137]. I think that this is simply the German word for the same architectural feature that the English socle names—a plinth used to support a pillar. This idea extends to similar structures in more general algebras and even to lattice theory (page 97).

In particular cases, which were often developed without reference to generalities, the terminology tends to become local, complicated and non-standard.

In ring theory, people often first define **semi-simple modules** as modules which are direct sums of simple modules and then define **semi-simple rings** as rings R for which the module ${}_R R$ is semi-simple [94]. One problem with this neat definition is that it results in simple rings which are not semi-simple, because semi-simple rings have a built-in finiteness condition (i.e. the sum part (I am not saying they are themselves finite!)).

They then responded to that particular hole by attempting to dig it out—with simple rings required to have Artinian (page 97) chain conditions on left or right ideals. I believe the Bourbaki group do this, and Lang [96]. Of course, now we have to rename simple rings, and these are triumphantly rechristened **quasi-simple**.

An alternative, popular, definition of semi-simplicity is that the *Jacobson radical* should be trivial [66]. I have seen that described as "Jacobson semi-simplicity"; but others instead use the term **semi-primitive ring**, because such rings are subdirect products of primitive rings [84]; and commutative Jacobson semi-simple rings are subdirect products of fields [66]. This is actually quite a useful idea because semi-primitive rings are the rings which have faithful semi-simple modules. Anyway, I do employ this terminology in Chap. 12.

It is perhaps more modern to look at the sum of the minimal submodules of a module and call that the socle. If there are no simple modules, we make the socle the trivial ideal [21].

In general, it seems to me that the terminology is poor. The real problem is another category error—semi-simplicity being defined in terms of modules and simplicity being defined in terms of rings. Our names for concepts grow organically; it is difficult to revisit terms that have become standardised, if contradictory. If one then seeks to apply these names into an extension of the subject, such as the theory of near-rings, one can end up compounding misadventures. At this point the reader might find it useful to look at Lemma 430.

1.10 Prime and Semi-prime Ideals

Notation (1.9) and the definitions on page 12 are useful here. Again, I stress that products of sets: $S \cdot T$ are simply sets of the products $(s \cdot t)$ where, $s \in S, t \in T$, and I don't go on to generate a further (additive) structure.

1.10.1 Prime Ideals

Definition 31 An ideal P of a near-ring is called **prime** if whenever we can find two ideals I and J for which $I \cdot J \subseteq P$, it must be the case that one of these two ideals is itself in P.

One defines a **prime near-ring** as one in which the trivial ideal $\{0\}$ is prime.

There are some trivial deductions one can make already from this. For example, if we have two ideals each properly containing P, then their product may not be contained in P; and more, if neither of the two ideals are contained in P, then their product can't be. Prime ideals containing another ideal map to prime ideals in the corresponding quotient of that ideal as one would expect.

Lemma 32 *These conditions are equivalent for an ideal P of a near-ring N.*

(i) The ideal P is prime.
(ii) Whenever $x,\ y \in N$ are elements not in P, the product $\left(\langle x\rangle_{(ideal)} \cdot \langle y\rangle_{(ideal)}\right)$ is itself not in P.

(*iii*) *Whenever two ideals properly contain P, their product is not contained in P.*
(*iv*) *Two ideals each not properly contained in P have a product which is not contained in P.*

Proof $(i) \Rightarrow (ii)$, $(i) \Rightarrow (iii)$, $(i) \Rightarrow (iv)$ are straightforward.

(iv) means that any product of two ideals that is contained in P involves one which is also contained in P, so $(iv) \Rightarrow (i)$.

$(ii) \Rightarrow (iii), (iv)$, again seems obvious.

Proving $(iii) \Rightarrow (iv)$ is trickier but really just involves moving from the ideals I, J mentioned in (iv) to the ideals $(I + P)$, $(J + P)$, catered for by (iii). □

Prime ideals and maximal ideals are important in commutative ring theory, where, of course, maximal ideals are prime.

Lemma 33 *If J is a maximal ideal of a near-ring N and J is not prime, then $N^2 \subseteq J$.*

Proof The near-ring $\frac{N}{J}$ has no proper non-trivial ideals; it is simple. Of course, this near-ring itself is an improper ideal, and it could happen that it is a trivial near-ring (Definition 6). That would be equivalent to $N^2 \subseteq J$. □

Corollary 34 *In unital near-rings, maximal ideals are prime.*

It's easy to see that an ideal J is prime if and only if the corresponding quotient near-ring is prime.

Even in the tame environs of commutative unital rings, it is not true that whenever you take an intersection of prime ideals, you get another one; and one can expect some ideals not to be prime because if one has a ring for which all proper ideals are prime, that ring is a field [83].

Definition 35 Given an ideal J of a near-ring N, there may be prime ideals containing J. If H is just such a prime ideal and it contains no further prime ideals which contain J, then H is called a **minimal prime ideal of J**.

Minimal prime ideals of $\{0\}$ are simply called **minimal prime ideals**.

In point of fact, the near-ring itself is a prime ideal, and an argument based on Zorn's lemma, applied to minimal, rather than maximal, structures, shows that all ideals have minimal prime ideals in the above sense.

The near-ring theory of prime and semi-prime ideals is the work of several mathematicians, among them, Carlton Maxson, Andries Van der Walt and Davuluri Ramakotaiah [128].

1.10.2 Semi-prime Ideals

Definition 36 An ideal J is **semi-prime** provided that whenever I is an ideal of the near-ring $(N, +, \cdot)$ such that $I^2 \subseteq J$, it must also be the case that $I \subseteq J$.

Semi-prime near-rings are just near-rings in which the trivial ideal is semi-prime.

Lemma 37

(i) The intersection of any chain of prime ideals is a prime ideal.
(ii) The intersection of any collection of prime ideals is a semi-prime ideal.

Proof

(i) This intersection is an ideal. Suppose I and J are ideals and $I \cdot J$ is contained in the intersection. Then this product is in each of the ideals of the chain, so, given an ideal of the chain, one of I or J lies in it. The same applies for any ideal of the chain "lower" than our one, but since our one contains that lower ideal, it must contain the choice made with respect to it.
(ii) Again the intersection is an ideal. Almost the same argument applies as in (i) except that now $I = J$.

□

1.10.3 Complements of Prime and Semi-prime Ideals

I use the ideal generation notation from page 14 and the complement notation from page 77. Recall any subset of a near-ring generates an ideal.

Definition 38 Near-ring theorists call a subset S of a near-ring N an **M-system** provided that whenever $x, y \in S$, we can find $w \in \langle x \rangle_{(ideal)}$ and $z \in \langle y \rangle_{(ideal)}$, such that $(w \cdot z) \in S$.

In other words the product of any two ideals generated by elements of S must contain elements of S. When P is a prime ideal, P^c (the complement) is an M-system. More than that, if the complement of an ideal is an M-system, then that ideal is prime. In any near-ring the whole near-ring and the empty set are both M-systems. If $n \in N$, then the set of all powers of n, that is, $\{n, n^2, n^3, \ldots\}$, is an M-system.

Pilz attributes this result both to Van der Walt and to Ramakotaiah.

Lemma 39 *If M is a non-empty M-system in the near-ring N and J is an ideal of N which is disjoint from M, then there is a prime ideal $P \supset J$ which is disjoint from M.*

Proof Consider the set of ideals containing J and still having empty intersection with M. Zorn's lemma gives us a maximal element to this set; call that P.

Take two ideals X and Y, properly containing P; these both contain elements of M.

So we may select $x \in (X \cap M)$ and $y \in (Y \cap M)$ and know that

$$\left(\left(\langle x \rangle_{(ideal)} \cdot \langle y \rangle_{(ideal)} \right) \bigcap M \right) \neq \phi$$

This means

$$\left(\langle X \cdot Y \rangle_{(ideal)} \bigcap M \right) \neq \phi$$

and

$$\left(\langle X \cdot Y \rangle_{(ideal)} \right) \subsetneq P \Rightarrow (X \cdot Y) \subsetneq P$$

□

This definition and the following lemma are due to Maxson.

Definition 40 Subsets S of near-rings are called **sp-systems** provided that whenever $x \in S$, we may find $s_1, s_2 \in \langle x \rangle_{(ideal)}$, with the property that $(s_1 \cdot s_2) \in S$.

Clearly this is a generalisation of a M-system.

If J is a semi-prime ideal of the near-ring N, and $S = J^c$, suppose $x \in S$ and $s_1, s_2 \in \langle x \rangle_{(ideal)}$.

Suppose $(s_1 \cdot s_2) \in J = S^c$ is always true. Then $(\langle x \rangle_{(ideal)} \cdot \langle x \rangle_{(ideal)}) \subset J$, which means $\langle x \rangle_{(ideal)} \subset J$. This is a contradiction. It means at least some of the elements $(s_1 \cdot s_2)$ must lie in S and that S is a sp-system. In other words, the complement of a semi-prime ideal is an sp-system.

Lemma 41 *Given an sp-system and any element of it, one can find an M-system containing that element and lying entirely within the original sp-system.*

Proof Let S be the sp-system, and suppose $x \in S$. From the definition of an sp-system, we can find $x_1, x_2 \in \langle x \rangle_{(ideal)}$ such that $(x_1 \cdot x_2) \in \langle S \rangle_{(ideal)}$.

But then, by the same argument, we can find $x_3, x_4 \in \langle x_1 \cdot x_2 \rangle_{(ideal)}$ such that $(x_3 \cdot x_4) \in \langle S \rangle_{(ideal)}$.

Iterating gives us a sequence of pairs, $\{\{x_k, x_{(k+1)}\} : k = (2n+1), n \in \mathbb{N}\}$, and a sequence of ideals

$$\langle x \rangle_{(ideal)} \supseteq \langle (x_1 \cdot x_2) \rangle_{(ideal)} \supseteq \langle (x_3 \cdot x_4) \rangle_{(ideal)} \supseteq \cdots$$

Then, define the subset of S

$$M = \{x, (x_1 \cdot x_2), (x_3 \cdot x_4), \ldots\}$$

This is an M-system. □

1.10.4 Prime and Semi-prime Ideals

This lemma is the work of a number of mathematicians [128].

Lemma 42

(i) A semi-prime ideal is the intersection of all the prime ideals that contain it.
(ii) A semi-prime near-ring is simply a subdirect product of prime near-rings.

Proof

(i) From Lemma 37 this intersection is itself a semi-prime near-ring. Suppose our semi-prime ideal were not equal to it. One could then find an element in the intersection but not in the ideal. The complement of our ideal is an sp-system, one that contains this element. From the previous lemma, there will be an M-system containing the element and lying in that complement, and from Lemma 39 there is a prime ideal containing our semi-prime ideal and disjoint from the M-system, therefore not containing the element. But then the intersection of the prime ideals containing our ideal does not contain the element either, which is a contradiction.
(ii) The notes in Sect. 1.7 applies here with the ideals $\{K_a\}_{a\in\alpha}$ being the prime ideals.

□

If J is an ideal, the intersection of the prime ideals containing it is called the **prime radical** of the ideal, $\mathfrak{p}(I)$ (see Sect. 12.3.6). This ideal is also the intersection of the semi-prime ideals containing it, from Lemma 1.10.4.

Lemma 43 *Suppose J is an ideal of the near-ring N. The quotient near-ring $\left(\frac{N}{J}\right)$ has no nilpotent ideals if and only if J is semi-prime.*

Proof We have the canonical mapping

$$\pi : N \longmapsto \left(\frac{N}{J}\right)$$

Suppose that the quotient has no nilpotent ideals. Suppose that for some ideal K, we have $K^2 \subset J$. Then, $\pi(K^2) = J$, which would make the ideal $\pi(K)$ nilpotent. This means $\pi(K) = J$, and so $K \subset J$. On the other hand, suppose J is semi-prime and that K is the inverse image, under π of a nilpotent ideal of the quotient. We know for some $n \in \mathbb{N}$ that $K^n \subset J$. Now, if J were prime, we could deduce that $K \subset J$, but it is semi-prime. However, from Lemma 1.7, when J is merely semi-prime, it is the intersection of the prime ideals containing it. Consequently, K is a subset of each of these prime ideals and a subset of their intersection, which is J.

□

Pilz lists a number of authors who have worked on these questions, other than the ones I have mentioned, and there has been further progress since his book was written [101].

1.11 Near-Fields

Definition 44 Suppose N is a near-ring.

(i) If the semi-group $(N - \{0\}, \cdot)$ is a group, then $(N, +, \cdot)$ is called a **near-field**.
(ii) If N is a near-field but not a field, then I shall say that N is a **proper** near-field.

Near-fields were the first proper, non-ring, near-rings to appear in scientific publications. Dickson worked with them in the early twentieth century, and they probably constitute the most well-known area of modern near-ring theory due to their links with important areas of group theory and geometry and their obvious intrinsic appeal.

Division Rings and Skew Fields
Rings in which the non-zero elements constitute a multiplicative group are usually called **division rings** or **skew fields**, but some authors reserve these terms for non-commutative cases (fields then not being division rings), and some authors distinguish between division rings and skew fields on the basis of commutativity—with fields being division rings, but not skew fields. Wedderburn's theorem [66] would then say that there are no finite skew fields but plenty of finite division rings. Cohn talks of general skew fields and says division algebras are skew fields which are finite dimensional over their prime subfields [27].

I shall simply use the terms "skew field" and "division ring" interchangeably and usually permit them to be fields. I would normally wish to prevent the two-element constant near-ring (Definition 78) from being classified as a near-field, and this should be understood in subsequent usage.

This is defined on the set $\{0, 1\}$ with product

$$(0^2) = 0, \ (0 \cdot 1) = 0, \ (1 \cdot 0) = 1, \ (1^2) = 1 \tag{1.24}$$

It is not zero-symmetric, and it has no multiplicative identity.

1.12 $\mathcal{A}$-Matrices

In algebra, functions are reliably single-valued, and mappings from a set to itself may sometimes be usefully represented using column-stochastic square matrices, when functions are applied on the left, or row-stochastic matrices, when applied on the right.

Variations of this technique occur throughout this text, with function composition being consistent with matrix products. One sees similar ideas in graph theory and stochastic processes, so this is nothing new. Nevertheless, I find a unifying term for the idea useful in this context of function composition.

I call the sort of matrices occurring, generically, **$\mathcal{A}$-matrices**. They are really a sort of incidence matrix. A more precise definition, for the particular case of groupoids, is given in Sect. A.2.

1.13 Functions and Function Composition

In keeping with my emphasis on right distributivity, functions are usually applied to the *left* of their arguments: $sin(x)$ rather than $(x)sin$. Unfortunately, situations arise in which I am forced to apply them on the right—the permutations in Sect. 8.9.2, for instance—$(x)sin$. I usually note this in some way.

Function composition often gives rise to the key near-ring products which concern us; but I don't always keep to the policy of explicitly noting the product where function composition is concerned, although I may use the composition symbol $\circ$ for clarity or emphasis, particularly when there are other products around, as in Chap. 7.

Thus, $(\theta \circ \phi)$ means *first* apply ϕ and *then* apply θ and will more usually be written $(\theta \cdot \phi)$.

1.14 The δ Operator and Phomomorphisms

1.14.1 The δ Operator

Definition 45

(i) Suppose that A and B are additive groups. A mapping $f : A^n \to B$ is **normalised** provided that $f(a_1, a_2, \dots .a_n) = 0$ whenever any of the a_j are zero. For general mappings of one variable, normalised mappings are simply the ones that preserve zero elements.

(ii) The set of all normalised mappings from A^n to B is denoted by $\mathbf{C^n(A, B)}$. It is convenient to use different notation in Chaps. 7, 11 and 8, where $\mathbf{C^1(A, B)}$ would be written $\mathrm{T}_0(A, B)$.

(iii) Suppose $\theta : A \to B$, where $(A, +)$ and $(B, +)$ are additive groups and θ is normalised.

- I define $\delta(\theta) : A \times A \to B$ by $\delta(\theta)(a, b) = \theta(b) - \theta(a + b) + \theta(a)$.
- More generally, for any normalised mapping $f : A^n \to B$, I define

$$\delta(f) : A^{(n+1)} \to B$$

by : $\delta(f)(a_0, a_1, a_2, \ldots a_n) =$

$$\Big(f(a_1, a_2, \ldots a_n) - f(a_0 + a_1, a_2, \ldots a_n) + f(a_0, a_1 + a_2, a_3, \ldots a_n) - \ldots$$

$$\ldots . + (-1)^n f(a_0, a_1, \ldots, a_{n-2}, a_{n-1} + a_n) + (-1)^{(n+1)} f(a_0, a_1, \ldots a_{n-1})\Big)$$

Lemma 46 *For any normalised mapping* $f : A^n \to B$

(i) $\delta(f)$ *is normalised.*
(ii) $\delta : C^n(Q, G) \to C^{n+1}(Q, G)$ *is a homomorphism modulo the commutator subgroup* $[C^{n+1}(Q, G), C^{n+1}(Q, G)]$.
(iii) $\delta(\delta(f)) = \delta^2(f) \in C^{n+2}(Q, [G, G]) = [C^{n+1}(Q, G), C^{n+1}(Q, G)]$ *(see Lemma 306).*

1.14.2 Phomomorphisms

Definition 47 I call normalised mappings between groups $\theta : (A, +) \to (B, +)$ which are homomorphisms modulo the commutator subgroup of the co-domain group $(B, +)$ **phomomorphisms** from $(A, +)$ to $(B, +)$.

Phomomorphisms $\theta : A \to B$ are exactly the zero-fixing mappings for which

$$\theta(a_1 + a_2) \equiv \theta(a_1) + \theta(a_2) \text{ (modulo}[B, B]) \quad (1.25)$$

This property has been described as characterising *pseudo-homomorphisms* in previous work [103, 104]. I call the collection of phomomorphisms from A to B, Phom(**A**, **B**).

All additive homomorphisms are phomomorphisms; the sum of any two phomomorphisms is a phomomorphism (making Phom(A, B) an additive group); and the composition of two phomomorphisms is a phomomorphism, because phomomorphisms preserve group commutators. These two closure properties, particularly additive closure, are the most striking feature of phomomorphisms.

Phomomorphisms from a group to itself preserve the cosets of the commutator subgroup. Phomomorphisms in which the co-domain is abelian are simply homomorphisms. It is possible for a bijective phomomorphism between two groups to have an inverse which is not a phomomorphism. However, bijective phomomorphisms from a group to itself which have inverses that are also phomomorphisms constitute a multiplicative sub-semi-group of the semi-group of phomomorphisms from the group to itself. It might be interesting to consider the additive structure they generate, but I don't do that here.

Phomomorphisms whose co-domains are simple non-abelian groups are simply zero-fixing mappings. Since all finite groups embed in simple groups (recall $S_n \subset A_{n+2}$), all finite near-rings embed in finite phomomorphism near-rings.

There is more information on phomomorphisms in Chap. 9.

1.15 Annihilators

Notation 48 Suppose $(N, +, \cdot)$ is a pre-near-ring.

(i) If S is any subset of N, then

$$\binom{\mathbf{S}}{\mathbf{0}}_{\mathcal{R}} \text{ is } \{x \in N : S \cdot x = 0\}$$

Similarly,

$$\binom{\mathbf{S}}{\mathbf{0}}_{\mathcal{L}} \text{ is } \{x \in N : x \cdot S = 0\}$$

here, $S \cdot x = \{s \cdot x : s \in S\}$, etc. These sets are, respectively, the right and left **annihilators** of the set S.

(ii) When $S = \{a\}$ is a singleton, I write $\binom{\{a\}}{0}$ as $\binom{a}{0}$.

(iii) On occasion, a near-ring N acts on a group Γ with Γ distinct from N. At need I might adopt the notation $\binom{S}{0}^{N}_{\mathcal{R}/\mathcal{L}}$ and $\binom{S}{0}^{\Gamma}_{\mathcal{R}/\mathcal{L}}$ to indicate explicitly which near-ring group (see Definition 20) is being considered. The superscript indicates the superset of the set of annihilated elements (here, N in the first example, then Γ, in the next).

An example of my use of this notation is given in (7.18).

(iv) I also write

$$\binom{N}{S}_{\mathcal{R}} = \{x \in N : N \cdot x \subseteq S\}$$

and

$$\binom{N}{S}_{\mathcal{L}} = \{x \in N : x \cdot N \subseteq S\}$$

In the next result, I shall adopt the temporary notation

$$S_{\mathcal{A}} = \binom{S}{0}_{\mathcal{R}}$$

$$S_{\mathcal{B}} = \binom{S}{0}_{\mathcal{L}}$$

Recall from Sect. 1.8 that a right sideal is a subset $H \subseteq N$ for which $H \cdot N \subseteq H$ (and a left sideal has $N \cdot H \subseteq H$).

Lemma 49 *Suppose $(N, +, \cdot)$ is a near-ring and $H \subseteq N$.*

(i) *$S_{\mathcal{A}}$ is a right sideal.*
(ii) *$S_{\mathcal{B}}$ is a left ideal of $(N, +, \cdot)$.*
(iii) $((S_{\mathcal{A}})_{\mathcal{B}})_{\mathcal{A}} = S_{\mathcal{A}}$.
(iv) $((S_{\mathcal{B}})_{\mathcal{A}})_{\mathcal{B}} = S_{\mathcal{B}}$

Lemma 50 *Suppose N is a near-ring and $S \lhd N$ is a left ideal.*

(i) *Then,*

$$\binom{N}{S}_{\mathcal{L}} = \{x \in N : x \cdot N \subseteq S\}$$

is an ideal of N.
(ii) *$\left(\frac{N}{S}\right)$ is a near-ring group under the obvious action.*
More,

$$\binom{\left(\frac{N}{S}\right)}{0}_{\mathcal{L}} = \binom{N}{S}_{\mathcal{L}} \tag{1.26}$$

One can extend the idea of annihilators to near-ring groups in the obvious way. The next results need two of the definitions given on page 380.

Lemma 51 *Suppose $_N\Gamma$ is a left near-ring group and Γ is monogenic, $\Gamma = \{N \cdot \gamma\}$.*

(i) *The annihilator of Γ,*

$$\binom{\Gamma}{0}_{\mathcal{L}} = \{n \in N : (n \cdot \omega) = 0, \text{ for all } \omega \in \Gamma\} \tag{1.27}$$

is a (two-sided) ideal of N.
(ii) *The mapping*

$$\theta : N \longmapsto \Gamma, \quad n \mapsto (n \cdot \gamma) \tag{1.28}$$

is a NR-linear mapping from the near-ring group $_N N$ to the near-ring group $_N\Gamma$. The mapping is an epimorphism and has kernel $\binom{\gamma}{0}_{\mathcal{L}}$, which is a left ideal of $(N, +, \cdot)$, and

$$\frac{N}{\binom{\gamma}{0}_{\mathcal{L}}} \cong \Gamma \tag{1.29}$$

(the isomorphism is one of N-groups).
(iii) *When S is a left sideal of N and a subgroup of $(N, +)$, $_N\{S.\gamma\}$ will be a left near-ring group contained in Γ.*

(iv) When ${}_N\Omega$ is a left near-ring group contained in Γ, $S = \{n \in N : (n \cdot \gamma) \in \Omega\}$ is a subgroup of $(N, +)$ and a left sideal of N.
(v) Suppose J is a left ideal of N. Then $\{J \cdot \gamma\}$ is a left mideal of Γ.
(vi) Suppose Γ is a near-ring group of type 0 (see Definition 413). Then, $\binom{\gamma}{0}_{\mathcal{L}}$ is maximal as a left ideal of $(N, +, \cdot)$.
(vii) Suppose Γ is a near-ring group of type 2 (see Definition 413). Then, $\binom{\gamma}{0}_{\mathcal{L}}$ is maximal as a left sideal of $(N, +, \cdot)$.

Proof

(v) $\{J \cdot \gamma\}$ is a subgroup of Γ.
Since $(-(n \cdot \gamma) + (j \cdot \gamma) + (n \cdot \gamma)) = (-n + j + n) \cdot \gamma$, it is a normal subgroup.

$$(n \cdot ((m \cdot \gamma) + (j \cdot \gamma)) - (n \cdot m \cdot \gamma)) = ((n \cdot (m + j) - (n \cdot m)) \cdot \gamma) \in \{J \cdot \gamma\}$$

(vi) Suppose J is a left ideal properly containing the annihilator left ideal. Then $\{J \cdot \gamma\}$ must be equal to Γ. If $J \neq N$, then for any $n \in N$, we may find $j \in J$ such that $(n - j) \cdot \gamma = 0$. But that means $n \in J$.
(vii) If a left sideal S properly contains the annihilator, then $\{S \cdot \gamma\}$ is a near-ring group in Γ and non-trivial. The argument then follows the one made in (vi).

□

Lemma 52 *Suppose $(N, +, \cdot)$ is a simple, zero-symmetric near-ring which possesses an additive subgroup which is also a minimal left sideal $S \neq \{0\}$.*
Then N is a two-primitive near-ring and S is a type 2 near-ring group.

Proof $\binom{S}{0}_{\mathcal{L}}$ is an ideal of N and so trivial. If $s \in S^*$, then the additive group $N \cdot s$ is non-trivial, and a left sideal, so it actually is S, which is therefore monogenic and type 2. □

1.16 Conjugacy and Annihilators

If $(N, +, \cdot)$ is a near-ring, and $a, x \in N$, then

$$\binom{x}{0}_{\mathcal{R}} = \binom{-a + x + a}{0}_{\mathcal{R}}$$

Thus, right annihilators of elements are equal to the right annihilators of their conjugates.

Lemma 53 *Suppose $(N, +, \cdot)$ is a near-ring and that the additive group $(N, +)$ has finite class number $(r + 1)$ (r non-trivial conjugacy classes). Suppose the conjugacy classes are $\{\mathcal{C}_j\}_{j=0,1,2,\ldots}$ with $\mathcal{C}_0 = \{0\}$. Write $\mathcal{A}_j = \begin{pmatrix} \mathcal{C}_j \\ 0 \end{pmatrix}_{\mathcal{R}}$.*

Then any right annihilator in the near-ring is a finite intersection of some of the sets $\mathcal{A}_j$.

Correspondingly, all finite intersections of the sets $\{\mathcal{A}_j\}_{j=1,2,\ldots r}$ correspond to the right annihilator of the union of the corresponding conjugacy classes.

This prescription accounts for all possible right annihilators in the near-ring.

Proof An annihilator, by definition, annihilates some set of elements. The annihilator also annihilates the conjugacy classes of all elements in the set and therefore annihilates the union of the conjugacy classes of these elements.

This is a finite union, and its annihilator is the intersection of the corresponding annihilators of the individual conjugacy classes. □

1.17 Sylow Subgroups

Suppose $(N, +, \cdot)$ is a finite near-ring and p is a prime divisor of its order.

The group has Sylow p-subgroups, and I suppose the collection of them to be $\{S_\alpha\}_{\alpha \in A}$.

If $x \in N$, then $S_\alpha \cdot x = \{a \cdot x \ : \ a \in S_\alpha\}$ will be properly contained in one of the Sylow subgroups of the group (by right distributivity and order considerations).

If this other Sylow subgroup is S_β, then, of course, I can find some $w \in N$ such that $S_\alpha = (S_\beta)^w$, and we have

$$S_\alpha \cdot x = (S_\beta \cdot x)^{(w \cdot x)}$$

I define the union of all the Sylow p-subgroups as

$$\mathcal{S} = \left(\bigcup_{\gamma \in A} S_\gamma \right)$$

This is precisely the collection of elements of the additive group $(N, +)$ that have additive orders some power of p, plus the zero element.

When more than one group or prime is being discussed, I may write this collection as (in this case)

$$\mathcal{S}_p^N \tag{1.30}$$

and I use the same notation when M is merely some *subset of* N, $\mathcal{S}_p^M$ (see (8.18)).

Lemma 54 *Using the notation just established, suppose $\alpha, \beta \in A$.*

(i)

$$\begin{pmatrix} S_\alpha \\ 0 \end{pmatrix}_{\mathcal{R}} = \begin{pmatrix} S_\beta \\ 0 \end{pmatrix}_{\mathcal{R}}$$

(ii)

$$\begin{pmatrix} S_\alpha \\ 0 \end{pmatrix}_{\mathcal{R}} = \begin{pmatrix} \mathcal{S} \\ 0 \end{pmatrix}_{\mathcal{R}}$$

(iii) $\begin{pmatrix} S_\alpha \\ 0 \end{pmatrix}_{\mathcal{R}}$ *is a (two-sided) sideal of $(N, +, \cdot)$.*

In the case of finite solvable groups, there are *Hall subgroups*, which involve collections of primes [132], and these are also conjugate; so the same result applies—the right annihilator of a Hall subgroup is also the right annihilator of all the Hall subgroups for that collection of primes and is also a two-sided sideal.

Returning to the Sylow p-subgroup case – suppose $x \in \begin{pmatrix} \mathcal{S} \\ 0 \end{pmatrix}_{\mathcal{R}}$. The set $N \cdot x = \{n \cdot x \ : \ n \in N\}$ is a subgroup of $(N, +)$ and its intersection with $\mathcal{S}$ is trivial.

I shall often use the notation $\boldsymbol{S_p}$ for a Sylow p-subgroup that interests me.

1.18 The Zeroiser Ideal

Definition 55 The two-sided ideal

$$\mathcal{Z}_{\mathbf{N}} = \{n \in N \ : \ n \cdot N = \{0\}\} = \begin{pmatrix} N \\ 0 \end{pmatrix}_{\mathcal{L}}$$

is called the **zeroiser ideal** of the near ring $(N, +, \cdot)$.

A near-ring equal to its own zeroiser ideal is called trivial (page 7). Near-fields (Definition 44) will have trivial zeroiser ideals, and so the wider class of near-rings in which this ideal is trivial is of particular interest.

In a general, and zero-symmetric, p.n.r., the collection of elements which left annihilate everything will be a two-sided sideal and a normal subgroup, but I cannot deduce it to be an ideal without the associativity law.

1.19 The Core of a Left Ideal

In this section, suppose $(N, +, \cdot)$ is a zero-symmetric, unital, near-ring and that J is a left ideal.

The annihilator

$$\binom{J}{0}_{\mathcal{L}}^{N}$$

is a two-sided ideal of N.

$$\binom{N}{J}_{\mathcal{L}}^{N} = \{x \in N \ : \ x \cdot N \subset J\}$$

is an ideal of N. If $x \in \binom{N}{J}_{\mathcal{L}}$, then $x = x \cdot 1 \in J$.

Motivated by the **normal core** construction, used in group theory (page 117), define the **core of the left ideal** J to be the **largest ideal of** N **contained in** J.

Obviously, $\binom{N}{J}_{\mathcal{L}}^{N}$ is contained in this core. But if x is one of the core's elements, then $x \cdot N \subset J$, so $x \in \binom{N}{J}_{\mathcal{L}}^{N}$

Lemma 56 *The core of the left ideal J is the ideal*

$$\mathcal{C}(\mathrm{J}) = \binom{N}{J}_{\mathcal{L}}^{N} \tag{1.31}$$

A similar idea construction occurs in ring theory [135].

Interestingly, even when the near-ring N is not unital, the structure $\binom{N}{J}_{\mathcal{L}}^{N}$ is actually a two-sided ideal. The identity element serves only to place it inside J; but there are situations in which this occurs without recourse to an identity, (see Lemma 105). Compare this idea with Lemma 19.

1.20 Anti-chains of Subgroups

In this section, $(N, +, \cdot)$ is a near-ring with identity e, and $(N, +)$ is a group with finite exponent n.

When I speak of an **anti-chain** of normal subgroups, I mean a collection of normal subgroups which has the property that no member of the collection is properly contained in any of the others.

$|e| = n$ as a direct consequence of right distributivity. I use the notation for highest common factors described in Notation 1 on page xvi.

If $1 \leq r \leq n$, then $\left(\frac{n}{(n,r)}e\right) \cdot re = 0$. Most of the arguments that follow are based on observing that one can distinguish annihilators of elements re on the basis of the highest common factors (n,r) and annihilating elements that lie within $\langle e \rangle$.

Lemma 57

(i) Suppose $r, s \in \{1, 2, 3, \ldots n\}$; *then*

$$\begin{pmatrix} re \\ 0 \end{pmatrix}_{\mathcal{L}} \subseteq \begin{pmatrix} se \\ 0 \end{pmatrix}_{\mathcal{L}}$$

requires $\frac{n}{(s,n)}$ *divides* $\frac{n}{(r,n)}$.

(ii) When (s, n) *and* (r, n) *are distinct, these two annihilators must also be distinct.*

Proof The argument takes place entirely within $\langle e \rangle$. It is clear that if $1 \leq t < n$, then $te.re = 0 \Rightarrow tr \equiv 0$ (modulo(n)). Thus, $\frac{n}{(r,n)}|(t, n)$.

From this, it is immediate that $\frac{n}{(s,n)}$ divides $\frac{n}{(r,n)}$.

If (r,n) and (s,n) are distinct, I may then write $k.(n, r) = (n, s)$ for non-trivial $k \in \mathbb{N}$.

Now, $\frac{n}{(n,s)}e \in \begin{pmatrix} se \\ 0 \end{pmatrix}_{\mathcal{L}}$ but $\frac{n}{(n,s)}e \notin \begin{pmatrix} re \\ 0 \end{pmatrix}_{\mathcal{L}}$ □

Notation 58 In the next lemma, $\tau(n)$ represents the number of distinct divisors of the number n.

Lemma 59 *The additive group* $(N, +)$ *has at least* $\tau(n)$ *distinct normal subgroups including the proper and the trivial normal subgroups.*

Lemma 60 *Suppose n has prime decomposition* $n = \prod_{i=1}^{r} p_i^{\alpha_i}$; *then the additive group* $(N, +)$ *possesses an anti-chain consisting of r proper, non-trivial, normal subgroups.*

The key point here is that each left annihilator ideal, $\begin{pmatrix} p_i^{\alpha_i} e \\ 0 \end{pmatrix}_{\mathcal{L}}$, contains elements not in any of the others.

1. From this result it is immediate that, say, alternating groups A_n $(n \geq 5)$ host no near-rings with identity.
2. The condition is necessary but insufficient. I prove later that D_{12} hosts no near-rings with identity; but it does have the required normal subgroup structure.
3. The special case $(\pi, n) = 1$ is of interest. Some power of π is congruent to 1 modulo n, and consequently the map $x \mapsto te \cdot x$ must be a bijection of the set N.

This work originates with John Krimmel [89].[3]

Dr Krimmel independently duplicated some part of the work that Marjorie Johnson did on near-rings with identity on dihedral groups, which I report later, and did other interesting work establishing host numbers for various classes of groups, as I understand it, using mainly combinatorial arguments (see Sect. 5.6).

1.21 Subsets

In this section, N is a zero-symmetric near-ring.

Notation 61

(i) I fix a positive integer m and define

$$\mathbf{P^m(N)} \;=\; \{n_1 \cdot n_2 \cdot \ldots \cdot n_m \;:\; n_j \in N\} \tag{1.32}$$

that is, the set of all products of m elements of N.

(ii) $\mathbf{r^{(m)}(N)} \;=\; \langle P^m(N)\rangle$ (subgroup closure).

Since

$$P^{(m+1)}(N) \;\subseteq\; P^m(N)$$

we have that

$$r^{(m+1)}(N) \;<\; r^{(m)}(N)$$

(iii) $\mathbf{R^{(m+1)}(N)}$ is defined as the normal closure of $r^{(m+1)}(N)$ in the additive group $(r^{(m)}(N), +)$.

That is

$$R^{(m+1)}(N) \;=\; \left\langle P^{(m+1)}(N)\right\rangle_N^{r^{(m)}(N)}$$

(iv)

$$\mathbf{j_m(N)} \;=\; \{x \in r^{(m)}(N) \;:\; x \cdot N = 0\} \;=\; \left(\binom{N}{0}_{\mathcal{L}} \bigcap \left(r^{(m)}(N)\right)\right)$$

[3] I do not have access to this. My account develops information obtained from my then-supervisor, Robert Laxton, many years ago. I believe that Dr Laxton was the external examiner for Dr Krimmel's thesis and I hope I have done Dr Krimmel's work justice here.

Naturally, when $(N, +, \cdot)$ has either a right or a left multiplicative identity, $P^m(N) = N$ and therefore $r^{(m)}(N) = N = R^{(m+1)}(N)$.

Lemma 62

(i) $N \cdot r^{(m)}(N) \subseteq r^{(2)}(N)$.
(ii) $r^{(m)}(N) \cdot N = r^{(m+1)}(N)$.
(iii) $r^{(m)}(N)$ *is a sub-near-ring of* N.
(iv) $R^{(m+1)}(N)$ *is an ideal of* $r^{(m)}(N)$.
(v) $j_m(N)$ *is an ideal of* $r^{(m)}(N)$.
(vi) $\frac{r^m(N)}{R^{(m+1)}(N)}$ *is a trivial near-ring, in the sense of Definition 6.*
(vii) Both $r^{(m)}(N)$ *and* $R^{(m+1)}(N)$ *are right sideals of* $(N, +, \cdot)$.
(viii) $R^{(m)}(N) \cdot N \subseteq R^{(m+1)}(N)$

Notation 63 We have the obvious sequence defined as

$$R_1(N) = N, R_2(N) = R^{(2)}(N), \ldots, R_{m+1}(N) = R^{(2)}(R_m(N))$$
$$m = 1, 2, 3, \ldots$$

except that, where possible, I drop the reference to $(N, +, \cdot)$, writing R_{m+1} for $R_{m+1}(N)$.

The sequence is

$$N = R_1 \rhd R_2 \rhd R_3 \rhd R_4 \ldots. \tag{1.33}$$

where each R_{m+1} is an ideal of each R_m and each $\frac{R_{(m)}}{R_{(m+1)}}$ is a trivial near-ring.

1.21.1 Generating Near-Rings

Suppose H is some subset of the near-ring $(N, +, \cdot)$. I am interested in the smallest sub-near-ring that contains H, and, of course, the intersection of two sub-near-rings is a sub-near-ring.

I shall partially extend Notation 61 so that it may apply to subsets and define a sequence of nested subgroups, $\{\mathbf{Q^{(k)}}\}_{\mathbf{k=1,2,\ldots}}$ of subgroups of $(N, +)$ by

(i) $Q^{(1)} = \langle P^1(H)\rangle = \langle H\rangle$
(ii) $Q^{(2)} = \langle P^2(Q^{(1)}) \bigcup Q^{(1)}\rangle$
(ii) $Q^{(n+1)} = \langle P^2(Q^{(n)}) \bigcup Q^{(n)}\rangle$
(iv) $\mathcal{Q}(H) = \bigcup_{k=1}^{\infty} Q^{(k)}$

Lemma 64 *$\mathcal{Q}(H)$ is the near-ring generated by the subset $H \subseteq N$.*

Proof $\mathcal{Q}(H)$ is a subgroup of $(N, +)$ containing H. Any two elements, $x, y, \in \mathcal{Q}(H)$ must lie in some $Q^{(k)}$, and so their product will lie in $Q^{(k+1)}$. Thus

$\mathcal{Q}(H)$ is additively and multiplicatively closed and is the smallest sub-near-ring containing H. □

1.21.2 Lifting Near-Rings

Suppose $(A, +)$ is an additive group and $(B, +, \cdot)$ is a near-ring, and there is a group homomorphism

$$\Theta : (A, +) \longrightarrow (B, +)$$

with kernel K.

Suppose further that there is a **lifting** or **section** of Θ—that is, a (normalised) mapping $\tau : B \to A$ such that $(\Theta \circ \tau)$ is the identity mapping on B—which implies Θ is surjective.

I define a product on the quotient group $\left(\frac{A}{K}, +\right)$ with the formula

$$(a_1 + K) * (a_2 + K) = (\tau(\Theta(a_1) \cdot \Theta(a_2)) + K) \tag{1.34}$$

Lemma 65 $\left(\frac{A}{K}, +, *\right)$ *is a right near-ring isomorphic to* $(B, +, \cdot)$.

The product extends to a near-ring product on A defined by

$$(a_1 + k_1) \overset{*}{*} (a_2 + k_2) = (a_1 * a_2 + 0)$$

Here, K is an ideal and the quotient near-ring is isomorphic to B.

1.22 Nil and Nilpotent Sets

Ring theorists often consider elements $x \in R$ such that $x^n = 0$ for some $n \in \mathbb{N}$. These are the **nilpotent elements** [66]. Subsets $S \subset R$ are described as **nil sets** provided they consist entirely of nilpotent elements.

We allow products of subsets of a near-ring as is described in (1.9), or (1.32)—except I am not now restricting to zero-symmetric cases and have extended the notation to general subsets of near-rings.

A subset S is described as **nilpotent** if we can find some positive integer m such that

$$P^m(S) = \left\{\left(s_1 \cdot s_2 \cdot \ldots s_{(m-1)} \cdot s_m\right) = 0 \,\forall\, s_i \in S, 1 \le i \le m\right\} \tag{1.35}$$

Clearly, nilpotent subsets are nil subsets. The opposite is rarely true.

The important cases involve nilpotent and nil ideals. In nilpotent ideals, as we have defined it, the issue as described above is whether the product of *any m* elements should be zero, rather than the $m^{th.}$ powers of elements as in the nil case.

Some texts define products of ideals by forming the usual product and then generating the smallest ideal containing it (see page 12), a procedure invoked in both rings and near-rings but involving appreciably more complicated mechanisms in the latter—simple sums of elements not sufficing. When this is done, there are occasions when such a power eventually becomes zero and is an appreciably stronger nilpotence condition than the one that concerns me in this book.

Specifically,

Definition 66

(i) If I and J are ideals of a near-ring, their **strong product** is the additive group

$$(I \times J) = \langle\{(i \cdot j) : i \in I, j \in J\}\rangle_{(add)} \tag{1.36}$$

(ii) An ideal J is **strongly nilpotent** provided that for some finite positive integer n, the strong product

$$\overbrace{(J \times J \times J \times \ldots \times J)}^{\text{(n times)}} = \{0\} \tag{1.37}$$

The strong product $(I \times J)$ forms an ideal in rings but not generally in near-rings where it forms a right sideal and a subgroup. When the near-ring is zero-symmetric, it forms a right near-ring group.

I do extend these definitions in the obvious way in order to talk about strong nilpotence in one-sided ideals and sideals.

1.22.1 Sums of Nil Ideals

Lemma 67

(i) Suppose N is a near-ring, J is a nil ideal and $H \subset N$ and H are a nil set. Then $(H + N)$ (see (1.8)) is a nil set.
(ii) The sum of any finite collection of nil ideals is a nil ideal.
(iii) The sum of any collection of nil ideals is a nil ideal.

Proof

(i) $(H + I)$ is actually an additive subgroup. The quotient $\frac{(H+I)}{I}$ is nil, and since I is too, the whole set is.
(ii) The named sum is the set of finite sums of elements from the ideals and is the smallest ideal containing them all. It is nil from the first part.

(iii) To cater for possibly infinite sums, we need only realise that any given element is in an ideal consisting of but a finite number of nil ideals and therefore is nilpotent from part (ii).

□

1.22.2 Sums of Nilpotent Ideals

Lemma 68 *Suppose I and J are nilpotent ideals of a near-ring N.*

(i) Then $(I + J)$ is a nilpotent ideal.
(ii) The sum of any ***finite*** *collection of nilpotent ideals is a nilpotent ideal.*

Proof

(i) Both I and J are nil ideals, so $(I + J)$ is nil. However, nilpotence informs us of fixed integers in I and J with products involving those terms resulting in zero. Arguments similar to the proof in Lemma 67 now ensure that $(I + J)$ is nilpotent.

□

We cannot go on to adduce that sums of infinite collections of nilpotent ideals would be nilpotent. The problem is that the fixed integers referred to in the previous proof may constitute a potentially unbounded set of integers.

Indeed, it is not true that the sum of any collection of nilpotent ideals will be nilpotent and there are good examples of this.

1.23 Cores

Definition 69 Suppose $(N, +, \cdot)$ is a bi-zero-symmetric, right pre-near-ring (see Sect. 1.4.2).

The **distributivity core** is defined as

$$\mathcal{C}_{\mathcal{D}}(N) = \{d \in N : d \cdot (a + b) - (d \cdot b) - (d \cdot a) = 0, \ \forall \, a, b, \in N\} \quad (1.38)$$

The **associativity core** is defined as

$$\mathcal{C}_{\mathcal{A}}(N) = \{a \in N : a \cdot (b \cdot c) = (a \cdot b) \cdot c, \ \forall \, b, c, \in N\}$$

For near-rings with identity, the **invertibility core** is defined as

$$\mathcal{C}_{\mathcal{I}}(N) = \{a \in N : \exists \, b \text{ such that } (a \cdot b) = 1 = (b \cdot a)\}$$

Observations 70

(i) The first two cores contain the zero element and any multiplicative identity that exists. Both contain the zeroiser ideal (Definition 55).

(ii) The associativity core is a subgroup of $(N, +)$. For this reason it has previously been referred to as **the associativity subgroup** of $(N, + \cdot)$ (see [100]).

(iii) The associativity core is multiplicatively closed and forms a sub-near-ring of the pre-near-ring $(N, + \cdot)$. It is important in my approach to the host problem (Chap. 2).

(iv) When the pre-near-ring product is associative (i.e. we have a near-ring), the distributivity core is multiplicatively closed and, of course,

$$\{((d \cdot b) - d \cdot (a + b) + (d \cdot a)) : d, a, b \in N\} \subseteq r^{(2)}(N).$$

(v) In near-rings, the distributivity core additively generates a distributively generated sub-near-ring of the original near-ring (see Definition 75).

(vi) Distributively generated near-rings have been studied since the 1950s [42, 98, 119].

(vii) In near-rings hosted by abelian groups, the distributivity core is actually a ring since d.g. near-rings hosted by abelian groups are rings (see page 49). Some authors refer to d.g. near-rings by stipulating the existence of a multiplicative semi-group of distributive elements which should additively generate the near-ring. This is inappropriate as a definition if one wishes to extend the idea to pre-near-rings (as I do). Any such set of generators produces a multiplicatively associative structure.

(viii) Two lemmata[4] on near-fields.

(a) **Lemma 71** *Suppose that $(N, +, \cdot)$ is a zero-symmetric near-ring with the property that for any $n \in N^*$, $N \cdot n = N$.*
Suppose further that $\mathcal{C}_{\mathcal{D}}(N))^* \neq \phi$.
Then, $(N, +, \cdot)$ is a near-field.

Proof Taking $a, b \in N^*$, we can find $c, d \in N$ such that $a = c \cdot b$, and $d \cdot a = c$.

Now,

$$d \cdot (a \cdot b) = (d \cdot a) \cdot b = (c \cdot b) = a$$

This implies $a \cdot b \neq 0$, and it applies to any pair of non-zero elements.

Now select $d \in \mathcal{C}_{\mathcal{D}}(N))^*$, and write $e \cdot d = d$. Clearly, $(d \cdot e - d) \cdot d = 0$, so $d \cdot e = d$.

[4] See the footnote on page 94.

Moreover, taking $n \in N$, $d \cdot (e \cdot n - n) = 0$, so $(e \cdot n - n) = 0$. e is a left identity on N.

But $(n \cdot e - n) \cdot d = 0$, so e is a right identity on N, and N is a near-field. □

(b) **Lemma 72** *Suppose $(N, +, \cdot)$ is a zero-symmetric near-ring with a left identity element, e (so, $e \cdot x = x$ for all x).*

Suppose the near-ring has no proper non-trivial subgroups which are themselves left sideals.

Then $(N, +, \cdot)$ is a near-field.

Proof N acts faithfully upon itself because the set $\{m \in N : m \cdot N = \{0\}\}$ is a left annihilator ideal and therefore a left sideal.

This means $N = N \cdot n$ for all $n \in N^*$. Obviously $e \in (\mathcal{C}_{\mathcal{D}}(N))^*$.

Now apply Lemma 71. □

(ix) The invertibility core is a multiplicative subgroup of the multiplicative semi-group in near-rings (associativity assumed) and suggests a natural right action on that semi-group: $x \mapsto x^a$. It seems generally less useful than the other two cores and is not usually additively closed, nor is its additive closure usually multiplicatively closed. For rings with identity, the invertibility core is a multiplicative group which is often called the **group of units** of the ring.

I suppose one might christen unital near-rings for which $\{N - \mathcal{C}_{\mathcal{I}}(N)\}$ is an ideal, **local near-rings** (page 280 of [94] covers local rings, as do many other sources).

Near-fields are, of course, local near-rings, although the required ideal is trivial.

Local rings have been extensively explored and have striking properties. For example, factored out by the Jacobson radical, they give division rings [74, 127]. I believe that several people have looked at *local near-rings*, perhaps using a rather more intelligent definition for them than I have supplied [128].

(x) In unital rings (ring with identity), people sometimes describe a ring automorphism $\phi : R \mapsto R$ as **inner** if there exists some invertible element a for which $\phi(x) = (a^{-1} \cdot x \cdot a)\ \forall x \in R$. If one extends this definition to near-rings with identity, one has

$$a^{-1} \cdot (x + y) \cdot a = \phi(x + y) = (\phi(x) + \phi(y)) = \left(a^{-1} \cdot x \cdot a + a^{-1} \cdot y \cdot a\right) = \left(a^{-1} \cdot x + a^{-1} \cdot y\right) \cdot a$$

which implies

$$a^{-1} \cdot (x + y) = \left(a^{-1} \cdot x + a^{-1} \cdot y\right)$$

So a^{-1} should be in $\mathcal{C}_{\mathcal{D}}(N)$.

Now, write $a \cdot (x + y) = (a \cdot x) + (a \cdot y) + d$ and get

$$\begin{aligned}(x + y) \cdot a &= a^{-1} \cdot (a \cdot x) \cdot a + a^{-1} \cdot (a \cdot y) \cdot a + a^{-1} \cdot d \cdot a \\ &= (x \cdot a) + (y \cdot a) + a^{-1} \cdot d \cdot a\end{aligned}$$

Consequently, $a^{-1} \cdot d \cdot a = 0 \Rightarrow d = 0$.

From that, $a \in \mathcal{C}_{\mathcal{D}}(N)$, and, for near-rings with identity:

Lemma 73

(i) If $x \mapsto \left(a^{-1} \cdot x \cdot a\right)$ *is a near-ring automorphism, then,*

$$a, a^{-1} \in \left(\mathcal{C}_{\mathcal{D}}(N) \bigcap \mathcal{C}_{\mathcal{I}}(N)\right) \tag{1.39}$$

(ii) The condition is both necessary and sufficient.

(iii) The group of distinct inner automorphisms of the near-ring is then

$$\left(\frac{(\mathcal{C}_{\mathcal{D}}(N) \bigcap \mathcal{C}_{\mathcal{I}}(N))}{(Cent(N) \bigcap \mathcal{C}_{\mathcal{I}}(N))}\right) \tag{1.40}$$

Lemma 73 is interesting because one can use it to determine the inner automorphism groups for all finite near-fields.

For Dickson near-fields, these are trivial—Lemma 188. The non-Dickson near-fields are given in Table 3.3, and the inner automorphisms are only non-trivial for the "fantastic four", where they are cyclic, with orders two, three, five and two, respectively.

Near-Ring Automorphisms

Automorphisms—bijective homomorphisms from a near-ring to itself—form a group under composition.

Suppose $a \in \mathcal{C}_{\mathcal{I}}(N)$ (N a near-ring with identity), and suppose θ is an automorphism of $(N, +, \cdot)$.

Then,

$$\theta(a^{-1}) \cdot \theta(a) = \theta(1) = 1 \Rightarrow \theta(a^{-1}) = (\theta(a))^{-1}$$

Now, again suppose θ is an automorphism and $i(x) = \left(a^{-1} \cdot x \cdot a\right)$ is an inner automorphism.

Then,

$$\left(\theta^{-1} \circ i \circ \theta\right)$$

maps

$$x \mapsto \theta^{-1}\left(a^{-1} \cdot \theta(x) \cdot a\right) = \left(\left(\theta^{-1}(a)\right)^{-1} \cdot x \cdot \left(\theta^{-1}(a)\right)\right)$$

Lemma 74 *The set of distinct inner automorphisms form a normal subgroup of the multiplicative group of all distinct automorphisms.*

There will be a corresponding *outer automorphism group* (see page xviii), though I don't know how useful that might be.

The Associator
The expression

$$(((x \cdot y) \cdot z) - (x \cdot (y \cdot z))) \tag{1.41}$$

is often called the **associator** in non-associative structures.

It is an important **ternary** operation (see [136] and Sect. C.1.3).

1.24 Classes of Near-Rings

The reader will be beginning to appreciate that near-rings represent a considerable generalisation of rings. The only strategy one might have in order to obtain some reasonable structure theorems seems to be to restrict attention to various judiciously chosen subclasses. I am particularly interested in near-rings with identity elements. Other, more restricted, subclasses seem to occur naturally and have always attracted researchers.

It seems obvious that we need clear examples of the near-rings that occur, both their structure and the groups that host them. This is the defence's case for the space I devote to the *host problem*, in what follows.

1.24.1 Distributively Generated and F-Near-Rings

Suppose that $(N, +, \cdot)$ is a right near-ring (see Definition 9).

Definition 75

(i) It is a **zero-symmetric** near-ring if $m \cdot 0 = 0 \quad \forall m \in N$.[5]

[5] This restates an earlier definition.

(ii) It is a (right) **F-near-ring** provided it is zero-symmetric and

$$m \cdot (n+k) \equiv (m \cdot n) + (m \cdot k) \ \ (\text{modulo } ([N, N]))$$

$\forall\, m, n, k, \in N$

(iii) It is **distributively generated** (d.g.), provided there is an additive set of generators of the group $(N, +)$ which are themselves left-distributive (i.e. generators d such that $d \cdot (x+y) = (d \cdot x) + (d \cdot y)$, $\forall\, x, y \in N$). All such near-rings are zero-symmetric.

> In this book I try to be fairly specific, but if there is any doubt, one should assume my near-rings to be zero-symmetric. The unqualified term "near-ring" should be understood to mean "right-distributive zero-symmetric near-ring" unless there is a prior statement to the contrary.

The multiplicatively opposite structure to a right-distributive near-ring is a **left-distributive, near-ring** (see Sect. 1.24.5).

I do occasionally need to consider left-distributive near-rings (e.g. throughout Chap. 16) and shall always refer to them in those terms, expecting the reader to make the required modifications (d.g. left-distributive near-rings, left-distributive F-near-rings, etc.).

The reader may reasonably expect me to stick to right- or left-distributive near-rings throughout. The reason I do not do this is that occasionally constructions occur in which one crosses the divide, with right near-rings spawning left ones and vice versa (see Sect. 2.1.3).

All these distributivity terms may also be applied to p.n.r. in the obvious way.

F-near-rings form a proper subclass of the class $\mathcal{F}$ of near-rings whose commutator subgroups are ideals (Sect. 1.24.2) and a proper superclass of the class of d.g. near-rings. They were named, by me, after Albrecht Fröhlich, a kind man, and one of the British pioneers of near-ring research.

Pseudo-Distributive Generation

If $(N, +, \cdot)$ is a right near-ring with a set of additive generators $\{n_j\}_{j=1}^{k}$, all of which have the property

$$(n_j \cdot (a+b)) \equiv \big((n_j \cdot a) + (n_j \cdot b)\big) \ (\text{modulo } ([N, N]))$$

then $(N, +, \cdot)$ will be a F-near-ring.

This is a sort of generalisation of the idea of distributive generation.

1.24.2 Class $\mathcal{F}$ Near-Rings

Class $\mathcal{F}$ near-rings are precisely the near-rings satisfying the weaker distributivity property that for all $a, b, \in N$ and for all $d \in [N, N]$

$$a \cdot (b + d) \equiv (a \cdot b) \text{ (modulo } [N, N]) \tag{1.42}$$

In these near-rings additive commutator subgroups are ideals. The class includes d.g. near-rings and F-near-rings but is larger than that. Simple near-rings from class $\mathcal{F}$ are either abelian or hosted by perfect groups (Definition 3). Appendix B gives examples of class $\mathcal{F}$ near-rings which are not actually F-near-rings. I use this idea to generalise the concept of a phomomorphism in Sect. 8.2.1.

Remark 76

(i) Although right distributivity guarantees $(-x) \cdot y = -(x \cdot y)$ in all near-rings, it is not generally true that $(x) \cdot (-y) = -(x \cdot y)$. The property does hold in near-fields.

(ii) My near-rings are not automatically assumed to have multiplicative identities. When they have them, I shall often use the symbol 1 to denote them.

(iii) Given a *near-ring homomorphism* $\Theta : N \longrightarrow R$, $\Theta(1)$ acts as the identity on $\Theta(N)$. When Θ is surjective, this maps identities to identities.

(iv) d.g. near-rings hosted by abelian groups are actually rings, and d.g. pre-near-rings on abelian groups are bi-distributive (see Definition 87 and Observations 70).

(v) Similarly, F-near-rings hosted by abelian groups are rings.

(vi) When $(N, +, \cdot)$ is a right-distributive stem with left multiplicative identity 1 $(1 \cdot x = x)$, then if 1 has finite additive order, that order will be equal to the exponent of the additive group. This immediately rules out many kinds of groups as hosts for right-distributive stems with multiplicative identities, such as D_5, because the group has exponent 10 and no element with that order; and it rules out non-abelian groups with square-free order.

In Chap. 4 (page 171), and in Chap. 5, I describe groups that do have elements with order equal to their own exponents as *I-appropriate groups*.

It is interesting that the additive group exponent could be used to extend the field theory idea of the **characteristic** to general near-rings, whether or not they were I-appropriate.

(vii) If $(N, +, \cdot)$ is a near-ring with (multiplicative) identity e hosted by a group $(N, +)$ which has finite exponent n, and n is composite (say n = m.r), then the near-ring possesses non-zero, zero divisors, because $(me) \cdot (re) = ne = 0$.

1.24.3 The $\mathcal{F}_j$ Cores, ($j = 1, 2, 3$)

Definition 77 Suppose $(N, +, \cdot)$ is a right-distributive near-ring.

(i) The $\mathcal{F}_1$-core is the set

$$\mathcal{C}_{\mathcal{F}_1}(\mathrm{N}) = \{n \in N : n \cdot (a+b) \equiv (n \cdot a) + (n \cdot b) \text{ (modulo } [N, N]), \text{ for all } a, b, \in N\} \quad (1.43)$$

(ii) The $\mathcal{F}_2$-core is the set

$$\mathcal{C}_{\mathcal{F}_2}(\mathrm{N}) = \{n \in N : n \cdot (a+d) \equiv (n \cdot a) \text{ (modulo } [N, N]) \text{ for all } a, \in N \text{ and all } d \in [N, N]\} \quad (1.44)$$

(ii) The $\mathcal{F}_3$-core is the set

$$\mathcal{C}_{\mathcal{F}_3}(\mathrm{N}) = \{n \in N : n \cdot [N, N] \subseteq [N, N]\} \quad (1.45)$$

Each of these cores is a sub-near-ring of $(N, +, \cdot)$ containing $[N, N]$; but we cannot claim that the $\mathcal{F}_1$-core is an $\mathcal{F}$-near-ring or that the $\mathcal{F}_2$-core is a class $\mathcal{F}$ one.

It might be interesting to investigate iterations of this process—taking the core of cores; and we do know

$$\mathcal{C}_{\mathcal{F}_1}(N) \subseteq \mathcal{C}_{\mathcal{F}_2}(N) \subseteq \mathcal{C}_{\mathcal{F}_3}(N) \quad (1.46)$$

1.24.4 Constant and Near-Constant Near-Rings

Definition 78

Structures which satisfy all the near-ring axioms except zero symmetry are called **non-zero-symmetric** near-rings.

A **constant near-ring** is one for which, $\forall\, x, y, \in N$ we have that $x \cdot y = x$.

I describe near-rings for which

$$x \cdot y = \begin{cases} x & \text{if } y \neq 0 \\ 0 & \text{otherwise} \end{cases}$$

as **near-constant**.

Non-trivial constant near-rings are not zero-symmetric. In fact, they are essentially the complement of zero-symmetric near-rings (see Lemma 80).

Any d.g. near-ring N is an F-near-ring. Moreover, for all x, y, z in N, we have that

$$z \cdot (x + y) \equiv (z \cdot x) + (x \cdot y) \ (modulo\ [r^2(N), r^2(N)]) \tag{1.47}$$

1.24.5 Opposites

Definition 79 If $(N, +, \cdot)$ is a (right-distributive) near-ring, $(\mathbf{N}, +, \cdot)^{\mathbf{opp}}$ is the opposing *left-distributive* near-ring with multiplication (:) given by

$$x : y = y \cdot x \tag{1.48}$$

It is called the corresponding **opposite** near-ring. Notice that the multiplication of $(N, +, \cdot)^{opp}$ is (:) rather than $(\cdot)$. That is, $(N, +, \cdot)^{opp} = (N, +, :)$ (it's tempting to think the multiplication of $(N, +, \cdot)^{opp}$ is $(\cdot)$).

I also speak of *opposite* stems in what follows.

1.24.6 Non-Zero-Symmetric Near-Rings

The next lemma discusses near-rings without the zero symmetry requirement.

Lemma 80

(i) Any near-ring $(N, +, \cdot)$ is of the form $(N_0 + N_c)$ where N_0 is the largest zero-symmetric sub-near-ring contained in N and N_c is the largest constant near-ring contained in N.
(ii) The group $(N, +)$ is a semi-direct product of the normal subgroup N_0 by the subgroup N_c, $N = N_0 \rtimes N_c$.
(iii) N_c is a (two-sided) sideal of $(N, +, \cdot)$, that is, a near-ring group.
(iv) N_0 is a left ideal of the near-ring $(N, +, \cdot)$.

Proof Write $N_0 = \{n \in N : n \cdot 0 = 0\}$. This is clearly a zero-symmetric near-ring. It is also the left annihilator of the element 0 and, as such, a left ideal (Lemma 49). Write $N_c = \{n \cdot 0 : n \in N\}$. This is a constant near-ring. If $x \in N$, then $x = x - x \cdot 0 + x \cdot 0$. Certainly, $(x - x \cdot 0) \in N_0$ and it is equally clear that $x \cdot 0 \in N_c$. So we have that $N = N_0 + N_c$. N_0 is all the zero-symmetric elements, so the term "largest" is obvious. Now suppose $R \subset N$ is a constant near-ring. We know that $r \cdot 0 = r$ since R, by hypothesis, is a constant near-ring. This means $r \in N_c$ and hence $R \subset N_c$. The mapping $N \Longrightarrow N \cdot 0 = \{n \cdot 0 : n \in N\}$ in which $n \mapsto n \cdot 0$ is a homomorphism of groups whose kernel corresponds to N_0 and whose image corresponds to N_c. □

Elements of $x \in N_c$ have the property that $x \cdot y = x$ for all $y \in N$. That is, all elements of the near-ring are right identities for the elements of N_c.

1.25 The Distributor and the Annular Ideal

Let N be a near-ring.

1.25.1 The Distributor Ideal

I define $\delta(\mathcal{N})(a, b, c) =$

$$b \cdot c - (a + b) \cdot c + a \cdot (b + c) - a \cdot b = -(a \cdot c) + a \cdot (b + c) - (a \cdot b)$$

The *additive* closure of all the expressions $\delta(\mathcal{N})(a, b, c)$ is a subgroup of $r^2(N)$.

Notation 81 I shall use the symbol $\mathcal{D}(N)$ to represent the normal closure of *that* subgroup within $(N, +, \cdot)$.

Definition 82 $\mathcal{D}(N)$ will be called the **distributor ideal** of $(N, +, \cdot)$.

Thus,

$$\mathcal{D}(N) = \langle\{\delta(\mathcal{N})(a, b, c) : a, b, c \in N\}\rangle_{\mathcal{N}}^{N}$$

and

$$\mathcal{D}(N) \lhd R^2(N) \lhd N$$

Notice

(i) $d \cdot 0 \in \mathcal{D}(N)$ for all $d \in N$.
(ii) $d \cdot (-x) + (d \cdot x) \in \mathcal{D}(N)$ for all $d, x \in N$.
(iii) $(d \cdot a) - d \cdot (a + b) + (d \cdot b) \in \mathcal{D}(N)$ for all $d, a, b \in N$.
(iv) $\{d \cdot \mathcal{D}(N)\} \subset \mathcal{D}(N)$ for all $d \in N$.

Lemma 83

(i) $\mathcal{D}(N)$ *is an ideal of* $(N, +, \cdot)$.
(ii) $\frac{N}{\mathcal{D}(N)}$ *is a bi-distributive near-ring.*

Proof Taking $a, b, c \in N$, we know

$$(a \cdot (b + c)) \equiv ((a \cdot b) + (a \cdot c))(\text{modulo } \mathcal{D}(N))$$

This, in turn, means that $N \cdot \mathcal{D}(N) \subseteq \mathcal{D}(N)$ and ensures that $\mathcal{D}(N)$ is a left ideal. That it is a right ideal is an immediate consequence of right distributivity. □

If $(F, +, \cdot)$ is a right-distributive near-field, then $\mathcal{D}(F) = F$.

1.25.2 The Multiplicative Centre

For any near-ring $(N, +, \cdot)$, we may define a **multiplicative centre**

$$\mathbf{Cent}(N) = \{n \in N : n \cdot r = r \cdot n \, \forall r \in N\} \tag{1.49}$$

This structure is multiplicatively closed and a subset of the distributivity core. In standard ring theory, the multiplicative centre is an important subring, but this is generally untrue in near-rings, though it is true for *most* finite near-fields. The multiplicative centre is of particular interest to me in Chap. 3, although, in much of that chapter, confusingly, I use the symbol $\mathcal{Z}$ for the multiplicative centre (page 117)—the excuse being that in that chapter the additive groups are usually abelian anyway, so additive centres are of no interest.

Consider the subset of Cent(N) consisting of its multiplicatively idempotent elements (i.e. $x^2 = x$). We obtain a sub-near-ring of N. In rings with identity, these **central idempotent elements** form a **Boolean algebra** (Sect. 2.5.2) with operations

$$\begin{aligned}(x \wedge y) &= (x \cdot y)\\ (x \vee y) &= (x + y - x \cdot y)\\ x' &= (1 - x)\end{aligned}$$

(I am using the definition on page 95: x' represents the complement operation I define there).

This result is given in [128], there attributed to the doctoral dissertation of Gerhard Betsch.

Lemma 84 *Suppose that e is a central idempotent of the near-ring $(N, +, \cdot)$ and that*

$$(N \cdot e) = \{(n \cdot e) : n \in N\} \text{ is an ideal of } N$$

Then,

$$N = \left((N \cdot e) \oplus \begin{pmatrix} \{e\} \\ 0 \end{pmatrix}_{\mathcal{L}}\right) \tag{1.50}$$

Proof Centrality means

$$\begin{pmatrix} \{e\} \\ 0 \end{pmatrix}_{\mathcal{L}} = \begin{pmatrix} (N \cdot e) \\ 0 \end{pmatrix}_{\mathcal{L}}$$

and this is an ideal.

Clearly,

$$\left((N \cdot e) \cap \left(\begin{pmatrix} \{e\} \\ 0 \end{pmatrix}_{\mathcal{L}}\right)\right) = \{0\}$$

□

Central idempotents are additively closed in near-rings, and the result is a Boolean near-ring with identity (see Sect. 2.5.2). According to Corollary 130, this will be a Boolean ring. In that case, we again obtain a Boolean algebra.

In Sect. 14.1 I generalise the idea of the multiplicative centre for F-near-rings.

Peirce Decomposition

Ordinary (i.e. not necessarily central) idempotents provide what some people have called a **Peirce decomposition for near-rings.**[6]

Suppose e is an idempotent and $n \in N$; then,

$$n = ((n - n \cdot e) + (n \cdot e)) \tag{1.51}$$

so any element is a sum of something in the left ideal $\begin{pmatrix} \{e\} \\ 0 \end{pmatrix}_{\mathcal{L}}$ and something in the subgroup and left sideal $\{N \cdot e\}$.

1.25.3 The Annular Ideal

Lemma 85 *In any near-ring N, the set*

$$\mathcal{A}_{\mathbf{N}} = (\mathcal{D}(N) + [N, N]) = \{(d + c) : d \in \mathcal{D}(N), c \in [N, N]\}$$

forms a two-sided ideal which is inside the kernel of any epimorphic image of N which is a ring.

[6] Two Peirces occur in this book. This one has been described as the "Father of Pure Mathematics in America". He is also the father of the other one.

Definition 86 $\mathcal{A}_N$ will be called the **annular ideal** of the near-ring N.

The annular ideal always contains the distributor ideal. In F-near-rings (Definition 75), the annular ideal is equal to the commutator subgroup of the additive group. In particular, this will be true for d.g. near-rings.

The annular ideal might appropriately be termed the **near-ring core** of the near-ring. Another possible name for it would be the **ring radical**; see Sect. 12.3.1.

1.26 Bi-distributive Stems

Definition 87 A **bi-distributive** stem is a pre-near-ring, N, which is also left-distributive.

Lemma 88 *When N is a bi-distributive near-ring, $r^2(N)$ is a ring.*

Proof $a \cdot c + a \cdot d + b \cdot c + b \cdot d = a \cdot (c + d) + b \cdot (c + d) = (a + b) \cdot (c + d) = (a + b) \cdot c + (a + b) \cdot d = a \cdot c + b \cdot c + a \cdot d + b \cdot d$, and so, $[b \cdot c, a \cdot d] = 0$. □

Corollary 89 *A bi-distributive near-ring with either a left or a right identity must be a ring.*

Corollary 90 *In any bi-distributive near-ring N,*

$$[N, N] \subseteq \mathcal{Z}_N$$

that is, the commutator subgroup is contained within the zeroiser ideal (Definition 55).

Observations 91 The commutator subgroup is actually an ideal in any bi-distributive near-ring, and the corresponding quotient structure will be a (possibly trivial) ring. In the example in Sect. 5.1, the quotient is the non-trivial ring on the cyclic group of order 2.

Lemma 92 *Suppose $(N, +, \cdot)$ is a d.g. near-ring and $r^2(N)$ is abelian. Then $(N, +, \cdot)$ is bi-distributive.*

Notation 93 When $(N, +, \cdot)$ is a pre-near-ring, we may define, for each $n \in N$, mappings

$$\mathcal{L}_\mathbf{n} : N \longrightarrow N \quad \mathcal{L}_\mathbf{n}(a) = (n \cdot a)$$

$$\mathcal{R}_\mathbf{n} : N \longrightarrow N \quad \mathcal{R}_\mathbf{n}(a) = (a \cdot n)$$

$$\mathcal{N} : N \times N \longrightarrow N \quad \mathcal{N}(a, b) = (a \cdot b)$$

Using the notation given in Definition 45, we have

$$\delta\mathcal{L}_n(a,b) = n \cdot b - n \cdot (a+b) + n \cdot a$$

and

$$\delta\mathcal{N}(a,b,c) = (b \cdot c) - ((a+b) \cdot c) - (a(b+c)) + (a \cdot b)$$

The mapping $\mathcal{R}_n$ induces another mapping, from the collection of conjugacy classes to itself. So, if $\mathcal{C}_j$ is a conjugacy class of the additive group $(N, +)$, then $\mathcal{R}_n\left(\mathcal{C}_j\right) \subseteq \mathcal{C}_i$, for some other conjugacy class, $\mathcal{C}_i$, of $(N, +)$.

Notation 94 When $(N, +, \cdot)$ is a right-distributive stem, I write

$$\mathcal{F}_{(N,\ +,\cdot)} = \{n \in N \ : \ \delta\mathcal{L}_n(a,b) \in [N,N]\,,\ \forall\, a, b, \in N\}$$

Near-rings for which $\mathcal{F}_{(N+\cdot)} = N$ are called F-near-rings (see Definition 75). In a general near-ring N, the set

$$\mathcal{F}_{(N\ +\ \cdot)} = \{d \in N \ : \ d \cdot (a+b) \equiv (d \cdot a) + (d \cdot b)\ (\text{ modulo}[N,N])\}$$

(defined as $\mathcal{C}_{\mathcal{F}_1}(N)$ in Definition 77) is merely a sub-near-ring, but in the case of transformation near-rings, this set has further significance (see Lemma 371).

Lemma 95

(i) $[N,N]$ *is an ideal of the sub-near-ring* $\mathcal{F}_{(N,\ +,\ \cdot)}$
(ii) $\frac{\mathcal{F}_{(N,\ +,\cdot)}}{[N,N]}$ *is a ring.*
(iii) $\mathcal{F}_{(N,\ +\ \cdot)} \lhd (N,\ +)$

One might call pre-near-rings in which $\delta\mathcal{N} : N^3 \longrightarrow [N,N]$, **F-pre-near-rings.**

Notation 96 I am particularly concerned with the near-ring Phom(**N**, **N**) of all phomomorphisms from an additive group $(N, +)$ to itself, and I usually write that near-ring as **Phom**(N), or even $\mathrm{P}_0(N)$.

Using the notation given in (93), supposing $(F, +, \cdot)$ is a F-near-ring, then the mapping

$$\begin{aligned} \mathcal{L} : N &\longrightarrow \text{Phom}(N) \\ n &\mapsto \mathcal{L}_n \end{aligned}$$

is a homomorphism of near-rings.

1.27 Subgroup Series

This section covers ideas which began in some early papers of Fröhlich [41–43] and uses Notations 63 and 81.

1.27.1 Weak Distributivity

I define

Notation 97

$$\mathcal{D}_m(N) = \mathcal{D}(R_m) \tag{1.52}$$

I would also write $\mathcal{D}_m$ for $\mathcal{D}_m(N)$ (dropping the reference to the near-ring) where possible.

Lemma 98

(i) Each $\mathcal{D}_m$ is an ideal of each R_m in the sequence (1.33).
(ii) Each quotient $\frac{R_m}{\mathcal{D}_m}$ is bi-distributive.
(iii) $\mathcal{D}_{m+1} \subseteq \mathcal{D}_m \subseteq R_{(m+1)}$.

It is apparent that we have a sequence

$$N \rhd \mathcal{D}_1 \rhd \mathcal{D}_2 \rhd \mathcal{D}_3 \rhd \dots \tag{1.53}$$

In which each term is an ideal in the one to its left.

Definition 99 A near-ring is **weakly distributive** provided that the sequence (1.53) terminates finally with some term $\mathcal{D}_m$ equal to $\{0\}$.

If $(N, +, \cdot)$ is weakly distributive, then the corresponding R_m is a bi-distributive near-ring. It will either be additively perfect (equal to its own commutator subgroup), and trivial, or have a homomorphic image which is a ring (see Lemma 91).

1.27.2 Annularity

Starting from a near-ring N and Definition 86, we may define a sequence

$$\mathcal{A}_1 \supseteq \mathcal{A}_2 \supseteq \mathcal{A}_3 \supseteq \dots$$

in which each $\mathcal{A}_{m+1}$ is an ideal of each $\mathcal{A}_m$ and $\frac{\mathcal{A}_m}{\mathcal{A}_{m+1}}$ is a ring, using the obvious prescription: $\mathcal{A}_1 = \mathcal{A}_N$, $\mathcal{A}_{m+1} = \mathcal{A}_{\mathcal{A}_m}$.

Definition 100 The near-ring N is **annular** provided that for some finite m, $\mathcal{A}_m = 0$.

So, annular near-rings are additively solvable.

1.27.3 $N_{(+\times)}$-Nilpotence

Returning to the sequence (1.33) on page 40, we have the associated subnormal (page 4) sequence

$$R_1 \rhd [R_1, R_1] \rhd [R_2, R_2] \rhd [R_3, R_3] \rhd [R_4, R_4] \ldots. \tag{1.54}$$

Observations 101 We know that $R_m \cdot N \subseteq R_{m+1}$ and conclude

$$[R_m, R_m] \cdot N \subseteq [R_{m+1}, R_{m+1}]$$

and that, in particular, R_m is a right sideal of N (see page 15).

Definition 102 The near-ring N is $N_{(+\times)}$**-nilpotent**, provided that (1.54) terminates finitely in 0. That is, for some finite integer m, $[R_m, R_m] = \{0\}$

1.28 Modular Ideals

Monogenic structures are important in rings and so it is with near-rings. This section is about the annihilators of monogenic generators—something that may seem to verge on the obscure in isolation but which is right at the centre of the work on primitive modules (see Chap. 11).

I present a definition which is standard in both rings and near-rings. It originates, I think, in the ring theory work of Nathan Jacobson [76], where he attributes the idea to Segal. He remarks that papers of his time used the term "regular" which he finds conflicts with the important term "quasi-regular" (Sect. 1.29). The ring theory book that I found most useful as a postgraduate [66] is loyal to the older term.

I need the notation on page 32 in this section.

Definition 103 A left ideal J of the near-ring $(N, +, \cdot)$ is **modular** if there is some $e \in N$ such that for all $n \in N$

$$(n - (n \cdot e)) \in J \tag{1.55}$$

Modularity is equivalent to the congruence

$$n \equiv (n \cdot e)\ (modulo\ J) \tag{1.56}$$

so e is a sort of "modular right identity". It is sometimes referred to as a **right identity modulo** J, or as a **modular right identity**.

The next lemma is at least partly due to Betsch.

Lemma 104

(i) Supposing J to be a modular left ideal, as above, then, for zero-symmetric near-rings,

$$e \in J \Leftrightarrow J = N \tag{1.57}$$

(ii) Suppose I is a modular left ideal; then any left ideal containing it is also a modular left ideal.

(iii) If the near-ring is zero-symmetric, then any proper modular left ideal is contained in a maximal proper left ideal which is also modular.

Proof

(iii) The proper left ideal has a modular right identity e which does not lie inside it. Left ideals containing it but not containing that modular right identity are themselves modular, and the union of an ascending chain of them is itself a modular left ideal not containing e. The set of these left ideals is inductive, and the set has maximal elements by Zorn's lemma.

□

The next lemma might explain why I have this section on modular left ideals.

Lemma 105

(i) Modular left ideals are exactly the annihilators of the monogenic generators of monogenic left near-ring groups.

(ii) If $J = \begin{pmatrix} \{\gamma\} \\ 0 \end{pmatrix}_{\mathcal{L}}$, *then* $J \supset \begin{pmatrix} N \\ J \end{pmatrix}_{\mathcal{L}}$

Proof

(i) If J is a non-trivial modular left ideal, the group $\left(\frac{N}{J}\right)$ is a left near-ring group, and the coset $\{e + J\}$ is a monogenic generator of it.

The left annihilator of this coset is the collection of elements $n \in N$ such that $(n \cdot e) \in J$; from (1.55) this is the set of elements that lie in J.

On the other hand, suppose Γ is a non-trivial, monogenic, left near-ring group. We can find some $e \in N$ such that $(e \cdot \gamma) = \gamma$ (where γ is the monogenic generator) and $J = \binom{\{\gamma\}}{0}_{\mathcal{L}}$ is certainly a left ideal. If $n \in N$, then $(n - (n \cdot e)) \in J$, so the left ideal is modular.

(ii)

$$J = \begin{pmatrix} \{\gamma\} \\ 0 \end{pmatrix}_{\mathcal{L}} \supset \begin{pmatrix} \{\Gamma\} \\ 0 \end{pmatrix}_{\mathcal{L}} = \begin{pmatrix} N \\ J \end{pmatrix}_{\mathcal{L}}$$

□

Suppose J is a modular left ideal with e a modular right identity. Then,

$$\begin{pmatrix} N \\ J \end{pmatrix}_{\mathcal{L}} = \begin{pmatrix} \{N \cdot e\} \\ J \end{pmatrix}_{\mathcal{L}} \tag{1.58}$$

These identical structures are actually two-sided ideals of N, and from Lemma 105 they are contained in the left ideal J; but, of course, any two-sided ideal in J must be inside $\begin{pmatrix} N \\ J \end{pmatrix}_{\mathcal{L}}$, so the structures represent maximal two-sided ideals lying in J. Compare this with Sect. 1.19.

1.29 Quasi-regular Left Ideals

Jacobson [76] selects an element e in a ring R and looks at the collection of elements[7]

$$\mathfrak{l}_e = \{(a - a \cdot e) : a \in R\} \tag{1.59}$$

In rings, this is always a left ideal—in fact, a *modular left ideal*—and e is a modular right identity. $\mathfrak{l}_e$ is the smallest modular right ideal having e as its modular right identity.

Jacobson defines an element e to be **left quasi-regular** provided $\mathfrak{l}_e = R$. This is, of course, equivalent to the condition that $e \in \mathfrak{l}_e$.

$$e \in \mathfrak{l}_e \Leftrightarrow \mathfrak{l}_e = R \tag{1.60}$$

If that is true, then, of course, $\pm e \in \mathfrak{l}_e$, and so, realising that $-e \in \mathfrak{l}_e$, there exists $e' \in R$ such that

$$\left(e + e'\right) = \left(e' \cdot e\right) \tag{1.61}$$

that is, for e to be left quasi-regular, there must be an element e' such that

$$\left(e + e' - (e' \cdot e)\right) = 0 \tag{1.62}$$

[7] Actually, he works with *right quasi-regular elements*; I adjusted things to fit in with subsequent near-ring definitions.

e' is then described as a **left quasi-inverse** for e. Thus, an element a is left quasi-regular if and only if either

$$(a - b + (b \cdot a)) = 0 \, for \; some \; b \in R \tag{1.63}$$

or

$$(a + b - (b \cdot a)) = 0 \, for \; some \; b \in R \tag{1.64}$$

because these are equivalent conditions. Equation (1.64) is precisely Herstein's definition for the element a to be accounted **left quasi-regular** [66].

1.29.1 Quasi-regularity in Rings

Jacobson defines a **circle operation** ($\circ$) on rings

$$(a \circ b) = (a + b - (a \cdot b)) \tag{1.65}$$

This makes $(N, \circ)$ into a semi-group, because the operation is associative. When the ring is unital, the semi-group is a monoid, with identity element 0. By the way, the original idea for this operation came in a 1942 paper by Perlis.

For unital rings the mapping $a \leftrightarrow (1 - a)$ forms an isomorphism of monoids. The element b is left quasi-regular provided $(a \circ b) = 0$ for some a; the element b is right quasi-regular provided we can find a such that $(b \circ a) = 0$. Finally a is **quasi-regular** provided we can obtain b such that

$$(a \circ b) = 0 = (b \circ a) \tag{1.66}$$

Suppose $(a \circ x) = 0 = (y \circ a)$. Then, $y = (y \circ 0) = (y \circ (a \circ x)) = ((y \circ a) \circ x) = (0 \circ x) = x$. The quasi-regular elements constitute a group under the ($\circ$) operation—it is isomorphic to the group of units (page 45) of the original ring. Kurosh [93] describes elements satisfying (1.66) as **radical elements** and calls the group of quasi-regular elements the **adjoint group** of the ring. This is terminology which still appears to be occasionally in use.

Lemma 106 *Suppose R is a ring and J is a left quasi-regular ideal. Then the elements in J are also right quasi-regular.*

Proof Take $a \in J$. We can find some $r \in R$ such that $(r \circ a) = 0$. The fact J is a left ideal ensures $r \in J$, so we may find $s \in R$ such that

$$(s + r - (s \cdot r)) = (s \circ r) = 0$$

s is right quasi-regular, and

$$a = (0 \circ a) = ((s \circ r) \circ a) = (s \circ (r \circ a)) = (s \circ 0) = s$$

which means a is right quasi-regular. □

There is an exercise which is usually attributed to Kaplansky about rings in which all but one element are left quasi-regular. Such rings turn out to be division rings [71].

Near-Rings

Meldrum warns that quasi-regularity has several equivalent characterisations in ring theory which devolve into different things in near-rings and that care is needed [119, p. 91]. The story is that in near-rings we are forced to manipulate accretions of elements rather than individual ones.

Starting with a near-ring $(N, +, \cdot)$, one can select $r \in N$ and look at the left ideal $\mathfrak{l}_r$ *generated* by the totality of expressions of the form

$$\{(s - (s \cdot r)) \ : \ s \in N\} \tag{1.67}$$

This gives us a collection of left ideals. An element $r \in N$ is held to be **left quasi-regular** iff $r \in \mathfrak{l}_r$. If S is a subset of the near-ring $(N, +, \cdot)$ in which all elements are left quasi-regular, the set *itself* is called left quasi-regular.

Lemma 107

(i) Nilpotent elements of near-rings are left quasi-regular.
(ii) Nil sets in near-rings are left quasi-regular.
(iii) Non-zero idempotent elements of near-rings cannot be left quasi-regular.
(iv) If J is a proper modular left ideal of N with modular right identity e, then e is not quasi-regular.

Proof

(i) Suppose $x \in N$ and $x^n = 0$. Then, for all positive integers j, and all $z \in N$,

$$\left(z \cdot x^j - z \cdot x^{(j+1)}\right) = \left((z \cdot x^j) - (z \cdot x^j) \cdot x\right) \in \mathfrak{l}_x$$

Now,

$$(z - z \cdot x) + \left(z \cdot x - z \cdot x^2 \cdot\right) + \left(z \cdot x^2 - z \cdot x^3\right) + \\ \ldots + \left(z \cdot x^{(n-1)} - z \cdot x^n\right) \in \mathfrak{l}_z$$

Consequently, $(z - z \cdot x^n) \in \mathfrak{l}_z \Rightarrow z \in \mathfrak{l}_z$. Thus, $\mathfrak{l}_z = N$.

(iii) Suppose $e \neq 0$ and $e^2 = e$. There is an additive homomorphism: $N \mapsto N \cdot e$ and this is non-trivial. The left ideal L_e is in the kernel of this homomorphism.
(iv) We know $(n - n \cdot e) \in J$ for all $n \in N$. If e is left quasi-regular, we can find s such that $(s - s \cdot e) = e$ and so e is in J. Take any $n \in N \cdot n. n = (n - n \cdot e) + n \cdot e \in J$, which makes J improper.

□

An element a of a near-ring is held to be **right quasi-regular** provided there is some $m \in N$ such that

$$a = (m - a \cdot m) \tag{1.68}$$

and a *left ideal* is right quasi-regular provided that it consists entirely of right quasi-regular elements. There are particular examples of left quasi-regular ideals in Sect. 7.7.7.

1.30 Pseudo-Rings

Definition 108 A **pseudo-ring** is a ring-like structure in which the additive group is not necessarily abelian, and the two distributivity laws and the associativity law of standard ring theory are replaced by congruences, modulo the additive commutator subgroup.

One can define ideals for pseudo-rings, and the commutator subgroup is one. Pseudo-rings have homomorphic images which are non-trivial rings. One can take $(n \times n)$ matrices over F-near-rings, the resulting structure being a pseudo-ring. It is also true that matrices over pseudo-rings are themselves pseudo-rings, so the process iterates.

There is a term "pseudo-ring" which is already used in ring theory and which should not be confused with my definition above. I describe these pseudo-rings in Appendix C, page 509, but do not mention them elsewhere in this document; nor shall I return to my pseudo-rings, here, or elsewhere. They seem to be well worth further study.

1.31 Propriety

1.31.1 "Left" and "Right" Confusion

The terms "left" and "right" distributivity are used in different ways in algebra, and there are important papers in near-rings whose definition of *left* distributivity corresponds exactly to what I have called *right* distributivity, and vice versa.

For me, and many others, right distributivity involves

$$((x + y) \cdot z) = ((x \cdot z) + (y \cdot z))) \tag{1.69}$$

and characterises p.n.r. whose elements distribute on the *right*—"right p.n.r".

However, things are not so simple. For Betsch [11] the above identity would characterise a *left* p.n.r. Additionally, it is not always clear that people following this prescription might describe z as a left-distributive element, above. After all, it is operating on the right. So, for consistency, one might then start talking about a right-distributive near-ring in which all elements were left-distributive....

All this probably stems from the mistake of adducing spurious spatial orientations into marks on paper (or screens),[8] but it is far too late to do anything about that, now.

For me, *right* F-near-rings are *right*-distributive, $((x + y) \cdot z = ((x \cdot z) + (y \cdot z)))$, with their modular distributivity occurring on the *left*, and each of their elements is *right-distributive*. This strikes me as an important contribution to the already rich possibilities for confusion in this terminology.

1.31.2 Proper Structures

I use the unfortunate term "proper near-ring" when attempting to stress that a structure in consideration is not also a ring; I even use "non-ring" on occasion. I also refer to "proper near-fields" when trying to emphasise that the structure in question is not actually a field. This terminology seems locally appropriate, but it might cause some confusion when sub-structures are involved if the emphasis is rather on their proper set-theoretic inclusion. I try to avoid this arising, where it might (e.g. page 145).

1.31.3 Transferred Epithets

I may sometimes transfer group properties to near-rings through slack language—speaking of "abelian" near-rings for near-rings hosted by abelian groups, "solvable" near-rings for near-rings hosted by solvable groups, etc.

[8] I am following Buliwyf's dictum, in the film "The 13th Warrior", that writing "draws sounds". Where would left and right hide if this book were translated into Arabic?

1.31.4 Problematic Terminology

I should own that the two terms **unitary** and **unital** may be poorly chosen. The offence compounds in that, on the face of it, they are easily confused and there are publications in which each means the other.

Jacobson [76] says that the standard term for ring identities acting as left identities on modules is "unital", and Meldrum follows suit [119]. Pilz [128], on the other hand, uses "unitary" as I do. I think our usage had historical precedences in Jacobson's time, but that Jacobson had solid geometric reasons for changing it, which he mentions on the very first page of his book!

I stick with Pilz, but the reader should regard my terms as severely local. My defence is that the practical consequences of confusion are small (it's usually obvious whether I am talking about a ring or a near-ring group) and that this book is well enough indexed to remove doubts, but I would hesitate to use these terms in a wider context than this immediate volume and don't really like either of them.

1.32 An Unsettling Homomorphism

It is possible to have a near-ring homomorphism

$$\Theta : (N, +, \cdot) \longrightarrow (R, +, \cdot) \tag{1.70}$$

where one of the two near-rings shown is right-distributive and the other left-distributive. In that case, the image near-ring, $\Theta(N)$, will be bi-distributive (Sect. 1.26); and the kernel of the homomorphism contains the appropriate distributor ideal (Sect. 1.25).

Chapter 2
Near-Ring Theory

2.1 Pre-Near-Ring Construction Conditions and the Associativity Core

The abbreviation "p.n.r." is explained in Sect. 1.4.2.

Suppose G is the finitely presented group

$$\langle x_1, x_2, \ldots x_r \;:\; R_1(x_1, x_2, \ldots x_n) \;=\; R_2(x_1, x_2, \ldots x_n) = \ldots$$
$$= \;\ldots = R_s(x_1, x_2, \ldots x_n) = 0\rangle \qquad (2.1)$$

(I shall use this notation to represent a general additive group of interest throughout this chapter.)

Definition 109 An r-tuple of elements from G, $(g_1, g_2, \ldots g_r)$ satisfies the **p.n.r. construction conditions for G**, provided that

$$R_j(g_1, g_2, \ldots, g_r) \;=\; 0, \;\text{for } 1 \leq j \leq s \qquad (2.2)$$

There always are r-tuples that satisfy these conditions, $(0, 0, \ldots 0)$, for instance.

Strictly, I should relate construction conditions to a particular presentation, and there are situations in which this can be important.

Lemma 110 *Suppose G is a group with presentation given above, and we associate with each $h = h(x_1, x_2, \ldots x_r) \in G$ some r-tuple $(h_1, h_2, \ldots, h_r)$ which satisfies the p.n.r. construction conditions for G associated with this presentation.*

Then the product $(*)$ defined on G by the prescription

$$w\,(x_1, x_2, \ldots x_r) * h\,(x_1, x_2, \ldots, x_r) \;=\; w\,(h_1, h_2, \ldots, h_r)$$

R. Lockhart, *The Theory of Near-Rings*, Lecture Notes in Mathematics 2295,
https://doi.org/10.1007/978-3-030-81755-8_2

($w = w(x_1, x_2, \ldots, x_r)$ being any word in the generators of the group) is a right *pre-near-ring product hosted by* $(G, +)$.

Moreover, all pre-near-ring products hosted by $(G, +)$ *do arise in this way.*

Proof It is an application of von Dyck's theorem, in representation theory, that the mapping

$$w\,(x_1, x_2, \ldots, x_r) \mapsto w\,(h_1, h_2, \ldots, h_r)$$

is a well-defined endomorphism of the additive group and, from that, the result follows. □

2.1.1 Host Determination Strategies

Working from Presentations

One strategy for determining the near-rings hosted by a finitely presented group is to first determine the non-isomorphic pre-near-rings, using the construction conditions, and then isolate the ones whose generating sets $\{x_1, x_2, \ldots, x_r\}$ lie in the *associativity core* of the resulting pre-near-ring (see Observations 70).

That is, for all $w, h, \in G$

$$\big(x_j \cdot (w \cdot h)\big) = \big((x_j \cdot w) \cdot h\big)$$

The assumption is that the additive structure as encoded by (2.1) is fully known. This viewpoint relates to the *multiplicative* host problem (see page 329, which also describes the *additive* host problem).

Working from Primitivity

Here, I need some of the notation and lemmata from Chaps. 7 and 11, where definitions, explanations and proofs are to be found. The reader new to the subject might find it easier to wait until they had assimilated Chaps. 7 and 11 before reading this next material.

Given a finite, zero-symmetric, near-ring with identity, one can factor out by maximal ideals, obtaining a quotient near-ring which is simple. This quotient near-ring will either be a simple non-ring, in which case it has the structure given in (7.200) and is a direct sum of isomorphic, left ideals, or it is a simple ring, in which case it is a full ring of matrices over a finite field (and one-dimensional matrices correspond to it actually *being* a finite field)—this is part of the Jacobson density theorem from ring theory.

The point of these observations relates to order considerations. Given a group, we may wish to determine the zero-symmetric near-rings with identity that it may host. If we consider simple near-rings, we require the group to have orders dictated by these constraints. If we consider non-simple near-rings, we know that, factored out by a maximal ideal, the near-ring again would have only a circumscribed set of orders and a limited range of additive structures.

The Group G_7
Consider the group G_7 from Sect. 4.1.1 and posit that it does in fact host a zero-symmetric, unital, near-ring. Note that G_7 is nilpotent and has order 2^4.

If the near-ring is to be simple, it would consist of a sum of isomorphic minimal left ideals and this is not the case. So it should have ideals. Factoring by a maximal ideal, one should get either a simple *ring* or a simple *near-ring*.

In the former case, the near-ring is either a finite field—and C_2, $(C_2 \oplus C_2)$ and $(C_2 \oplus C_2 \oplus C_2)$ all possibilities—or it should be a full ring of matrices over a finite field, the formula for the order of which is given in (7). The characteristic of the field would have to be two, and the field order must be $n = 2^k$ for some $k \leq 3$, but in all possible orders obtained, one would have odd prime divisors (from the $(N^m - 1)$ term), so this is impossible.

In the near-ring case, the near-ring is a direct sum of copies of a minimal left ideal, as in (7.202), whose order would have to be a power of two. From the structure of G_7 itself, we can say that these minimal left ideals could only have order 2. Only one non-trivial quotient of the group gives anything other than a direct sum of copies of C_2, and that quotient is simply Q_8, but in any case we know that Q_8 does not host zero-symmetric near-rings with identity (Lemma 207). This means that in the near-ring case, the non-trivial ideal must contain the commutator subgroup of G_7 and be abelian.

To summarise, the suggestion that G_7 hosts a zero-symmetric near-ring with identity implies that that near-ring is not simple and that it factors out to a simple near-ring hosted by one of C_2, $(C_2 \oplus C_2)$ or $(C_2 \oplus C_2 \oplus C_2)$ which may or may not be a ring. The commutator subgroup of G_7 is in the ideal. But we cannot have a simple non-ring hosted by either $(C_2 \oplus C_2)$ or $(C_2 \oplus C_2 \oplus C_2)$ because in such structures the number of summands is circumscribed and two and three are ruled out. So, if we did have either of these cases, we would be dealing with a quotient which was a finite field.

2.1.2 *Distributive Generation*

The elements of a p.n.r. hosted by the group given by (2.1) consist of words $w(x_1, \ldots x_r)$ in the generators $\{x_1, x_2, \ldots x_r\}$.

I deal mainly with right-distributive p.n.r. and am interested in elements from the corresponding distributivity core and whether the p.n.r. could be distributively generated (see Definition 75).

Selecting $w = w\,(x_1, x_2, ..x_r) \in G$ and writing

$$\overleftarrow{w_i} = w\,(x_1, x_2, ..x_r) \cdot x_i = w\,(x_1 \cdot x_i, x_2 \cdot x_i, \ldots, x_r \cdot x_i)$$

one might consider whether the construction conditions

$$R_j\left(\overleftarrow{w_1}, \overleftarrow{w_2}, \ldots, \overleftarrow{w_r}\right) = 0, \text{(for } 1 \leq j \leq s) \tag{2.3}$$

are satisfied.

Equation (2.3) is a necessary condition for $w \in \mathcal{C}_{\mathcal{D}}(G)$. If, additionally, it is the case that for all $h(x_1, x_2, \ldots x_r)$ we have

$$w(x_1, x_2, \ldots, x_r) \cdot h(x_1, x_2, \ldots, x_r) = h\left(\overleftarrow{w_1}, \overleftarrow{w_2}, \ldots, \overleftarrow{w_r}\right)$$

then $w(x_1, x_2, \ldots, x_r)$ actually is in $\mathcal{C}_{\mathcal{D}}(G)$.

This will be the technique used to check for distributive generation in some of the examples in later chapters. To illustrate this, in Table B.1 it will be seen that

$$b \cdot (4a) = b + b + b + b = 0$$
$$b \cdot (2b) = b + b = 0$$
$$b(2(a + b)) = b + b + b + b = 0$$

so the element b satisfies (2.3) in this case. However, from the table, $(2a + b) = b \cdot (2a + b) \neq b \cdot (2a) + b \cdot b = 0 + b = b$, so $b \notin \mathcal{C}_{\mathcal{D}}(D_4)$.

2.1.3 *Co-structures: A Sort of Duality*

I now discuss a technique which may be used to associate a right near-ring with a left near-ring to which it may not necessarily be *anti-isomorphic* (definition—page 479). It's a sort of "syntactic symmetry". I maintain the notation introduced in Sect. 2.1.2, here.

Suppose that the collection $\{x_1, x_2, \ldots, x_r\}$ of additive generators satisfies (2.3) for all possible choices of $w(x_1, x_2, \ldots, x_r)$. One can then define a *left* p.n.r. hosted by G with the prescription

$$w(x_1, x_2, \ldots, x_r) \cdot h(x_1, x_2, \ldots, x_r) = h(\overleftarrow{w_1}, \overleftarrow{w_2}, \ldots, \overleftarrow{w_r})$$

and this might, in some sense, be regarded as a co-structure to the original left p.n.r.

Of course there are potentially many distinct such co-structures to be found; and I do need to be precise about exactly which set of additive generators is being used.

Elements from $x \in \mathcal{C}_{\mathcal{D}}(G)$ (original, right-distributive structure) will have the same values for left products: $x \cdot y$, in both the original p.n.r. and any co-structure. In particular, left identities map to left identities, and the zeroiser ideal (Definition 55) is mapped into the collection of elements in the co-structure which left annihilate everything (since the co-structure is potentially non-associative, I cannot describe this collection of elements itself as a zeroiser ideal).

To illustrate this, I consider the near-ring hosted by D_4 given by Table B.7 on page 500; and I use the presentation of this group given in (5.4) on page 195. Elements of the group are words $w(a, b)$, and $w(a, b) \cdot b$ is always 0 or b. $w(a, b) \cdot a$ is of course just $w(a, b)$ and it will be seen that for all possible choices of $w(a, b)$

$$2\,(w(a, b) \cdot a + w(a, b) \cdot b) = 0$$

The conditions for the definition of a co-structure are met. The resulting *left* p.n.r. is given by Table 2.1.

The distributivity core for this left p.n.r. is $\{0, a, b\}$ and it is multiplicatively closed. This is a near-ring; it is anti-isomorphic to the near-ring given in Table B.2.

Each of the seven near-rings Table B.1–B.7 given in Appendix B has a set of additive generators for which a co-structure can be defined. The generating sets $\{a, b\}$ and $\{a, 2a + b\}$ are not suitable in cases Table B.3, B.4, B.5 and B.6. With that caveat, we can use $\{a, b\}$, $\{a, a + b\}$, $\{a, 2a + b\}$ and $\{a, 3a + b\}$ in all seven cases.

Consider Table B.1, using the generators $\{a, a+b\}$. The anti-isomorphism of the co-structure is given by Table 2.2—it is, of course, a *right* p.n.r. Using the notation of Appendix B, $\mathcal{C}_{\mathcal{D}}(2.2) = \{0, a, a + b\}$. $\mathcal{N} = \{0, 2a\}$. $\mathcal{I} = \{0, a, 3a, b, a + b, 2a + b\}$.

Table 2.1 Co-structure to Table B.7

$*$	0	a	$2a$	$3a$	b	$a+b$	$2a+b$	$3a+b$
0	0	0	0	0	0	0	0	0
a	0	a	$2a$	$3a$	b	$a+b$	$2a+b$	$3a+b$
$2a$	0	$2a$	0	$2a$	0	$2a$	0	$2a$
$3a$	0	$3a$	$2a$	a	b	$3a+b$	$2a+b$	$a+b$
b	0	b	0	b	b	0	b	0
$a+b$	0	$a+b$	0	$a+b$	0	$a+b$	0	$a+b$
$2a+b$	0	$2a+b$	0	$2a+b$	b	$2a$	b	$2a$
$3a+b$	0	$3a+b$	0	$3a+b$	0	$3a+b$	0	$3a+b$

Table 2.2 Anti-isomorphism of a co-structure to Table B.1

$*$	0	a	$2a$	$3a$	b	$a+b$	$2a+b$	$3a+b$
0	0	0	0	0	0	0	0	0
a	0	a	$2a$	$3a$	b	$a+b$	$2a+b$	$3a+b$
$2a$	0	$2a$	0	$2a$	0	0	0	0
$3a$	0	$3a$	$2a$	a	b	$a+b$	$2a+b$	$3a+b$
b	0	b	$2a$	$2a+b$	b	0	$2a+b$	$2a$
$a+b$	0	$a+b$	0	$a+b$	0	$a+b$	0	$a+b$
$2a+b$	0	$2a+b$	$2a$	b	b	0	$2a+b$	$2a$
$3a+b$	0	$3a+b$	0	$3a+b$	0	$a+b$	0	$a+b$

With these characteristics, the p.n.r. can potentially only be isomorphic to Table B.2, and the isomorphism would have to be

$$\Phi : a \mapsto a$$

$$\Phi : b \mapsto a + b$$

It will be seen that the product

$$(x * y) = \Phi^{-1} (\Phi(x) \cdot \Phi(y))$$

does indeed map Tables 2.2 to B.2 and so this p.n.r. is a near-ring isomorphic to Table B.2.

Lemma 111 *Suppose that $(G, +, \cdot)$ is a (right) near-ring hosted by (2.1) and that $(G, +, *)$ is a (left) pre-near-ring which is a co-structure to it, based on the additive generators $\{x_1, x_2, \ldots, x_r\}$. Suppose, additionally, that these additive generators are multiplicatively closed in the sense*

$$x_i \cdot x_j = x_k \text{ (for some } k\text{)}$$

for all possible choices of i and j (k varying with the choices). Then this p.n.r. is a near-ring—that is, the product is associative.

Proof

$$(w * h) * x_i = h\,(w \cdot x_1, w \cdot x_2, \ldots, w \cdot x_r) * x_i =$$
$$h\,(w \cdot x_1 \cdot x_i, w \cdot x_2 \cdot x_i, \ldots, w \cdot x_r . x_i) =$$
$$w * h\,(x_1 \cdot x_i, x_2 \cdot x_i, \ldots, x_r \cdot x_i) =$$
$$w * (h * x_i)$$

□

The assertion that

$$h\,(w \cdot x_1 \cdot x_i, w \cdot x_2 \cdot x_i, \ldots, w \cdot x_r \cdot x_i) = w * h\,(x_1 \cdot x_i, x_2 \cdot x_i, \ldots, x_r \cdot x_i)$$

is uncertain, unless the set $\{x_1, x_2, \ldots, x_r\}$ is multiplicatively closed.

A non-associative structure may be produced when the relation does not hold; and the relation does not hold for the generating set $\{a, 2a+b\}$ in Table B.2, and for the generating set $\{a, a+b\}$, in Table B.6. In all other cases, we can be confident of obtaining a near-ring.

In Table B.2 and generators $\{a, 2a + b\}$, the *left* p.n.r. obtained is given by Table 2.3.

Table 2.3 A non-associative co-structure to Table B.2

$*$	0	a	$2a$	$3a$	b	$a+b$	$2a+b$	$3a+b$
0	0	0	0	0	0	0	0	0
a	0	a	$2a$	$3a$	b	$a+b$	$2a+b$	$3a+b$
$2a$	0	$2a$	0	$2a$	0	$2a$	0	$2a$
$3a$	0	$3a$	$2a$	a	b	$3a+b$	$2a+b$	$a+b$
b	0	b	0	b	b	0	b	0
$a+b$	0	$a+b$	0	$a+b$	$2a$	$3a+b$	$2a$	$3a+b$
$2a+b$	0	$2a+b$	0	$2a+b$	b	$2a$	b	$2a$
$3a+b$	0	$3a+b$	0	3a + b	$2a$	$a+b$	$2a$	$a+b$

Table 2.4 A non-associative co-structure to Table B.6

$*$	0	a	$2a$	$3a$	b	$a+b$	$2a+b$	$3a+b$
0	0	0	0	0	0	0	0	0
a	0	a	$2a$	$3a$	b	$a+b$	$2a+b$	$3a+b$
$2a$	0	$2a$	0	$2a$	$2a$	0	$2a$	0
$3a$	0	$3a$	$2a$	a	$2a+b$	$a+b$	b	$3a+b$
b	0	b	0	b	$2a+b$	$2a$	$2a+b$	$2a$
$a+b$	0	$a+b$	0	$a+b$	$2a$	$3a+b$	$2a$	$3a+b$
$2a+b$	0	$2a+b$	0	$2a+b$	b	2a	b	2a
$3a+b$	0	$3a+b$	0	3a + b	0	$3a+b$	0	$3a+b$

Since

$$2a = (2a*(a+b)) = (((a+b)*(2a+b))*(a+b))$$

whereas

$$((a+b)*((2a+b)*(a+b))) = ((a+b)*2a) = 0$$

the product is not associative. It is, however, a left-distributive p.n.r. in which the element a is a multiplicative identity element.

Similarly, the generating set $\{a, a+b\}$ produces this non-associative left p.n.r. from Table B.6 (Table 2.4). The two non-associative left p.n.r. are isomorphic under the isomorphism

$$\Phi(a) = a, \quad \Phi(b) = (a+b)$$

There are 18 possible near-rings to be produced as co-structures from Tables B.1 to B.7 and the symmetries listed. In all cases one obtains something which, bearing in mind the change from right to left distributivity, is essentially isomorphic to Table B.2; and that near-ring is distributively generated—a phenomenon to which I return.

Table 2.5 A non-associative co-structure to Table 2.4

$*$	0	a	$2a$	$3a$	b	$a+b$	$2a+b$	$3a+b$
0	0	0	0	0	0	0	0	0
a	0	a	$2a$	$3a$	b	$a+b$	$2a+b$	$3a+b$
$2a$	0	$2a$	0	$2a$	0	0	0	0
$3a$	0	$3a$	$2a$	a	b	$a+b$	$2a+b$	$3a+b$
b	0	b	0	b	$2a+b$	$2a$	$2a+b$	$2a$
$a+b$	0	$a+b$	0	$3a+b$	$2a$	$3a+b$	0	$a+b$
$2a+b$	0	$2a+b$	0	$2a+b$	$2a+b$	$2a$	$2a+b$	$2a$
$3a+b$	0	$3a+b$	0	$a+b$	$2a$	$3a+b$	0	$a+b$

Other co-structures may be formed. For example, in Table B.1 we may use the generating set $\{3a, b\}$. This is not multiplicatively closed and the left p.n.r. produced is not associative. Nor it is isomorphic to Table 2.3—the p.n.r. has a left, but not a right, identity. Finally, of course, this process iterates.

Looking at Table 2.4, we see that we may devise a co-structure to it, based on the generating set $\{a, b\}$. This right-distributive p.n.r. is given by Table 2.5.

Lemma 112 *When the conditions of Lemma 111 apply, the additive generators* $\{x_1, x_2, \ldots, x_r\}$ *are in the distributivity core of this (left) near-ring and distributively generate it—that is, the near-ring is d.g.*

Proof The elements $\{x_1, x_2, \ldots, x_r\}$ are distributive over the other elements. Since they are multiplicatively closed, the near-ring is d.g. (see Observations 70). □

2.1.4 Reduced Free Groups and Another Sort of Duality

Finitely generated reduced free groups are characterised by the property of having a presentation in which each relation is a law in the group[1] (see [124] or [132] for more on this).

These are precisely the groups possessing a generating set with the property that each mapping of that set into the group itself can be extended to an endomorphism of the group—a "reduced" form of the property which characterises free groups and one that is exploited in *type two addition* (page 323).

[1] A law is just an identity that is true throughout the group—for example, that any two elements commute, as is the case in the variety of abelian groups. The corresponding reduced free groups are the free abelian groups. Other varieties include such things as the variety of nilpotent groups and so on. See page xx.

I suppose (2.1) is just such a presentation and consider a bi-zero-symmetric *stem* product hosted by $(G, +)$ (Definition 5). Following (2.3), it is clear that for all $w(x_1, x_2, \ldots, x_r) \in G$, if we write

$$\overleftarrow{w}_i = w(x_1 \cdot x_i, x_2 \cdot x_i, \ldots, x_r \cdot x_i)$$

and

$$\overrightarrow{w}_i = w(x_i \cdot x_1, x_i \cdot x_2, \ldots, x_i \cdot x_r)$$

then both

$$R_j(\overleftarrow{w_1}, \overleftarrow{w_2}, \ldots, \overleftarrow{w_r}) = 0 \quad \text{for } 1 \leq j \leq s$$

and

$$R_j(\overrightarrow{w_1}, \overrightarrow{w_2}, \ldots, \overrightarrow{w_r}) \quad \text{for } 1 \leq j \leq s$$

These two relations are the construction conditions for, respectively, a *left* p.n.r. hosted by $(G, +)$ and a *right* one. In this way, each stem product spawns two p.n.r. I believe that this is a new sort of symmetry and that there are many issues that might be investigated regarding it.

2.1.5 Construction Conditions and F-Near-Rings

I again work with the presentation (2.1). My interest is in F-near-rings (Definition 75). Suppose we have a (right) p.n.r. hosted by G. One can ask whether, for any additive generating set $\{x_1', x_2', \ldots .x_k'\}$, it is the case that

$$R_j\left(x_t' \cdot x_1, x_t' \cdot x_2, \ldots, x_t' \cdot x_r\right) \equiv 0 \text{ (modulo } [G, G]) \tag{2.4}$$
$$(\text{ for } 1 \leq j \leq s \text{ and } 1 \leq t \leq k)$$

and one can ask whether it might also be true that for each $1 \leq t \leq k$ and all $h(x_1, x_2, \ldots, x_r) \in G$

$$x_t' \cdot h(x_1, x_2, \ldots, x_r) \equiv h\left(x_t' \cdot x_1, x_t' \cdot x_2, \ldots, x_t' \cdot x_r\right) \text{ (modulo } [G, G])$$

If both questions can be answered affirmatively, the p.n.r. is an F-near-ring. Sect. 5.4 relies heavily on these ideas. They are implicit in the work of a number of authors, such as Meldrum [119].

2.1.6 *Bounds on Associativity Checking*

When presented with a p.n.r. hosted by a group with presentation (2.1), the strategy outlined in Sect. 2.1.1 involves us in a little fewer than $r.|G|^2$ separate checks in order to verify associativity. And all these checks must be done for *each* possible p.n.r. if we wish to identify the near-rings.

We essentially must verify that

$$x_k \cdot (a \cdot b) \;=\; (x_k \cdot a) \cdot b$$

for all $1 \leq k \leq r$ and all $a, b \in G$ (although, of course, some values of a and b, such as $a = 0$, do not need checking).

In some circumstances fewer checks than this are needed; and it is sometimes possible to see that a product cannot be associative for other reasons (e.g. if the distributive core is not multiplicatively closed); also, there may be situations where isomorphic near-rings are easily identifiable, rendering some checks otiose.

My host determination strategy (page 68) can easily involve more checks than is feasible, even when automatic computation is employed, and this can happen for quite small groups.

An example of this is given in Sect. 4.1.1. The group G_7 has sufficient structure to reduce the number of possible p.n.r. definable, but even then I found myself having to search for associativity from among (7×2^{46}) possible p.n.r. Unsurprisingly, the computer program I wrote to do this ran for days on my PC without result. I attempted a different search, selecting p.n.r. at random, but that too never uncovered a near-ring of the required type. The group, inescapably, has three generators, and it was beyond my skill to obtain restrictions to sufficiently reduce the number of p.n.r. that I had to consider. Nor could I decide probabilistically whether my lack of success was indicative of a general negative result.

Special Cases

These special cases may be worth detailing. If the p.n.r. is distributively generated by $\{x_j\}_{j=1}^r$, one only need verify

$$\left(x_i \cdot x_j\right) \cdot a \;=\; x_i \cdot \left(x_j \cdot a\right) \quad \text{for } 1 \leq i, j, \leq r \;\text{ and } a \in G.$$

That is, approximately $r^2.|G|$ checks. If the (right) p.n.r. is actually left distributive (i.e. a *bi-distributive* p.n.r.), one need only verify

$$\left(x_i \cdot x_j\right) \cdot x_k \;=\; x_i \cdot \left(x_j \cdot x_k\right) \quad \text{for } 1 \leq i, j, k \leq r$$

That is, approximately r^3 checks.

Verifying associativity is usually the most laborious part of solving the host problem. It would be useful to reduce the amount of checking needed, and it is likely that existing bounds could be further reduced (by taking isomorphism into account, for instance).

2.2 Coupling and Dickson Near-Rings

Suppose $(N, +)$ is an additive group, and recall that End$(N, +)$ is the associated monoid of additive endomorphisms (Notation 2). Suppose we have a normalised mapping (see Definition 45)

$$\Phi : N \to \text{End}(N, +) \quad (n \mapsto \phi_n) \tag{2.5}$$

When $(N, +, \cdot)$ is a stem, we can define a new stem product $(*)$ hosted by the additive group $(N, +)$ with the formula

$$m * n = \phi_n(m) \cdot n \tag{2.6}$$

(i) When $(N, +, \cdot)$ is right zero-symmetric (i.e. $0 \cdot x = 0$), $(N, +, *)$ is bi-zero-symmetric (Definition 5).
(ii) When $(N, +, \cdot)$ is left zero-symmetric (i.e. $x \cdot 0 = 0$), so is $(N, +, *)$.
(ii) When $(N, +, \cdot)$ is a right p.n.r., so is $(N, +, *)$.

Displacement products of this sort are common in near-ring theory. The reader might like to compare (2.6) with the various notions of *isotopy* which crop up in this book (see, for instance, Appendix A or page 203).

I shall now assume that $(N, +, \cdot)$ is a (right) pre-near-ring. We have that

$$m * (n * k) = m * (\phi_k(n) \cdot k) = \phi_{(\phi_k(n)\cdot k)}(m) \cdot (\phi_k(n) \cdot k)$$

and

$$(m * n) * k = \phi_k\ (\phi_n(m) \cdot n) \cdot k$$

The structure $(N, +, *)$ is a near-ring precisely when

$$\phi_{(\phi_k(n)\cdot k)}(m) \cdot (\phi_k(n) \cdot k) = \phi_k\ (\phi_n(m) \cdot n) \cdot k \tag{2.7}$$

Suppose $(N, +, \cdot)$ is already multiplicatively associative, that is, a near-ring. Suppose that the mapping Φ maps N into the subset of $End(N, +)$ consisting of group endomorphisms which are actually near-ring endomorphisms of $(N, +, \cdot)$. Then, in particular, Eq. (2.7) is satisfied when the logically distinct mapping condition

$$\phi_{(\phi_k(n)\cdot k)} = (\phi_k \circ \phi_n) \tag{2.8}$$

applies. Equation (2.8) is to apply for all possible choices of n and k in N. It is sometimes called a **coupling condition**.

2.2.1 $\mathcal{D}$-Near-Rings

Definition 113 The near-ring $(N, +, *)$ will be described as the **$\mathcal{D}$-near-ring** of $(N, +, \cdot)$, or, when the sense is clear, the **associated $\mathcal{D}$-near-ring**. I mainly use this term in the work on near-fields, speaking there of "$\mathcal{D}$-near-fields". The two near-rings $(N, +, \cdot)$ and $(N, +, *)$ will be described as being **coupled**.

Coupling is a formalisation of a process used by Dickson and later Zassenhaus in their work on near-fields. Maxson [117] extended the idea to near-rings. Clay calls the process that I have just described the **Dickson** process [23]. It is the mechanism used to construct *most* finite near-fields from fields $(N, +, \cdot)$; and the near-fields it produces are known as **Dickson near-fields** (see Sect. 3.4). Dembowski calls these near-fields **regular near-fields**, as do other authors [32]. In infinite cases, one may extend the coupling process to division rings rather than fields, and one still gets near-fields from it.

Equation (2.8) suggests a product which is an alternative to (2.5). It is

$$(k \overset{*}{*} n) = (\phi_k(n) \cdot k) = (n * k) \tag{2.9}$$

Of course, this is just the opposite stem to (2.5) (page 51), and associativity is preserved in opposites. Consequently the associativity condition (2.8) must be equivalent to the similar

$$\phi_{(\phi_k(n) \cdot k)} = (\phi_n \circ \phi_k) \tag{2.10}$$

as we shall see again, shortly.

2.3 Affine Near-Rings

My next definition uses Definition 69 and the notation established in Lemma 80. I shall also need the idea of a *sideal* in this section (see Sect. 1.8).

Definition 114 Suppose N is a near-ring and that it is not zero-symmetric. I shall say that it is an **affine near-ring** provided that:

- The near-ring $(N_0, +, \cdot)$ is the distributivity core of $(N, +, \cdot)$.
- The additive group $(N_c, +)$ is a normal subgroup of $(N, +)$.

Suppose $(b \cdot 0 + a)$ and $(n \cdot 0 + m)$ are elements of the affine near-ring $(N, +, \cdot)$, where $a, m \in N_0$ and $b, n \in N$. We know that

$$(b \cdot 0 + a) \cdot (n \cdot 0 + m) = (b \cdot 0 + a \cdot (n \cdot 0 + m)) = \tag{2.11}$$
$$(b \cdot 0 + a \cdot n \cdot 0 + a \cdot m)$$

Both N_c and N_0 are normal subgroups of $(N, +)$. Their intersection is trivial and we have the *additive* direct sum

$$N = N_c \oplus N_0$$

N_c is both a two-sided ideal of $(N, +, \cdot)$ and a two-sided sideal of it.

If the original near-ring $(N, +, \cdot)$ has a multiplicative identity, this is of course a distributive element and a member of N_0, which would mean that N_0 would be a ring. Suppose J is a right sideal of the affine near-ring $(N, +, \cdot)$. Elements of J have the form $(b \cdot 0 + a)$ where $a \in N_0$. We have that $(b \cdot 0 + a) \cdot 0 = (b \cdot 0) \in J$ and this implies that $a \in J$ too. That is, both a and $b \cdot 0$ are individually elements of J. More than that,

$$(J \cap N_c) = \{b \cdot 0 : (b \cdot 0 + a) \in J\} \tag{2.12}$$

and

$$(J \cap N_0) = \{a : (b \cdot 0 + a) \in J\} \tag{2.13}$$

and J is additively of the form $(M_c + M_0)$, where M_0 is a subgroup of N_0 and M_c is a subgroup of N_c. M_c is, of course, a right sideal of N. Indeed, any subgroup of N_c is a right sideal of N. We can also say that M_0 must be a right sideal of N_0.

Suppose, instead, that J is a left *sideal* of $(N, +, \cdot)$. The general observation that $(x \cdot 0) \cdot y = (x \cdot 0)$ guarantees that $N_c \subset J$. Again, each coordinate of J is separately a member of the set J, so (2.12) and (2.13) again apply. In this situation, J is additively of the form $(N_c + M_0)$, where M_0 is a subgroup of N_0, and, of course, M_0 is a left sideal of N_0.

Lemma 115 *All two-sided sideals of the affine near-ring N have the form $(N_c + M_0)$, where M_0 is a two-sided sideal of N_0.*

Now consider a subgroup J of $(N, +, \cdot)$ which has the weak left ideal property (see Eq. (1.23)). Taking, $b, c \in N, a \in N_0$ and $j \in J$, we have

$$\begin{aligned}(b \cdot 0 + a) \cdot (c + j) - ((b \cdot 0 + a) \cdot c) &= b \cdot 0 + a \cdot c + a \cdot j - ac - b \cdot 0 \\ &= (a \cdot j)^{-a \cdot c - b \cdot 0} \in J\end{aligned}$$

When $(N, +, \cdot)$ has a multiplicative left identity element, this means that J will be normal in $(N, +)$ and therefore a left ideal in its own right. If we set b and c equal to zero above, we obtain that $N_0 \cdot J \subseteq J$ and, indeed, $N \cdot J \subseteq J$. So, J turns out to be a left sideal of $(N, +, \cdot)$ and therefore has the form $(N_c + J_0)$ as we discovered above. We note that J_0 has the weak left ideal property in N_0.

Lemma 116 *Ideals of the affine near-ring N have the form $(N_c, \oplus J)$, where J is an ideal of the near-ring N_0.*

Affine near-rings first arose when people considered the near-ring of affine mappings on vector space. Affine transformations involve a linear mapping followed by a translation, and their composition is a further affine mapping, as is their sum. When one restricts to invertible linear transformations, (2.12) expresses a multiplicative semi-direct product in the **affine group**.

The near-rings that affine transformations generate are abelian, and not zero-symmetric. They were considered in the 1950s in some of the first papers on near-rings, and there have been several generalisations of them since (see (3.75)).

2.4 Near-Rings Hosted by Semi-direct Products

One might ask about whether there would be a natural way to go from near-rings on the constituents of additive semi-direct products to near-rings on the semi-direct product itself. This certainly cannot be done with respect to near-rings with identity (see page 172).

Suppose $(N, +, \cdot)$ and $(K, +, \cdot)$ are both bi-zero-symmetric stems and that there is an exact sequence of groups

$$N \rightarrowtail G \twoheadrightarrow K \tag{2.14}$$

which splits. That is, $G = N \rtimes K$. Elements of $(G, +)$ are pairs (n, k), with $n \in N$ and $k \in K$. There is the usual group homomorphism, corresponding to an *action* of K on N

$$\begin{aligned} \theta : K &\to \mathrm{Aut}(N, +) \\ k &\longmapsto (n \mapsto n^k) \end{aligned}$$

and we have

$$(n_1, k_1) + (n_2, k_2) = (n_1 + n_2^{(-k_1)}, k_1 + k_2) \tag{2.15}$$

We can define an additional product $(\times)$ on G by

Notation 117

$$(n_1, k_1) \times (n_2, k_2) = (\alpha(n_1, n_2, k_2) + \beta(k_1, n_2, k_2), k_1 \cdot k_2) \tag{2.16}$$

Equation (2.16) is not offered as an expression of *any* possible stem product on $(G, +)$. It builds in the assumption that N should be both a left sideal and a weak left ideal (recall Eq. 1.23). If any resulting near-ring had a multiplicative right identity, it could not lie within the subgroup N, and so this product rules out known classes of near-rings, for example, near-rings with identity on dihedral groups.

In rings, or, indeed, near-rings, $(G, +)$ could be a direct sum of the groups $(N, +)$ and $(K, +)$, and we might set

$$(\alpha(n_1, n_2, k_2) + \beta(k_1, n_2, k_2)) = (n_1 \cdot n_2)$$

which would give us the near-ring direct sum of $(N, +, \cdot)$ and $(K, +, \cdot)$.

In near-rings, we might again choose G to be the direct sum of the groups $(N, +)$ and $(K, +)$ and take $(K, +, \cdot)$ to be any zero-symmetric near-ring and $(N, +, \cdot)$ to be the constant near-ring. Then, motivated by Eq. (2.12), we might set

$$\alpha(n_1, n_2, k_2) = n_1$$

and

$$\beta(k_1, n_2, k_2) = \beta(k_1, n_2, 0)$$

That is

$$(n_1, k_1) \times (n_2, k_2) = (n_1 + \beta(k_1, n_2, 0), k_1 \cdot k_2) \tag{2.17}$$

Imposing right distributivity on (2.17) gives

$$n_2 + \beta(k_1 + k_2, n_3, 0) = \beta(k_1, n_3, 0) + n_2 + \beta(k_2, n_3, 0)$$

and associativity implies

$$\beta(k_1, n_2 + \beta(k_2, n_3, 0), 0) = \beta(k_1, n_2, 0) + \beta(k_1 \cdot k_2, n_3, 0)$$

One possible solution to these two conditions is to set $\beta(k, n, 0) = 0$, giving the near-ring product

$$(n_1, k_1) \times (n_2, k_2) = (n_1, k_1 \cdot k_2)$$

With this product (using the notation of Lemma 80)

$$G_c = G \cdot 0 = \{(n, 0) : n \in N\}$$

and

$$G_0 = \{g \in G : g \cdot 0 = 0\} = \{(0, k) : k \in K\}$$

When K is bi-distributive, it will be the distributivity core of G (or at least, the set $\{(0, k) : k \in K\}$ is that core); and the resulting near-ring is affine (see Definition 114).

The standard near-ring of affine mappings on a vector space is not of this form. In that near-ring, the product is the usual composition of affine transformations, and so n_1 would be replaced by $n_1 + k_1(n_2)$ with k_1 a matrix operator and n_1 and n_2 vectors.

2.4.1 Near-Rings Hosted by $\mathbf{D}_n$

In more general cases, we shall not assume a group direct sum. The question that interests us is whether we can obtain some sort of general near-ring product. Now, $(N, +, \cdot)$ and $(K, +, \cdot)$ are supposed to be bi-zero-symmetric right-distributive near-rings.

We look for near-rings with addition given by (2.15) and multiplication given by

$$(n_1, k_1) \times (n_2, k_2) = (n_1 \cdot n_2, k_1 \cdot k_2)$$

It is immediate that the above product is associative. Right distributivity corresponds to the condition

$$\left(n_2^k\right) \cdot n_3 = (n_2 \cdot n_3)^{kk_3} \tag{2.18}$$

holding for all values of the variables. This in turn is equivalent to the conditions

$$\left(n_2^k\right) \cdot n_3 = (n_2 \cdot n_3) = (n_2 \cdot n_3)^{kk_3}$$

and these conditions certainly apply in central extensions, but then the near-ring operations will be those of a typical direct product, and we gain nothing new. $D_4 = \langle a, b : 4a = 2b = 2(a+b) = 0\rangle$ is a split extension of C_4 by C_2, that is, $D_4 = C_4 \rtimes C_2$ and

$$C_4 \rightarrowtail D_4 \twoheadrightarrow C_2$$

C_4 is the cyclic group $\langle a\rangle$ and C_2 is $\langle b\rangle$.

The group action is

$$(a)^b = -a = 3a$$

Table 2.6 describes one of the three possible non-isomorphic rings hosted by C_4.

Table 2.6 A ring multiplication on C_4

·	0	a	2a	3a
0	0	0	0	0
a	0	2a	0	2a
2a	0	0	0	0
3a	0	2a	0	2a

The action fits in with the product in the way described by (2.18) and therefore gives us a near-ring hosted by D_4.

The other two rings hosted by C_4 include the trivial ring $x \cdot y = 0$ and the ring in which $a^2 = 3a$. The latter does not satisfy (2.18) because, taking $k = b = k_3 = k$, we have $3a = (3a \cdot a) \neq (3a)^b = +a$. The former, a trivial ring, does fit the condition but is uninteresting.

If we turn to general dihedral groups, D_n with $n > 2$, the presentation would be

$$\langle a, b \ : \ na = 2b = 2(a + b) = 0 \rangle \tag{2.19}$$

and the sequence is

$$C_n \rightarrowtail D_n \twoheadrightarrow C_2$$

with action

$$(a)^b = -a$$

giving addition

$$(ma, rb) + (ka, sb) = ((m + (-1)^r k)a, (r + s)b)$$

where $m, k \in \{0, 1, \ldots .(n - 1)\}$ and $r, s \in \{0, 1\}$.

Then, (2.18) will imply

$$(-a) \cdot a = a^2$$

(taking $n_2 = n_3 = a$ and $k = b, k_3 = 0$).

This means that unless the additive generator of C_n squares to zero and C_n is a trivial ring, we must have that the additive group $C_n = \langle a \rangle$ is of even order and $a^2 = -(a^2)$. This latter condition does relate to a ring on C_n and consequently gives us a non-trivial near-ring on D_n, but only when n is even. In that case, $a^2 = \frac{n}{2}a$ and the product is

$$(ma, rb) \times (ka, sb) = \left(\frac{mkn}{2} a, rsb \right) \tag{2.20}$$

Lemma 118 *Consider the dihedral group $D_{2\eta}$ (i.e. $n = 2\eta$) with $\eta \geq 2$. Relation (2.20) represents a non-trivial near-ring product hosted by $D_{2\eta}$ and $C_{2\eta}$ is an ideal in this near-ring. This near-ring is commutative and bi-distributive, and the ring $r^2(D_{2\eta})$ is additively isomorphic to the Klein group (see Lemma 88). In fact, $r^2(D_{2\eta}) = \langle \eta a, b \rangle$ and*

$$(\eta a)^2 = \begin{cases} 0 & \text{when } \eta \text{ is even} \\ \eta a & \text{otherwise} \end{cases}$$

The zeroiser ideal (Definition 55) is $\langle 2a\rangle$. This is also the commutator subgroup of the group and the annular ideal (Definition 86). Its quotient is additively a direct sum of two copies of the ring of integers C_2. In the case that η is odd, the quotient near-ring is the direct sum of two copies of the ring of integers modulo 2. When η is even, the quotient near-ring itself has a proper non-trivial zeroiser ideal, and its quotient is the ring of integers modulo 2.

2.5 Ideas from Mathematical Logic and Universal Algebra

Much of my subject matter involves finite structures, but I do need **Zorn's lemma** on a few occasions (see [96]).

I am conscious of my own deficiencies in this section. I have far less expertise here than I would wish. The subject matter is quite beautiful and I hope I have not misrepresented it.

2.5.1 *Equational Products*

Given a word $w(x, y)$ in two symbols, I now investigate products on an arbitrary group $(G, +)$ defined using the formula

$$(a * b) = w(a, b) \tag{2.21}$$

The twin considerations of right distributivity and associativity can be expressed by the two conditions

$$w(x + y, z) = w(x, z) + w(y, z) \tag{2.22}$$

$$w(w(x, y), z) = w(x, w(y, z)) \tag{2.23}$$

which themselves correspond to a set $\mathcal{W}$ of two words in the free group on two symbols. These words generate a corresponding *variety* of groups, $V(\mathcal{W})$ (see Notation 3).

I use Notation 61 in the next lemma.

Lemma 119

(i) Equation (2.21) represents a near-ring product for any group $G \in V(\mathcal{W})$.
(ii) In this near-ring, $r^{(2)}(G)$ is the verbal subgroup corresponding to our original word $w(x, y)$ in (2.21) and, as such, is itself fully invariant in $(G, +)$.
(iii) This last point reinforces the trivial fact that not all near-ring products can be equational.

The variety $V(\mathcal{W})$ is associated with a **reduced free group on two symbols** (see [124] or [132]). I shall denote this reduced free group by $\mathcal{F}$. There is a corresponding variety of groups generated as quotients of $\mathcal{F}$.

Lemma 120 *Prescription (2.21) represents a near-ring hosted by $\mathcal{F}$. The near-rings of the corresponding variety defined using the same prescription are near-ring epimorphic images of that near-ring.*

Proof Other groups of the variety are epimorphic group images of $\mathcal{F}$

$$\Phi : \mathcal{F} \twoheadrightarrow G \tag{2.24}$$

$$\Phi\left((a * b\right) = \Phi\left(w(a, b)\right) = w\left(\Phi(a), \Phi(b)\right) = \left(\Phi(a) * \Phi(b)\right)$$

This means that Φ is an epimorphism of near-rings. □

Example 121 Perhaps the simplest example of a word I can offer is the standard commutator

$$w(x, y) = [x, y] = -x - y + x + y$$

The product becomes

$$(a * b) = [a, b]$$

and the two words corresponding to right distributivity and associativity are

$$[x + y, z] - [y, z] - [x, z] \tag{2.25}$$

$$[[x, y], z] - [x, [y, z]] \tag{2.26}$$

The corresponding variety of groups $\mathcal{W}$ is the variety of nilpotent groups of class 2.

It is immediate that the prescription

$$w(x, y) = [y, x] = -[x, y]$$

must also represent a near-ring product on these groups.

There are obvious and direct generalisations, which would be suggested by any text on **universal algebra** [24, 124].

One can identify a class of groups as a variety precisely when it is known to be closed under subgroups, direct products and homomorphic images [132]—and this result is more generally true and usually associated with Garrett Birkhoff, who was also one of the initiators of lattice theory.

2.5.2 *Boolean Algebras and Boolean Rings*

The power set of a set X is the set of all subsets, $\mathfrak{P}(X)$. There are natural operations to be performed on $\mathfrak{P}(X)$—unions, intersections, complementation—and there are distinguished subsets, the whole set itself and the empty set. These ingredients form an algebra called a **Boolean algebra**. Boolean algebras are vastly important and have an interesting theory which connects strongly with mathematical logic and set theory and which crops up in many guises [9, 52]. A Boolean algebra is really a kind of lattice (see page 96).

One striking example of their ubiquity involves the *clopen* sets of a topological space. Using standard unions and intersections, these form a Boolean algebra. Not only that, but every Boolean algebra arises in this way—a result usually referred to as the **Stone representation theorem**.

Boolean algebras connect intimately to power sets. Any finite subset of $\mathfrak{P}(X)$ which is closed under complementation and finite is a Boolean algebra, for instance—and a subalgebra of the power set algebra itself. Algebras of this type are often, distressingly, referred to as being a **field of sets in X**. **Boolean rings** [27] are rings with identity elements with the property that for all elements x, $x^2 = x$. In other words, all elements are multiplicative idempotents. It's easy to prove that these rings are commutative and have characteristic 2. It turns out that the theory of Boolean rings is equivalent to the theory of Boolean algebras—from one, you can get the other.

2.5.3 *Boolean Near-Rings*

Definition 122 A (zero-symmetric, right-distributive) near-ring N is **Boolean** provided that $\forall\, x \in N$, we have that $x^2 = x$.

Near-constant near-rings (see Definition 78) are Boolean, so any group at all hosts a Boolean near-ring.

Remark 123 It is well-known that Boolean rings $(R, +, \cdot)$ have these properties.

(i) $x + x = 0, \ \forall\, x \in R.$
(ii) $x \cdot y = y \cdot x, \ \forall\, x, y, \in R$

I shall now discuss the more general, near-ring case. In this section $(N, +, \cdot)$ represents a typical Boolean near-ring. Notice that the near-ring has no non-trivial multiplicatively nilpotent elements.

Lemma 124 *Suppose* $x, y \in N$.

(i) $x \cdot y = 0 \Leftrightarrow y \cdot x = 0.$
(ii) $x \cdot y = 0 \Rightarrow \langle x\rangle \cdot \langle y\rangle = \{0\} \Rightarrow \langle y\rangle \cdot \langle x\rangle = \{0\}.$

Proof

(i) $y \cdot x = y \cdot x \cdot y \cdot x = y \cdot (x \cdot y) \cdot x = 0$.
(ii) $x \cdot y = 0 \Rightarrow \langle x \rangle \cdot y = 0 \Rightarrow y \cdot \langle x \rangle = 0 \Rightarrow \langle y \rangle \cdot \langle x \rangle = 0$.

□

Lemma 125 *Suppose* $x, y \in N$.

(i) $x \cdot (x - y) = (x - y) + y \cdot (x - y)$
ii) $x \cdot (x + y) = (x + y) - y \cdot (x + y)$
(iii) $x \cdot y = x \cdot y \cdot x$
(iv) $x \cdot (-x) = x$

Proof

(i) $(x+y)\cdot(x-y) = x\cdot(x-y)+y\cdot(x-y)$ $(x+y)\cdot(x-y) = (x-y+y+y)\cdot(x-y) = (x-y)+y\cdot(x-y)+y\cdot(x-y)$. Consequently, $x \cdot (x-y) = (x-y)+y\cdot(x-y)$
(ii) Simply replace y with -y in (i).
(iii) $(x \cdot y - x \cdot y \cdot x) \cdot x \cdot y = 0$. Thus, $0 = x \cdot y \cdot (x \cdot y - x \cdot y \cdot x) = (x \cdot y - x \cdot y \cdot x) + x \cdot y \cdot x \cdot (x \cdot y - x \cdot y \cdot x) = (x \cdot y - x \cdot y.x)$.
(iv) $(-x) = (-x)^2 = -(x \cdot (-x))$

□

Lemma 126 *Any Boolean near-ring with a multiplicative identity element has additive characteristic equal to 2 and is abelian.*

Proof From Lemma 125 we have that $(-1) = 1 \cdot (-1) = 1$. This means $x + x = (1 + 1) \cdot x = 0 \cdot x = 0$ and so $|x| = 2$. Groups with exponent 2 are abelian. □

Lemma 127 *Suppose J is a left ideal of the Boolean near-ring N. Then J is also a right ideal.*

Proof Take $j \in J$ and $m \in N$. Then $j \cdot m = j \cdot m.j \in J$. □

Now, suppose $a \cdot x = 0$. Then, $a \cdot (x + y) = a \cdot (x + y) \cdot a = a \cdot y \cdot a = a \cdot y$ which means

Lemma 128 $a \cdot x = 0 \Rightarrow a \cdot (x + y) = a \cdot (y + x) = a \cdot y$.

Lemma 129 *Select* $a, x, y \in N$.

(i) $a \cdot x \cdot y = a \cdot y \cdot x$ *(this is sometimes referred to as "weak commutativity").*
(ii) If $a \cdot x = y \cdot a$*, then* $a \cdot y = a \cdot x$ *and* $a \cdot y = y \cdot a$.

Proof

(i) $y\cdot(a-a\cdot y) = 0 \Rightarrow y\cdot(a-a\cdot y)\cdot x = 0 \Rightarrow y\cdot(a\cdot x-a\cdot y\cdot x) = 0 \Rightarrow (a\cdot x-a\cdot y\cdot x)\cdot y = 0 \Rightarrow (a\cdot x\cdot y-a\cdot y\cdot x\cdot y) = 0 \Rightarrow (a\cdot x\cdot y-a\cdot y\cdot x) = 0 \Rightarrow a\cdot x\cdot y = a\cdot y\cdot x$
(ii) $a \cdot x = a \cdot a \cdot x = a \cdot y \cdot a = a \cdot y$

□

Weak commutativity means that any product of elements from a Boolean near-ring can be expressed uniquely as a product involving single occurrences of any of the symbols appearing.

Corollary 130 *Any Boolean near-ring with an identity is multiplicatively commutative and a ring.*

Proof We have already established the near-ring is abelian. We have $1 \cdot x \cdot y = 1 \cdot y \cdot x$. □

Lemma 131 *Suppose N is a Boolean near-ring and $a \in N$. Define $a \cdot N = \{a \cdot n : n \in N\}$.*

(i) $a \cdot N \subseteq N \cdot a = \{n \cdot a : n \in N\}$.
(ii) $a \cdot N$ is a Boolean ring with identity element a.

Proof

(i) $a \cdot n = a \cdot n \cdot a \in N \cdot a$, consequently, $a \cdot N \subseteq N \cdot a$.
(ii) a acts as a two-sided identity on $a \cdot N$, which is therefore a Boolean ring. □

Remark 132 Right distributivity gives us

$$N \cdot a = \frac{N}{\begin{pmatrix} a \\ 0 \end{pmatrix}_{\mathcal{L}}}$$

Lemma 133 *Taking any $a \in N$,*

(i) The set

$$\begin{pmatrix} a \\ 0 \end{pmatrix}_{\mathcal{L}} = \{x \in N : x \cdot a = 0\}$$

is a two-sided ideal of N.
(ii) The set $N \cdot a = \{n \cdot a : n \in N\}$ forms a Boolean near-ring, and a subring of $(N, +, \cdot)$.
(iii) a acts as a right identity on the set $N \cdot a$.

Now, suppose, with the notation just established, that $(a + a) \in a \cdot N$. We have $(a+a) = (a+a)^2 = a \cdot (a+a) + a \cdot (a+a)$. We know that a acts as a left identity on $a \cdot N$ and conclude $(a + a) = (a + a) + (a + a)$. This means $(a + a) = 0$, as happens generally in the ring situation.

2.5.4 *Finite Boolean Near-Rings*

In this subsection suppose that $(N, +, \cdot)$ is a finite Boolean near-ring. Consider the set $a \cdot N = \{a \cdot n : n \in N\}$. If $x \in a \cdot N$, then $x \cdot N \subseteq a \cdot N$. The set $x \cdot N$ cannot be $\{0\}$ when x is non-zero. When $a \cdot N$ is of minimal size, $x \cdot N = a \cdot N$, and so, $a \cdot r = x$ and $x \cdot s = a$ for some $r, s, \in N$. Now, $(a - x) \cdot s \cdot r = (a - ar) \cdot s \cdot r = (a \cdot s \cdot r - a \cdot r \cdot s) = 0$. We have $a = a \cdot r \cdot s$ and $x = x \cdot s \cdot r$ and conclude $(a - x) = (a \cdot s \cdot r - x \cdot s \cdot r) = (a - x) \cdot s \cdot r = 0$. This means that $a \cdot N$ is a doubleton set, $\{0, a\}$. More than that, the near-ring and $N\cdot$ a has the property

$$x \cdot y = \begin{cases} x & \text{if } y \neq 0 \\ 0 & \text{otherwise} \end{cases}$$

that is, a near-constant, near-ring (see Definition 78). This is because $(m \cdot a) \cdot (n \cdot a) = (m \cdot a) \cdot (a \cdot n)$ and both $n \cdot a$ and $a \cdot n$ are non-zero for the same values of n, which is when $a \cdot n = a$. I have already observed that $N_1 = \{x : xa = 0\} = \begin{pmatrix} a \\ 0 \end{pmatrix}_{\mathcal{L}}$ is a two-sided ideal of N; but the exact sequence of groups

$$N_1 \rightarrowtail N \twoheadrightarrow Na \tag{2.27}$$

splits additively, since $ma + na = (m + n)a$. Notice N_1 is another Boolean near-ring.

Lemma 134 *Suppose N is a finite Boolean near-ring and $a \in N$ has the property that $a \cdot N = \{a, 0\}$.*

(i) We have an exact sequence of near-rings

$$N_1 \longmapsto N \twoheadrightarrow \{N \cdot a\} \tag{2.28}$$

which splits additively.

(ii) $N \cdot a$ is a near-constant near-ring.

(iii) $N_1 = \begin{pmatrix} a \\ 0 \end{pmatrix}_{\mathcal{L}}$ is a Boolean near-ring.

(iv) If two arbitrary elements of N are written as $(n \cdot a + j_1)$ and $(m \cdot a + j_2)$, where j_1 and j_2 are elements of N_1, then

$$(n \cdot a + j_1) \cdot (m \cdot a + j_2) = (n \cdot a + (j_1 \cdot j_s))$$

Proof The only part of this needing proof is probably the last and that uses Lemma 128 like this $(n \cdot a + j_1) \cdot (m \cdot a + j_2) = (n \cdot a) \cdot (m \cdot a + j_2) + (j_1) \cdot (m \cdot a + j_2) = (n \cdot a) \cdot (m \cdot a) + (j_1) \cdot (j_2)$. □

The above lemma means that questions about the near-ring product on N are essentially reduced to questions about the near-ring product on N_1. Of course, the decomposition iterates. First, N_1 decomposes into $N_2 \longmapsto N_1 \twoheadrightarrow N_1 \cdot b$, then N_2

Table 2.7 The Boolean algebra of subsets of a three-element set

$\cdot$	0	1	2	3	4	5	6	7
0	0	0	0	0	0	0	0	0
1	0	1	0	0	1	1	0	1
2	0	0	2	0	2	0	2	2
3	0	0	0	3	0	3	3	3
4	0	1	2	0	4	1	2	4
5	0	1	0	3	1	5	3	5
6	0	0	2	3	2	3	6	6
7	0	1	2	3	4	5	6	7

decomposes, and so on. We finally reach a near-ring $M = N_n$ for which, choosing $m \in M$ so that $m \cdot M = \{m, 0\}$ is minimal gives us $M \cdot m = M$ so no further descent is possible. That is, one expects finally to end with a near-constant near-ring.

I shall illustrate this with a specific example. Suppose we consider the Boolean algebra of subsets of the three-element set $\{A, B, C\}$. The group addition is set symmetric difference, the multiplication, set intersection. The subsets are $\{\phi, \{A\}, \{B\}, \{C\}, \{A, B\}, \{A, C\}, \{B, C\}, \{A, B, C\}\}$. I shall represent them with the symbols: $\{0, 1, 2, 3, 4, 5, 6, 7\}$.

The multiplication table is represented in Table 2.7.

We look at the sets $a \cdot N$ for various choices of a and get that $a \cdot N$ is minimal for $a = 1, 2, 3$. Choose a = 3. Then $N \cdot a = \{0, 3\}$, and $\begin{pmatrix} 3 \\ 0 \end{pmatrix}_{\mathcal{L}} = \{0, 1, 2, 4\}$—earlier, we used the symbol N_1 for this set. $N \cdot a$ is certainly a near-constant near-ring. The Boolean near-ring $N_1 = \{0, 1, 2, 4\}$ involves us in a search for elements a for which $a \cdot N_1$ is minimal. Either of the elements $\{1, 2\}$ do the job. Choosing $a = 2$, we obtain $N_1 \cdot a = \{0, 2\}$ and the corresponding annihilator ideal is $\{0, 1\}$. Again, $N_1 \cdot a$ is near-constant as indeed is $\{0, 1\}$.

As a further example, consider finite symmetric groups S_n. These host Boolean near-rings which are near-constant and near-rings in which the alternating group, A_n, annihilates a transposition and forms a near-constant near-ring in its own right.

Connections with Propositional Logic

Boolean rings are important in propositional logic, ring addition being identified with the exclusive or (XOR) and multiplication with conjunction.

In computer science, one sometimes generalises conjunction in terms of left-to-right evaluations. So, $A \wedge B$ would be interpreted as involving the evaluation of A, and the subsequent evaluation of B, when A proves to be true (if false, there being no point in proceeding with the evaluation since the ultimate truth value is known). With this interpretation, $A \wedge B$ and $B \wedge A$ can be regarded as distinct. For example, when A is true and B false, the former involves two evaluations, the latter merely one. The two expressions $((A \wedge B) \wedge C)$ and $(A \wedge (B \wedge C))$ are still formally the same under this regime. Moreover, the left distributivity law

$$(A \wedge (B \text{ XOR } C)) = (A \wedge B) \text{ XOR } (A \wedge C)$$

applies. However the corresponding *right* distributivity rule

$$((A \text{ XOR } B) \wedge C) \;=\; ((A \wedge C) \text{ XOR } (B \wedge C))$$

is now no longer valid, for, when A, B and C are all true, the left-hand side involves us in the evaluation of A and B only (since their truth makes the whole expression false) whereas the right-hand side involves us in the evaluation of A, C and then B.

All this corresponds to having a left-distributive, Boolean near-ring under these operations.

2.5.5 Partially Ordered Sets

My **partially ordered sets** ($S \leq$) satisfy the three axioms of **reflexivity** ($x \leq x$), **transitivity** ($x \leq y \leq z \Rightarrow x \leq z$) and **anti-symmetry** ($x \leq y, y \leq x \Rightarrow x = y$). Many authors call such sets **"posets"**; but some use the same term for sets for which the third axiom does not apply.[2]

People habitually extend this notation in the obvious way, writing $y \geq x$ when if $x \leq y$. A **totally ordered set** is a partially ordered set in which any two elements are comparable—that is, either $a \leq b$ or $b \leq a$. A **directed set** is a pre-order with the property that any pair of elements has an upper bound—that is,

$$x, y \in S \Rightarrow \exists\, b \in S, \textit{ such that both } x \leq b \textit{ and } y \leq b$$

This assortment of definitions is part of an attempt to introduce the analytical ideas of sequence convergence and limits to a more general, topological and algebraic, setting. The central notion is that of a directed set, which was a constant in much of this work, although not the categorial part.

It all started with Moore-Smith convergence, defined for general topology using **nets**, in the 1920s. Nets replace the integers used in sequences with arbitrary directed sets—one sees the same sort of thing in algebra with direct and projective limits (described below). An equivalent approach due to Henri Cartan (1937) was based on *filters*, which I cover later (page 100). It turned out that most topological ideas can be expressed using these simple generalisations of limits [82, 87]. In group theory, for instance, suppose $(S, \leq)$ is a directed set and we have a mapping

$$\Phi : S \longmapsto \{G_s\}_{s \in S} \quad s \mapsto G_s$$

[2] A bit of grumpiness here! Those of us who are not brilliant mathematicians can spend hours trying to figure out which definition an author is using, and in a surprising number of cases, this remains wilfully inexplicit. The general story is that **"pre-orders"** are appropriate to limiting processes, these being sets satisfying only the first two axioms, and that, as I see it, it is regrettable that they are sometimes called partial orders—which, for many of my generation, are pre-orders satisfying the third! See Sect. C.1.10.

where $\{G_s\}_{s\in S}$ is some family of groups. Further suppose there are homomorphisms $\theta_{(s_1,s_2)} : G_{s_1} \mapsto G_{s_2}$ for all $s_1 \leq s_2$, for which

(i) $\theta_{s,s} = ID_{G_s}$.
(ii) $s_1 \leq s_2 \leq s_3 \Rightarrow \theta_{(s_1,s_3)} = (\theta_{(s_2,s_3)} \circ \theta_{(s_1,s_2)})$

This is called a **direct system of groups**. From it, one may construct a group

$$D = \varinjlim_{s\in S} G_s \tag{2.29}$$

and homomorphisms

$$\Theta_s : G_s \longmapsto D \tag{2.30}$$

The set $\{D, \{\Theta_s\}_{s\in S}\}$ is called the **direct limit** of the original system.

The general idea is that in D elements of the groups $\{G_s\}$ are identified with all their possible images under mappings $\{\theta_{(s,t)}\}_{t\in S}$. We insist that the groups $\{G_s\}$ should be regarded as disjoint, so D will actually be a *disjoint union* below.

We construct D by forming $\bigcup_{s\in S} G_s$ and then taking equivalence classes from this set. The equivalence relation used is that if $g_{s_j} \in G_{s_j}$, $j = 1, 2$, then .

$$g_{s_1} \sim g_{s_2} \tag{2.31}$$

means that there is some $s \geq s_1, s \geq s_2$ such that

$$\theta_{(s_1,s)}(g_{s_1}) = \theta_{(s_2,s)}(g_{s_2})$$

and then,

$$D = \varinjlim_{s\in S} G_s = \left(\frac{\bigcup_{s\in S}(G_s)}{\sim}\right)$$

The maps Θ_s map elements of G_s to their equivalence class under this relation.

As to the group operation on D, we consider equivalence classes $[g_{s_1}], [g_{s_2}]$ and know we can find some $s \geq s_1, s_2$. That means $\theta_{(s_1,s)}(g_{s_1}), \theta_{(s_2,s)}(g_{s_2}) \in G_s$ (although they may well be distinct elements), and it therefore makes sense to define

$$[g_{s_1}] + [g_{s_2}] = \left[\theta_{(s_1,s)}(g_{s_1}) + \theta_{(s_2,s)}(g_{s_2})\right]$$

(here the first "+" is being defined; the second is the group operation in G_s). This definition makes D into a group and the mappings $\Theta_s : G_s \mapsto D$ are homomorphisms.

The main properties of direct limits are:

(i) $D = \bigcup_{s \in S}(\Theta_s(G_s))$
(ii) If $s_1 \leq s_2$, then $\Theta_{s_1}(G_{s_1}) \leq \Theta_{s_2}(G_{s_2})$.
(iii) If the original set of homomorphisms, $\{\theta_{(s_1,s_2)}\}$, were all monomorphisms, so are the homomorphisms $\{\Theta_s\}_{s \in S}$.

Full details are available in [132]. Direct limits provide us with a mechanism for regarding any sequence of groups

$$\{G_1 \subseteq G_2 \subseteq \ldots\}$$

as embedded in a larger group. Direct limits apply to the common algebraic structures, and there are existence theorems relating to the properties a category must have for limits to exist. In the case of near-rings, the mappings mentioned above should be near-ring homomorphisms. Selecting g_{s_1} and g_{s_2}, and supposing $g_{s_3} \in [g_{s_1}]$, $g_{s_4} \in [g_{s_2}]$, where $g_{s_j} \in S_j$ $(j \in \{1, 2, 3, 4\})$, and (2.31) applies. We know we can find $s \geq s_1, s_2, s_3, s_4$ and such that

$$\begin{aligned} \theta_{(s_1,s)}(g_{s_1}) &= \theta_{(s_3,s)}(g_{s_3}) \\ \theta_{(s_2,s)}(g_{s_2}) &= \theta_{(s_4,s)}(g_{s_4}) \end{aligned}$$

The definition of the product is

$$[g_{s_1}] \cdot [g_{s_2}] = [\theta_{(s_1,s)}(g_{s_1}) \cdot \theta_{(s_2,s)}(g_{s_2})]$$

which is clearly well-defined with respect to the two equivalence class representatives we have chosen. The question is whether it is well-defined with respect to *any* two. Imagining two more representatives of exactly the same cosets implies the existence of $s' \in S$, which exceeds both s the indices for the two new representatives. We know

$$\theta_{(s,s')}\left(\theta_{(s_1,s)}(g_{s_1}) \cdot \theta_{(s_2,s)}(g_{s_2})\right) = \left(\theta_{(s_1,s')}(g_{s_1}) \cdot \theta_{(s_2,s')}(g_{s_2})\right)$$

from which, well-definedness follows. This relies crucially on these mappings being near-ring homomorphisms.

There are dual structures, called **inverse limits**, and one may construct inverse limits of near-rings too. So far as I am aware, no-one has looked seriously at these structures within the context of near-ring theory, or, more specifically, within the context of Dickson near-fields.

Direct limits are related to the categorial notion of **"co-limits"** and are sometimes called **"inductive limits"**. Inverse limits relate to the categorial **"limit"** and are sometimes called **"projective limits"** ($\varprojlim_S G_s$). **Profinite groups** are inverse limits of finite groups (Profinite ≡ "Projective Limit of Finite Groups") [88]. The

Nottingham group (page 338) is a profinite group and there is much active research into profinite groups.

2.5.6 *Lattices*

Once one has a partially ordered set, one may define the **supremum** of two elements $\{x, y\}$ as an element z bigger than both, and smaller than any other element with that property. Partial order properties ensure that any supremum is unique.

That is

$$x \leq z, y \leq z, \text{ and } x \leq w, y \leq w \Rightarrow z \leq w$$

That is not to say that such things automatically exist.

Dually, one defines the **infinum** of two elements as an element smaller than both but bigger than anything else with that property. In analysis the supremum is often called the "least upper bound" and the infinum the "greatest lower bound". A partially ordered set with the property that any two elements do have both a supremum and an infinum is called a **lattice**.

In finite lattices, all subsets have infina and suprema,[3] but this does not always happen in infinite cases. Those lattices which do have this property are called **complete lattices**. They possess **greatest elements** and **least elements**, usually denoted by 1 and 0, respectively. More generally, lattices having least and greatest elements are called **bounded lattices**. The classic example of a lattice involves $\mathbb{N}$, with $a \leq b \equiv a|b$. The supremum of a and b is their least common multiple; the infinum is their highest common factor.

In lattice theory, the supremum of two elements is often called their **"join"** and written $(a \vee b)$. Their infinum is often called their **"meet"** and written $(a \wedge b)$. As one might expect, there is an extensive literature on lattices; one good introduction is [57], and [31] is another. [91] is always beautifully clear, on lattices, and everything else, though I guess it would now be thought old-fashioned.

Lattices occur in basic algebra as collections of subgroups, normal subgroups, modules, etc. For subgroups, inf corresponds to intersection and sup to the subgroup generated by the ones considered, and the same is done for modules over rings.

A **modular lattice** is one satisfying

$$(a \vee (b \wedge c)) = ((a \vee b) \wedge c), \text{ for all } a, b, c \text{ for which, } a \leq c \qquad (2.32)$$

[3] The Pedanta will tell you that the plural of lemma is lemmata, not lemmas, and that the plural of supremum is "suprema", not supremums. .

Modularity is something that excites algebraists, though it is hardly a catchy property. The reason is that lattices of submodules of a module over a ring, lattices of normal subgroups and lattices of ideals of a ring are all modular.

Famously, lattices of subgroups of a group are not generally modular. The groups for which the subgroup lattice is modular are called **Iwasawa groups** (sometimes, **modular groups** although that term has various, conflicting uses and when people speak of **the modular group** they usually mean $\mathrm{PSL}(2, \mathbb{Z})$).

Kurosh gives this neat characterisation of modular lattices in his book [91]. It is useful when verifying modularity for lattices, of ideals, left ideals, etc.—something which, later on, I leave to the reader.

Theorem 135 *A lattice is modular if and only if for all elements a, b, c satisfying*

$$a \leq b, \quad (a \wedge c) = (b \wedge c), \quad (a \vee c) = (b \vee c) \tag{2.33}$$

we can deduce that $a = b$.

Robinson gives an exercise in [132] to the effect that when a and b are subgroups, above, and c is a normal subgroup, then the deduction in (2.33) holds. Not immediately relevant here, this applies when considering modularity in ideal lattices.

Modularity in algebras usually involves **quasi-normal** (otherwise, **permutable**) subgroups, which are simply subgroups that commute with other subgroups, as normal subgroups do. When dealing with the various flavours of ideals, one has additive normality, so modularity is more or less immediate from Robinson's exercise, though I spell it out in Lemma 16. A **distributive lattice** is one in which meet distributes over join and vice versa (there is a duality in lattices—the one implies the other). That is

$$(a \ \wedge \ (b \vee c)) \ = \ ((a \wedge b) \vee (a \wedge c))$$

or

$$(a \ \vee \ (b \wedge c)) \ = \ ((a \vee b) \wedge (a \vee c)) \tag{2.34}$$

These are the lattices which may be identified as subsets of some power set [31]. The lattice of ideals of a ring is both complete and modular.

Last, a **complemented lattice** is a lattice with a greatest element, 1, and a least element, 0, in which every element has a **complement**. That is, for all elements a, there is an element b such that $(a \vee b) = 1$ and $(a \wedge b) = 0$. Boolean algebras are exactly the complemented, distributive, lattices [27] (see Sect. 2.5.2).

I now record the basic isomorphism theorems for modules over rings; I have simply copied them from [7], page 75.

Theorem 136 *Let M be a left module over a ring R, and let N be a submodule of it.*

(i) If K is any submodule of M, then

$$\frac{(N+K)}{K} \cong \frac{N}{(N \cap K)} \tag{2.35}$$

(ii) If $K \subseteq L$ are submodules of M, then

$$\frac{(\frac{M}{K})}{(\frac{L}{K})} \cong \frac{M}{L} \tag{2.36}$$

(iii) If $K \subseteq L$ are submodules of M, then

$$(L \cap (K+N)) = (K + (L \cap N)) \tag{2.37}$$

Beachy calls this "the modular law" [7]. It is often associated with Richard Dedekind.

Hall on Lattices

Marshall Hall covers lattices [62]. Given a lattice, he defines a **quotient sublattice** for each pair of elements $\{a, b\}$ as

$$\frac{a}{b} = \{x \, : \, a \geq x \geq b\} \tag{2.38}$$

The two lattices

$$\frac{(a \vee b)}{b} = \frac{a}{(a \wedge b)}$$

are then said to be **perspective** to each other.

He then gives a theorem to the effect that in modular lattices perspective quotients are isomorphic to each other (with the obvious definition of isomorphic lattices). This is strikingly similar to Theorem 136 (i). Hall goes on to define chains (page 97), dimension and chain lengths. From these ideas he obtains versions of some of the typical algebraic isomorphism theorems and a sort of Jordan-type result on maximal chains (page 4)—all of this without recourse to underlying algebra.

I bring this fascinating story up here because in near-rings there is a plethora of definitions for structures generalising modules, with accompanying isomorphism results usually proved using the algebra to hand. Many of them are probably just examples of this sort of work and do not need to be repeatedly reproved. The key requirement is always modularity (2.32).

I had thought that the sort of recurring relations given in Theorem 136 were episodic—recurrently rediscovered as people extended the algebras that interested them—and that, seeing a common pattern, people abstracted these ideas into lattice theory. But this seems to be the wrong way round, and in fact these variants on "modular laws" were uncovered by Dedekind in great generality, in a series of papers written in his retirement. My opinion was probably an artefact of the way that people choose to present special cases, and the individualised nature of university courses. Lattice theory is a much earlier development than I had realised. Grätzer lists the founders as Peirce,[4] Schröder and Dedekind and dates it to late nineteenth-century times [57].

The Socle

Another example of a standard idea in general algebra actually being related to abstract lattice theory is the **socle** (page 23). This applies to **complete, modular lattices**. One defines an element of the lattice to be **essential** provided it has non-trivial intersection with non-zero elements. That is

$$b \neq 0 \Rightarrow (a \wedge b) \neq 0 \tag{2.39}$$

The socle of the lattice is then defined as

$$\bigwedge (a \ : \ a \text{ is essential}) \tag{2.40}$$

These ideas occur in the theory of radicals and **antiradicals** [58]. I had intended to cover them in Chap. 12 but ran out of space and time.

2.5.7 *Finiteness Conditions: Chains, Intersections, Generators*

Finite near-rings preoccupy me in this book. When dealing with infinite structures, algebraists often look for gadgets that bring finite-like properties to particular classes of infinite structures, and chains provide such gadgets.

Definition 137 Suppose S is a partially ordered set; then a **chain** in S is simply a totally ordered subset of S.

One considers **ascending chains**, $A_1 \leq A_2 \leq \ldots$, and **descending chains**: $A_a \geq A_b \geq \ldots$. The finiteness gadgets then restrict to situations in which infinite, strictly *descending* chains cannot exist (the D.C.C.—descending chain condition—a device often associated with Emil Artin; so a ring that has the D.C.C. for left ideals would be called "left Artinian") or restrict to situations in which infinite, strictly *ascending* chains cannot exist (the A.C.C.—ascending chain condition—a device often associated with Emmy Noether; so a ring that has the A.C.C. for

[4] There are two Peirces in this book. This one is sometimes called "the Father of Pragmatism". Luminaries such as Bertrand Russell and Karl Popper regarded him as an outstanding philosopher. He is actually the *son* of the other one. .

left ideals would be called "left Noetherian"). Ring theory is replete with theorems about variations on these two conditions and about how they may relate to each other and about situations in which one holds but not the other. Of course, so too is the theory of near-rings (see Sect. 12.3.3). As an example, the ring of integers is a module over itself, and the A.C.C. holds for its modules, mainly because they are each generated by a single element. The module generated by an integer a is often written (a), where

$$(a) = \{m.a : m \in \mathbb{Z}\} \tag{2.41}$$

and so

$$(a) \subset (b) \Rightarrow b|a \tag{2.42}$$

which indicates that only a finite number of modules are contained in (b). On the other hand, the D.C.C. does not hold for these modules. Here is an infinitely descending chain of them.

$$(2) \supsetneq (2^2) \supsetneq (2^4) \supsetneq (2^8) \supsetneq \ldots \supsetneq (2^{2^\alpha}) \supsetneq \ldots \tag{2.43}$$

Complexities multiply in near-rings because so do the sub-structures to which they might be applied. For instance, there are various generalisations of ring modules available, I give several in Chap. 11, and within a single near-ring, some of these may obey some of these chain conditions, while others do not.

There are other finiteness gadgets that are quite common. One is to insist on rings having minimal left or right ideals, for instance, and the same idea does crop up in near-rings (see [119]), where one might also insist on minimal left or right sideals (see Lemma 52).

There is room for a more systematic account of these ideas than I have space for. Instead, I offer a tiny, and rather overloaded, amount of information about chains, both here and in Chaps. 7 and 11.

Sub-structures

One frequently wants to discuss sub-structures and almost invariably intersections provide examples. So, for illustration, the intersection of two ideals is an ideal contained in both and so on.

Sometimes, these sub-structures are infinite, and one considers sub-structures that are minimal with respect to containing some *finite* collection of elements. In that case, the sub-structures are usually termed **finitely generated**. This extends naturally to the original structure itself—which would be *finitely generated* provided there is a set of finite elements for which no proper sub-structure is properly contained in the original structure.

Often finite generation can be viewed in another guise. In group theory the smallest subgroup containing a given set of elements consists of the collection of all finite sums and subtractions of elements from the set. Rather than worrying about intersections, we then concern ourselves with algebraic structure.

Thus, the additive group of integers $(\mathbb{Z}, +)$ is finitely generated by the element $\{1\}$, whereas the additive group of rational numbers, $(\mathbb{Q}, +)$, is not finitely generated at all. However, the *field* of rational numbers, $(\mathbb{Q}, +, .)$, is finitely generated by the set $\{1\}$—no, proper, subfield contains that set.

Lemma 138 *If the near-ring $(N, +, \cdot)$ is finitely generated, then the union of any ascending chain of proper left ideals is a proper left ideal.*

Proof Suppose N is finitely generated by a set $\{n_1, n_2, \ldots, n_k\}$.

Suppose we have a chain of left ideals

$$L_1 \subset L_2 \subset L_3 \subset \ldots$$

Suppose

$$J = \bigcup_{i=1}^{\infty} (L_i)$$

J is itself a left ideal of N.

Suppose $J = N$.

Then, each n_j must lie in one of the left ideals C_t. But then N is a union of a *finite* number of the C_t, which implies that the biggest of these is not proper.

The conclusion is that $J \subsetneq N$. □

Lemma 139 *Suppose $(N, +, \cdot)$ is a near-ring in which the union of any ascending chain of proper left ideals is a finitely generated proper left ideal. Then the near-ring is finitely generated.*

Proof Consider the collection of finitely generated left ideals in the near-ring, ordered by inclusion. The hypothesis about ascending chains ensures that this is *inductive* in the sense of Zorn's lemma (page 390). So, the set contains maximal elements—these being finitely generated left ideals not contained in any other such finitely generated left ideal.

If M is one such maximal element, and not itself equal to N, then there is some $n \in \{N - M\}$, and so the left ideal generated by $\{M \cup \{n\}\}$ must be equal to N, which is therefore finitely generated. □

Lemma 140 *The following conditions are equivalent for a near-ring $(N, +, \cdot)$.*

(i) There is no infinite strictly ascending chain of left ideals

$$\{0\} \subsetneq I_1 \subsetneq I_2 \subsetneq I_3 \subsetneq \ldots \tag{2.44}$$

(ii) Any collection of left ideals contains a maximal member.
(iii) Any left ideal is finitely generated.

Proof

(i) ⇒ (ii) The proof flirts with Zorn's lemma. Suppose we have a collection of left ideals. Select any left ideal in it; if it is maximal, we are done; if it is not, we can select another left ideal strictly containing it.

In this way we obtain a strictly ascending sequence of left ideals. It breaks off finitely and when it does we have our maximal left ideal in the collection.

(ii) ⇒ (i) Taking any strictly ascending chain of left ideals, the collection of left ideals in the chain has a maximal member of the chain, so the chain cannot be infinite.

(iii) ⇒ (i) The union of left ideals in any chain is a left ideal. It is finitely generated, so each of its generators occurs at some point in the chain and in every left ideal thereafter. There must be a finite point at which all the generators have been accounted for, and at that point the chain stops ascending.

(i) ⇒ (iii) If we had a non-finitely generated left ideal, we could construct an infinite ascending chain of proper left ideals working with a posited infinite generating set.

□

This lemma is proved in almost the same way.

Lemma 141 *The following conditions are equivalent for a near-ring* $(N, +, \cdot)$.

(i) There is no infinite strictly descending chain of left ideals.
(ii) Every non-empty collection of left ideals has a minimal member.

I have expressed these lemmata in terms of left ideals, but they are, of course, more widely applicable, and I exploit that in Chap. 11 without further explanation. The anxious reader can find fuller proofs in [11, 128] and in [119].

2.5.8 Ultra-Products

Ultra-products are algebraic constructions with properties relating to model-theoretic considerations [6, 8, 9]. As such, they preserve some algebraic properties in unexpected ways.

They are more usually called "ultraproducts" (no hyphen) these days. Amitsur introduced them into ring theory. My account is based on that of Herstein [66].

Definition 142 Suppose S is some non-empty set. A collection of subsets of S is a **filter** provided that it does not include the empty set, it does include all finite intersections of members, and whenever a subset of S contains one of the subsets in the filter, that subset is also in the filter.

There is a partial order on filters based on inclusion of the constituent sets. An **ultra-filter** is a filter that is maximal under this partial ordering.

Any filter can be enlarged into an ultra-filter (this uses Zorn's lemma). A simple lemma characterises ultra-filters as filters for which any subset is either in the filter, or its complement is.

Filters are closely related to nets and result in the same sort of convergence theories. Suppose we have a collection of near-rings $\{N_\alpha\}_{\alpha \in A}$. Form their direct product

$$\mathfrak{D} = \prod_A (N_\alpha)$$

which we can think of as the set of distinct functions from A to $\bigcup_A (N_\alpha)$ for which $f(\alpha) \in N_\alpha$.

Suppose we have a filter S on A. We can define an equivalence relation on the direct product, viewed as functions, by saying

$$f \sim g \text{ means } \{\alpha \in A \,:\, f(\alpha) = g(\alpha)\} \in S.$$

Then we have a set representing the equivalence class of the zero mapping.

$$\mathfrak{I} = \{f \,:\, f \sim 0\} = \{f \,:\, \{a \in A \,:\, f(a) = 0\} \in S\}$$

Lemma 143 *$\mathfrak{I}$ is an ideal of the near-ring $\mathfrak{D}$.*

Proof Suppose $f, g \in \mathfrak{I}$. The function $(f - g)$ has a set of elements which it maps to zero which is a superset of the set of elements which both f and g map to zero. So it is a set in the filter A and is in $\mathfrak{I}$.

Additive conjugates of elements of $\mathfrak{I}$ are similarly seen to be in $\mathfrak{I}$, so $\mathfrak{I} \lhd (\mathfrak{D}, +)$. Suppose $f \in \mathfrak{I}$ and $\theta \in \mathfrak{D}$. The product $(f \cdot \theta)$ is not a composition of functions: it is the function produced by multiplying individual respective coordinate values. When the f coordinate is zero, the product at that coordinate will be too. Consequently, $(f \cdot \theta) \in \mathfrak{I}$. If we investigate $\theta \cdot (\phi + f) - (\theta \cdot \phi)$, we see that the function maps all coordinates that f takes to zero (as well as possibly other coordinates). Thus the function is also in $\mathfrak{I}$. □

The quotient structure $\mathfrak{U} = \frac{\mathfrak{D}}{\mathfrak{I}}$ is a near-ring.

Definition 144 When the filter S is an ultra-filter, the quotient structure $\frac{\mathfrak{D}}{\mathfrak{I}}$ is called **an ultra-product** of the near-rings $\{N_\alpha\}_{\alpha \in A}$.

Ultra-products may be defined for other algebraic structures: groups, vector spaces, modules and rings, for instance. As I understand it, one needs the index set, A, to be infinite and the ultra-filter to have the property that it contains all **co-finite** subsets of A (these being subsets whose complements are finite). If that is not the case, the ultra-product is apparently isomorphic to a product of some of its constituents and is called **principal** (see [24]). Ultra-powers are simply ultra-products in which all factors are the same.

The key result is usually called Łoš theorem, which is a technical result in mathematical logic relating the logical properties of the constituent algebras to the logical properties of the ultra-product. It is a useful device for going from a collection of finite things to a larger structure, which may be infinite, preserving some of the properties that may be of interest.

I defined ultra-filters in terms of collections of subsets of a set; but it is possible to instead work with general lattices. One defines filters as non-empty subsets of lattices which are closed under meets of element pairs and which contain all elements larger than individual elements. Then there is a dual idea, that of an **ideal** which involves subsets of a lattice closed under joins and which contains all elements smaller than individual elements. More details are given in [8]. The generalisation may be less profound than one might think because there are theorems associated with Stone and with Birkhoff, which relate lattices and Boolean algebras to power sets.

2.6 Adjoining an Identity

There is a standard and well-known way in which an identity element may be adjoined to a ring that lacks one.

Starting with the ring $(\mathcal{R}, +, \cdot)$, one forms the additive group $\mathcal{R} \oplus \mathbb{Z}$ and defines

$$((r_1, z_1) \cdot (r_2, z_2)) = ((r_1 \cdot r_2) + z_2 r_1 + z_1 r_2, z_1 z_2) \tag{2.45}$$

The original ring $\mathcal{R}$ is embedded in the new one under the embedding

$$r \mapsto (r, 0)$$

and is an ideal of the new ring—which has infinite size. I believe this is sometimes referred to as the "Dorroh extension". In the case of rings hosted by groups with finite exponent, the infinite cyclic group may be replaced by the corresponding finite cyclic group—C_m. This has the virtue of not always extending to an infinite group.

One might try to do the same thing for near-rings but hit problems. When the near-ring is abelian, we do obtain a p.n.r. with identity, but to obtain a full-blown near-ring in this way, we really need *both* multiplicative associativity and bi-distributivity (excluding possible pathologies). The way in which these laws influence each other in this context is explored more fully in Chap. 15 Product Theory. I also consider an alternative way to embed a group into a near-ring with identity (Sect. 7.3.8).

2.7 Planarity

A near-field (page xvii) which is a quasi-field (page 508) is called a **planar near-field**. These structures are important in the theory of coordinates in projective geometries. They are near-fields alternatively described as being near-fields $(F, +, \cdot)$ for which, given $a, b, c \in F$, with $a \neq b$, there is always a unique $x \in F$, such that

$$a \cdot x = (b \cdot x + c)$$

This is always true of finite near-fields, but there are infinite near-fields for which it is not; however the reader will see that any near-ring with this property is automatically a near-field.

In general near-rings, $(N, +, \cdot)$, one may define an equivalence relation $(\sim)$ by setting

$$(a \sim b) \Leftrightarrow (a \cdot x = b \cdot x) \textit{ for all } x \in N$$

Then, the near-ring $(N, +, \cdot)$ is said to be **planar** provided that:

(i) The equivalence relation has at least three equivalence classes
(ii) The equation $a = (b \cdot x + c)$ has a unique solution x whenever $(a \not\sim b)$

Planar near-rings have attracted quite a lot of attention, recently. They are zero-symmetric, and any planar near-ring with identity is actually a near-field. Much more information on them can be found in [23, 40]. The reviewers of this book felt I might reasonably have given them much more prominence than I did—a criticism I fully accept. The following construction appears in [25] where it is attributed to Ferrero.

2.7.1 *The Ferrero Construction*

Suppose that $(G, \circ)$ is some group of fixed-point-free automorphisms of the *finite* additive group $(A, +)$ (so only the identity element leaves any elements other than zero fixed). Choose some non-empty collection of the orbits of these automorphisms of the form $\{A_i\}_{i \in I}$, none of these zero. Choose $\{a_i \in A_i\}_{i \in I}$ as distinct orbit representatives. Select $x, y \in A^*$ and suppose $x \in A_j$. We can find some unique $g_j \in G$ such that $g_j(a_j) = x$ and we set

$$(x \wedge y) = (g_j(y))$$

In the event that $x \neq \left(\bigcup_{i \in I}(A_i)\right)$, define $x \wedge y = 0$. The structure $(A, +, \wedge)$ is a zero-symmetric near-ring hosted by $(A, +)$. It turns out to be planar, and all planar near-rings arise in this way (see [23]). In infinite cases we additionally must assume that for all non-identity maps $\phi \in G$, the mapping $(\phi + ID_A)$ is surjective (as always happens in finite cases).

Compare these ideas with Sect. 11.4.6.

Chapter 3
Near-Fields

Nearly all near-fields in this chapter are finite and right-distributive.

This chapter requires some of the standard theory of finite fields. Lang [96] is a general reference for that material and [123] is a very accessible account of it.

I am almost entirely concerned with finite near-fields here, though some of what I say has more general application. I also cover the work of a number of authors. Principal sources include [32, 62, 128].

Much of the modern theory of near-fields is only available in German, reflecting the key authorities in this area: Karzel, Wähling, et al. I only have an outline knowledge of this work and have certainly not done it justice here, not having obtained some of the most basic references.[1] I believe at least some of the material given here is original to me but would not be surprised if that were not the case and hope that if I have misrepresented the work of better mathematicians they will understand that this is unintentional.

In this chapter, I shall often refer to the group of generalised quaternions, Q_{2^n} (3), and, in turn, some of the information given, and some of the notation developed, surfaces again, in Chap. 15. The reader is reminded that for me the class of generalised quaternion groups is identical to the class of quaternion groups and, non-standardly, includes Q_8.

I use Gothic "N" on page 120 ($\mathfrak{N}$), principally to avoid confusion with N^*. $\mathfrak{K}$ and $\mathfrak{F}$ both appear in Sect. 3.8.

In this chapter and Chap. 16, S_2 usually means the Sylow 2-subgroup.

[1] I did go to some lengths to try to get hold of the book on near-fields by Wähling [141]. It would probably be accessible through university libraries and does not seem to be commercially available, even second hand.

R. Lockhart, *The Theory of Near-Rings*, Lecture Notes in Mathematics 2295,
https://doi.org/10.1007/978-3-030-81755-8_3

The Frobenius Automorphism
I shall often refer to the **Frobenius automorphism** of a finite field with characteristic p. This is the map

$$x \mapsto x^p \tag{3.1}$$

(which is a field automorphism).

3.1 Near-Fields

Recall that a near-ring N is a **near-field** provided that the non-zero elements form a multiplicative group (see Definition 44) and the near-ring is not a constant near-ring (Definition 78). I sometimes refer to near-fields that are not themselves fields as "proper" near-fields (Definition 44). As I remarked on page 64, this is unfortunate terminology.

Simple Properties
Suppose $(N, +, \cdot)$ is a near-field in which 1 is the multiplicative identity. We know $(-x) \cdot y = -(x \cdot y)$ in all near-rings, so $(-1)^2 = +1$.

If $(N, +)$ has an element, x, with prime, additive, order, p, then right distributivity ensures that $x \cdot N = \{x \cdot n \ : \ n \in N\}$ consists of the additive identity, and elements with additive order p. This being the whole group, we deduce that every non-zero element of $(N, +)$ has additive order p. The conclusion is that when $(N, +)$ is not torsion-free, it will be a p-group with exponent p, for some fixed prime.

Both left and right multiplication by any fixed non-trivial element of the near-field induce bijective mappings of its elements. Right multiplication $(x \mapsto x \cdot y)$ induces group automorphisms. The (additive) commutator subgroup is a right ideal in any near-ring; so, in near-fields, multiplicative invertibility implies that it must either be trivial or equal to $(N, +)$. Moreover, the host group of a near-field $(N, +)$ must be both characteristically simple (contains no proper non-trivial subgroups invariant under all automorphisms) and fully invariantly simple (has no proper, non-trivial, subgroups fixed by all endomorphisms), since no proper, non-trivial subgroup is invariant under all possible right multiplications.

In summary, $(N, +)$ could only be non-abelian when perfect; and unless it is torsion-free, it has exponent p, for some prime. The group is fully invariantly simple and characteristically simple.

Lemma 145 *All finite near-fields are hosted by elementary abelian p-groups.*

If $(N, +)$ has exponent 2, then it is, of course, abelian. I shall call the exponent of the additive group of the near-field its **characteristic** as one does in field theory. This fits in with my earlier suggestion for characteristics of near-rings (Remark 76).

The additive group hosting a near-field has unusual properties. If torsion-free, then, for each positive integer r, each element (1 + 1 + ... 1) (r times) is multiplicatively invertible, with inverse, m. So $rm = m + m + \ldots m = 1$ and for any $a \in N$ we have that $a = 1 \cdot a = r(m \cdot a)$, which means that the group is additively divisible (though not yet shown to be abelian). Simple manipulations reveal that the multiplicative identity of a near-field generates a **prime subfield**. In finite cases, the near-field will be a vector space over this prime subfield, C_p, and consequently have order equal to p^α for some integer $\alpha > 1$.

3.1.1 Near-Fields Not of Characteristic 2

In this subsection, near-fields will be assumed to have characteristic not equal to 2.

Lemma 146 *Suppose $(N, +, \cdot)$ is a near-field with multiplicative identity, 1. Then the multiplicative group $(N^*, \cdot)$ of non-zero elements has exactly one element with order 2 and this is (-1).*

Proof Clearly, (-1) does have multiplicative order 2. Suppose $n \in N$ and $n^2 = 1$. Define $m = \big((1+1)^{-1} \cdot (n-1) + 1\big)$. Then,

$$m \cdot n = \Big((1+1)^{-1} \cdot (1-n) + (n-1) + 1\Big) =$$
$$\Big((1+1)^{-1} \cdot (-1) + 1\Big) \cdot (n-1) + 1 \tag{3.2}$$

Now,

$$\Big((1+1)^{-1} \cdot (-1) + 1\Big) \cdot (-(1+1)) = \Big((1+1)^{-1} \cdot (1+1) - (1+1)\Big) = -1$$

which means $(1+1)^{-1} = \big((1+1)^{-1} \cdot (-1) + 1\big)$, and, from Eq. (3.2), we obtain

$$m \cdot n = \Big(\big((1+1)^{-1}\big) \cdot (n-1) + 1\Big) = m$$

This identity implies that n is 1 when m is non-zero. For n to be distinct from 1, we require $(n-1) = -(1+1)$, that is, $n = -1$. □

Select $x \in N$ and form $y = x^{-1} \cdot (-1) \cdot x$. Clearly, $y^2 = 1$. y can only be equal to 1 if the characteristic is 2, forbidden in this subsection, so from Lemma (146), we conclude $y = -1$ and that $x^{-1} \cdot (-1) = -x^{-1}$. That is to say, $x \cdot (-1) = -x$ for all $x \in N$.

3.1.2 General Near-Fields

Lemma 147 *In any near-field*

$$x \cdot (-y) = -(x \cdot y)$$

Selecting, $x, y \in N$, we obtain $(-x+y) = (x-y)\cdot(-1) = (y-x)$ and conclude

Lemma 148 *The additive group of a near-field is abelian.*

In general near-fields with odd prime characteristic, the mapping $x \mapsto (-1) \cdot x$ is a bijection. The mapping interchanges $+1$ and -1 and bijectively associates elements with odd order with elements with order congruent to 2 modulo 4. To see this, notice that if x has order $(2 + \psi.4)$, then $(x)^{(1+\psi.2)}$ has order 2 and is therefore equal to -1. This means (-x) has order $(1 + \psi.2)$ The remaining elements are the elements with order congruent to 0 modulo 4. These the bijection preserves.

Lemma 149 *In finite near-fields with odd characteristic, there are as many elements with odd order as there are elements with order congruent to 2, modulo four.*

Lemma 150 *Suppose* $(N, +, \cdot)$ *is a general near-field. Then the distributivity core,* $\mathcal{C}_{\mathcal{D}}(N)$, *(see Definition 69) is a division ring within* $(N, +, \cdot)$, *one which contains the prime subfield.*

Proof The fact that the additive group is abelian forces $\mathcal{C}_{\mathcal{D}}(N)$ to be an additive subgroup of the near-field. It obviously contains the prime subfield. The distributive elements are multiplicatively closed and if $d \in N^*$ is such an element, then

$$x = d^{-1} \cdot (a+b) - (d^{-1} \cdot b) - (d^{-1} \cdot a) \Rightarrow d \cdot x = (a+b) - b - a = 0$$

Hence $x = 0$ and so $d^{-1} \in \mathcal{C}_{\mathcal{D}}(N)$ □

The Kernel of a Near-Field

Pilz follows other authors in calling $\mathcal{C}_{\mathcal{D}}(N)$ the **kernel** of the near-field [128], a term relating to their close connection with Frobenius groups. In the case of *finite* near-fields, $(\mathcal{C}_{\mathcal{D}}(N), +, \cdot)$ is a subfield of $(N, +, \cdot)$ (Wedderburn's theorem). We know that $(N, +, \cdot)$ has order p^α for some prime p and that any subfield has order p^β for some $\beta \leq \alpha$; consequently, the subset of non-zero distributive elements is a cyclic subgroup of $(N^* \cdot)$, one with order $(p^\beta - 1)$, dividing $(p^\alpha - 1)$. Of course, any bi-distributive near-ring (Definition 87) whose non-zero elements are, multiplicatively, a cyclic group is a field.

Lemma 151 *Suppose* $q, \beta, \alpha, \in \mathbb{N}$, *with* $q > 1$, *and suppose that*

$$(q^\beta - 1) \mid (q^\alpha - 1) \textit{ (divides)} \tag{3.3}$$

Then, β *divides* α.

Proof We may write $\alpha = (m\beta + r)$ with $0 \leq r < \beta$. Then, $q^\alpha = q^{m\beta}.q^r$. Working modulo $(q^\beta - 1)$, we have

$$q^\alpha \equiv 1 \equiv q^r$$

This is only possible if $r = 0$, and we deduce that whenever (3.3) holds, then $\beta \mid \alpha$. □

This result places restrictions on the finite near-fields that can occur and is used repeatedly in this chapter.

Writing $\alpha = \beta.\delta$, we have

$$(p^\alpha - 1) = (p^\beta - 1).\left(p^{(\delta-1).\beta} + p^{(\delta-2).\beta} + \ldots + p^\beta + 1\right) \tag{3.4}$$

Hence, if q is some prime dividing $(p^\beta - 1)$ (i.e. $p^\beta \equiv 1$ (*modulo* q)), then

$$\left(p^{(\delta-1).\beta} + p^{(\delta-2).\beta} + \ldots + p^\beta + 1\right) \equiv \delta\ (modulo\ q) \tag{3.5}$$

This is a useful result in the analysis of Sylow subgroups. For instance, a Sylow q-subgroup of $(\mathcal{C}_\mathcal{D}(N), \cdot)$ will be a Sylow q-subgroup of $(N^*, \cdot)$ unless q divides δ; and if the multiplicative group of non-zero elements of our near-field is to have Sylow 2-subgroups which are generalised quaternion, we should need that δ be even (see Lemma 165). A further interesting observation is that in the odd prime case, the condition that $(p + 1)$ divides $(p^\alpha - 1)$ is that α be even. This follows since $(p + 1)$ divides $(p^2 - 1)$ which, in turn, divides $(p^\alpha - 1)$ exactly when α is even.

Alternative argument:

$$(p^\alpha - 1) \equiv \begin{cases} (p-1) & \text{if } \alpha \text{ is odd} \\ (p^2-1) & \text{if } \alpha \text{ is even} \end{cases} \quad (modulo\ (p+1))$$

The Multiplicative Centre

Consider the multiplicative centre, Cent(N) (see page 53), of the *general* near-field $(N, +, \cdot)$. We already know Cent$(N) \subseteq \mathcal{C}_\mathcal{D}(N)$. The multiplicative identity is in this set, so we know that it contains both 0 and 1. The multiplicative centre is a normal subgroup of the multiplicative group of non-zero elements of the near-field. Standard group theory dictates that the corresponding quotient cannot be cyclic in non-field cases. The multiplicative centre is not an additive subgroup in general near-rings. This is related to the failure of left distributivity. However, we do have

$$\text{Cent}(N) \subseteq \text{Cent}(\mathcal{C}_\mathcal{D}(N))$$

Lemma 152 *When $(N, +, \cdot)$ is a near-field:*

(i) Cent($\mathcal{C}_{\mathcal{D}}(N)$) *is a commutative subring of $\mathcal{C}_{\mathcal{D}}(N)$, and therefore a subfield containing the prime subfield*
(ii) N is a vector space over the field Cent($\mathcal{C}_{\mathcal{D}}(N)$)
(iii) $\mathcal{C}_{\mathcal{D}}(N)$ is a sub-vector space of that vector space
(iv) $\mathcal{C}_{\mathcal{D}}(N)$ is an additive direct summand of $(N, +)$ (see 16.24 on page 458)
(v) When N is finite, $\mathcal{C}_{\mathcal{D}}(N) = \mathrm{Cent}(\mathcal{C}_{\mathcal{D}}(N))$

If $a \cdot (-1) = b$, then

$$(a+b) \cdot (-1) = (b + b \cdot (-1)) = (b+a) = (a+b)$$

Now,

$$x \cdot (-1) = x \neq 0 \Rightarrow (-1) = +1$$

so either the characteristic of the near-field is two, or $(a + b) = 0$. In either case, $a \cdot (-1) = -a = (-1) \cdot a$. Thus, when the characteristic of the near-field is not equal to two, Cent(N) has at least two non-zero elements—something that follows more directly from Lemma 147, of course. These two elements are ± 1. They are in the centre of all finite non-fields with odd characteristic and are actually equal to the centre in exactly four cases which I shall identify in this chapter.

Select $a, b \in N^*$, $a \neq 0$. Let us suppose that a has finite multiplicative order $r \in \mathbb{N}$, that is, $a^r = 1$. Then,

$$\left(b + b \cdot a + b \cdot a^2 + \ldots + b \cdot a^{(r-1)}\right) \cdot a = \left(b \cdot a + b \cdot a^2 + b \cdot a^3 + \ldots + b\right)$$

If now, $(b + b \cdot a + b \cdot a^2 + \ldots + b \cdot a^{(r-1)}) \neq 0$, we may cancel, obtaining $a = 1$.

Lemma 153 *Suppose a and b are elements of the near-field $(N, +, \cdot)$. Suppose a has multiplicative order $r \in \mathbb{N}$, with $2 \leq r < \infty$; then*

$$\left(b + b \cdot a + b \cdot a^2 + \ldots + b \cdot a^{(r-1)}\right) = \left(b + a \cdot b + a^2 \cdot b + \ldots + a^{(r-1)} \cdot b\right) \tag{3.6}$$

$$= 0$$

This result provides an easier proof of Lemma 146 (the r=2 case). Generally, if $(1+a+a^2+\ldots.+a^t) = 0$, then $-1 = (1+a+\ldots+a^{(t-1)}) \cdot a$, and so $a^t = a^{-1}$.

Corollary 154 *Each of the r expressions*

$$\{1, (1+a), (1+a+a^2), \ldots (1+a+a^2+\ldots a^{(r-1)})\} \tag{3.7}$$

represents a distinct element of the near-field. Moreover

$$\left(1 + a + a^2 + \ldots + a^{(r-1)}\right) = 0 \tag{3.8}$$

Lemma 155 *Suppose H is a finite, non-trivial, (multiplicative) subgroup of $(N^*, \cdot)$ and $H = \{h_0 = 1, h_1, \ldots, h_{(r-1)}\}$. Suppose $a \in N^*$. Then,*

$$\left(a + a \cdot h_1 + a \cdot h_2 + \ldots + a \cdot h_{(r-1)}\right) = 0$$

3.2 Commutators and the Sub-near-Field $\mathcal{L}$

The group $(N^*, \ \cdot)$ is non-abelian in proper near-fields and has *multiplicative* commutator subgroup $[N^*, N^*]$. We have that $a \cdot b = b \cdot a \cdot [a, b]$ and so, taking $a, b, g \in N^*$

$$\begin{aligned} a \cdot (b + g) = (b + g) \cdot a \cdot [a, (b + g)] &= (b \cdot a + g \cdot a) [a, (b + g)] \\ &= ((a \cdot b \cdot [b, a] \cdot [a, (b + g)]) + (a \cdot g \cdot [g, a] \cdot [a, (b + g)])) \\ &= \left(a \cdot b \cdot ((a^{-1})^b \cdot a^{(b+g)}) + a \cdot g \cdot ((a^{-1})^g \cdot a^{(b+g)})\right) \end{aligned} \tag{3.9}$$

Quite generally,

$$a \cdot (b + c) = (a \cdot b) \cdot l_1 + (a \cdot c) \cdot l_2 \tag{3.10}$$

where $l_1, l_2 \in [N^*, N^*]$. The elements of the multiplicative commutator generate an *additive* subgroup of $(N, \ +)$. Its elements are all finite sums of finite products of multiplicative commutators.

Definition 156

$$\mathcal{L} = \langle [N^*, N^*] \rangle_{(add)} \tag{3.11}$$

In what follows, I shall occasionally use the term $\mathcal{L}$ in the above sense when the near-field referred to is clear.

Lemma 157 *$\mathcal{L}$ is a sub-near-field of $(N, \ +, \ \cdot)$, and $(\mathcal{L}^*, \ \cdot) \lhd (N^*, \ \cdot)$.*

Proof Suppose $C_1, C_2, C_3, C_4 \in [N^*, N^*]$. Then,

$$\begin{aligned} (C_1 - C_2) \cdot (C_3 - C_4) &= (C_1 \cdot (C_3 - C_4)) - (C_2 \cdot (C_3 - C_4)) \\ &= (C_1 \cdot C_3 \cdot c') + (C_1 \cdot C_4 \cdot c'') + (C_2 \cdot C_3 \cdot c''') + (C_2 \cdot C_4 \cdot c'''') \end{aligned}$$

for some $c', c'', c''', c'''' \in [N^*, N^*]$. This is an element of $\mathcal{L}$. Normality follows since $\mathcal{L}^* \supseteq [N^*, N^*]$. □

There are examples of proper near-fields in which $\mathcal{L}$ is a proper sub-near-field (Sect. 3.3.1), and there are examples where it is the entire near-field (pages 132 and 162).

Write $N = \mathcal{L}_0, \mathcal{L} = \mathcal{L}_1$. We may apply this same commutator construction to the near-ring $\mathcal{L}_1$, obtaining the chain of near-fields

$$N = \mathcal{L}_0 \supseteq \mathcal{L}_1 \supseteq \mathcal{L}_2 \supseteq \dots \tag{3.12}$$

One might describe near-fields for which the above chain terminates finitely at $\{0\}$ as **$\mathcal{L}$-solvable**. The near-field in Sect. 3.3.1 is $\mathcal{L}$-solvable. The near-field in Sect. 3.4.3 is not.

Lemma 158 *If H is a multiplicative subgroup of $(N^*, \cdot,)$, then $\langle \mathcal{L}^* \cdot H \rangle_{(add)}$ is a sub-near-field of $(N, +, \cdot)$.*

Proof The proof comes down to verifying that the additive subgroup is multiplicatively closed. We should consider expressions of the form

$$(l \cdot h) \cdot (l_1 \cdot h_1 + \dots + l_n \cdot h_n)$$

And closure follows from (3.10) plus multiplicative normality of $\mathcal{L}^*$. □

3.2.1 The Sub-near-Field $\mathcal{F}$

Define

$$\mathcal{F} = \{x \in N \,:\, (x \cdot (a + b)) = ((x \cdot a) + (x \cdot b)) \textit{ for all } a, b \in \mathcal{L}\} \tag{3.13}$$

Then $\mathcal{C}_\mathcal{D}(N) \subseteq \mathcal{F}$.

Lemma 159

(i) $\mathcal{F}$ is a sub-near-field of N.
(ii) When $\mathcal{L}$ is a field, we can say $\mathcal{L} \subseteq \mathcal{F}$.

Proof $\mathcal{F}$ is additively closed. Suppose $x, y \in \mathcal{F}$, and recall that $\mathcal{L}^* \lhd N^*$. Then,

$$\begin{aligned}(x \cdot y) \cdot (a + b) = (x \cdot (y \cdot a + y \cdot b)) = \\ \left(x \cdot (y \cdot a \cdot y^{-1} + y \cdot b \cdot y^{-1}) \cdot y\right) = \\ ((x \cdot y) \cdot a + (x \cdot y) \cdot b)\end{aligned}$$

□

3.3 Finite Near-Fields

> I am concerned primarily with finite near-fields in the remaining sections of this chapter, although infinite cases are discussed briefly in Sect. 3.9.

Finite near-fields have orders of the form p^{α}, where p is prime and $\alpha \geq 2$. The first two possible numbers are both powers of two, and posited near-fields would have multiplicative groups which were cyclic of prime orders 3 and 7, respectively; so they would be fields.

3.3.1 The Smallest Proper Near-Field

The next possible number is $3^2 = 9$. A near-field with this order which was not a field would have to have its multiplicative group either the quaternions, Q_8, or the dihedral group with degree four, D_4. The characteristic is three. We can reject D_4 because it has more than one element with order 2 (Lemma 146). I investigate the quaternion possibility, using the notation in (5.3). That notation uses a symbol, -1, which represents the sole element with order 2 in Q_8. We already know that the near-field should have such a unique element and that it is the negative of the multiplicative identity, so we have no notational conflict.

I attempt to construct an addition table for the elements; this is an exercise in the *additive host problem* (see Remark 332). The elements are the *so-called* Hamiltonian quaternions

$$\{0, 1, -1, i, \quad i, j, -j, k, -k\} \tag{3.14}$$

and we know that $(x+x) = -x$ is true generally. The element $(1+i)$ cannot be equal to any of the elements $\{0, 1, i, -i\}$ without obvious conflict. From the symmetries inherent in the quaternions, we may set

$$(1 + i) = j$$

From this, obtaining

$$j = (1 + i) \quad \text{and} \quad k = (1 - i)$$

This fixes the addition completely. The additive group is $C_3 \oplus C_3$ and its Cayley table is Table 3.1.
The corresponding structure is right-distributive and is a near-field.

For example,

$$(i + j) \cdot k = k^2 = -1 = (-j + i) = (i \cdot k + j \cdot k)$$

Table 3.1 Addition: a direct sum of two cyclic groups

+	0	1	−1	i	-i	j	-j	k	-k
0	0	1	−1	i	-i	j	-j	k	-k
1	1	−1	0	j	k	-k	-i	-j	i
−1	−1	0	1	-k	-j	i	k	-i	j
i	i	j	-k	-i	0	k	-1	1	-j
-i	-i	k	-j	0	i	1	-k	j	-1
j	j	-k	i	k	1	-j	0	−1	-i
-j	-j	-i	k	−1	-k	0	j	i	1
k	k	-j	-i	1	j	−1	i	-k	0
-k	-k	i	j	-j	−1	-i	1	0	k

This structure is the smallest *proper* near-field (i.e. it is not actually a field). Its multiplicative centre is the set $\{0, -1, 1\}$, and that is also the set of all left-distributive elements–the distributivity core (page 43). It is the finite field with three elements. We have $[i, j] = -i \cdot -j \cdot i \cdot j = -1$, and so the multiplicative commutator subgroup is isomorphic to C_2. As expected,

$$(a + a) = -a$$

Furthermore, $\mathcal{L} \cong C_3$, and $\mathcal{L}^* \triangleleft N^*$, and $\mathcal{L} = \mathcal{C}_{\mathcal{D}}(N)$. Not only that, for all $a \in N^*$ and for all $l_1, l_2 \in \mathcal{L}^*$, we have

$$(a \cdot (l_1 + l_2)) = ((a \cdot l_1) + (a \cdot l_2)) \tag{3.15}$$

Frobenius Automorphisms
Notice that

$$-1 = (-1)^3 = (i - j)^3 \neq (i^3 + (-j)^3) = (-i + j) = 1^3 = 1$$

so the familiar "Frobenius automorphism" of field theory (3.1) does not automatically apply in finite near-fields.

Polynomial Equations
The polynomial equation

$$(x^2 + 1) = 0$$

is satisfied by six distinct elements of the near-field, and the polynomial identity

$$(x + 1)^2 \equiv (x^2 + 2x + 1)$$

does not hold true for polynomials over this near-field, in the sense that the polynomials on each side of the equivalence evaluate to distinct values at $x = i$ (j on the left and $-i$ on the right).

Table 3.2 Identifying quaternion symbols

The field	π	π^2	π^3	π^4	π^5	π^6	π^7	π^8
The group	i	j	k	-1	-i	-j	-k	1

A Dickson Near-Field

This near-field is a Dickson near-field (see Sect. 3.4). It can be built from the Galois field with order 9 $(N, +, \cdot)$ in a way which I shall now describe (the reader may wish to look at Sect. 2.2 to make sense of what follows). If π is a primitive element for that near-field, its non-zero elements form the set $\{\pi^k\}_{k=1,2,\ldots,8}$. I now define a coupling mapping

$$\Phi : N \to \mathrm{Aut}(N, + \cdot) \quad (n \mapsto \phi_n)$$

where

$$\phi_{\pi^r}(x) = \begin{cases} x^3 & \text{if } r \text{ is odd} \\ x & \text{if } r \text{ is even} \end{cases}$$

We can identify the non-zero field members with the standard quaternion symbols by means of Table 3.2.

The near-field product is, of course, $(a * b) = (\phi_b(a) \cdot b)$.
For future reference (see (3.51))

$$\mathcal{K}^* = \{j, j^2 = -1, -j, 1\} \tag{3.16}$$

and this is not a subfield since it is not additively closed.

It is confusing that this example should represent really the first proper near-field that one can get to grips with because this particular near-field is atypical in so many ways. It seems to me to be the most unusual of the finite Dickson near-fields, and the fact that it was the easiest to grapple with may have set people off on the wrong track.

For instance, it is the only proper finite near-field whose multiplicative group of non-zero elements is a p-group. To see this, represent the near-field as having order p^α, with $(p^\alpha - 1) = q^\beta$, for some integers $\alpha, \beta \in \mathbb{N}$, and primes, p, q. The case $\alpha = 1$ cannot occur, such near-fields would have p odd, and their distributivity cores should equal the whole near-field. Assume, then, that $\alpha > 1$. If q were odd, then, from Lemma 165, the group N^* is cyclic, and we have a field. When q is 2, we would require that $(p^\alpha - 1) = 2^\beta$. We have

$$(p^\alpha - 1) = (p - 1) \cdot (p^{(\alpha-1)} + p^{(\alpha-2)} + \ldots + p + 1)$$

The left-hand side of this identity involves a product with α terms, each of them odd. Since it should be a power of two, we require α to be even, say $\alpha = 2 \cdot \gamma$. But then

$$(p^\alpha - 1) = (p^\gamma - 1) \cdot (p^\gamma + 1)$$

and both factors are powers of two. The only possibility is that $p = 3$ and $\gamma = 1$, which is the case we have just covered.

One can make some deductions from this, for example, that there is no proper near-field with order 17 (its Sylow 2-subgroup would be the generalised quaternion group with order 16) and, similarly, that there is no proper Dickson near-field with order 257.

Dembowski says it is the only proper finite near-field with the property that all non-zero elements not equal to ± 1 have squares equal to -1 [32]. This particular near-field reappears in Sect. 16.4, and the next smallest possible proper near-field would have order 25 (Sect. 3.4.3), and, interestingly, it will turn out that there is more than one near-field with that order, including a non-Dickson example.

Our near-field is called **the miniquaternion system** in [133] and gives rise to some *projective planes* which are not *field planes* (see Sect. C.2). It now becomes important to consider what would be the *next* smallest Dickson near-field with quaternion Sylow 2-subgroups. It turns out that all other finite Dickson near-fields have multiplicative groups which are 3-groups, so there are none—a further indication of how special this small near-field is.

3.3.2 General Cases

I establish some notation which shall apply mainly in the remainder of this chapter, and in Chap. 15.

Notation 160

(i) $(N, +, \cdot)$ is a finite near-field with characteristic p and order p^α. Later, this will actually represent a field, and there will be a related near-field, $(N, +, \wedge)$.
(ii) $\mathcal{C}_\mathcal{D}(N)$ is represented by the single symbol $\mathcal{D}$. This is a commutative field (Lemma 150). It has order p^β, $(\beta \leq \alpha)$, and $\alpha = (\beta.\delta)$. I don't entirely reserve β to $\mathcal{D}$, though that is its main use. I sometimes use it in other subfields.
(iii) Cent(N) is represented by the single symbol $\mathcal{Z}$ (don't confuse this with the additive centre, which is irrelevant here).
(iv) As usual, starred symbols represent subsets of non-zero elements: N^*, $\mathcal{D}^*$, $\mathcal{Z}^*$ ... (as in Notation 7).
(v) I define the symbol $\mathcal{C}$ in Notation 162, which is closer to where it is actually used.
(vi) Since near-fields are hosted by abelian groups, all uses of the standard notation for commutators refer to *multiplicative* rather than additive structure. In

situations of ambiguity, I will sometimes suffix structures with (add) or $(mult)$ as described in Notation (1); and, when more than one near-field is around, I may append the near-field product to notation in which there may be confusion: $[N^*, N^*]^{(\wedge)}$.

In Sect. 3.3.1, we have values: $p = 3, \alpha = 3, \beta = 1, \delta = 3$. We have the chain of multiplicative subgroups

$$\mathcal{Z}^* \subseteq \mathcal{D}^* \subseteq N^*$$

and know $\mathcal{D}^*$ is cyclic. Each of the expressions $\{1, (1+1), (1+1+1), \ldots\}$ lies in $\mathcal{D}$, if not in $\mathcal{Z}$ itself. If π is a primitive root modulo p (the near-field characteristic), then

$$\pi = \pi.1 = \overbrace{(1+1+1\ldots+1)}^{(p \text{ times})} \tag{3.17}$$

is in this set and has multiplicative order $(p-1)$. The conclusion is that $\mathcal{D}^*$ has order at least as large as $(p-1)$. Moreover, $\langle \mathcal{Z} \rangle_{(add)}$ is a field containing $\mathcal{Z}$ and contained in $\mathcal{D}$. Its non-zero elements, $\left(\langle \mathcal{Z} \rangle_{(add)}\right)^*$, form a cyclic subgroup of $\mathcal{D}^*$. In our near-field with order 9, $\mathcal{Z}^* \cong C_2$ and $\mathcal{D} \cong C_3$.

The question as to whether $\langle \mathcal{Z} \rangle_{(add)}$ is ever strictly larger than $(\{0\} \cup \mathcal{Z})$, or indeed whether $\pi \cdot 1 \in \mathcal{Z}$, is of some interest, as is the related question of under what conditions $\mathcal{Z} \subsetneq \mathcal{D}$. It turns out that the two subgroups are equal in finite Dickson near-fields (page 126) and that there are also non-Dickson near-fields in which distributive elements are always central (Sect. 3.7). The near-field $\mathcal{L}$ has order p^τ, and $\alpha = \tau.\omega$. The additive quotient group $(\frac{N}{\mathcal{L}}, +)$ has order

$$\frac{p^\alpha}{p^\tau} = p^{(\tau.(\omega-1))}$$

and the multiplicative quotient group $(\frac{N^*}{\mathcal{L}^*}, .)$ has order

$$\frac{(p^\alpha - 1)}{(p^\tau - 1)} = \left(p^{(\tau.(\omega-1))} + p^{(\tau.(\omega-2))} + \ldots + p^\tau + 1\right)$$

3.3.3 The Normal Core of $\mathcal{D}^*$

The largest normal subgroup of N^* contained in $\mathcal{D}^*$,(the *normal core* [132, 139]) is

$$\bigcap_{x \in N^*} (\mathcal{D}^*)^x$$

Multiplicatively, this is a cyclic subgroup of $\mathcal{D}^*$, one which contains $\mathcal{Z}^*$, and is certainly non-trivial, at least for odd-prime characteristics. I shall call it $\langle \xi \rangle$. Taking any $x \in N^*$, we can say

$$\xi \cdot x = x \cdot \xi^a \quad (a \in \mathbb{N}) \tag{3.18}$$

and since (multiplicative) conjugation is a multiplicative automorphism of groups, we know that

$$\langle \xi \rangle_{(mult)} = \langle \xi^a \rangle \tag{3.19}$$

Fixing a, we form the set

$$\mathcal{C}_a^* = \{x \in N^* \ : \ \xi \cdot x = x \cdot \xi^a\} \tag{3.20}$$

Additively, this generates a subgroup $(\mathcal{C}_a, +)$ in which only the zero element does not lie in the original set. If ξ is not central, then each set $\mathcal{C}_a$ has order some power of the near-field characteristic, p, and is a proper subgroup of $(N, +)$. We know $\xi \in \mathcal{C}_1$; and

$$\left(\mathcal{C}_{a_i} \cdot \mathcal{C}_{a_j}\right) = \mathcal{C}_{(a_i.a_j)}$$

In particular;

$$(\xi \cdot \mathcal{C}_a) = \mathcal{C}_a = (\mathcal{C}_a \cdot \xi)$$

The non-zero elements of the group $(\mathcal{C}_1, +)$ form the multiplicative centraliser, in N^*, of ξ, a set which contains $\mathcal{D}^*$. This is actually a near-field containing $\mathcal{D}$. If $x \in \mathcal{C}_a$, then $\mathcal{C}_a = \mathcal{C}_1 \cdot x$, so the two subgroups $\mathcal{C}_a$ and $\mathcal{C}_1$ are additively isomorphic, and all the $\mathcal{C}_a$ are additively isomorphic. Moreover,

$$\left(\mathcal{C}_a \bigcap \mathcal{C}_b\right) = \{0\} \, (for\ a \neq b) \tag{3.21}$$

Lemma 161 *Suppose N is a finite near-field. Then,*

$$\bigcap_{x \in N^*} (\mathcal{D}^*)^x = \mathcal{Z}^* \tag{3.22}$$

In other words, $N = \mathrm{C}_1$*, and* $a \neq 1 \Rightarrow \mathcal{C}_a^* = \{\phi\}$

This proof is offered in [32].

Proof Suppose $t \notin \bigcap_{x \in N^*} (\mathcal{D}^*)^x$. We may find $z, z_1, z_2 \in \bigcap_{x \in N^*} (\mathcal{D}^*)^x$ such that

$$z \cdot t = t \cdot z_1$$

$$z \cdot (t + 1) = (t + 1) \cdot z_2$$

But then,

$$(t \cdot z_1 + z) = (z \cdot t + z) = (z \cdot (t+1)) = ((t+1) \cdot z_2) = (t \cdot z_2 + z_2)$$
$$\Rightarrow (t \cdot (z_1 - z_2)) = (z_2 - z)$$
$$\Rightarrow t \in \bigcap_{x \in N^*} (\mathcal{D}^*)^x \quad \text{or} \quad (z_1 - z_2) = 0$$

To avoid contradiction, we deduce $z = z_1 = z_2$, and hence

$$\bigcap_{x \in N^*} (\mathcal{D}^*)^x \subseteq \mathcal{Z}^*$$

□

Standard group theory says that if a subgroup has finite index n, then the index of its core is no larger than $n!$, [84].

3.3.4 The Multiplicative Centre

The multiplicative centre plays an important role in the standard structure theorems for associative algebras, and it does so here, although that role bifurcates into the usual centre, and a larger subgroup of multiplicatively distributive elements, which turns out to be actually larger than the centre in only a very few special cases, which I call "the fantastic four" (see page 164).

We can only admit the *possibility* of centreless cases when $p = 2$. The multiplicative group then has order $(2^\alpha - 1)$ and is solvable, but there is still considerable structural latitude in such cases, for example, groups with order 255 and groups with order 511 are actually cyclic (so, no proper near-fields with order 256 or 512). In general, $\mathcal{Z}^*$ is a cyclic group, and, as I have already observed, its quotient cannot be cyclic when $(N^*, \cdot)$ is non-abelian. We also know

$$\mathcal{Z}^* = \mathcal{D}^* \Leftrightarrow \mathcal{D}^* \lhd N^* \tag{3.23}$$

and would wish to say more about the structure of the group $(N^*, \cdot)$.

In fact, in finite Dickson near-fields, $\mathcal{Z}^*$ and $\mathcal{D}^*$ are identical (Lemma 188); and in all $\mathcal{D}^*$ is multiplicatively cyclic. Suppose it to be generated by Λ. In general, if there are elements $x, y \in N^*$ such that $x^y \in \mathcal{L}$, then since $[x, y] \in \mathcal{L}$, we deduce that $x^y = x \cdot l$, for some $l \in \mathcal{L}$. From this, $x^y \in \mathcal{L} \Rightarrow x \in \mathcal{L}$.

Notation 162

$$\mathcal{C}^* = \{n \in N^* : n \cdot d = d \cdot n \, \forall d \in \mathcal{D}\} \tag{3.24}$$

$\mathcal{C}^*$ is multiplicatively, closed but also, when $a, b \in \mathcal{C}^*$

$$(a - b) \cdot d = (a \cdot d - b \cdot d) = (d \cdot a - d \cdot b) = d \cdot (a - b) \qquad (3.25)$$

so $\mathcal{C}$ is a sub-near-field of N, one which contains the field $\mathcal{D}$. It is actually the centraliser of any cyclic generator of $\mathcal{D}^*$, such as Λ. In the same way

$$\mathfrak{N}^* = \{n \in N^* \; : \; (\mathcal{D}^*)^n = (\mathcal{D}^*)\} \qquad (3.26)$$

is the normaliser of $\mathcal{D}^*$. It is a subgroup of N^* containing $\mathcal{C}^*$ and consists of the elements of $\mathcal{N}^*$ that map cyclic generators to each other via inner automorphisms. The index $|\frac{N^*}{\mathfrak{N}^*}|$ is the number of distinct conjugates of $\mathcal{D}^*$. In most finite near-fields, $\mathcal{D}^*$ is actually normal and $\mathcal{C} = (N, +, \cdot)$. There is more about this normaliser in non-Dickson settings in Chap. 16 and the above symbol reappears there.

3.3.5 *The Multiplicative Group Structure of Finite Near-Fields*

Almost all my remaining work in this chapter relates to finite near-fields.

We have established that when the near-field characteristic is an odd prime, $\mathcal{Z}^*$ is non-trivial. In that situation, there is a unique element with multiplicative order 2, and when N^* contains Q_8, this element is in the commutator subgroup.

Lemma 163 *Let q and s be primes, not necessarily distinct. Then any subgroup of* (N^*) *having order* $(q \times s)$ *is cyclic.*

Proof First, treat the $q = s$ case and suppose we have a non-cyclic subgroup with order q^2, generated by elements x and y, which, of course, commute. Taking any non-zero element a, consider the sequence of elements which form the $q \times q$ matrix $(a_{i,j})$, where

$$a_{i,j} = a \cdot y^{(i-1)} \cdot x^{(j-1)}$$

This matrix consists of entirely distinct elements of N^*. From Corollary (154), we can say that the sum of the elements in any row or any column of the matrix is zero; so the sum of *all* of the elements of the matrix is zero. In fact, for any fixed i and j not both zero, we have

$$a + a \cdot (y^i \cdot x^j) + a \cdot (y^i \cdot x^j)^2 + a \cdot (y^i \cdot x^j)^3 + \ldots + a \cdot (y^i \cdot x^j)^{q-1} = 0$$

This expression involves q distinct terms, and the set of all $(q^2 - 1)$ such expressions (as i and j vary) contains $(q - 1)$ repetitions of each particular expression, with the terms after the first appropriately re-ordered as each of the last $(q - 1)$ terms takes its turn at the second term slot. So we actually distinguish $(q + 1)$ distinct expressions.

These involve each term of our matrix $(a_{i,j})$ but on the first term, $a_{1,1} = a$, appear more than once—it appears $(q + 1)$ times. The sum of all such expressions is equal to the sum of all the terms of our matrix plus qa and we deduce

$$qa = 0$$

But our near-field has characteristic p and q is coprime to that, so this equality cannot possibly be correct. To escape contradiction we deduce that our subgroup has to be cyclic. Second, treat the $q \lneq s$ case and apply the Sylow theorems, assuming the group to be non-abelian. We can say that there is a unique cyclic and normal Sylow subgroup with order s, and exactly s distinct subgroups with order q. We suppose these $(s + 1)$ subgroups are generated by elements $\{y_0, y_1, y_2, \ldots, y_s\}$, the first having order s. Select a non-zero element $a \in N^*$ and consider the $(s + 1)$ expressions

$$z_i = \left(a + a \cdot y_i + a \cdot y_i^2 + \ldots + a \cdot y_i^{(t_i - 1)}\right)$$

where $t_i = |y_i|$ (which is s when i = 0 and q otherwise). Each z_i is equal to zero; and the individual terms of each sum involve distinct elements of N^*, except for the very first, which is a and appears once in each expression. Thus, a total of $(s + sq)$ terms appear, and these involve $(s + 1)$ repetitions of the element a, and each of the elements $a \cdot h$ where h is an element of our subgroup not equal to 1. The sum of all of the terms appearing, which is $(z_0 + z_1 + \ldots + z_s)$, is zero; but from Lemma 155, this sum must also be $sa + 0$. As in the previous case, we know that p and $|a|$ are coprime and deduce our group is cyclic. □

Corollary 164 *Any abelian subgroup of the multiplicative group* (N^*) *is cyclic.*

Proof The fundamental theorem for abelian groups induces us to consider direct sums of cyclic subgroups with prime power order. For the result to be false, there would have to be two subgroups C_{q^α} and C_{q^β} for which $\langle C_{q^\alpha}, C_{q^\beta} \rangle_{(mult)} \cong (C_{q^\alpha} \oplus C_{q^\beta})$, with $(\alpha + \beta) > 2$. Such a subgroup would contain a non-cyclic subgroup with order q^2, which is impossible. □

I imagined I was the first to notice this lemma, but it is actually an obvious consequence of what Joseph Wolf calls the p^2 conditions (his Theorem 5.3.2, [144]).[2] In the situation in which the multiplicative group $(N^*, \cdot)$ is solvable, Wolf lists four possible families of groups to which $(N^*, \cdot)$ belongs ([144] page 179). This situation actually corresponds to finite *Dickson near-fields* and to all but three of the finite *non-Dickson near-fields* (Lemma 181, Sect. 3.7). Finite p-groups containing no non-cyclic subgroup with order p^2 are known to be cyclic themselves,

[2] In fact, I experienced something familiar to most mathematicians, obtaining Wolf rather late in the development of this book and then discovering that many results I had laboured over were actually well-known (this does not apply to the results that were wrong!).

when p is odd, and either cyclic or generalised quaternion (see (3)), when p is even (see [62]).

Lemma 165 *The Sylow subgroups of* (N^*) *are cyclic, for odd primes, and either cyclic or generalised quaternion, for even primes.*

Interestingly, it is known that finite subgroups of the multiplicative group of non-zero elements in a division ring will have Sylow subgroups which are either cyclic or generalised quaternion. Isaacs [72] gives a theorem to the effect that whenever a finite group has a cyclic Sylow p-subgroup, it will be the case that that subgroup is either contained in the commutator subgroup or has only trivial intersection with it. Consequently, in finite near-fields, the only prime that could divide both $|[N^*, N^*]|$ and $\left|\frac{N^*}{[N^*,N^*]}\right|$ will be 2, and that can occur only when the Sylow 2-subgroup is generalised quaternion. When the Sylow subgroup S_2 is quaternion, it has the form (3) and we can say that $y^2 = -1 = x^{2^{(n-2)}}$. Such groups contain copies of Q_8 and using the notation in (5.3), we have $[i, j] = (-i \cdot -j \cdot i \cdot j) = k^2 = -1 \in \mathcal{Z} \subseteq \mathcal{D}$.

3.3.6 Presentations for Finite Near-Fields with S_2 Cyclic

In the case in which all Sylow subgroups are cyclic, Robinson [132] reports that the group has a presentation with the form

$$N^* = \left\langle a, b \,:\, a^m = 1,\ b^n = 1,\ a^b = a^r \right\rangle \tag{3.27}$$

where m is an odd integer, $1 < r < m$, $r^n \equiv 1$ (modulo m) and $(m, n(r-1)) = 1$. These are the $\mathfrak{Z}$**-groups** (page xviii) (from the German, "Zyklische gruppe").[3]

I do not claim that *all* such groups are the multiplicative groups of near-fields (see Sect. 3.4.7). After all, it may well not be the case that $nm = (p^\alpha - 1)$; and, in any case, for proper near-fields we should require that n be composite. There are actually necessary and sufficient condition on the parameters n, r and m for finite non-fields [62].

For near-fields on these groups, the factor group $\frac{N^*}{[N^*,N^*]}$ is isomorphic to C_n and $[N^*, N^*] \cong C_m$. Consequently, $-1 \notin [N^*, N^*]$. Moreover, the group is supersolvable and $b^{\binom{n}{2}} = -1$ (odd prime characteristics). Robinson [132] gives a theorem due to Isaai Schur which says that in finite groups the index of the group centre is a multiple of the exponent of the commutator subgroup. In this case that exponent is m, and we conclude that the group centre has only trivial intersection with $\langle a \rangle$ (see (3.34)). We can say

$$\mathcal{D}^* \cap [N^*, N^*] \equiv C_t$$

[3] The term is not universal, which is why I use $\mathfrak{Z}$ rather than "Z". Robinson himself uses "Z-group" to denote a quite distinct class of groups.

for some divisor t of m. Of course, for $x \in N^*$,

$$(C_t)^x \subseteq [N^*, N^*]$$

and since the commutator subgroup is cyclic, we conclude

$$(C_t)^x = C_t$$

Then, from Lemma 161, $C_t \subseteq \mathcal{Z}^*$, and so

$$\big(\mathcal{D} \cap [N^*, N^*]\big) = \{1\} \tag{3.28}$$

Lemma 166 *Suppose N is a finite near-field.*

(i) *$\mathcal{D}^*$ is cyclic.*
(ii) *Whenever $[N^*, N^*]$ is cyclic, (3.28) applies.*

3.3.7 $\mathfrak{Z}$-Group Properties

$\mathfrak{Z}$-groups have many interesting properties. For example, a theorem due to Taunt says that their centres are hypercentres [132] and they possess subgroups with orders equal to all divisors of the group order, all such subgroups being conjugate [72] (and see Lemma 172). In finite $\mathfrak{Z}$-groups, all *subnormal subgroups* (page 4) are actually normal. The centre of (3.27) is the cyclic group

$$\langle b^s \rangle_{(mult)} \tag{3.29}$$

where s is defined as the order of r modulo m (so, $r^s \equiv 1$ $(modulo\ m)$); and, of course, s is coprime with m, it divides $n = st$, and it divides $\frac{n}{2}$ (odd prime characteristics).

Wolf ([144]—Theorem 5.4.1) reports a result of Burnside to the effect that when a $\mathfrak{Z}$-group has the cyclicity property described in Lemma 163, then $t = \frac{n}{s}$ is divisible by all prime divisors of s, the condition being both necessary and sufficient. If $(N, +, \cdot)$ is a proper near-field in which the group N^* has a presentation of the form (3.27), we can say that these number-theoretic conditions apply.

$$mn = mst = \big(p^\alpha - 1\big) \tag{3.30}$$

$$t = (p^\beta - 1) \tag{3.31}$$

Finite near-fields in which the Sylow 2-subgroup S_2 is abelian have these properties:

(i) N^* has the presentation (3.27).
(ii) $[N^*, N^*] = \langle a \rangle_{(mult)} \cong C_m$.

(iii) $\mathcal{Z}^* = \langle b^s \rangle_{(mult)}$, where s is the order of r modulo m, and $n = st$.
(iv) $\mathcal{L} \supsetneq \langle a \rangle$ (because $-1 \in \mathcal{L}$).
(v) The number-theoretic conditions (3.31) apply.
(vi) $\alpha = (\beta.\delta)$, and the Hall-Zassenhaus conditions apply too (page 131).
(vii) The multiplicative group (N^*) has elements of order $(p-1)$. It will have elements with order $(p+1)$ precisely when α is even.

Elements of N^* can be uniquely expressed in the form $x = b^i \cdot a^j$, and

$$(b^i \cdot a^j) \cdot (b^w \cdot a^v) = \left(b^{(i+w)} \cdot a^{(r^w j + v)}\right) \tag{3.32}$$

which expresses the near-field multiplication in a very wide range of cases. Subgroups of N^* containing $[N^*, N^*]$ must have the form

$$\left\langle a, b^k \right\rangle$$

for some divisor k of n, and their elements have the form $(b^{ku} \cdot a^j)$. Now,

$$\left(b^k \cdot a^j\right)^m = \left(b^{km} \cdot a^x\right)$$

where

$$x = r^{((m-1)k+(m-2)k+\ldots+r^k+1)} = \left(\frac{(r^{km}-1)}{(r^k-1)}\right)$$

We know

$$\mathcal{L}^* = \langle a, b^l \rangle_{(mult)} \tag{3.33}$$

where $n = lv$. $\mathcal{L}$ is a field precisely when $\mathcal{L}^*$ is cyclic, which occurs precisely when the two generators commute, that is, $r^l \equiv 1$ (modulo m). Equivalently, $\mathcal{L}$ is a field precisely when the element b^l is central (but see page 146, because it turns out that this structure is either a commutative field or the whole of the near-field).

Lemma 167 *N^* contains a Sylow 2-subgroup which is quaternion if and only if $-1 \in [N^*, N^*]$.*

Trivially, this particular case requires the characteristic of the near-field to be an odd prime. If the Sylow 2-subgroup is the generalised quaternion group with order 2^n, then $[N^*, N^*]$ will have order $2^{(n-2)}.m$, with m odd (page xviii). Isaacs [72] mentions a theorem (5.3 in his numbering), to the effect that a finite group G has non-abelian Sylow p-subgroups for all primes dividing $|\,([G, G] \cap Cent(G))\,|$. The proof of this theorem uses transfer theory. In our context, this can only mean that

$$|\left([N^*, N^*] \cap \mathrm{Cent}(N^*)\right)| = 2^\tau \tag{3.34}$$

where

$$\tau = \begin{cases} 1 & \text{if } S_2 \textit{ is generalised quaternion.} \\ 0 & \text{if } S_2 \textit{ is cyclic.} \end{cases}$$

M-Groups and SL(**2**, **3**)

Like all supersolvable groups, $\mathfrak{Z}$-groups are also **M-groups**, which means that their irreducible complex representations are induced from linear ones and only involve monomial matrices, which are matrices produced by multiplying an invertible diagonal matrix by a permutation matrix (see [62] or [139]). Wolf [144] gives very specific information on this (see his Theorem 5.5.1).

The group SL(2, 3) (page xviii) is of some later interest to us. Robinson gives it as an example of a solvable finite group which is not an M-group [132]. More generally, Karl Gruenberg, page 6 [61], mentions a theorem of Suzuki dating from 1955 which says that a finite group in which all Sylow p-groups have unique subgroups with order p, and which is non-solvable, must have a normal subgroup H_1 with index either 1 or 2, and such that $H_1 = (\mathfrak{Z} \oplus L)$, where $\mathfrak{Z}$ is a (of course, solvable) $\mathfrak{Z}$-group, and $L \cong \mathrm{SL}(2, p)$ (presumably, $p > 3$).

Presumably this offers a proof that finite non-solvable near-fields have quaternionic Sylow 2-subgroups. In fact there are only three such near-fields, and they are given in Table 3.3; and the result is obvious anyway since $\mathfrak{Z}$-groups are solvable.

I suppose that $\mathfrak{Z}$ should have odd order in our situation—we only want one element with order 2; and, indeed, it turns out that there are exactly three finite near-fields with non-solvable multiplicative groups. They comprise the last three cases of Table 3.3. In all three cases, the index of this normal subgroup, H_1, is one. In the first case, $\mathfrak{Z}$ is trivial, in the second, cyclic with order 7, and in the third, cyclic with order 29. L is always SL(2, 5).

3.3.8 The Product of All the Non-zero Elements

The product of the non-zero elements of a finite field is always -1. In the case of finite proper near-fields for which N^* is a $\mathfrak{Z}$-group, we may apply a theorem due to Dénes and Hermann, which appeared in the *Annals of Discrete Mathematics*, in 1982. There will be a set of these products in non-commutative cases. The theorem says that this set is a coset of the commutator subgroup. This applies to all finite groups. Since one possible arrangement of these products pairs elements with their inverses, and comes to -1, the coset must be

$$\{[N^*, N^*]_{(mult)} \cdot (-1)\} = \{-a^s \ : \ 1 \leq s < m\} \tag{3.35}$$

Of course, for finite fields, the commutator subgroup is trivial. In non-Dickson cases, the theorem still applies, and we can conclude that the set of these products will actually be the commutator subgroup (Sect. 3.7). In the smallest proper non-field, the set of products must be the commutator subgroup itself, which is C_2 (page 113).

3.4 Finite Dickson Near-Fields

Nearly all near-fields in this chapter are finite.

I described Dickson near-fields in Sect. 2.2, page 77. Dickson near-fields are not necessarily finite (see Sect. 3.9), but I really only look at finite cases here, and the reader should assume all further near-fields in this chapter are finite unless specifically advised to the contrary. I do find a very small amount to say about infinite Dickson near-fields in the final section of this chapter (page 167).

Recall that the centre of a generalised quaternion group is cyclic with order 2. The multiplicative centre of a Dickson near-field is actually the group $\mathcal{D}^*$ (Lemma 188). Let us suppose that we have a Dickson near-field in which the Sylow 2-subgroups are generalised quaternion. Applying Sylow theory, we can deduce that a posited cyclic subgroup with order 4 would have to lie in the centre of one of the Sylow 2-subgroups, which is impossible. We deduce that four cannot divide $|\mathcal{D}^*|$; and, from Theorem 178, it must follow that four does not divide δ. So, $\tau = 1$, above.

Lemma 168 *If, in a finite Dickson near-field, we have that* $[N^*, N^*] \subseteq \text{Cent}(N^*)$, *then it must be the case that the Sylow 2-subgroup of* N^* *is* Q_8.

Proof Two is the only prime that divides $|[N^*, N^*]|$, so S_2 is quaternion. But the group centre may not have order divisible by four in quaternion cases. So the commutator subgroup has order equal to 2. So S_2 is Q_8. □

This situation actually occurs in the smallest possible proper near-field. It must also follow that four must divide $\left|\frac{N^*}{\mathcal{D}^*}\right|$. If the near-field order is $p^{\alpha=\beta.\delta}$, where $|\mathcal{D}| = p^{\beta}$, we can deduce that $p \equiv 3$ (*modulo 4*), that β is odd, and that α is two times an odd number, say $\alpha = (2.m)$.

Lemma 169 *The proper Dickson near-field with order* $p^{\beta.\delta}$ *and multiplicative centre with order* $(p^{\beta} - 1)$ *can only have Sylow 2-subgroups that are generalised quaternion if:*

(i) δ *is two times an odd number*
(ii) $p \equiv 3$ *(modulo 4)*
(iii) β *is odd*

We know

$$\left(p^{2mp} - 1\right) = \left(\left(p^{mp} + 1\right)\left(p^{mp} - 1\right)\right)$$

so the (possible) quaternion Sylow 2-subgroups depend on the precise power of two dividing $(p^{mp} + 1)$. The conditions of the lemma are met in Sect. 3.3.1, where $\delta = 2, p = 3, \beta = 1$. This is a result about finite Dickson near-fields though part of it is true more generally. In situations in which $\mathcal{D}^*$ is not multiplicatively central, the congruence relation can fail; see Table 3.3, cases A and F.

The first possibility for a Dickson near-field with Sylow 2-subgroup generalised quaternion seems to be $p = 7, \beta = 1, \delta = 2$, representing a possible near-field with order 49. From Theorem 178, we know there are Dickson near-fields with this order. There are groups with order 48 having Sylow 2-subgroups equal to Q_{16}; but my own GAP analysis seems to show that in any Dickson near-field, the Sylow 2-subgroups will be cyclic. The moral is that (169) represents necessary, but insufficient, conditions for the quaternion case. There is a non-Dickson near-field with this order, and it does have generalised quaternion Sylow 2-subgroups (Table 3.3 case C).

The group $G = N^*$ of (3.27) (all Sylow subgroups cyclic case) has additional properties. We can say that the group has order $mn = (p^\alpha - 1)$ (p odd) and that $b^{\frac{n}{2}} = -1$. Consequently, $r^{\frac{n}{2}} \equiv 1$ (modulo m). Finally, because the group is supersolvable, the elements of odd order are characteristic [132], and there is an epimorphism

$$\Phi : N^* \twoheadrightarrow S_2 \cong C_{2^a} \tag{3.36}$$

for some fixed $a \geq 4$.

Normal Complements

The Schur-Zassenhaus theorem tells us that N^* is of the form $H \cdot K$, where H is a normal subgroup with odd order and K is a cyclic group with order 2^a—that is, K has a **normal complement** in N^* (see [132]).

Metacyclic Groups

Groups which are cyclic extensions of cyclic groups are usually called **metacyclic** groups.[4] On that basis, (3.27) is metacyclic—a special sort of metacyclic group—which is called a *split metacyclic group* in [122]. The location of the various Sylow q-subgroups is of interest. Suppose $q \mid (p^\beta - 1)$ (q prime). From relation (3.5), the condition that S_q is not entirely contained in D is that $q \mid \delta$. When that is not the case, the subgroup is cyclic and unique, therefore normal. Now turn to the cases in which the Sylow 2-subgroups of N^* are generalised quaternion, and this happens in the smallest possible near-field, described above.

Following the terminology just introduced, we would say that these groups have **normal 2-complements** provided they have the form $H \cdot S_2$, where H is normal and of odd order and S_2 is a Sylow 2-subgroup. In the case of the near-field hosted by $(C_3 \oplus C_3)$, this happens, but H is trivial.

---

[4] Older authors sometimes restrict the term to groups having cyclic commutator subgroups with cyclic quotients. Q_8 does not have this property, but it is metacyclic, in modern parlance.

The reader may need to look at Notation (2), page xvii, in what follows. A theorem of Burnside says that if q is the smallest prime dividing the order of a finite group, and the Sylow q-subgroup, S_q is cyclic, then q has a normal q-complement [72].

The Involvement of SL(2, 3)
Gorenstein defined a group K being **involved** in a group G to mean that K is an epimorphic image of some subgroup H of G. It would perhaps be more common, and modern, to describe K as being a **section** of the group G. Gorenstein [55] offers an exercise in Chap. 7 of his book to the effect that whenever G is a group having a generalised quaternion Sylow 2-subgroup, that group either has a normal 2-complement, or else the group SL(2, 3) is *involved* in G. The theorem means that SL(2, 3) should be involved in all near-fields in which the multiplicative group is non-solvable. I should point out that the group SL(2, 3) is not supersolvable.

3.4.1 Coupling Maps and Dickson Near-Fields

I now want to discuss finite Dickson near-fields in some detail. It is well-known that most finite near-fields can be produced by a variant of the Dickson process (Sect. 2.2). One starts with a finite field $(N, +, \cdot)$ and its group of field automorphisms $\mathrm{Aut}(N, +, \cdot)$, which replaces the additive endomorphisms of (2.5). The mapping $n \mapsto \phi_n$ applies to N^* rather than N, and I define $\phi_0 = 0 = C_0$ (the constant mapping—(7.15)). As before we define a binary product, now called $\wedge$, with

$$m \wedge n = \phi_n(m) \cdot n$$

and when

$$(\phi_m \circ \phi_n) = \phi_{\phi_m(n) \cdot m} \tag{3.37}$$

the resulting structure, $(N, +, \wedge)$ will be found to be a right-distributive near-field.

Near-fields obtained in this way are called **Dickson near-fields**, and following previous terminology (page 78), $(N, +, \wedge)$ will be called the **associated $\mathcal{D}$-near-field**. Informally, I shall sometimes express this idea notationally as

$$\Phi : (N, +, \cdot) \longmapsto (N, +, \wedge) \tag{3.38}$$

This process preserves the field additive and multiplicative identities, and, of course, the field and near-field characteristics are the same. The field automorphisms $\{\phi_n\}_{n \in N^*}$ are members of the Galois group of $(N, +, \cdot)$ over the prime subfield—a cyclic group with order α. As such, they are powers of the Frobenius automorphism (3.1) and they commute (see (3.48)).

Notational ambiguities are encountered when discussing these structures and I adopt the following policy.

Notation 170

(i) When ϕ is a field automorphism, the symbol ϕ^{-1} denotes the inverse automorphism.
(ii) N^* denotes both the non-zero elements of the field $(N, +, \cdot)$ and the non-zero elements of the near-field $(N, +, \wedge)$, for they are the same.
(iii) When $a \in N^*$, the symbol a^{-1} denotes the inverse field element only, that is, $(a \cdot a^{-1}) = 1$.
(iv) The corresponding inverse near-field element will be denoted by $a^{\wedge(-1)}$. that is: $(a \wedge a^{\wedge(-1)}) = 1$.

I also use this notation to represent multiplicative powers in $(N^*, \wedge)$, so

$$(a^{\wedge(s)}) = \overbrace{(a \wedge a \wedge \ldots \wedge a)}^{(s \text{ times})}$$

Always, a^s will mean the power of a using field multiplication.

(v) I normally write the multiplicative order of a field element x as $|x|$. If I wish to stress the order of that element as an element of the related near-field $(N, +, \wedge)$, I might write $|x|^\wedge$, to maintain clarity (but see Corollary 174).
(vi) Multiplicative commutator notation should be understood to refer to the near-field multiplication, since the field is, of course, commutative. However I do write $[N^*, N^*]^\wedge$ to stress this, on occasion.

Observations 171

(i) An alternative expression for (3.37) or (2.8) is $\phi_a \circ \phi_b = \phi_{(b \wedge a)}$. In near-fields, this exactly corresponds to associativity in the associated $\mathcal{D}$-near-field.
(ii) $\phi_{(a \wedge b)} = \phi_{(b \wedge a)}$ (because these field automorphisms commute).
(iii) $\phi_a^{-1} = \phi_{a^{\wedge(-1)}}$.
(iv) $a^{\wedge(-1)} = \phi_a^{-1}(a^{-1})$.
(v) The collection of *distinct* field automorphisms $\{\phi_a\}_{a \in N^*}$ forms a cyclic group $(\mathcal{A}\ \circ)$ under composition. I call this **the coupling group**.
(vi) $(a \cdot b) \wedge c = \big((a \wedge c) \cdot (c^{-1}) \cdot (b \wedge c)\big)$.
(vii) If $\phi_x(a) = a^{p^r}$ and $s \in \mathbb{N}$, then $x^{\wedge(s)} = x^w$, where $w = \frac{(p^{sr}-1)}{(p-1)}$
(viii) $(a^r \wedge c) = \big(a \wedge c)^r \cdot (c^{(r-1)})\big)$.
(ix) $(a \cdot b) = (\phi_b^{-1}(a) \wedge b)$.

Points (ii) and (v) are the ones that seem to need commutativity of $(\cdot)$; the others will apply more generally, when $(\cdot)$ is not commutative.

Our near-field contains the prime subfield C_p, and there is a multiplicative pth root of unity, which I denote by π—to clarify

$$\pi = \overbrace{(1 + 1 + \ldots + 1)}^{\pi \text{ times}} \tag{3.39}$$

We do know that $\phi_n(\pi) = \pi$ for all $n \in N^*$. Suppose H is some multiplicative subgroup of $(N^*, \cdot)$, which will, of course, be cyclic. Selecting $h_1, h_2 \in H$, we have

$$(h_1 \wedge h_2) = (\phi_{h_2}(h_1) \cdot h_2) = h_1^k \cdot h_2 \in H$$

(where k indicates a power of the element h_1).

Lemma 172 *When $(N, +, \cdot)$ is a field, all subgroups $(H, \cdot) < (N^*, \cdot)$ give rise to subgroups $(H, \wedge) < (N^*, \wedge)$.*

Of course, $(N^*, \cdot)$ is itself cyclic, so this pairing of subgroups does not generally result in isomorphisms; and, in principal, we have no reason to suppose that the correspondence works in reverse.

Corollary 173 *Suppose $(S, +, \cdot)$ is a subfield of the field $(N, +, \cdot)$. Then, $(S, +, \wedge)$ is a sub-near-field of $(N, +, \wedge)$.*

Proof We already know that the set of non-zero elements of S constitute a subgroup of $(N^*, \wedge)$. Adding two of these elements results either in 0 or in some non-zero element of S. Thus,

$$\langle S \rangle_{(add)} = (S \cup \{0\})$$

which is therefore a sub-near-field. □

Corollary 174 *Suppose $x \in N^*$, and $|x|$ is prime. Then $|x|^\wedge$ is the same prime. Thus, the Dickson process preserves elements with prime multiplicative order.*

The same result applies for elements of order 4 because there are two groups with this order but only the cyclic one has a single element with order 2.

Corollary 175 $|x| = 4 \Longrightarrow |x|^\wedge = 4$

The reader may wish to look back at Sect. 3.3.1. The field involved has exactly two elements with order 4, and both have order 4 in the corresponding near-field; but that near-field has other elements with order 4 which do not arise from field elements with that order.

$\mathcal{K}_y$ **and** $\mathcal{J}_y$

Suppose $(N, +, \wedge)$ is a Dickson near-field coupled with the field $(N, +, \cdot)$. Selecting $y \in N^*$, define

$$\mathcal{K}_y = \{x \in N \ : \ (x \wedge y) = (x \cdot y)\} \tag{3.40}$$

(see (3.51)). Take $x_1, x_2, \in \mathcal{K}_x$, then

$$((x_1 - x_2) \wedge y) = ((x_1 \cdot y) + (x_2 \cdot y)) = ((x_1 - x_2) \cdot y)$$

and

$$((x_1 \cdot x_2) \wedge y) = (x_1 \wedge y) \cdot (y^{-1}) \cdot (x_2 \wedge y) = ((x_1 \cdot x_2) \cdot y)$$

Lemma 176 *$(\mathcal{K}_y, +, \cdot)$ is a subfield of $(N, +, \cdot)$.*

It will emerge later that $\mathcal{K}_y$ is always non-trivial (see Lemma 189). Symmetrically, we can define

$$\mathcal{J}_y = \{x \in N \ : \ (y \wedge x) = (y \cdot x)\} \tag{3.41}$$

If $x_1, x_2 \in \mathcal{J}_y$, then

$$(y \wedge (x_1 \wedge x_2)) = \Big((y \wedge x_2) \cdot (x_2^{-1}) \cdot (x_1 \wedge x_2)\Big) = (y \cdot (x_1 \wedge x_2))$$

Lemma 177 *$(\mathcal{J}_y^*, \wedge)$ is a non-trivial (multiplicative) subgroup of $(N^*, \wedge)$.*

3.4.2 A Theorem Reported by Marshall Hall

Marshall Hall [62] gives a theorem entirely characterising Dickson near-fields (see Sect. 3.4.1). He offers no proof of it, referring the reader to a paper of Zassenhaus for that.[5]

The Hall-Zassenhaus Conditions

First, I need some number-theoretic conditions—the **Hall-Zassenhaus** conditions. I fix a prime number, p, and a positive integer, β. I then select a further positive integer, δ, with the following properties:

(i) All prime factors of δ are prime factors of $(p^\beta - 1)$.
(ii) $(p^\beta - 3) \equiv 0 \text{ (modulo 4)} \Rightarrow \delta \not\equiv 0 \text{ (modulo 4)}$

Theorem 178 *One may produce a Dickson non-field with $p^{(\beta.\delta)}$ elements from the Galois field with the same order, and the multiplicative centre of this near-field is the Galois field with order p^β. All finite Dickson near-fields have this form.*

This is my version of Hall's theorem. His account includes rather more information, and he actually gives a formula for the near-field multiplication too (but see 3.59). More than that, the theorem says that one cannot have two Dickson near-fields with the same order and centre but having Sylow 2-subgroups quaternion and cyclic, respectively.

[5] The literature contains references to Zassenhaus's book on group theory [145] for this proof; but it does not seem to appear in the English edition of that work. It seems to me that this is one of several chance happenings that have rendered finite near-field theory less accessible to non-German-speaking audiences.

It took me a while to realise that the theorem did not include the Dickson near-field in Sect. 3.3.1 but does include all the other finite Dickson non-fields. It takes a bit of work, but from it, one can figure out that finite Dickson near-fields have multiplicative groups which are 3-groups and that near-fields having such groups as their multiplicative groups are Dickson. As I say, this does not exactly leap out of the page.

Isomorphism Classes
Once one has numbers $\{p, \delta\}$ satisfying the Hall-Zassenhaus conditions (and I believe some authors call such numbers a **Dickson pair**), one may produce $\frac{\Phi(n)}{k}$ non-isomorphic Dickson near-fields, using Hall's prescription—he actually gives a recipe for this. Here Φ is the Euler function and k is the order of p modulo δ [128]. The few books that deal with near-rings usually include an account of this theorem [23, 128]).[6]

Using the relation (3.5), one can say that a prime dividing δ and also dividing $(p^\beta - 1)$ must further divide $\left|\frac{N^*}{D^*}\right|$. Consequently, if such a prime exists then its square divides $|N^*|$. If the prime is 2, and 4 divides δ, then 8 divides $|N^*|$; but, of course, by basic number theory, when p is odd, $(p^{2m} - 1) \equiv 0$ *(modulo 8)*, so the result is scarcely surprising. I return to this on page 148.

3.4.3 The Smallest Proper Near-Field Having All Sylow Subgroups Cyclic

It seems worthwhile to discuss the smallest proper Dickson near-field all of whose Sylow subgroups are cyclic. I appeal to Hall's Theorem 178 and Lemma 169 to do this. This is also the smallest proper near-field not containing the multiplicative group Q_8, and the smallest proper near-field whose group of non-zero elements is a 3-group (page xviii).

We look at powers of primes and associate them with the process Hall outlines. Our first candidate would have order $5^2 = 25$. Using the notation given in Hall's theorem, we can say $h = 1, v = 2$, and the group $(N^*, \wedge)$ has order $24 = (2^3 \times 3)$. There will be a Dickson near-field with this order. Since its centre has order divisible by four, we can rule out the quaternion possibility and conclude we have a 3-group. The presentation for N^* must be

$$G = \left\langle a, b\ :\ a^3 = 1, b^8 = 1, a^b = a^2 \right\rangle$$

[6] A version of it does also appear in [35] but, frankly, it puzzles me. In that book, condition (ii), above, is replaced by

$$(p^\beta - 1) \equiv 0 \ (\textit{modulo 4}) \ \Rightarrow \delta \not\equiv 0 \ (\textit{modulo 4}) \tag{3.42}$$

(page 237). These authors are not the sort of writers to make simple errors, so all I can presently do is report it.

and we can say

$$[N^*, N^*] = \langle a \rangle \cong C_3, \qquad \mathcal{D}^* = \langle b^2 \rangle \cong C_4$$

$\mathcal{L}$ has order either 5 or 25, but it contains $\langle a \rangle$, so in this case $\mathcal{L} = N$.

There is another proper near-field with order 25, and it does have Sylow 2-subgroups (of the multiplicative group) which are quaternion, but it is not Dickson (see [62] and also Table 3.3). Each of the finite non-Dickson near-fields has quaternion Sylow 2-subgroups (Sect. 3.7); and the smallest proper Dickson near-field has $N^* = Q_8$ (Sect. 3.3.1).

3.4.4 The Algebra of the Dickson Process

$(N, +, \wedge)$ is the associated $\mathcal{D}$-near-field of $(N, +, \cdot)$ (with respect to the defined automorphisms). One might wonder whether the Dickson process is reversible.

This would entail our having near-field automorphisms $\{\theta_n\}_{n \in N^*}$ for which

$$(m \cdot n) = \theta_n(m) \wedge n$$

which, in turn, would mean

$$(m \cdot n) = \phi_n(\theta_n(m)) \cdot n$$

that is

$$m = (\phi_n \circ \theta_n)\,(m)$$

In other words

$$\phi_n = \theta_n^{-1}$$

For this to work, we should seem to need (2.8)

$$(\theta_m \circ \theta_n) = \theta_{(\theta_n(m)(n) \wedge n)}$$

but the condition corresponds exactly to associativity of $(\cdot\ )$, which is already known; so no additional restriction is imposed. However, it is not clear that θ is an automorphism of the near-field $(N, +, \wedge)$. We do know that $\theta_n = \phi_n^{-1} = \phi_{n \wedge (-1)}$, so we need only consider mappings ϕ_a.

The condition

$$\phi_a(b \wedge c) = (\phi_a(b) \wedge \phi_a(c)) \tag{3.43}$$

is only true when

$$\phi_c = \phi_{\phi_a(c)} \tag{3.44}$$

is generally true, which seems likely never to be the case (this is the requirement that the automorphisms ϕ_a should preserve cosets of $\mathcal{K}^*$ in (3.55)).

3.4.5 A Generalisation of the Dickson Process

I now consider structures produced by coupling from general near-fields $(N, +, \cdot)$. The collection of mappings $\{\phi_n\}_{n \in N^*}$ is still a collection of additive group automorphisms, but they are not specifically assumed a priori to be automorphisms of the near-field. I shall, however, insist on the normalisation condition (applying to all $n \in N^*$)

$$\phi_n(1) = 1 \tag{3.45}$$

From (2.8), we can say that associativity corresponds entirely to the condition

$$\phi_c\,(\phi_b(a) \cdot b) = \phi_{(b \wedge c)}(a) \cdot \phi_c(b) \tag{3.46}$$

where as before, $(a \wedge b) = \phi_b(a) \cdot b$. The structures $(N, +, \wedge)$ are seen to be near-fields because

$$(1 \wedge a) = (\phi_a(1) \cdot a) = a$$

so 1 is a left identity; and when $\phi_a(m) = a^{-1}$ (which may be arranged since the mappings are bijections)

$$(m \wedge a) = (a^{-1} \cdot a) = 1$$

so we always have left inverses, and this means $(N^*, \wedge)$ is a multiplicative group.

Consider situations of the form (see (3.38))

$$(N, +, \cdot) \xrightarrow{\Phi} (N, +, \wedge) \xrightarrow{\Theta} (N, +, *)$$

involving near-fields, rather than fields, not overtly insisting on near-field automorphisms, but requiring normalisation, and that all products be multiplicatively associative. We know that

$$(a * b) = (\theta_b(a) \wedge b) = ((\phi_b \circ \theta_b)(a) \cdot b)$$

and that $(\Phi \circ \Theta)$ is a collection of normalised group automorphisms.

Lemma 179 *This generalised coupling process is an equivalence relation on the near-fields hosted by a given elementary abelian p-group.*

Suppose $(N, +, \cdot)$ and $(N, +, \wedge)$ are coupled near-fields. Then there are group automorphisms $\{\phi_n\}_{n \in N^*}$ such that

$$(a \wedge b) = (\phi_b(a) \cdot b)$$

and so

$$\phi_b(a) = (a \wedge b) \cdot b^{-1}$$

for which,

$$\phi_c(a \wedge b) = \phi_{(b \wedge c)}(a) \cdot \phi_c(b)$$

Two near-fields $(N, +, \cdot)$ and $(N, +, \wedge)$ are coupled when $(a \wedge b) = (\phi_b(a) \cdot b)$ for suitable mappings Φ. Then

$$\phi_b(a) = (a \wedge b) \cdot b^{-1}$$

a prescription which always defines normalised bijective group automorphisms $\{\phi_n\}_{n \in N^*}$. Since, by hypothesis, both near-field products are associative, we already know that (3.46) must be valid; consequently:

Lemma 180 *All near-field products hosted by a given finite elementary abelian group are coupled, in this extended sense.*

3.4.6 The Historical Dickson Process

I now return to my original view of the Dickson process and insist that $(N, +, \cdot)$ be a commutative field and that the mappings $\{\phi_n\}_{n \in N^*}$ are field automorphisms. The question now is whether all near-fields are coupled to fields in this more restrictive situation. As we have already observed, automorphisms ϕ_n are iterations of the Frobenius automorphism (3.1). Taking π as a primitive root modulo p (see (3.17)), we have

$$\phi_x(\pi) = \pi \tag{3.47}$$

for all $x \in N^*$

The Coupling Group
Consider the collection of field automorphisms $\{\phi_n\}_{n \in N^*}$. We already know that the *distinct* members of this set constitute a multiplicatively cyclic group $(\mathcal{A}, \circ)$

under mapping composition (see (Observations 171)). We also know that the Galois group of the field $(N, +, \cdot)$ over its prime subfield is cyclic and generated by the Frobenius automorphism. We have $|N| = p^{\alpha}$ and that the order of the Galois group of field automorphisms is α, so $\mathcal{A}$ consists of powers of some power r of the Frobenius automorphism, where $1 \leq r \leq \alpha$ is fixed, and is a cyclic group with order dividing α (its order is $\frac{\alpha}{(\alpha,r)}$, using highest common factor notation). Set $\zeta(x) = x^{p^r}$

$$\mathcal{A} = \langle \zeta \circ \rangle \tag{3.48}$$

Fixed points of these automorphisms are field elements x for which

$$x^{p^r} = x$$

that is, solutions to the polynomial equation

$$\left(x^{p^r} - x\right) = 0$$

These form a subfield $\mathcal{F}$ of $(N, +, \cdot)$ one containing the prime subfield. The field has order $p^{(r,\alpha)}$. Suppose $(r, \alpha) = \tau$. Then, $\tau = a.r + b.\alpha$, so, if $x \in N^*$,

$$x^{p^r} = x \Leftrightarrow x^{p^{\tau}} = x$$

From which, it follows

$$\{x : x^{p^r} = x\} = \{x : x^{p^{\tau}} = x\}$$

and this is the Galois field of order p^{τ}, which has p^{τ} elements and is a subfield of $(N, +, \cdot)$. The elements $x \in \mathcal{F}$ have the property that $\phi_a(x) = x$, for all $a \in N^*$. Consequently, for all $a \in N^*$

$$(x \wedge a) = \phi_a(x) \cdot a = (x \cdot a)$$

It follows that

$$\mathcal{F} = \{x \in N : (x \wedge a) = (x \cdot a), \text{ for all } a \in N\} \tag{3.49}$$

and that $(\mathcal{F}, +, \wedge)$ is a subfield of the near-field $(N, +, \wedge)$ and that it is also the Galois field with order p^{τ}. It must also be the case that

$$\mathcal{F} \subseteq D$$

so $\beta = \tau.\omega$ for some integer ω dividing α. I introduce a mapping

$$\varphi : N^* \longrightarrow \mathrm{C}_{(\frac{\alpha}{\tau})} \tag{3.50}$$

where $\phi_m = \zeta^s \Rightarrow \varphi(m) = [s]_{\frac{\alpha}{\tau}}$ (using Notation 3 for cyclic groups, here expressed additively, and the square brackets to indicate a congruence class, modulo $\frac{\alpha}{\tau}$). It will be seen that since $\phi_m \circ \phi_n = \phi_{(m \wedge n)}$, we have

$$\varphi(m \wedge n) = (\varphi(m) + \varphi(n))$$

(the addition being modular and taking place within $C_{(\frac{\alpha}{\tau})}$). This mapping φ is an epimorphism of groups. The kernel of this mapping is

$$\mathcal{K}^* = \{m \in N^* \ : \ \phi_m = Id\}$$

It is a proper, non-trivial, normal subgroup of $(N^*, \ \wedge)$ and it contains $[N^*, N^*]^\wedge$.

We have

$$\mathcal{K}^* = \{a \in N^* \ : \ (x \wedge a) = (x \cdot a) \ \forall \, x \in N\} \tag{3.51}$$

and

$$\mathcal{Z}^* \supseteq \left(\mathcal{F}^* \bigcap \mathcal{K}^*\right)$$

We now know $[N^*, N^*]^{(\wedge)}$ has the same multiplication as its elements have in $(N^*, \ \cdot)$. It is a subgroup of this cyclic group, therefore itself cyclic. The same reasoning tells us that $(\mathcal{K}^*, \ \wedge)$ is also cyclic.

Lemma 181 *The commutator subgroup of the multiplicative group of a Dickson near-field is cyclic. The group itself is metacyclic—a cyclic extension of a cyclic group.*

Corollary 182 *The multiplicative group of a finite Dickson near-field is supersolvable, and therefore Lagrangian (page 4).*

Metacyclic groups are always supersolvable, so the multiplicative group of a Dickson near-field is supersolvable, and consequently cannot be *involved* with SL(2,3) (see page 128), because if it were, SL(2, 3) would have to be supersolvable (subgroups and factor groups of supersolvable groups are supersolvable [62]).

From that, we can deduce that when p is an odd prime, the proper Dickson near-field $(N, \ +, \ \wedge)$ has a normal 2-complement

$$O \longmapsto N^* \twoheadrightarrow S_2$$

When all Sylow subgroups are cyclic and the characteristic is an odd prime, we have an exact sequence of groups

$$C_m \longmapsto N^* \twoheadrightarrow (S_2 \cong C_{2^n}) \tag{3.52}$$

where m is some odd number and n is a positive integer.

The condition that an element $a \in N^*$ is central in $(N^*, \wedge)$ is that for all $x \in N^*$

$$(x \wedge a) = (a \wedge x)$$

That is,

$$\phi_x(a) = \phi_a(x) \cdot a \cdot x^{-1}$$

From which, for all $x, y \in N^*$,

$$\phi_{(x \cdot y)}(a) = \phi_a(x \cdot y) \cdot a \cdot y^{-1} \cdot x^{-1} = \phi_a(x) \cdot a \cdot x^{-1} \cdot \phi_a(y) \cdot a \cdot y^{-1} \cdot a^{-1}$$

So,

$$\phi_{x \cdot y}(a) = \phi_x(a) \cdot \phi_y(a) \cdot a^{-1} \tag{3.53}$$

From (3.47) we have that this relation is satisfied when $a = \pi$.

Lemma 183 *The prime subfield lies within the multiplicative centre of all Dickson near-fields.*

The condition that $d \in D^*$ is that for all $x \in D^*$,

$$(d \wedge (x + 1)) = ((d \wedge x) + d) \tag{3.54}$$

From which,

$$(\phi_{(x+1)}(d) - \phi_x(d)) = \left(d - \phi_{(x+1)}(d)\right) \cdot x^{-1}$$

x is the only term here that is not overtly in D, so this would mean either that $N \subseteq D$, which is impossible, or that for all $x \in N^*$, we have

$$\phi_x(d) = d$$

This means $(d \wedge x) = (d \cdot x)$, and so $d \in \mathcal{F}$.

Strictly, the argument given so far, and its conclusion, is only true for $x \notin D$. Suppose then $x \in D$. We know that $\phi_x(d) = d^a$, and choosing $y \in N^*$ such that $(y \wedge x) \notin D^*$, we have

$$d = \phi_{y \wedge x}(d) \; = \phi_x(\phi_y(d)) \; = \phi_x(d)$$

(since $y \notin D^*$).

Lemma 184 $(D, +, \wedge) = (\mathcal{F}, +, \wedge) = (\mathcal{F}, +, \cdot)$

Corollary 185 *Using the notation we have developed so far, we can say $\beta = \tau$ and that we have an exact sequence*

$$\mathcal{K}^* \rightarrowtail N^* \overset{\varphi}{\twoheadrightarrow} C_\delta \tag{3.55}$$

One consequence of this is that δ divides $(p^\alpha - 1)$, and $\alpha = (\beta.\delta)$.

This raises a problem because the fact that δ divides $(p^\alpha - 1)$ should therefore be a consequence of Hall's Theorem 178, which is by no means obvious to me.

Corollary 186 *The cyclic group $(\mathcal{A}, \circ)$ of automorphisms defined in (3.48) consists of powers of the automorphism*

$$\psi \equiv x \longmapsto x^{p^\beta} \tag{3.56}$$

so $\mathcal{A} = \{\psi, \psi^2, \dots, \psi^\delta \equiv I\}$ (where $\psi^r \equiv x \mapsto x^{p^{(\beta.r)}}$). Moreover, $(\mathcal{A}, \circ) \cong C_\delta$ and we have an exact sequence

$$\mathcal{K}^* \rightarrowtail N^* \overset{\varphi}{\twoheadrightarrow} \mathcal{A}$$

Corollary 187 *In Dickson near-fields, $\mathcal{D} = \mathcal{Z}$.*

Proof We know that $\phi_x(d) = d$ for all $d \in D^*$, so from (3.53) we have $D = \mathcal{Z}$. □

Since $\mathcal{Z}^*$ is cyclic, and $(N^*, \wedge)$ non-abelian, we must have that $\mathcal{Z}^* \subsetneq \mathcal{K}$.

I summarise, producing what is possibly the main result of this chapter.

Lemma 188 (Sofia)

Suppose $(N, +, \wedge)$ is a finite proper Dickson near-field.

(i) D is both a subfield of $(N, +, \cdot)$ and a subfield of $(N, +, \wedge)$, and within D the near-field multiplication and the field multiplication are identical.
(ii) D is identical to the field $(\mathcal{F}, +, .)$ defined in (3.49).
(iii) D is identical to $\mathcal{Z}$.
(iv) D^* is contained in the normal subgroup $\mathcal{K}^*$ defined in (3.51).
(iv) $\mathcal{K}^*$ is a cyclic normal subgroup of $(N^*, \wedge)$; it contains the commutator subgroup of $(N^*, \wedge)$ (which is therefore cyclic).
(v) The factor group $\frac{N^*}{\mathcal{K}^*}$ is cyclic, and so $(N^*, \wedge)$ is metacyclic.
(vi) Both $\mathcal{F}^*$ and $\mathcal{K}^*$ are proper subgroups of both $(N^*, \cdot)$ and $(N^*, \wedge)$.
(vii) $\mathcal{K}^*$ has order $\left(\frac{(p^\alpha-1)}{\delta}\right)$. It contains elements which have precisely this multiplicative order, both in $(N, +, \cdot)$ and in $(N, +, \wedge)$, and if Π is a cyclic generator of the cyclic group of non-zero field elements, then $\mathcal{K}^* = \langle \Pi^\delta \rangle_{(mult)}$.
(viii) $\langle \mathcal{Z} \rangle^*_{(add)} = \mathcal{Z}$.
(ix) $D^* = \mathcal{Z}^* \subsetneq \mathcal{K}^*$. If the near-field is proper, the quotient $\frac{N^*}{\mathcal{Z}^*}$ cannot be cyclic (basic group theory).

Notice from Lemma 168 that if $\frac{N^*}{\mathcal{D}^*}$ should be abelian, then the Sylow 2-subgroups of $(N^*, \wedge)$ will be isomorphic to Q_8, and the corresponding quotient group cannot be cyclic. This actually occurs in Sect. 3.3.1. We do know $(|\mathcal{K}^*|.\delta) = (p^\alpha - 1)$. By Zsigmondy's theorem (page 519), we know of a prime divisor of $(p^\alpha - 1)$ which does not divide $(p^\beta - 1)$, and by Hall's theorem (page 131), we know that all such primes cannot divide δ. This means we have further proof that $\mathcal{D} \subsetneq \mathcal{K}$. Referring back to (3.40) and (3.49), we have

Lemma 189

$$\mathcal{D} = \bigcap_{y \in N^*} (\mathcal{K}_y)$$

(3.16) identifies $\mathcal{K}$ in the case of the smallest proper near-field (Sect. 3.3.1). Notice that $\mathcal{K}$ is certainly not a subfield here; it is not additively closed.

The Dickson process associates each $b \in N^*$ with an automorphism ϕ_b of the field $(N, +, \cdot)$. The multiplicative centre of the near-field is $\mathcal{D}^*$, and since its corresponding factor group is isomorphic to the group of its inner automorphisms, one might wonder whether non-trivial group inner automorphisms occur among the set $\{\phi_b\}_{b \in N^*}$ of field automorphisms. The answer is that they don't. It's fairly easy to see that any such automorphism would have to involve members of $\mathcal{D}^*$ and would therefore be trivial (see page 46).

When N^* is non-nilpotent the group $\mathcal{K}^*$ is its Fitting subgroup.

3.4.7 *When N^* Is a $\mathfrak{Z}$-Group*

The reader should perhaps refer back to Sect. 3.3.6, because I use the notation developed there.

Supposing N^* has the presentation of (3.27). From (3.55) we can say that

$$b^\delta \in \mathcal{K}^*$$

and, consequently, b^δ commutes with a, so $b^\delta \in \mathcal{Z}^*$, and $s|\delta$.

The multiplicative subgroup $\langle a \cdot b^s \rangle$ is cyclic and has order mt. It contains $[N^*, N^*]$ and $\mathcal{D}^*$ and is contained in $\mathcal{K}^*$.

The conclusion is

$$\begin{aligned} s &= \delta \\ t &= (p^\beta - 1) \\ p^\alpha = (t+1)^\delta &= (mst + 1) \\ \mathcal{K}^* &= \langle a, b^\delta \rangle \end{aligned}$$

and from that, Hall's second condition in Theorem 178 is apparent.

The elements $\{b, b^2, b^3, \ldots b^{s=\delta}\}$ constitute a complete set of coset representatives for $\mathcal{K}^*$ in N^*.

Can there be a proper subfield $(F, +, \cdot)$ of $(N, +, \cdot)$ which contains $\mathcal{K}$?

We have that

$$|\mathcal{K}^*| = mt$$

where

$$n = st$$

$$t = (p^\beta - 1), \quad \alpha = (\beta.s)$$

and

$$|r|_m = s$$

$$(p^\alpha - 1) = (m.t.s)$$

We also know that all primes dividing s must divide $(p^\beta - 1)$.

The field F should have order (p^γ), where

$$(p^\gamma - 1) = (m.t.s_1), \quad s = (s_1.s_2)$$

Zsigmondy's theorem gives us a prime q dividing s_2 but not dividing $(p^\gamma - 1)$, that is, dividing s but not dividing $(p^\beta - 1)$. This means that the sub-field F cannot exist. The sub-near-field $\mathcal{L}$ is additively generated by the multiplicative commutator $[N^*, N^*]$. If $\mathcal{L}^*$ lies entirely within $\mathcal{K}^*$ (and so $\mathcal{L}$ is a commutative field), then we may suppose that $|\mathcal{L}| = p^\gamma$, and $\mathcal{L}^* = \langle a, b^{\delta.t_1} \rangle$, where $\alpha = (\gamma.w)$, and $t = (t_1.t_2)$. Then,

$$(p^\alpha - 1) = (m.\delta.t_1.t_2)$$

$$(p^\gamma - 1) = (m.t_2)$$

$$(p^\beta - 1) = (t_1.t_2)$$

and the constraints of Zsigmondy's theorem ensure that neither t_1 nor t_2 is equal to 1. The alternative is that $\mathcal{L} = (N, +, \cdot)$.

3.4.8 *Multiplication in Finite Dickson Near-Fields*

In view of (3.50) it seems that finite Dickson near-fields are very close indeed to being fields. Suppose $a, b \in N^*$, and $k \in \mathcal{K}^*$. Then,

$$(a \wedge (b \wedge k)) = \left(\phi_{(b \cdot k)}(a) \cdot b \cdot k\right)$$

and

$$((a \wedge b) \wedge k) = (\phi_b(a) \cdot b \cdot k)$$

Therefore (using Observations 171),

$$\phi_b = \phi_{(b \cdot k)} = \phi_{(b \wedge k)} = \phi_{(k \wedge b)} \tag{3.57}$$

which means that we can specify the near-field multiplication provided that we can specify the behaviour of a complete set of δ coset representatives for $\mathcal{K}^*$ in (3.55). If $\{x_1, x_2, \ldots, x_\delta\}$ is such a set of representatives, we can arrange things so that (in the notation of Corollary 186)

$$\phi_{x_j} = \psi^j$$

and, moreover, that

$$x_{(j+1)} = (x_1 \wedge x_j) = (x_1)^{\wedge(j+1)} \tag{3.58}$$

(*interpreting* x_0 *as 1*)

Since $(x_1)^{\wedge(2)} = (\phi_{x_1}(x_1) \cdot x_1) = (x_1)^{(p^\beta+1)}$, we conclude that near-field multiplication is very similar to the cyclic multiplication of fields. General elements of N^* have the form $(x_j \cdot k)$, for some $k \in \mathcal{K}^*$, and

$$((x_i \cdot k_1) \wedge (x_j \cdot k_2)) = \psi^j(x_i \cdot k_1) \cdot (x_j \cdot k_2) = (x_i^{p^{(\beta.j)}} \cdot k_1^{p^{(\beta.j)}}) \cdot (x_j \cdot k_2) \tag{3.59}$$

This completely determines the near-field product.

Notice that $\mathcal{K}^*$ is a subgroup of the multiplicative group of the field and of the near-field. It is therefore unique. The near-field product depends heavily on the multiplicative centre $\mathcal{D}$, the term $\mathcal{D}$—near-field being chosen to reflect this (page 78). Concomitantly, suppose we have a field with order $p^{\alpha=(\beta.\delta)}$ and that its multiplicative group $(N^*, \cdot)$ has a cyclic normal subgroup $\mathcal{K}^*$ with index δ. We may select a primitive element Π of the field $(N, +, \cdot)$ such that the set

$$\{\Pi, \Pi^q, \Pi^{q^2}, \ldots, \Pi^{q^{(\delta-1)}}\}$$

(these the *conjugates* of Π, in the field sense) is a basis for the vector space $(N, +)$ over the field $\mathcal{D} + \cdot$, where $q = p^\beta$ and $q^j = p^{j.\beta}$ (a so-called primitive normal basis—see [123]). Field theory then tells us that the related set

$$\{\Pi, (\Pi \cdot \Pi^q), (\Pi \cdot \Pi^q \cdot \Pi^{q^2}), \ldots, (\Pi \cdot \Pi^q \cdot \Pi^{q^2} \cdot \ldots \cdot \Pi^{q^{(\delta-1)}})\} \tag{3.60}$$

will also be a basis for this vector space precisely when the determinant

$$\begin{vmatrix} \Pi & (\Pi \cdot \Pi^q) & \dots & \left(\Pi \cdot \Pi^q \cdot \Pi^{q^2} \cdot \dots \cdot \Pi^{q^{(\delta-1)}}\right) \\ \Pi^q & (\Pi \cdot \Pi^q)^q & \dots & \left(\Pi \cdot \Pi^q \cdot \Pi^{q^2} \cdot \dots \cdot \Pi^{q^{(\delta-1)}}\right)^q \\ \Pi^{q^2} & (\Pi \cdot \Pi^q)^{q^2} & \dots & \left(\Pi \cdot \Pi^q \cdot \Pi^{q^2} \cdot \dots \cdot \Pi^{q^{(\delta-1)}}\right)^{q^2} \\ \dots & \dots & \dots & \dots \\ \Pi^{q^{(\delta-1)}} & (\Pi \cdot \Pi^q)^{q^{(\delta-1)}} & \dots & \left(\Pi \cdot \Pi^q \cdot \Pi^{q^2} \cdot \dots \cdot \Pi^{q^{(\delta-1)}}\right)^{q^{(\delta-1)}} \end{vmatrix} \tag{3.61}$$

is non-zero (again [123]—and note that we are dealing with *field* multiplication here). This expression may be simplified, for example, we have that

$$\Pi^{q^\delta} = \Pi^{(p^\beta)^\delta} = \Pi^{p^{\beta.\delta}} = \Pi$$

In the general 6×6 case, we would end up contemplating the determinant

$$\begin{vmatrix} a & ab & abc & abcd & abcde & abcdef \\ b & bc & bcd & bcde & bcdef & bcdefa \\ c & cd & cde & cdef & cdefa & cdefab \\ d & de & def & defa & defab & defabc \\ e & ef & efa & efab & efabc & efabcd \\ f & fa & fab & fabc & fabcd & fabcde \end{vmatrix} \tag{3.62}$$

where

$$a = \Pi, b = \Pi^q, c = \Pi^{q^2}, d = \Pi^{q^3}, e = \Pi^{q^4}, f = \Pi^{q^5},$$

and the reader will see the general pattern. When $\delta = 2$, the determinant is

$$\begin{vmatrix} \Pi & \Pi \cdot \Pi^q \\ \Pi^q & (\Pi^q \cdot \Pi) \end{vmatrix}$$

with value $\left(\Pi^2 \cdot \Pi^q - \left(\Pi \cdot (\Pi^q)^2\right)\right)$. This will be zero precisely when $\Pi = \Pi^{p^\beta}$, and that never happens. Observe that we already know that p must be an odd since, by hypothesis, δ divides $\left(p^{(\beta.\delta)} - 1\right)$. The conclusion is that we can produce the basis (3.60) in the $\delta = 2$ case, the characteristic is odd, and, by the way, the hypotheses of Theorem 178 are upheld. In fact, the number-theoretic hypotheses of this theorem are automatically valid whenever δ is power of an odd prime—say, $\delta = d^a$, for, the hypothesis, δ divides $(p^{\delta.\beta} - 1)$, requires $p \neq d$; and since $p^d \equiv p$ (*modulo* d), we have $p^\delta \equiv p$ (*modulo* d) and can deduce $p^\alpha \equiv p^\beta$ (*modulo* d). Theorem 178 is true, of course, more generally than this. Ideally, we should like to evaluate (3.61) for all values of δ—that is, all possible sizes of matrix. The rationale for the present analysis is that we wish to define a Dickson near-field using the field in question. From (3.57) we need to associate a distinct automorphism of the field with each coset representative of $\mathcal{K}^*$, and primitive roots provide a convenient way in which this might be achieved. One might guess that the occasions in which our determinant (3.62) is non-zero correspond precisely to the number-theoretic conditions given in Hall's theorem. If this suggestion is flawed, it may because our choice of vector space *basis* is extreme, when all we really need are coset representatives for $\mathcal{K}^*$. And the set (3.60) is actually a transversal for $\mathcal{K}^*$. To see this, first observe that field automorphisms will map transversals to transversals and will map $\mathcal{K}^*$ to itself. Were the set not a transversal, we could expect, for some r and t, and some $k \in \mathcal{K}^*$, that

$$\left(\Pi \cdot \Pi^q \cdot \Pi^{q^2} \cdot \ldots \cdot \Pi^{q^{(r+t)}}\right) = \left(\Pi \cdot \Pi^q \cdot \Pi^{q^2} \cdot \ldots \Pi^{q^r}\right) \cdot k$$

which means

$$\left(\Pi^{q^{(r+1)}} \cdot \Pi^{q^{(r+2)}} \ldots \cdot \Pi^{q^{(r+t)}}\right) \in \mathcal{K}^*$$

or

$$\left(\Pi \cdot \Pi^q \ldots \cdot \Pi^{q^{(t-1)}}\right)^{q^{(r+1)}} \in \mathcal{K}^*$$

From the observation about automorphisms, this would mean

$$\left(\Pi \cdot \Pi^q \ldots \cdot \Pi^{q^{(t-1)}}\right) \in \mathcal{K}^*$$

Applying an automorphism, this means, for some $k \in \mathcal{K}^*$

$$\left(\Pi \cdot \Pi^q \ldots \cdot \Pi^{q^{(t-1)}}\right) = \left(\Pi^q \cdot \Pi^{q^2} \ldots \cdot \Pi^{q^{(t)}}\right) \cdot k$$

that is,

$$\Pi = \Pi^{q^t} \cdot k$$

which is impossible. When we do have a transversal, we may build our near-field, by setting

$$\phi_\Pi \equiv \left\{x \mapsto x^{p^\beta}\right\}$$

This is, of course, an automorphism of the field $(N, +, \cdot)$. Now, for $j = 1, \ldots (\delta - 1)\}$,

$$\left(\Pi \cdot \Pi^q \cdot \Pi^{q^2} \cdot \ldots \Pi^{q^j}\right) = \left(\phi_\Pi(\Pi \cdot \Pi^q \cdot \Pi^{q^2} \ldots \Pi^{q^{(j-1)}}) \cdot \Pi\right)$$

a condition that will make the elements of (3.60) successive powers of Π in the near-field. We next define field automorphisms

$$\phi_{\Pi \cdot \Pi^q \cdot \Pi^{q^2} \cdot \ldots \Pi^{q^j}} = (\phi_\Pi)^{(j+1)}$$

for $j = 1, 2, \ldots, (\delta - 1)$ and make the obvious definitions within cosets of $\mathcal{K}^*$. The resulting structure has been built to satisfy the coupling conditions and gives a Dickson near-field hosted by $(N, +)$.

3.4.9 Isomorphism in Finite Dickson Near-Fields

Dickson near-fields hosted by a given finite abelian group which have the same multiplicative centre will be isomorphic. This raises the question as to whether all proper Dickson near-fields with the same order are isomorphic. Appealing directly to the theorem of Marshall Hall (page 131), we may set $\alpha = 60$ and consider near-fields with prime characteristic $p = 3$ and order 3^{60}.

(i) There will be a Dickson near-field in which $\delta = 2$ and $\beta = 30$. Its centre is the Galois field with order 3^{30}.
(ii) There will be a Dickson near-field in which $\delta = 3$ and $\beta = 20$. Its centre is the Galois field with order 3^{20}.
(iii) There will be a Dickson near-field in which $\delta = 5$ and $\beta = 12$. Its centre is the Galois field with order 3^{12}.

Consequently, there will be several distinct isomorphism classes of Dickson near-fields with order 3^{60}. Hall's theorem means that these results do depend on the characteristic chosen. For instance, there is no Dickson near-field with order 5^{60} whose centre is the Galois field with order 5^{12}.

3.4.10 Sub-near-Fields

In this subsection, the term "proper" means "properly contained in", and is not used as described in Sect. 1.31.2.

Our standard Dickson near-field has order $p^{\alpha=(\beta.\delta)}$. There will be a sub-near-field with order $p^{\mathcal{A}}$ for each divisor $\mathcal{A}$ of α (see Corollary 173). I refer to these as **divisor** near-fields. Non-Dickson near-fields have exactly one proper non-trivial sub-near-field and that is C_p (Sect. 3.7). They do not properly contain non-Dickson sub-near-fields. Finite Dickson near-fields cannot have non-Dickson sub-near-fields either, because that would involve supersolvable groups having non-supersolvable subgroups. So, possible sub-near-fields of Dickson near-fields must themselves either be finite fields or they must be Dickson near-fields.

Suppose $(N, +, \wedge)$ is a finite Dickson near-field. We know $\mathcal{K}^* \subsetneq N^*$. Is it possible that the near-field (N) contains a proper sub-near-field, $(M, +, \wedge)$, which is a non-field and which itself contains $\mathcal{K}^*$? Appealing to Zsigmonde's theorem (Appendix E), we could then expect a prime q dividing $(p^\alpha - 1)$ which does not divide $|M^*|$. Such a prime would divide δ but not $(p^\beta - 1)$, contradicting Hall's theorem. So, no proper sub-near-fields of $(N, +, \wedge)$, which are not fields, can contain $\mathcal{K}^*$; and, in fact, the same argument proves both that no proper subfields of the near-field can contain $\mathcal{K}^*$ and that $\mathcal{K}$ cannot itself be a subfield.

Suppose, now, that $(N, +, \wedge)$ is our standard finite Dickson near-field and that $(M, +, \wedge)$ is a proper sub-near-field of it. Suppose this sub-near-field is a non-field. N has a collection of structures associated with it: $\mathcal{D}^*$, $\mathcal{K}^*$, $\mathcal{A}$ and so on. Being Dickson, M has similar structures: $\mathcal{D}_M^*, \mathcal{K}_M^*, \mathcal{A}_M, \ldots$. It seems clear that $\mathcal{A}_M$ is a cyclic subgroup of $\mathcal{A}$. M has order p^{α_M} and there is an exact sequence of groups

$$\mathcal{K}_M^* \mapsto M^* \overset{\varphi}{\twoheadrightarrow} \mathrm{C}_{\delta_M} \tag{3.63}$$

and we know δ_M divides δ. M^* is a multiplicative subgroup of N^*, so $(p^{\alpha_M} - 1)$ divides $(p^\alpha - 1)$, and from Lemma 151 we can say that α_M divides α. The near-field M is coupled with a field with order $|F_M| = p^{\alpha_M}$. This field will be a subfield of F. The sub-near-field is not itself a commutative field and must contain non-zero elements lying outside the cyclic group $\mathcal{K}^*$ identified in (3.51). M contains an element $y \notin \mathcal{K}^*$ such that $\phi_y \equiv \{x \mapsto x^{p^{\beta.\delta_1}}\}$, where $\delta = (\delta_1.\delta_2)$ and $1 < \delta_1 < \delta$, and ϕ_y is a cyclic generator of $\mathcal{A}_M$. For notational consistency, $\beta_M = (\beta.\delta_1)$, and $\delta_M = \delta_2$. There is a subfield

$$\mathcal{D}_M = \{x \in N^* \ : \ (x)^{p^{\beta_M}} = x\} \supsetneq \mathcal{D}$$

This is a subfield of $(N, +, \cdot)$. It has order p^{β_M}. The near-field $(M, +, \wedge)$ itself has order $p^{\alpha_m} = (p^{\beta_M})^{\beta_1} = p^{\beta.\delta} = p^\alpha$ and so is improper. It follows that our original Dickson near-field has no proper sub-near-fields which are non-fields and that proper sub-near-fields are commutative fields.

Lemma 190 *The only sub-near-fields of a finite Dickson near-field are themselves subfields of the original field. All proper subfields of the original field occur in this way.*

Interestingly, the number-theoretic conditions required by Hall's theorem may apply to what one would have thought would be possible sub-near-fields. There may well be Dickson near-fields with these orders, but they are not sub-near-fields of our one. The structure $\mathcal{L}$ is a sub-near-field (Lemma 157). It follows that either $\mathcal{L}$ is a commutative field contained in $\mathcal{K}$ (as happens in Sect. 3.3.1) or that it is an improper sub-near-field—that is, $\mathcal{L} = N$. The same applies to the near-field $\mathcal{F}$ (page 112). Either it is a field containing $\mathcal{L}$, or it is the near-field $(N, +, \wedge)$.

3.4.11 Number-Theoretic Issues

In this subsection it is occasionally necessary to distinguish the field $(N, +, \cdot)$ and its coupled Dickson near-field, $(N, +, \wedge)$. We have that $|\mathcal{D}^*| = (p^\beta - 1)$, and $|\mathcal{K}^*| = (\frac{(p^\alpha-1)}{\delta}$, and $\alpha = (\beta.\delta)$. In Dickson cases, all prime divisors of δ divide $(p^\beta - 1)$. We also know that $\mathcal{D}^*$ is a subgroup of the cyclic group $\mathcal{K}^*$, and the latter is generated by π^δ, where π is a multiplicative generator of the cyclic group of non-zero field elements, $(N^*, \cdot)$. We may write

$$(p^\alpha - 1) = ((p^\beta - 1).\delta.\sigma)$$

σ is the index $|\mathcal{K}^* : \mathcal{D}^*|$, and $|\mathcal{K}^*| = (p^\beta - 1).\sigma$. From (3.4)

$$(\delta.\sigma) = \left(p^{(\delta-1).\beta} + p^{(\delta-2).\beta} + \ldots + p^\beta + 1\right)$$

and from (3.5) we can say that for all primes q dividing $(p^\beta - 1)$,

$$(\delta.\sigma) \equiv \delta \ (modulo\ q)$$

This means either that q divides δ, or it means that $\sigma \equiv 1$ (modulo q). Appendix E gives a theorem that suggests that if there should be a subfield of $(N, +, \cdot)$ with order p^τ with β a divisor of τ, and, of course, τ a divisor of α, then we can find a prime, q dividing $(p^\alpha - 1)$ which does not divide $(p^\tau - 1)$ (Zsigmondy's theorem). Obviously, q cannot divide $((p^\beta - 1).\delta)$, so it must divide σ. In fact, all primes dividing $(p^\alpha - 1)$ but not $(p^\beta - 1)$ must divide σ. This argument was employed in Sect. 3.4.10 to establish that $\mathcal{K}^*$ could not be contained in a proper subfield of $(N, +, \cdot)$. We know there must be a subfield of $(N, +, \cdot)$ with order p^τ for every integer τ which is divisible by β and which divides α, and we know that all such subfields of $(N, +, \cdot)$ must contain $\mathcal{D}^*$. We can say that

$$\tau = (\beta.\omega) \ where\ \delta = (\omega.\rho)$$

3.4.12 Near-Field Automorphisms

If θ is a near-field automorphism, then

$$\theta\,(a \wedge b) = (\theta(a) \wedge \theta(b))$$

so

$$\theta\,(\phi_b(a) \cdot b) = \left(\phi_{\theta(b)}\,(\theta(a)) \cdot \theta(b)\right)$$

which means

$$\theta (a \cdot b) = \left(\left(\phi_{\theta(b)}(\theta\left(\phi_b^{-1}(a)\right)\right) \cdot \theta(b)\right)$$

Now, this will also be a field automorphism when

$$\theta(a) = \phi_{\theta(b)}\left(\theta\left(\phi_b^{-1}(a)\right)\right) \tag{3.64}$$

for all $a, b, \in N^*$. But when θ is already a field automorphism

$$\left(\theta\left(\phi_b^{-1}(a)\right)\right) = \left(\phi_b^{-1}(\theta(a))\right)$$

and (3.64) becomes

$$\phi_{\theta(b)} \circ \phi_b^{-1} = Id$$

that is

$$\phi_b = \phi_{\theta(b)} \tag{3.65}$$

for all $b \in N^*$.

Lemma 191 *Suppose θ is a field automorphism. Then, (3.65) gives the condition that it is also an automorphism of the coupled near-field $(N, +, \wedge)$.*

Clearly, Dickson near-fields with characteristic 2 are multiplicatively $\mathfrak{Z}$-groups. There are Dickson near-fields with odd characteristic for which this is true—that is, having cyclic Sylow 2-subgroups in the multiplicative group. One example of this is the Dickson near-field with order $3^{(3\times 13)}$ and multiplicative centre $C_{(3^3-1)}$. Simple congruence arguments show that $(3^{(39)} - 1)$ is not divisible by 4 so cannot contain a generalised quaternion subgroup. But the conditions of Theorem 178 are met in these figures. The conclusion is that multiplicatively this must be a $\mathfrak{Z}$-group.

3.4.13 Prime Divisors of δ: Hall's Theorem

Suppose that we have a proper Dickson near-field and that there is a prime q dividing δ. Can it happen that q does not divide $(p^\beta - 1)$, that is, that the field $\mathcal{D}$ has no element with multiplicative order q?

From (3.55), we can say that q does divide $(|N^*| - 1)$, and we can select an element $x \in (N, +, \cdot)$ (the field) with multiplicative order q. Since q is prime, by Corollary 174 we know x has the same multiplicative order in the field as it has in the near-field. Suppose no element of $\mathcal{D}$ has multiplicative order q, or, what comes to the same thing, that q does not divide $(p^{\beta} - 1)$. Field theory tells us that there is an extension field of $\mathcal{D}$, a subfield of $(N, +, \cdot)$, which has order $p^{(\beta \cdot q)}$. The field contains all elements of $(N, +, \cdot)$ whose multiplicative order divides $\left(p^{(\beta \cdot q)} - 1\right)$ and so contains x. The simple algebraic extension $\mathcal{D}(x)$ is a subfield of this field, but since q is prime, it is improper. This means that the minimal polynomial of x is of degree q and is actually $(x^q - 1)$. We have a polynomial basis ([123]) for the vector space $\mathcal{D}(x)$. It is $\{1, x, x^2, \ldots, x^{q-1}\}$. But this would mean that the equality $(1 + x + x^2 + \ldots + x^{q-1}) = 0$ could not apply (see (3.6)). We conclude that q does divide $(p^{\beta} - 1)$, exactly as Theorem 178 claims. Suppose now that 4 divides δ but that it does not divide $(p^{\beta} - 1)$. We can say that

$$p \equiv 3 \ (modulo\ 4)$$

and that β is odd. Moreover,

$$p^m = \begin{cases} 1 \ (modulo\ 4) \text{ if } m \text{ is even} \\ 3 \ (modulo\ 4) \ if\ m\ is\ odd \end{cases}$$

Our field contains an element x with order 4, which, from Corollary 175, has the same order in the near-field. If $a \in (N, +, \wedge)$, then $(x \wedge a)$ has the form $(x^{p^j} \cdot a)$, this involving a power of an odd prime. Consequently,

$$(x \wedge a) = \mp(x \cdot a)$$

However,

$$(x \wedge a) = (-x \cdot a) \Rightarrow (x^2 \wedge a) = (x \wedge a)^2 \cdot x^{-1} = (x \cdot a^2)$$

but

$$(x^2 \wedge a) = (-a)$$

which would imply

$$(-1) = x \cdot a$$

which cannot generally be true. We conclude that

$$(x \wedge a) = (x \cdot a)$$

and this is true for all $a \in N^*$. From (3.49), this implies $x \in \mathcal{D}$, verifying the number-theoretic conditions in Hall's theorem. With respect to elements of the *field* with order 4 (which are then elements of the near-field with the same order), the argument just given permits us to say more.

Lemma 192 *Suppose $x \in (N, +, \cdot)$ has order 4 (δ being divisible by four), i.e. $|x| = 4$. Then $x \in \mathcal{D}$.*

Sect. 3.3.1 provides an example (with $\delta = 2$) of a near field with elements with order 4 with not one of them lying in $\mathcal{D}$. It has elements with order 4 that have that order in both the corresponding field and the near-field, and near-field elements with order 4 which do not correspond to order 4 elements of the field.

3.4.14 $\mathcal{L}$ and N

Suppose that $(N, +, \wedge)$ is a finite Dickson near-field and that $(\mathcal{W}, +, \wedge)$ is a *sub-near-field* for which $\mathcal{W}^* \lhd N^*$, $\frac{N^*}{\mathcal{W}^*}$ is abelian, and for all $n \in N^*$ and all $C_1, C_2 \in \mathcal{W}^*$, we have

$$(n \wedge (C_1 + C_2)) = ((n \wedge C_1) + (n \wedge C_2)) \tag{3.66}$$

Obviously, $(\mathcal{W}, +, \cdot)$ is a field, so its non-zero elements constitute a cyclic group. $\mathcal{W}$ must contain the near-field $\mathcal{L}$ (defined page 111), and I shall just assume $\mathcal{W} = \mathcal{L}$ in what follows and drop the term $\mathcal{W}$ (see Sect. 3.3.1). The reader may wish to look back to Lemma 159 at this point and perhaps (3.51). Suppose that $\mathcal{L}$ has order p^λ and that the near field has order p^α, where $\alpha = \omega.\lambda$. Referring back to (3.55), (3.51) and Lemma 188, we can say that $[N^*, N^*]^\wedge \subseteq \mathcal{K}^*$ and so, if $C_1, C_2 \in [N^*, N^*]^\wedge$,

$$n \wedge (C_1 + C_2) = (n \wedge C_1) + (n \wedge C_2) = (n \cdot C_1) + (n \cdot C_2) = n \cdot (C_1 + C_2)$$

From this, additive closure, we conclude $\mathcal{L}^* \subseteq \mathcal{K}^*$.

3.4.15 An Intrinsic Characterisation of Dickson Near-Fields

The reader may care to look at Sect. 2.2 and Observations 171 before reading the following.

Suppose $(N, +, \wedge)$ is a finite Dickson near-field coupled with the field $(N, +, \cdot)$. We know, from Sect. 3.4.1, that

$$(a \wedge b) = \phi_b(a) \cdot b$$

Starting, instead, from any finite proper near-field $(N, +, \wedge)$, we know there is a unique Galois field with the same order $(N, +, \cdot)$ and might ask whether the mapping

$$\phi_b(a) = (a \wedge b) \cdot b^{-1}$$

constitutes a field automorphism. That it is an additive group automorphism is trivial. The condition that $\phi_b(a \cdot c) = (\phi_b(a) \cdot \phi_b(c))$ is actually that

$$((a \cdot c) \wedge b) = \Big((a \wedge b) \cdot b^{-1} \cdot (c \wedge b)\Big) \tag{3.67}$$

(compare with Observations 171). The reader will realise that this condition is sufficient for the structures $(N, +, \cdot)$, $(N, +, \wedge)$ to be coupled. Associativity of $(\wedge)$ imposes

$$(\phi_a \circ \phi_b) = \phi_{(b \wedge a)} \tag{3.68}$$

which requires

$$\Big(\Big(\Big((c \wedge b) \cdot b^{-1}\Big) \wedge a\Big) \cdot a^{-1}\Big) = \Big((c \wedge b \wedge a) \cdot (b \wedge a)^{-1}\Big)$$

and, assuming condition (3.67), that condition becomes

$$a^2 = \Big((b^{-1} \wedge a) \cdot (b \wedge a)\Big)$$

But this condition is true if (3.67) holds, because

$$a = ((b^{-1} \cdot b) \wedge a) = \Big(b^{-1} \wedge a) \cdot a^{-1} \cdot (b \wedge a)\Big)$$

which, I suppose, we knew anyway. This gives us an intrinsic characterisation of finite Dickson near-fields.

Lemma 193 *The finite near-field $(N, +, \wedge)$ is Dickson and coupled with the field $(N, +, \cdot)$ precisely when (3.67) holds.*

3.5 Group Structure of N^*

This section continues to use the notation developed so far. So $|N| = p^\alpha$, $|\mathcal{D}| = p^\beta$, and $\alpha = \delta.\beta$. Furthermore, S_2 represents the Sylow 2-subgroup of N^* (odd prime characteristics). The structure of the multiplicative group of a finite near-field is quite restrictive.

General Finite Near-Fields
All finite near-fields have these properties:

(i) When the near-field characteristic is an odd prime, the group has precisely one multiplicative involution, the element (-1). This is central.
(ii) Subgroups with order equal to a product of two primes (whether distinct or not) are cyclic.

(iii) Abelian subgroups are cyclic.
(iv) Sylow subgroups are cyclic, for odd primes, and either cyclic or generalised quaternion otherwise.
(v) The only prime that could possibly divide the orders of both the group centre and the group commutator subgroup is 2, and that happens exactly when the Sylow 2-subgroup is quaternion. When it happens, the group centre has order $2x$, where x is some odd number; and the order of the commutator subgroup is divisible by four.

Finite Dickson Near-Fields
And these properties are exhibited by finite Dickson near-fields.

(i) The group $(N^*, \cdot)$ is supersolvable and metacyclic. It has a cyclic subgroup $\mathcal{K}^*$, which has a cyclic quotient group C_δ.
(ii) The commutator subgroup is cyclic.
(iii) The quaternion case can only possibly occur when the characteristic is congruent to 3 modulo 4 and δ is two times an odd number, and β is an odd number. These conditions are necessary but insufficient.
(iv) The multiplicative centre, $\mathcal{Z}^*$, is cyclic and has order $(p^\beta - 1)$. The near-field itself has order p^α, where $\alpha = (\beta.\delta)$.

In the Dickson case, we know N^* is supersolvable. So there is a normal subgroup $\mathcal{H}$ with odd order such that

$$\mathcal{H} \mapsto N^* \overset{\varphi}{\twoheadrightarrow} S_2 \tag{3.69}$$

where S_2 is the Sylow 2-subgroup.

3.5.1 *Presentations for Solvable Near-Fields with* S_2 *Quaternionic*

It turns out that one can obtain presentations for finite solvable groups with these characteristics. This is work that dates back to Zassenhaus, and I use the best account of it I know, which is due to Joseph Wolf [144], and which covers the multiplicative groups of all but three of the finite near-fields (Table 3.3). The conditions being so restrictive, one imagines similar presentations could be found for the non-solvable cases, if that has not been done already (and see page 128).

Wolf's proofs are clear and largely non-technical, but reproducing them would use up more space than seems appropriate, and take me too far from my central project. I simply quote the results, referencing the numbering Wolf uses, on the basis that in this book, known group theory may be taken as a given, particularly when available in a real classic such as [144]. His results often appear as lists of subcases. The group-theoretic properties of $(N^*, \cdot)$, listed above, permit me to select only such subcases as apply to near-fields, and my account of Wolf's theorems has modified them in that respect. Recall that S_2 is the Sylow 2-subgroup of the group—it is of special importance.

Wolf's Subcases

(A) First, Wolf's Lemma 5.4.5 says that when S_2 is cyclic, the quotient of the commutator subgroup is a cyclic group. This situation is the 3-group case and has already been noted, in Sect. 3.3.6, where a presentation is given (3.27). This means we need only consider cases in which S_2 is quaternionic.

(B) Next, Wolf's Lemma 6.1.9 says that if $|N^*| = 2^a.t$, with t odd, then there is a normal subgroup $\mathcal{I}$ which has cyclic Sylow 2-subgroups, and for which $\frac{N^*}{\mathcal{I}}$ is isomorphic to one of C_2, A_4 or S_4.[7] Now, neither A_4 nor S_4 is supersolvable, so the Dickson near-field case must be C_2. Looking at Table 3.3, we see cases A, B, C and D are the solvable non-Dickson near-fields. In the first two, the quotient $\frac{N^*}{\mathcal{I}}$ is A_4, and in the last two, it is S_4.

(C) Finally, via an intermediary Lemma 6.1.10, Wolf offers a Theorem 6.1.11 about finite solvable groups in which every abelian subgroup is cyclic.

Adapting Wolf's Theorem 6.1.11 to our particular case, I obtain a presentation for the multiplicative group of a finite Dickson near-field in which S_2 is quaternion based on Wolf's Type II which looks like this

$$\Big\langle a, b, c\ :\ a^m = 1, b^n = 1, (b^{-1}.a.b) = a^r, c^2 = b^{\frac{n}{2}} = -1, \\ (c^{-1}.a.c) = a^w, (c^{-1}.b.c) = b^k \Big\rangle \qquad (3.70)$$

This group has order $2.m.n$. $m \geq 1$, is odd, $n \geq 1$, and $(n.(r-1), m) = 1$, and $r^n \equiv 1$ (modulo m). Moreover, $w^2 \equiv r^{(k-1)} \equiv 1$ (modulo m); and $n = 2^u.v$, $u \geq 2$, v odd; and $k \equiv -1$ (modulo 2^u); and $k^2 \equiv 1$ (modulo n). Finally, if s is the order of r in the multiplicative group of residues modulo m, that is

$$r^s \equiv 1 \ (modulo\ m), \ \ (with\ s\ minimal)$$

then $d = (\frac{n}{s})$ is divisible by every prime divisor of s, and $n = (s.d)$.

The case in which this group is Q_8 requires $m = 1, n = 4, v = 1, k = 3$ (see Sect. 3.3.1).

I do not claim that all such groups are the multiplicative groups of near-fields. Indeed, a first necessary condition would be that the order 2 mn should be one less than a power of a prime.

Exploring (3.70) as Representing a Dickson Near-Field

The Sylow 2-subgroups are generalised quaternion, so the group centre must have order $2x$, with x some odd number.

[7] Wolf's formulation of the lemma is a little different, in that he specifically requires the group to have an element with order $2^{(a-1)}$. Since we already know this to be the case for N^*, I miss that out.

b^s is the generator of the subgroup of $\langle b \rangle$ which commutes with a. The group centre must be a subgroup of that, since, additionally, it must commute with c. The condition that b^t commutes with c is that

$$t.(k-1) \equiv 0 \ (modulo\ n)$$

So, the subgroup of $\langle b \rangle$ generated by elements commuting with c must be $\langle b^{(k-1)} \rangle$. Writing the order of $b^{(k-1)}$ as f, and $n = f.g$, this subgroup is $\langle b^g \rangle$. The group centre will then be $\langle b^l \rangle$, where l is the least common multiple of s and g, and $n = l.z$, and z divides d. We know that for some prime p, $p^\alpha = (2mn+1)$, and we know that the group centre has order $(p^\beta - 1)$, where $\alpha = (\beta.\delta)$. The commutator subgroup of the group is cyclic. It contains $a^{(r-1)}$, which means it contains a. It also contains $\langle b^{(k-1)} \rangle$. We conclude that it is

$$\left\langle a, b^{(k-1)} \right\rangle = \langle a.b^g \rangle \tag{3.71}$$

This has order mf. The integer g is clearly some multiple of s because the commutator is abelian. Therefore, $l = g$, and $f = z$. The group centre is $\langle b^g \rangle$. As I observed, it has order $2x$ where x is odd. The commutator subgroup will additionally have a contribution from the quaternion Sylow subgroups, so it will include an element with order $2^{(u-1)}$. The element b^g must have order some multiple of $2^{(u-1)}$. We conclude that $2^{(u-1)} = 2$, and $S_2 = Q_8$.

Lemma 194 *If* $(N, +, \cdot)$ *is a finite Dickson near-field, and* S_2 *is quaternion, then* $S_2 = Q_8$ *and in (3.70),* $u = 2$.

The group centre lies in the commutator subgroup. This means that its order is some power of two (see Isaacs' theorem on page 124). It's order is of the form $(p^\beta - 1)$ and it must actually be 2.

Lemma 195 *The prime p is equal to three, and* β *is equal to one.* N^* *has order* $(8.m.v)$*, where m and v are both odd. Finally* $\mathcal{Z} \cong C_3$.

In the case of the smallest proper near-field (page 113), $m = v = 1$. From Lemma 169, we can say $\alpha = (2.h) = \delta$, where h is some odd number. Then,

$$(1 + 8.m.v) = p^\alpha = 9^h \equiv 9 \ (modulo\ 16)$$

which is correct. So, the search for finite Dickson near-fields in which S_2 is quaternion may be restricted to situations in which $p = 3, \beta = 1, \alpha = (2.h)$, where $h = (m.v)$ is odd. From Hall's theorem (178), we must conclude $v = 1$. The following lemma presupposes the results stated in Sect. 3.7.

Lemma 196 *There are exactly eight proper finite near-fields having Sylow 2-subgroups which are quaternion. Seven of these are non-Dickson, with two having Sylow 2-subgroups with order 16 and five having Sylow 2-subgroups isomorphic to*

Q_8. *The sole Dickson near-field of this kind is the proper near-field with order 9 given in Sect. 3.3.1.*

So, all but one Dickson near-ring have N^* isomorphic to a $\mathfrak{Z}$-group, which has profound consequences for the structure of these creatures. For example, multiplicative subgroups of the same order are conjugate (page 123) in $\mathfrak{Z}$-groups.

3.5.2 Presentation for Non-Dickson Solvable Cases

Table 3.3 applies. $\mathcal{D}$ is multiplicatively central in all cases except C. The following list refers to Table 3.3 but uses results from case (B), page 153, and gives $\frac{N^*}{\mathcal{I}}$.

(i) $\frac{N^*}{\mathcal{I}} = \frac{\mathrm{SL}(2,3)}{C_2} \cong A_4$ (case A).
(ii) $\frac{N^*}{\mathcal{I}} = \frac{(\mathrm{SL}(2,3) \oplus C_5)}{C_{10}} \cong A_4$ (case B).
(iii) $\frac{N^*}{\mathcal{I}} = \frac{G_1}{C_2} \cong S_4$ (case C).
(iv) $\frac{N^*}{\mathcal{I}} = \frac{(G_1 \oplus C_{11})}{C_2} \cong S_4$ (case D).

It seems to me that one may identify the presentations for case A and case B based on Wolf's Theorem 6.1.11 Type III, as

$$\Big\langle a, b, c, d \,:\, a^m = 1, b^n = 1, b^{-1}.a.b = a^r, c^4 = 1, c^2 = d^2 = (c.d)^2,$$
$$(a^{-1}.c.a) = c, (a^{-1}.d.a) = d, (b^{-1}.c.b) = d, (b^{-1}.d.b) = (c.d)\Big\rangle \qquad (3.72)$$

Then, $m \geq 1, n \geq 1, (n(r-1), m) = 1, r^n \equiv 1$ (*modulo* m) and $n \equiv 1$ (*modulo* 2), $n \equiv 0$ (*modulo* 3). The group has order 8 mn, this being 24 in case A and 120 in case B. In these cases, the Sylow 2-subgroups are Q_8 and it is normal. It seems to me that the presentation for case C and case D corresponds to Theorem 6.1.11 Type IV and is

$$\Big\langle a, b, c, d, e \,:\, a^m = 1, b^n = 1, b^{-1}.a.b = a^r, c^4 = 1, c^2 = d^2 = (c.d)^2,$$
$$(a^{-1}.c.a) = c, (a^{-1}.d.a) = d, (b^{-1}.c.b) = d, (b^{-1}.d.b) = (c.d), e^2 = c^2,$$
$$(e.c.e^{-1}) = (d.c), (e.d.e^{-1}) = d^{-1}, (e.a.e^{-1}) = a^w, (e.b.e^{-1}) = b^k\Big\rangle \qquad (3.73)$$

Then, $m \geq 1, n \geq 1, (n(r-1), m) = 1, r^n \equiv 1$ (*modulo* m) and $n \equiv 1$ (*modulo* 2), $n \equiv 0$ (*modulo* 3), as in Type 111, and, additionally, $k^2 \equiv 1$ (*modulo* n), $k \equiv -1$ (*modulo* 3) and $r^{k-1} \equiv w^2 \equiv 1$ (*modulo* m). The group has order 16 mn, which is 48 in case C and 528 in case D. In these near-fields the Sylow 2-subgroups are Q_{16} and there are three conjugate ones.

The key non-solvable example in Table 3.3 is the binary icosahedral group (case E).[8] This has but one proper, non-trivial, normal subgroup, its centre, and that is isomorphic to C_2. The corresponding factor group is A_5—taking us out of Wolf's domain. Arguably, this represents the most exceptional of the finite non-Dickson near-fields, although a strong case for near-field F might be made, because that is the only finite near-field in which both $\mathcal{D}^* \supsetneq \mathcal{Z}^*$ and $\mathcal{Z}^* \supsetneq C_2$ (see page 470).

3.6 Frobenius Groups

In this section, all groups are finite.

3.6.1 *Basics*

Permutation groups have a central position in modern group theory, and theorems by Thompson, Zassenhaus and others on their structure when they possess restricted kinds of fixed points were pivotal to the technical work behind the determination of the finite simple groups [121]. Frobenius groups are a rather important subset of the permutation groups and are directly related to near-fields. Many of the multiplicative properties of near-fields are actually multiplicative properties of *Frobenius complements*, which I shall describe, shortly.

My groups are finite. The story for, say, infinite, sharply 2-transitive groups is much more complicated and, so far as I understand it, largely unknown [17]. Our permutation groups, G, act on a set Ω on the *right*

$$\Omega \times G \mapsto \Omega \quad \alpha \mapsto \alpha.g$$

Some of the ideas I cover were introduced on page xviii, but the reader would not comprehend much of what follows without some prior knowledge of permutation groups. I don't find the space to cover this here but believe that it is accessible, in [126], or in [143], or in [17], for instance, or even on the Internet.

Definition 197 A **Frobenius group** is a (in this section—finite) permutation group which is transitive without being fixed-point-free and with the property that only the group identity fixes more than one element.

They are alternatively characterised as groups $(G, .)$ possessing subgroups H with the property

$$g \notin H \Rightarrow \left(H \bigcap H^g\right) = \{Id_G\} \tag{3.74}$$

[8] There is also a group commonly known as the **full icosahedral group**. It is quite distinct from the binary icosahedral group but does have the same order because it is $(A_5 \oplus C_2)$—how to make a difficult subject harder!

(this property is sometimes described as characterising a *malnormal* subgroup). In this situation

$$K = \left(G - \bigcup_{g \in G} H^g \right) \cup \{1\}$$

is actually a nilpotent normal subgroup.[9] H is called a **Frobenius complement** of G and K is its corresponding **Frobenius kernel**. The Frobenius kernel turns out to be the set of fixed-point-free elements, plus the identity. Moreover, it is actually the **Fitting subgroup** of the group G[10] and so, unique; and K is a *regular normal subgroup* and elementary abelian. The group G is a semi-direct product of K by H

$$G = (K \rtimes H)$$

The Frobenius complement, H, is a subgroup of elements of the permutation group G which fix a specific letter α—written G_α in most texts on permutation groups (and referred to as **point stabilisers**). Other such subgroups are simply conjugates of H, that is, if $\beta = \alpha.g$, then $G_\beta = (G_\alpha)^g$ (G being transitive). They are the alternative Frobenius complements, and disjoint, no element fixing two members of Ω.

Transitivity implies that there are as many Frobenius complements as there are members of Ω, and that, by definition, is the degree of the permutation group. So $|G| = n.|H|$ and $|K| = n$. Further, if $H = G_\alpha$, then $\{\Omega - \alpha\}$ is a disjoint union of its orbits. This implies that $|H|$ divides $(n - 1)$; so the largest Frobenius group has order not exceeding $n(n - 1)$. As Dixon and Mortimer observe, Frobenius groups are actually rather small subgroups of the symmetric group on Ω [35].

G consists of the n elements of K, plus n sets which are essentially copies of $\{H - 1\}$. Since $\alpha.K$ has n distinct values in Ω, we deduce K acts transitively on Ω. It is known that when $|H|$ is even, K will be abelian and have odd order. When it is odd, the whole group G is solvable. In both situations the centre of G is trivial. Normal subgroups of G that are not contained in the kernel K must contain it (all this in [132]). The permutation representation of G on Ω is *equivalent* to the representation afforded by the cosets of H, in which

$$H.x \mapsto H.x.g$$

(this is an elementary theorem of permutation groups—see [17], Theorem 1.3). Here, the elements of K offer a complete set of coset representatives for H—a *transversal*. This equivalence means that the mapping $k \mapsto k^h$ where $h \in (H - 1)$ and $k \in (K - 1)$ is fixed-point-free. So, all non-trivial elements of

---

[9] Proving this is not remotely straightforward—as is true for many of the "facts" stated here.

[10] That is, the normal subgroup generated by the normal nilpotent subgroups. In finite cases, as we have here, this is the unique maximal normal nilpotent subgroup of G.

H induce, by conjugation, fixed-point-free automorphisms of K. In the argot of permutation groups, one would say **H is a regular group of automorphisms of the group K**. Conversely, suppose H is some regular group of automorphisms of the group K; then the associated semi-direct product $(K \rtimes H)$ will be found to be a Frobenius group with kernel K and complement H. Sylow subgroups of Frobenius complements are either cyclic or generalised quaternion (compare Lemma 165). Even order Frobenius complements have unique involutions; odd order Frobenius complements have unique subgroups with order p for all primes dividing the group order [72]. Moreover, subgroups with order p^2 or pq (p,q, primes) are cyclic in these groups; but the reverse is untrue—the non-solvable group SL(2,13) has this property without being a Frobenius complement. It is, however, true that the above property characterises finite solvable Frobenius complements, which is what concerns us here (see [126]).

From this, and Lemmata 163 and 181, we deduce that in finite Dickson near-fields, the subgroup of non-zero elements is a Frobenius complement. As it happens, the group is a Frobenius complement in non-Dickson cases. That is clearly the case for solvable non-Dickson cases, but non-solvable cases can occur (see [62]), and these turn out to involve non-solvable Frobenius complements (see Sect. 3.6.3). Frobenius complements are not always solvable: SL(2, 5) is a Frobenius complement which isn't [134]. In fact, there is a theorem of Zassenhaus which says that SL(2, 5) is the only Frobenius complement which is perfect [72]. Dixon and Mortimer observe that in the non-solvable case, the group must have but a single non-abelian composition factor,[11] and that is A_5, ([35]; see page 163). The dihedral groups which are Frobenius are precisely the ones with order 2 m, where m is an odd number [132].

3.6.2 *Sharply 2-Transitive Groups*

Suppose G is a permutation group acting on a set Ω, and suppose that G is sharply 2-transitive (see page xviii); choose $\alpha \in \Omega$, and set $H = G_\alpha$. Then, if $g \notin G_\alpha$, we have

$$\left(G_\alpha \cap (G_\alpha)^{g^{(-1)}}\right) = \left(G_\alpha \cap G_{(\alpha.g)}\right) = \left(G_{\alpha,\alpha.g}\right) = \{Id_G\}$$

making H a Frobenius complement. This means that sharply 2-transitive permutation groups are the same thing as doubly transitive Frobenius groups. According to [35, Theorem 3.4b], in these groups K is elementary abelian. It may be helpful to observe that Frobenius groups are not necessarily 2-transitive. In fact, Wielandt [143] showed that finite $\frac{3}{2}$-transitive permutation groups were *either* primitive or Frobenius, and it's a standard exercise in permutation groups to show that doubly

[11] Recall the Jordan-Hölder theorem from group theory.

transitive groups are primitive.[12] It's also a standard exercise to show that for transitive groups primitivity is equivalent to the point stabilisers, $\{G_\alpha\}_{\alpha\in\Omega}$, being maximal subgroups.

Otto Schmidt gave the first example of a Frobenius group in which K was not abelian (the group therefore being imprimitive). His group has order 1029, with kernel of order 343. The group is described more fully in [132, Exercises 8.5]. More examples of such groups are given in [35]. There are infinite Frobenius groups but most of these ideas don't apply to them. For instance, K does not have to be nilpotent, or even a subgroup [35].

All reasonably advanced texts on group theory cover Frobenius groups, Gorenstein [55], or Robinson [132], or Suzuki [139], or Cameron [17], for instance. I have found it quite hard to get a satisfyingly complete story from any one source, though, retrospectively, one can usually appreciate that it has in fact been covered in any of them.

3.6.3 Affine Groups

Finite near-fields are directly paired with 2-transitive Frobenius groups in a way that I shall now describe. I use some of Notation 3 in what follows. First, recall that a sharply 2-transitive permutation group is the same thing as a 2-transitive Frobenius group (observed above, noted in [35, page 88]).

3.6.4 Near-Fields to Sharply 2-Transitive Groups

Start with a finite near-field $(F,\ +,\ \cdot)$. Then, the collection of all mappings of the form

$$x \longmapsto ((x \cdot A) + B) \tag{3.75}$$

where $A, B \in F$, *and* $A \neq 0$, forms a group of **affine mappings** (compare (D.1), page 517). This group acts *sharply 2-transitively* on the set F (Notation 3). For choosing $x, y, v, w, \in F$, we have

$$A = \frac{(v - w)}{(x - y)}$$

$$(x - vA) = B = (y - wA)$$

[12] $\frac{3}{2}$-transitive permutation groups are transitive groups G acting on sets Ω in such a way that for each $\alpha \in \Omega$, $G_\alpha \neq \{\mathrm{Id}\}$, and the subgroup $\{G_\alpha\}$ acts transitively on $\{\Omega - \alpha\}$.

The group is known as the **one-dimensional affine group over F** (see page 517). To be explicit

We have gone from a field to a sharply 2-transitive permutation group.

The set acted on by the group is F. One can do the same thing for notionally weaker structures—*near-domains* (Sect. C.1.8); but in finite cases, at least, these are just near-fields anyway. In the literature the notation for this and related groups is uniformly horrible.

3.6.5 Further Affine Groups

I shall use notation based on that of [35] here, and in what follows, calling this first group $\mathbf{AGL}_1(F)$. Now consider linear transformations $\theta \in \mathrm{GL}(m, F)$ vectors $v \in F^m$ and **affine transformations** in m-dimensional vector space

$$\tau_{(\theta,v)} : u \longmapsto (\theta(u) + v) \tag{3.76}$$

These transformation constitute the m**-dimensional affine group over** F,

$$\mathbf{AGL}_m(F) = \{\tau_{(\theta,v)}\}_{\{\theta,\in GL(m,F),v\in F^m\}} \tag{3.77}$$

The group is a 2-transitive subgroup of the group $\mathrm{Symm}(F^m)$ (defined page xviii). It is a split extension of a regular normal subgroup of the form $\{\tau_{(I,v)}\}_{\{v\in F^m\}}$ (I the identity transform) by $\mathrm{GL}(m, F)$ [35]. Each field automorphism, σ, induces a permutation of F^m of the form

$$u = (u_1, u_2, \ldots, u_m) \overset{\tau_\sigma}{\longmapsto} (\sigma(u_1), \sigma(u_2), \ldots, \sigma(u_m)) = \tau_\sigma(u) \tag{3.78}$$

and the set

$$\{\tau_\sigma\}_{\{\sigma\in\mathrm{Aut}(F)\}} \tag{3.79}$$

is a subgroup of $\mathrm{Symm}(F^m)$ isomorphic to Aut(F). I define

$$\mathrm{A\Gamma L}_m(F) = \left\langle \{\tau_\sigma\}_{\{\sigma\in\mathrm{Aut}(F)\}}, \mathrm{AGL}_m(F) \right\rangle \tag{3.80}$$

and call this the group of **affine semi-linear transformations**. It consists entirely of permutations of F^m taking the form

$$\tau_{(\theta,v,\sigma)} : u \longmapsto (\theta(\tau_\sigma(u)) + v) \tag{3.81}$$

where, $\theta \in \mathrm{GL}(m, F)$, $v \in F^m$, $\sigma \in \mathrm{Aut}(F)$. The reader wishing to know more about affine groups would be well advised to acquire [35].

3.6.6 Sharply 2-Transitive Groups to Near-Fields

From the other side, suppose we have a finite sharply 2-transitive permutation group G acting on a set $N = \{n_0, n_1, \ldots, n_{(n-1)}\}$. I write $0 = n_0$, $1 = n_1$ and form the subset $\mathcal{A} \subset G$ consisting of the identity permutation plus all permutations that have no fixed points. Of course, this is simply the Frobenius kernel of the Frobenius group. Suppose the permutation

$$\begin{bmatrix} 0\,1 \ldots x \ldots n_{(n-1)} \\ b\,c \ldots y \ldots \quad z \end{bmatrix}$$

is in $\mathcal{A}$. Then define

$$y = x + b$$

Form the subset $\mathcal{H} \subset G$ consisting of permutation elements fixing 0, which can be identified with the Frobenius complement of the Frobenius group. Suppose

$$\begin{bmatrix} 0\,1 \ldots x \ldots n_{(n-1)} \\ 0\,c \ldots y \ldots \quad z \end{bmatrix}$$

is in $\mathcal{H}$. Then define

$$y = (x \cdot c)$$

These prescriptions will be found to define a finite near-field $(N, +, \cdot)$. (See [62] for more details.) The additive group is the Frobenius kernel and the multiplicative group the Frobenius complement.

These two processes are dual to one another. Finite near-fields emerge from finite sharply 2-transitive Frobenius groups, and vice versa. The determination of the finite near-fields is equivalent to the determination of the finite sharply 2-transitive permutation groups. This result is expressed in [35] as Corollary 7.6A. The work was done in the early twentieth century. It was initiated by Dickson, and completed by Zassenhaus, and it probably kick-started the serious study of near-fields and near-rings as objects in their own right. The published accounts that I have seen usually view this from the perspective of group theory. The best ones I know are in [126] and in [75] and I found them hard going. There is room for a near-ring-based account if someone would like to attempt it. I suspect such an account would be easier to understand!

3.6.7 *Dickson and Non-Dickson Near-Fields*

Finite near-fields are directly related to the groups $\mathrm{AGL}_1(F)$. Dixon and Mortimer [35] state:

(i) The Dickson near-fields are the ones whose corresponding sharply 2-transitive groups lie in $\mathrm{A\Gamma L}_1(F)$
(ii) The non-Dickson near-fields have sharply 2-transitive groups lying in $\mathrm{A\Gamma L}_2(F)$

This is, again, their Corollary 7.6A.

3.7 Finite Non-Dickson Near-Fields

It is known that there are exactly seven non-Dickson finite proper near-fields—a result finally settled by Zassenhaus in the 1930s [23]. Dembowski calls them the **irregular near-fields** [32]. The Dickson near-fields are then the **regular near-fields**. Their multiplicative groups have order $(p^2 - 1)$, where the characteristic, p, can be 5, 11, 7, 23, 11, 29 or 59. Each has a Sylow 2-subgroup which is quaternion, with Q_8 occurring for primes $p = 5$, 11, 29 and 59 and S_{16} for $p = 7$ and 23. In each case, $\mathcal{D}$ is the Galois field with order p. I summarise them in Table 3.3.

An Analysis of These Cases

The near-rings featured here were numbered as cases 1–7, in Clay [23], who states that he is following the original numbering of Zassenhaus. The reader may wish to refer back to Sect. 3.5.2.

Case A SL(2, 3) is known as the binary tetrahedral group. The multiplicatively central elements form a cyclic group with order 2. They additively generate a subfield with order 5 which must be identical to $\mathcal{D}$. Some elements of $\mathcal{D}$ are not multiplicatively central, and the subgroup $\mathcal{D}^*$ is not normal in N^*. The Sylow 2-subgroups are isomorphic to Q_8 and are normal in N^* (so there is one of them). The commutator subgroup is Q_8; so $\mathcal{L} = \mathrm{SL}(2, 3)$.

Table 3.3 The finite, proper, non-Dickson near-fields

Case	Char.	Group	Solvable	Centre	Order	S_2
A	p=5	SL(2, 3)	Yes	C_2	24	Q_8
B	p=11	$(\mathrm{SL}(2, 3) \oplus C_5)$	Yes	C_{10}	120	Q_8
C	p=7	G_1 = binary octahedral	Yes	C_2	48	Q_{16}
D	p=23	$(G_1 \oplus C_{11})$	Yes	C_{22}	528	Q_{16}
E	p=11	SL(2, 5)	No	C_2	120	Q_8
F	p=29	$(\mathrm{SL}(2, 5) \oplus C_7)$	No	C_{14}	840	Q_8
G	p=59	$(\mathrm{SL}(2, 5) \oplus C_{29})$	No	C_{58}	3480	Q_8

Case B The multiplicative centre is cyclic with order 10. This additively generates a subfield with order 11 which must be identical to $\mathcal{D}$, so all of its elements are multiplicatively central. The commutator subgroup is Q_8 again, and since 8 does not divide 10, $\mathcal{L}$ must be the whole group.

Cast C The multiplicative centre is cyclic with order 2. Additively it generates a subfield with order 7 which must be $\mathcal{D}$. Not all elements of $\mathcal{D}$ will be central, and $\mathcal{D}^*$ is not normal in N^*. The Sylow 2-subgroups are quaternion with order 16 and occur as three conjugates. Q_8 is a normal subgroup of N^*. In fact, this group has exactly three normal subgroups: C_2 and Q_8 and the binary tetrahedral group, with order 24, which is the commutator subgroup of the group. $\mathcal{L}$ must be the whole group.

Case D The multiplicative centre is cyclic with order 22. It additively generates a subfield with order 23 that is identical to $\mathcal{D}$, so all distributive elements are central. $\mathcal{L}$ must be the whole group.

Case E SL(2, 5) is the binary icosahedral group.[13] Again the multiplicative centre additively generates the field $\mathcal{D}$, and not all distributive elements are central, so $\mathcal{D}^* \ntrianglelefteq N^*$. The Sylow 2-subgroups are Q_8; there are five of them. This is a perfect group—its derived subgroup is the whole group, so, obviously, this is also equal to $\mathcal{L}$. The group has a single, proper, non-trivial, normal subgroup, its centre, and the corresponding quotient group is isomorphic to A_5 (see page xviii).

Case F The multiplicative centre is cyclic with order 14. It additively generates $\mathcal{D}$ which is a subfield with order 29. $\mathcal{D}^* \cong (C_7 \oplus C_4)$, and this is not a normal subgroup of N^*. Not all distributive elements are central. Order considerations imply $\mathcal{L}$ is the whole group.

Case G The multiplicative centre is cyclic with order 58. It additively generates the subfield $\mathcal{D}$ which has order 59. All distributive elements are central. $\mathcal{L}$ must be the whole group.

Group Centres

There is a clear pattern to the centres of these groups involving the three collections: $\{A, B\}, \{C, D\}, \{E, F, G\}$. For the first collection, the corresponding quotients form isomorphic non-abelian groups with order 12. This is the group A_4. For the second, we have isomorphic non-abelian groups with order 24, and the quotient is D_4. For the final collection, $\{E, F, G\}$, the quotient has order 60 and is A_5.

[13] I think it is also known as the **binary dodecahedral group** and is a unique stem extension (a special sort of central extension) with normal subgroup C_2 and quotient A_5. More recently, it has also been called the **icosian group** in [29] and other publications of John Horton Conway, whose death, from Covid-19, was announced on the day I wrote this. The Irish mathematician, W. R. Hamilton, invented an **Icosian game**, which was sold commercially in the 1860s and involved finding a **Hamiltonian cycle** around the edges of a dodecahedron, represented physically as a peg board.

Each group in Table 3.3 has S_2 quaternion. So, in each case, the group centre would have to the form $2w$, where w is an odd number, and so that centre has the form: $(C_2 \oplus A)$, where A is an abelian group with odd order. I consider this further on page 470.

The Fantastic Four
In these non-Dickson cases, the second prescription of Lemma 169 fails. The characteristic prime is congruent to 1 modulo 4 in cases A and F (the distributivity elements are not all central in those cases).

Notice that $\mathcal{D}^* \lhd N^*$, for cases B, D and G, and $\mathcal{D}^* \ntriangleleft N^*$ for cases A, C, E and F—the only four cases in which $\mathcal{Z}^* \neq \mathcal{D}^*$ among the finite proper near-fields (see Sect. 16.7).

I refer to the near-fields in cases A, C, E and F, as **the fantastic four** in what follows. They may be alternatively characterised as exactly the finite near-fields that have non-trivial inner automorphism groups (page 46).

Distinctiveness
None of the above groups are supersolvable, three not even being solvable, providing an intrinsic distinction between Dickson and non-Dickson near-fields because in Dickson near-fields N^* is supersolvable (Corollary 182).

Another, related, intrinsic distinction is that the non-Dickson near-fields are exactly the proper (finite) near-fields for which the multiplicative semi-group of non-zero elements has a quaternion Sylow 2-subgroup, but no normal 2-complement.

Finally, the finite non-Dickson near-fields are precisely the finite near-fields for which the commutator subgroup is non-cyclic (Sect. 3.3.6 would imply from that that the Sylow 2-subgroups of such near-fields cannot be cyclic).

The group SL(2, 3) is a subgroup of SL(2, 5) and a normal subgroup of the binary octahedral group. Technically, it is *involved* (page 128) in every group given in Table 3.3.

3.7.1 A Classification Lemma

It seems that the key determinant for finite proper near-fields is the commutator subgroup.

Lemma 198 *Suppose* $(N, +, \cdot)$ *is a finite, proper near-field.*

(i) The near-field is Dickson if and only if the multiplicative commutator subgroup, $[N^*, N^*]_{(mult)}$, *is cyclic.*
(ii) The near-field has cyclic Sylow p-subgroups for all primes p, if and only if $[N^*, N^*]_{(mult)}$ *has odd order.*

(iii) *When the order of* $[N^*, N^*]_{(mult)}$ *is even, all odd prime Sylow subgroups are cyclic, but* S_2 *is generalised quaternion.*
(iv) *The near-field is Dickson if and only if its multiplicative group is metacyclic.*

3.7.2 Element Orders

In Dickson near-fields with order p^α and α even, there is an element with multiplicative order equal to $(p+1)$ (Sect. 3.3.7). There are some non-Dickson near-fields with characteristic p which do not contain elements with order $(p+1)$. This is true for cases B, D, E, F and G, in Table 3.3.

3.8 General Finite Non-fields

In this section $(N, +, \wedge)$ represents *any* finite non-field. I extend some of the definitions from Sect. 3.4.6 to this more general situation, and I use Notations 160 and 170.

First, I introduce the field $(N, +, \cdot)$ which has the same size as my near-field, and I arrange that the same element of N^* is the multiplicative identity in each. Next, copying (3.49), I define

$$\mathfrak{F} = \{x \in N \ : \ (x \wedge n) = (x \cdot n), \ \forall\, n \in N\} \tag{3.82}$$

and copying (3.51), I define

$$\mathfrak{K} = \{x \in N^* \ : \ (n \wedge x) = (n \cdot x) \ \forall\, n \in N\} \tag{3.83}$$

Lemma 199

(i) $\mathfrak{F}$ *is a commutative field and if* $x \in \mathfrak{F}^*$, *then* $x^{-1} = x^{\wedge(-1)}$.
(ii) $(\mathfrak{F}, +, \cdot) = (\mathfrak{F}, +, \wedge)$.
(iii) $\mathfrak{F} \subseteq \mathcal{D}$.
(iv)

$$\left(\mathfrak{F}^* \cap \mathfrak{K}^*\right) \subseteq \mathcal{Z}^* \subseteq \mathcal{D}^* \tag{3.84}$$

(v) $\mathfrak{K}^*$ *is a non-empty, multiplicative, cyclic, group. It is a subgroup both of* $(N^*, \cdot)$ *and* $(N^*, \wedge)$.

Proof

(i) Working with $x, y \in \mathfrak{F}$

$$\begin{aligned} ((x \cdot y) \cdot n) &= (x \cdot (y \cdot n)) = (x \cdot (y \wedge n)) \\ &= (x \wedge y \wedge n) = ((x \cdot y) \wedge n) \end{aligned}$$

making $(\mathfrak{F}, \cdot)$ a commutative monoid.

$$((x + y) \wedge n) = ((x \wedge n) + (y \wedge n))$$
$$= ((x \cdot n) + (y \cdot n)) = ((x + y) \cdot n)$$

giving additive closure.

(v) $1 \in \mathfrak{K}^*$. If $x, y \in \mathfrak{K}^*$, then

$$((n \wedge x) \wedge y) = ((n \cdot x) \cdot y) = (n \cdot (x \cdot y)) = (n \cdot (x \wedge y))$$

which means $(\mathfrak{K}^*, \cdot)$ is a cyclic subgroup of the cyclic group $(N^*, \cdot)$.

□

In principle, the centre of the group $(N^*, \wedge)$ contains two sorts of element: first, those x for which $(x \wedge a) = (a.x) = (a \wedge x)$ and, second, those x for which $(x \wedge a) = (a \wedge x)$ but, for some $a \in N^*$, $(a \cdot x) \neq (a \wedge x)$. These last elements, if they exist, will not be in the intersection given in (iv), above.

The group $\mathfrak{K}^*$ does not necessarily contain $[N^*, N^*]_\wedge$. When it does, that group is cyclic, and we have examples such as case E of Table 3.3 in which this is not the case. We do know that the group is non-trivial, at least in odd prime characteristics.

This decomposition applies to any finite non-field. Fixing the corresponding field to have order p^α, I fix the order of $\mathcal{D}$ as p^β where $\alpha = (\beta.\delta)$ and then $\mathfrak{F}$ is a sub-field of that. $\mathcal{D}^*$ contains the multiplicative centre, $\mathcal{Z}^*$, which has even order and is its normal core. The two structures, $\mathcal{D}$ and $\mathcal{Z}$, are equal precisely when $\mathcal{D}^* \lhd (N^*, \wedge)$ (Lemma 161). The order of $\mathfrak{K}^*$ is a divisor of $(p^\alpha - 1)$. I shall set it at k and write $(p^\alpha - 1) = k.d$.

We can choose a cyclic generator Π of the cyclic group $(N^*, \cdot)$. This has order $(p^\alpha - 1)$, and the cyclic group $\mathfrak{K}^*$ is generated by Π^d. Elements of N^* are uniquely expressible as powers of Π, and

$$\left(\Pi^t \wedge \Pi^{d.r}\right) = \Pi^{(t+d.r)} \tag{3.85}$$

The subgroup $\mathfrak{K}^*$ offers two partitions of N^*—the left and the right cosets. The $(\wedge)$ product is described by specifying

$$(\Pi^a \wedge k_1) \wedge (\Pi^b \wedge k_2) \tag{3.86}$$

for all possible cases.

It will be seen that that requires us merely to specify

$$\Pi^s \wedge \Pi^b \tag{3.87}$$

for $1 \leq s \leq (p^\alpha - 1)$ and for $1 \leq b < d$.

Of course, the cyclic group $\mathfrak{K}^*$ may be very small in non-Dickson cases, so this may not be terribly helpful.[14] In fact, from Table 3.3, in case E, it is equal to C_2.

3.9 Infinite Near-Fields

I am mostly concerned with finite structures in this book but should at least mention infinite near-fields, and I do that here.

Pilz [128] attributes the following construction to Zemmer. Start with a commutative field, $(F, +, \cdot)$, and let $F(x)$ be the corresponding rational function field. Define a map

$$\Phi : F(x) \longrightarrow \mathrm{Aut}(F(x))$$
$$\Phi\left(\frac{f(x)}{g(x)}\right) \longmapsto \phi_{\frac{f(x)}{g(x)}}$$
$$\phi_{\frac{f(x)}{g(x)}}\left(\frac{p(x)}{q(x)}\right) = \frac{p\,(x + f(x) - g(x))}{g\,(x + f(x) - g(x))} \tag{3.88}$$

This is a *coupling map*, and from it one can produce an infinite Dickson near-field. There is an infinite non-Dickson near-field for each prime characteristic [59].

Theo Grundhöfer also proved, in 1989, that an infinite near-field, finite dimensional over its "kernel" (page 108), and with a solvable multiplicative group, is Dickson. All finite near-fields are planar. Clay says that Zemmer was the first to discover an infinite near-field which is not [51].

3.9.1 Characteristic Zero

Unlike finite near-fields, infinite near-fields may have characteristic zero. If $(N, +, \cdot)$ is just such a near-field, and $n \in \mathbb{N}$, then $\overbrace{(1+1+1\ldots+1)}^{\text{(n times)}}$ is invertible, and so, given any $x \in N$, we can find some $y \in N$ such that $ny = x$. This means that the additive group $(N, +)$ is **divisible**. It is also **torsion-free**. Scott, and most other basic group theory texts, says that a torsion-free divisible abelian group is a direct sum of copies of the additive rational numbers, $\mathbb{Q}$ [137].

[14] Berliners might describe it as a *schlimmbesserung*—an improvement that makes things worse!

3.10 A Continuing Story

I felt I should treat near-fields in this book and got rather submerged in them for several years. As I worked on them, it became clear to me that a series of chance events had made them less accessible than might have been the case, and I hope this chapter helps a bit, because I suffered from that myself. There is clearly more to be done on this topic, but the possibilities for a clear and straightforward account seem high.

Professor Pilz kindly informed me that there is a new and simple determination of the finite near-fields using (what else!) Zsigmondy's theorem. I have not had time to assimilate that, but the reader interested in near-fields should find it available, free of charge, on the Internet [60].

Part II
Near-Rings Hosted by Classes of Groups

Chapter 4
Near-Rings on Groups with Low Order

Near-rings are finite, right-distributive and zero-symmetric in this chapter.

Any ring is a near-ring, and intuitively, the more abelian a group is, the greater number of isomorphism classes of near-rings it may host[1] because normal subgroup richness may impose fewer restrictions on annihilators.

This chapter is almost exclusively concerned with near-rings hosted by non-abelian groups. As previously remarked (see Sect. 1.20), groups hosting such near-rings need to have elements whose additive orders are equal to the group exponent (the condition is necessary but insufficient). In this chapter, I shall refer to such groups as **I-appropriate groups**. I shall also demand that our near-rings be zero-symmetric (Definition 5).

Not all I-appropriate groups host near-rings with identities. In particular, all finite p-groups are I-appropriate, but Q_8, for example, hosts no unital near-rings (Lemma 207). Obviously, if H and K are groups hosting near-rings with identity, then $H \oplus K$ is also one. The structure of transformation near-rings means that any group-theoretic property which might debar unital, zero-symmetric, near-rings must be one, such as I-appropriateness, which is not preserved by sufficiently many direct summands. The next lemma follows since right near-ring multiplication induces additive endomorphisms.

Lemma 200 *Suppose $(G, +)$ hosts a near-ring with left identity e and that H is a proper, fully invariant, subgroup of G. Suppose, for some positive integer m, that $me \in H$. Then:*

(i) For all $x \in G$, we have $mx \in H$.
(ii) If e has finite additive order, then the order of me is strictly smaller than that.

[1] A piece of folklore that may well be more quantifiable .

R. Lockhart, *The Theory of Near-Rings*, Lecture Notes in Mathematics 2295,
https://doi.org/10.1007/978-3-030-81755-8_4

I have sometimes used Version 4.4.12 of the GAP computer package to determine groups and exponents. When I list the various non-isomorphic groups of a given order, the order given usually conforms to that offered by GAP although my numbering may be different from that of GAP because I may already have rejected or accepted some groups (Sects. 4.1 and 5.14).

4.1 Small Non-abelian Groups

Table 4.1 lists the number of I-appropriate non-abelian groups with order up to 30. There is exactly one non-abelian group with order 6, S_3. Its exponent is 6 and it is not I-appropriate. There are two non-abelian groups with order 8—the quaternion group with order 8 and D_4. Both are I-appropriate. I show in this chapter that the former hosts no near-rings with identity (Lemma 207), whereas the latter hosts seven non-isomorphic such near-rings (Lemma 209). D_5 is the sole non-abelian group with order 10. It is not I-appropriate (see Sect. 5.5.2). There are three non-abelian groups with order 12. One has exponent 12 and so is not I-appropriate (it is the dicyclic group Dic_3). The others are A_4 and D_6 and both have exponent 6. A_4 has elements with order 1, 2, 3 or 4 and is not I-appropriate. That leaves D_6 which is I-appropriate and which hosts exactly one near-ring with identity (Lemma 211). The exponent 12 group is a semi-direct product of C_3 by C_4: $(C_3 \rtimes C_4)$. Both constituent groups host rings with identities, so this is not a property carried over automatically to semi-direct products. There is one non-abelian group with order 14. This is D_7; it has exponent 14 and is not I-appropriate. There are nine non-abelian groups with order 16. Four of them have exponent 8 and the remaining five have exponent 4. All are I-appropriate. However, one of this list is the dihedral group D_8 which is known not to host zero-symmetric near-rings with identity (Lemma 210). I therefore list eight of these groups below, in GAP order but not with GAP numbering. There are three non-isomorphic non-abelian groups with order 18. One has exponent 18 (it is D_9) and is not I-appropriate. The other two both have exponent 6. Only one of these is I-appropriate: it is the group $(S_3 \oplus C_3)$ (otherwise: $(D_3 \oplus C_3)$). There are three non-abelian groups with order 20. These have exponents 20, 20 and 10. Only one is I-appropriate, that is, D_{10}, and it hosts exactly one near-ring with identity, up to isomorphism (Lemma 211). There is one non-abelian group with order 21. Its exponent is 21, so the group is not I-appropriate. Similarly there is only one non-abelian group with order 22. It is D_{11}, has exponent 22 and is not I-appropriate.

There are 12 non-abelian groups with order 24. One has exponent 24, eight have exponent 12, and three have exponent 6. Five of the exponent 12 groups are I-

Table 4.1 Non-abelian I-appropriate groups with order less than 32

Order	6	8	10	12	14	16	18	20	21	22	24	26	27	28	30
Number	0	2	0	1	0	9	1	1	0	0	7	0	2	1	0

appropriate and two of the exponent 6 groups are. One of the I-appropriate exponent 6 groups is ($D_6 \oplus C_2$), which certainly hosts near-rings with identity. One of the I-appropriate exponent 12 groups is D_{12}, which does not host near-rings with identity (Lemma 210). The single non-abelian group with order 26 is D_{13}. It has exponent 26 and is not I-appropriate. There are two non-abelian groups with order 27. These have exponents 3 and 9, respectively, and are both I-appropriate. There are two non-abelian groups with order 28. One has exponent 28 and is not I-appropriate. The other is D_{14}. It has exponent 14 and (Lemma 211) hosts exactly one near-ring with identity, up to isomorphism. There are three non-abelian groups with order 30. All three have exponent 30 and are therefore not I-appropriate.

We now have eight groups of order 16, one of order 18, five of order 24 and two of order 27 which could conceivably host near-rings with identity and whose status in that regard is undetermined—a total of 16 groups, in all. Incidentally, most of the group orders between 31 and 39 involve square-free numbers. So, their status with respect to hosting unital proper near-rings is known. Only groups with order 32 or 36 are problematic.

4.1.1 Groups with Order 16

There are eight groups to consider. I label them $\{G_j\}_{j=1}^{8}$.

First Group

Our first non-abelian group with order 16 has exponent 4 and presentation $G_1 =$

$$\langle a, b, c \ : \ 4a = 2b = 2c = 0, a + b = b + a, b + c = c + b, c + a - c = a + b\rangle$$

Elements of this group may be uniquely expressed in the form

$$(ma + nb + tc)$$

where $0 \leq m < 4, 0 \leq n < 2, 0 \leq t < 2$. The centre of the group is the subgroup

$$\{0, b, 2a, (2a + b)\}$$

The commutator subgroup of the group is $\langle b\rangle_{(add)}$, and this would be a right ideal in any near-ring. We know that

$$\begin{aligned}(c + a + c) &= (a + b)\\ (rc + ma + rc) &= (ma + mrb)\end{aligned}$$

The collection of elements with order 4 is

$$\{(ma + nb + rc) \ : \ m \equiv 1 \ (modulo\ 2)\ , 0 \leq n \leq 1, 0 \leq r \leq 1\}$$

There are eight such elements. I attempt to define a zero-symmetric near-ring in which a is the identity element with the prescription

$$(d \cdot x) = \begin{cases} 0 & \text{if } |x| = 2 \\ d & \text{if } |x| = 4 \end{cases} \tag{4.1}$$

where $d \in \{b, c\}$. I believe this is indeed a near-ring, actually, a distributively generated near-ring (see Definition 75), and that verifying associativity is straightforward (Sect. 2.1.1). On that basis, I conclude that the group does host near-rings with identity.

Second Group

Our second non-abelian group with order 16 has exponent 4 and presentation

$$G_2 = \langle a, b : 4a = 4b = 0, b + a - b = 3a \rangle$$

The group is a semi-direct product of C_4 with itself. Group elements may be expressed uniquely in the form

$$(ua + vb) \tag{4.2}$$

where $0 \leq u, v < 4$. The group centre is $\{0, 2a, 2b, (2a + 2b)\}$. The commutator subgroup is $\{0, 2a\}$. There are three elements with order 2 and they are central. Each of the remaining, non-trivial, elements has order 4, and

$$(b + a) = (-a + b)$$

Suppose a is the multiplicative identity. Then,

$$(2a \cdot b) = 2b$$

but

$$(2a \cdot b) \in \{0, 2a\} \; (\textit{commutator subgroup})$$

So this is impossible.

I define a product with the prescription

$$(a \cdot (ua + vb)) = va \tag{4.3}$$

and assume b to be the multiplicative identity. I believe this is a zero-symmetric near-ring. On that basis I assert the group does host near-rings with identity. Again, the near-ring is d.g. It is not bi-distributive.

Third Group
Our third non-abelian group with order 16 has exponent 8 and presentation

$$G_3 = \langle a, b : 8a = 2b = 0, b + a - b = 5a\rangle$$

It is a nilpotent member of the Krimmel class (Sect. 5.6). The commutator subgroup is $\langle 4a\rangle_{(add)}$. According to Theorem 215, the group does host zero-symmetric near-rings with identity a.

Fourth Group
Our fourth non-abelian group with order 16 has exponent 8 and presentation

$$G_4 = \langle a, b : 8a = 2b = 0, b + a - b = 3a\rangle$$

It is sometimes referred to as a semi-dihedral group. It is also known as the generalised semi-linear group of degree one over the Galois field of order 9 (see Appendix D, or [4], or [125]) and, as such, has the form

$$K^* \rtimes \mathrm{Aut}(K)$$

where, in this case, K is the Galois field of order 9 and Aut(K) is the group of field automorphisms. According to Sect. 5.10, this group does not host near-rings with identity.

Fifth Group
Our fifth non-abelian group with order 16 has exponent 8 and presentation

$$G_5 = \langle a, b : 8a = 0, 2b = 4a, a + b + a = b\rangle$$

This is a generalised quaternion group. As such, it hosts no near-rings with identity (see Sect. 5.7).

Sixth Group
Our sixth non-abelian group with order 16 has exponent 4.
It is a direct sum of D_4 and C_2:

$$G_6 = D_4 \oplus C_2$$

Both constituents of this direct sum host near-rings with identity, so the group itself does.

Seventh Group
Our seventh non-abelian group with order 16 has exponent 4. It is a direct sum of Q_8 and C_2 and is a Hamiltonian group (page 208):

$$G_7 = Q_8 \oplus C_2$$

We might write the presentation as

$$G_7 = (\langle e, y : 4e = 0, 2e = 2y, (-y + e + y) = -e\rangle \oplus \langle z : 2z = 0\rangle)$$

The elements with order 2 or 1 form the set

$$S = \{2e, z, (2e + z), 0\} \tag{4.4}$$

and this is the additive centre of G. There is no loss in generality in looking only for zero-symmetric near-rings in which e is the identity element. In all such near-rings, the additive centre of the group would be a (two-sided) sideal, S, consisting of the four group elements which do not have order 4. The other 12 non-trivial elements all have order 4. We know

$$(e + y) = (y - e)$$

The commutator subgroup of the group is $\langle 2e\rangle_{(add)}$. This group possesses a considerable amount of structure and is quite hard to manipulate. I was forced to resort to programming but still could not find any zero-symmetric unital near-rings. My initial attempt at a program exploited what I knew about the structure of possible near-rings and attempted an exhaustive search—something I came to realise was way beyond the capabilities of the PC I had. I then wrote a program that attempted random sampling of the possible p.n.r. products. I found no near-rings, however, the asymptotics mentioned in Sect. 6.11 simply do not exist, my program ran for days without terminating, and I have little idea whether a random searching policy of this kind could be expected to answer the question with any reasonable degree of confidence. I confess that, probably superstitiously, after 5 days I began to doubt that one could be found. It is unfortunate that this is the case. The group is quite small but requires three additive generators. The constraints imposed on searching for p.n.r. do not seem sufficiently restrictive to make the search for near-rings practicable with the resources I have (see page 76). On that basis, the status of this group remains undetermined by these direct methods.

Information from page 68 suggests that if G_7 does host a zero-symmetric unital near-ring, then it contains a proper non-trivial, maximal ideal J and the options are:

(i) $J = \langle 2e\rangle$ and $\frac{G_7}{J} \cong (C_2 \oplus C_2 \oplus C_2)$.
(ii) $J = \langle 2e, z\rangle$ and $\frac{G_7}{J} \cong (C_2 \oplus C_2)$.
(iii) $J = \langle y, z\rangle$ and $\frac{G_7}{J} \cong (C_2)$.

Analysis:

(i) By hypothesis, $J = \{0, 2e\}$ is a maximal two-sided ideal. The annihilator of the two-sided sideal $S = \{0, 2e, z, (z+se)\}$ is a two-sided ideal and it contains J, so it is J. This means $(z \cdot 2e) \neq 0$, but, of course, $(z \cdot 2e)^2 = 0$. We obtain $(z \cdot 2e) = 2e$. Then, $(z - e) \cdot 2e = 0$, but $(z - 2e) \notin J$. This case cannot occur.

(ii) Now, $J = \{0, 2e, z, (z + 2e)\}$ and is maximal. The additive group $\{G_7 \cdot z\}$ is a subgroup of S and has order 2 or 4. Suppose it has order 2. Then, $\{0, z\}$ is a left sideal, so its annihilator is an ideal, and that ideal has order 8. That annihilator certainly contains $2e$. If it does not contain z, then it must be Q_8, but since $e \cdot z \neq 0$, this is impossible. Therefore, that annihilator contains J, but J is maximal. The conclusion is that $\{G_7 \cdot z\} = J$. Now, the annihilator of z has order 4 and we know it contains $2e$. If it happens that $z \notin \binom{z}{0}_{\mathcal{L}}$, then this annihilator is cyclic with order 4. Without loss of generality, we may assume that this left ideal is generated by y. We have two possibilities.

$z^2 \neq 0$ Then, $\binom{z}{0}_{\mathcal{L}} = \{0, y, 2e, (y+2e)\}$. In this case, $\{J \cdot z\}$ is a left sideal and subgroup with order 2. Its annihilator is a two-sided ideal with order 8. This ideal cannot be Q_8, so it includes z, which is a contradiction.

$z^2 = 0$ Then, $\binom{z}{0}_{\mathcal{L}} = \{0, z, 2e, (z + 2e)\}$. The factor near-ring $\frac{G_7}{J}$ is a *field* hosted by the additive group $(C_2 \oplus C_2)$ (page 69). This means its non-zero elements are power of a single primitive element. In this case we could take that element to be the coset $\{y + J\}$. Then,

$$y^2 = (y + e) + j \textit{ for some } j \in J$$

Consequently,

$$(y + e + j) = (e + y + e) \cdot y = (y + y + e + j + e)$$

which is nonsense.

(iii) In this situation, the maximal ideal J is

$$\{y, 2e, (y + 2e), z, (z + 2e), (z + y), (z + y + 2e), 0\}$$

Consider the additive subgroup $H = G_7 \cdot y$. This contains y and has order at least 4. It lies in J and so has order no more than 8. We know

$$H \cong \left(\frac{G_7}{\binom{y}{0}_{\mathcal{L}}} \right)$$

If H has order 8, then the annihilator of y must have order 2. It does not contain $2e$, so, without loss of generality, it must be $\{0, z\}$. Since $z \in H$, there must be some element $g \in G_7$ such that $g \cdot y = z$. Furthermore, this element must have order 4. It seems this element is either y or $(y + 2e) = -y$. In that latter case, one deduces that $y^2 = (z + 2e)$. We know $(e + y + e) = y$. If $y^2 = z$, this implies

$$(y + z + y) = z$$

If $y^2 = (z + 2e)$, the implication is

$$(y + (z + 2e) + y) = (z + 2e)$$

Both are impossible. We are left to conclude that H has order 4. It must be that

$$H = \{y, 2e, (y + 2e), 0\}$$

The annihilator of y must have order 4 (and be a left ideal). The element $2e$ is not in this annihilator, but it has to have an element with order 2, so, without loss of generality, it must contain z. To have order 4 in this group and not contain $2e$ requires the annihilator to have an element with order 4. But then it would again have to contain $2e$. This case is impossible.

It seems to me that G_7 does not host zero-symmetric, unital, near-rings.[2]

Eighth Group
Our eighth non-abelian group with order 16 has exponent 4 and presentation $G_8 =$

$$\langle a, b, c \ : \ 4a = 2b = 0, 2a = 2c, b + a - b = -a, b + c = c + b, a + c = c + a\rangle$$

It is sometimes described as a central product of D4 and C4. The additive commutator subgroup of G_8 is $\langle 2a\rangle$ and the additive centre is the subgroup

$$\{0, c, 2c, 3c\} = \{0, c, (2a + c), 2a\}$$

We have that

$$(b + a) = (-a + b)$$

and that elements of G_8 may be uniquely written in the form

$$(ma + nb + rc)$$

where $0 \leq m < 4, 0 \leq n < 2, 0 \leq r < 2$. Such elements have order 4 precisely when $b = 0$, $m \equiv 1$ *(modulo 2)*. If G_8 does host near-rings with identity, we lose no generality in assuming that identity to be a.

Then,

$$-a = (b + a - b) \Rightarrow -c = (b.c + c - (b.c)) = 0$$

so G_8 does not host near-rings with identity.

[2] I'd like to think of this as proof by sheer exhaustion!

4.1.2 *Groups with Order 18*

We may write the group that we need to investigate as

$$G = D_3 \oplus C_3$$

where D_3 has the presentation given in (2.19) (with n = 3), and $C_3 = \langle y \rangle$. We regard elements of G as pairs of elements (w, v), $w \in D_3$, $v \in C_3$. We may assume that G hosts a near-ring $(G, +, \cdot)$ in which the identity is $e = (b, y)$, and $-e = (b, 2y)$. The group has commutator subgroup given by

$$\{(0, 0), (a, 0), (2a, 0)\}$$

Notice that the set $\{(b, y), (a, 0)\}$ additively generates the group $(G, +)$. We know, by hypothesis, that $e \cdot x = x$ *for all* $x \in G$. We can determine the near-ring by fixing the set $\{(a, 0) \cdot x : x \in G\}$, and since commutator subgroups are right ideals (see Remark 14), we know that set is contained in $\langle (a, 0) \rangle$. I define:

$$(a, 0) \cdot x = \begin{cases} (a, 0) & \text{if } x = \pm e \\ (0, 0) & \text{otherwise} \end{cases}$$

and

$$(0, y) \cdot x = 4x$$

$$(b, 0) \cdot x = 3x$$

This represents a near-ring with identity hosted by $(G, +)$.

4.1.3 *Non-abelian Groups with Order 21*

The smallest non-abelian group with odd order has order 21. This is a solvable, non-nilpotent group. Its exponent is equal to its order, so it is not I-appropriate. The group has trivial centre. Six of its elements have order 7, and 14 have order 3. Its commutator subgroup is its unique proper non-trivial normal subgroup, and that has order 7. It has seven conjugate cyclic subgroups with order 3. It has the presentation

$$\mathcal{G} = \langle a, b : 3a = 7b = 0, -a + b + a = 2b \rangle$$

The seven Sylow 3-subgroups have the form $\langle a + rb \rangle_{r=0,1,2\ldots6}$ and

$$\langle a + rb \rangle^{-b} = \langle a + (r + 1)b \rangle$$

Elements of the group may be uniquely expressed in the form

$$(ma + nb)$$

where $0 \leq m \leq 2$ and $0 \leq n \leq 6$. The group clearly hosts no near-rings with identity and the commutator subgroup $\langle b \rangle$ is a right ideal in any near-ring which it does host. With current techniques, it is computationally difficult to determine *all* the isomorphism classes of near-ring hosted by a particular group, and one is usually forced into further restrictions, such as demanding a multiplicative identity. Even strong restrictions of this sort typically lead to lengthy error-prone calculations. One imagines that these computations would be relatively easier in groups with very simple normal subgroup structure, as is the case with this group.[3] I shall not determine all the near-rings hosted by $\mathcal{G}$, but I shall look at some of them on the basis that the group is itself interesting, so determining some of its near-rings is of interest, and that the flavour of the calculation involved may be worth setting down. It may be that these ideas could be generalised to encompass the entire class of non-abelian groups with order pq.[4] Although there is not the space to attempt such a calculation here, it does not seem entirely intractable.

If the commutator subgroup should be an ideal of the near-ring, then for $0 \leq r \leq 6$

$$\mathcal{G} \cdot rb \subseteq \langle b \rangle$$

and a simple application of Lagrange's theorem shows that

$$\mathcal{G} \cdot \langle b \rangle = \{0\} \tag{4.5}$$

that is, the commutator subgroup right annihilates the near-ring. I investigate zero-symmetric near-rings hosted by $\mathcal{G}$ which possess an element x such that, additively,

$$\mathcal{G} \cdot x \cong C_3$$

In such situations the left annihilator of the element x is exactly the subgroup $\langle b \rangle$, and so (4.5) applies. From that, if follows that if $z \in \mathcal{G}$ has additive order equal to 3, then $z \cdot x \neq 0$. From (4.5) one concludes that x is an element with order 3. In particular, its powers are non-zero. We may write

$$\mathcal{G} \cdot x = \langle y \rangle$$

[3] This dictum was often offered by Robert Laxton, when I worked with him (see "folklore" on page 171).

[4] Recall that there will be a non-abelian group of order pq with $p > q$ precisely when $p \equiv 1 \pmod{q}$ and that any two such groups are isomorphic.

where $y \in G$ has additive order 3. We know that $\{\mathcal{G} \cdot x \cdot x\} \neq \{0\}$ and infer

$$\langle y \rangle \cdot x = \langle y \rangle$$

Since $\mathcal{G} \cdot \mathcal{G}$ contains elements with order 3 (such as y), we know

$$0 \neq \mathcal{G} \cdot \mathcal{G} \cdot x \subseteq \mathcal{G} \cdot x = \langle y \rangle$$

and conclude

$$\mathcal{G} \cdot \langle y \rangle = \langle y \rangle$$

Without loss of generality, one may assume $x^2 = y$. An immediate consequence of this is that

$$x \cdot y = y \cdot x \neq 0$$

With the above assumption

$$\mathcal{G} \cdot y = \langle y \rangle$$

and $x^3 \in \{y, 2y\}$. I investigate the possibility that

$$\mathcal{G} \cdot (2y) = \{0\}$$

and only consider near-rings of this sort. For such near-rings, $x \cdot y \cdot x \neq 0 \Rightarrow y \cdot x = x \cdot y = y$; and we know $y^2 \in \{y, 2y\}$. If $y^2 = 2y$, then

$$y = (2y) \cdot y = y^2 \cdot y = y \cdot y^2 = 0$$

and one has to conclude that $y^2 = y$ and so $y^3 = y$. Because $2y = 2x \cdot y$, it has to be the case that no element in $\mathcal{G} \cdot (2x)$ can have additive order 3 (such elements cannot left annihilate y). This means

$$\mathcal{G} \cdot (2x) = \{0\}$$

We seek to determine all values of $a \cdot w$ and of $b \cdot w$, as w ranges through $\mathcal{G}$. The first set of values represents group elements with additive order 3 or 0 and the second group elements with order 7 or 0. The elements of $\mathcal{G}$ having order 3 are of two types: those of the form $2x+rb$ and those of the form $x+rb$, (where $r \in \{0, 1, 2, \ldots 6\}$).

[A] For the form $w = (2x + rb)$, we have

$$a \cdot w = a \cdot (2x + rb) \equiv a \cdot (2x) \pmod{\langle b \rangle}$$

This means $a \cdot w \in \langle b \rangle$ and, from order considerations, that $a \cdot w = 0$. Then

$$(-a + b + a) \cdot w \;=\; (2b) \cdot w \Rightarrow (b \cdot w) \;=\; (2b) \cdot w \Rightarrow (b \cdot w) \;=\; 0$$

The form $w = (2x + rb)$ is completely resolved:

$$\mathcal{G} \cdot (2x + b) \;=\; \{0\}$$

[B] For the form for $w = (x + rb)$, we have

$$a \cdot w \;\equiv\; a \cdot x \pmod{\langle b \rangle}$$

and we know that $a \cdot x \in \{y, 2y\}$. If $a \cdot x = 2y$, then $a^2 \cdot x = 0$, and so $a^2 = 0$, and the left annihilator of a must be the whole of the group, that is,

$$\mathcal{G} \cdot a \;=\; \{0\}$$

From this it will be seen that

$$\begin{aligned} a \cdot x = 2y \;&\Leftrightarrow\; a \;\equiv 2x \;\equiv 2y \pmod{\langle b \rangle} \\ &\Leftrightarrow\; x \;\equiv\; y \;\equiv\; (2a) \pmod{\langle b \rangle} \end{aligned}$$

If $a \cdot x = y$, then $a^2 \cdot x = a \cdot y \neq 0$, so $a^2 \neq 0$, and $\mathcal{G} \cdot a \neq \{0\}$. This gives two subcases:

(i) $a \cdot x = y$, $a^2 \neq 0$, $\mathcal{G} \cdot a \neq \{0\}$, and $x \equiv y \equiv a \pmod{\langle b \rangle}$
(ii) $a \cdot x = 2y$, $\mathcal{G} \cdot a = \{0\}$, and $x \equiv y \equiv 2a \pmod{\langle b \rangle}$

Let us suppose that we can find some integer r such that

$$\mathcal{G} \cdot (x + rb) \;=\; \mathcal{G} \tag{4.6}$$

The construction conditions (page 67) tell us

$$-(a \cdot (x + rb)) + b \cdot (x + rb) + (a \cdot (x + rb)) \;=\; (2b) \cdot (x + rb)$$

This rules out subcase (ii), and so $a \cdot x = y$.

Lemma 201 *Suppose* $\theta : \{0, 1, 2, 3, 6\} \to \{0, 1, 2, 3, 4, 5, 6\}$ *is idempotent, in the sense that all image elements are fixed points. Then the group* $\mathcal{G}$ *hosts a non-trivial near-ring defined by the relations:*

(i)

$$b \cdot w \;=\; 0 \text{ for all } w \in \mathcal{G}$$

(ii)

$$a \cdot w = \begin{cases} (a + \theta(r)b) & \text{for } w = (a + rb) \\ 0 & \text{otherwise} \end{cases}$$

Proof The prescription is obviously that of a right-distributive, zero-symmetric, p.n.r. hosted by $\mathcal{G}$. The associativity core (Definition 69) of the p.n.r. certainly contains b. We really only need show that the element a is also in the associativity core, that is,

$$(a \cdot ((a + rb) \cdot (a + sb))) = ((a \cdot (a + rb)) \cdot (a + sb))$$

which follows from the fact that $\theta^2(s) = \theta(s)$. □

Two distinct near-rings formed using the above prescription will be isomorphic provided a simple congruence relation is satisfied. The interested reader will be able to derive that for themselves and will realise that the above prescription involves a subset of the near-rings hosted by $\mathcal{G}$. It is likely that there would be accessible generalisations of this result, first, to non-abelian groups with order 3p, perhaps.

4.1.4 Groups with Order 24

We need to check on five non-abelian groups.

First: The Dicyclic Group with Order 24
This has exponent 12. According to Sect. 5.8, it does not host near-rings with identity.

Second: $(S_3 \oplus \mathbf{C}_4)$
This has exponent 12. It is I-appropriate and therefore has a cyclic normal subgroup with index 2. Applying results from Sect. 5.6, it seems that this group does host zero-symmetric near-rings with identity.

Third: $(D_4 \oplus \mathbf{C}_3)$
This has exponent 12. As a direct sum of groups which host near-rings with identity, it too hosts near-rings with identity.

Fourth: $(\mathbf{Q}_8 \oplus \mathbf{C}_3)$
This has exponent 12. It is I-appropriate and therefore has a cyclic normal subgroup with index 2. Applying results from Sect. 5.6, it seems that this group does not host zero-symmetric near-rings with identity (see also Sect. 5.9).

Fifth: $(\mathbf{A}_4 \oplus \mathbf{C}_2)$
This has exponent 6. We know

$$\mathrm{A}_4 = \langle x, y \,:\, 3y = 0 = 2x, 3(x + y) = 0 \rangle$$

A_4 is a centreless group with commutator subgroup isomorphic to the Klein group. Precisely three elements have order 2, these being the distinct permutations of the form $(a, b)(c, d)$, where a, b, c, d are all different. These elements, plus the additive identity, 0, constitute the normal subgroup with order 4 which is the sole proper non-trivial normal subgroup of the group. In the presentation given, x may be taken as any order 2 element and y as a permutation of the form (a, b, c). Specifically, if we take $x = (1, 2)(3, 4)$ and $y = (1, 2, 3)$, it is possible to obtain simple identities for each of the two remaining elements with order 2 (which are $x_1 = (-y + x + y)$ and $x_2 = (y + x - y)$, respectively) and each of the remaining seven elements with order 3 (which are $2y$, $(x_1 + y)$, $(x_2 + y)$, $(y + x_1)$, $(x_1 + 2y)$, $(x + 2y)$ and $(2y + x_1)$, respectively). We define

$$G = (A_4 + C_2)$$

We may assume a multiplicative identity would have the form $e = (y + z)$, where z is the cyclic generator of C_2. So, 3e = z, and 2e = 2y = -y. These relations are sufficient to determine that, for any $w \in G$,

$$z \cdot w = 3w$$

$$y \cdot w = -2w$$

We also know $0 = 3(x + y) \cdot w = 3((x \cdot w) - 2w)$, which means that if $|w| = 2$,

$$x \cdot w = 0$$

and, obviously, $x \cdot w \in [G, G], \textit{for all } w \in G$, with $x \cdot (y + z) = x$. It seems clear that $(G, +)$ hosts p.n.r. with identity. We need only specify the value of $x \cdot w$ for $|w| > 2$ and apply the theory of the associativity core in Sect. 2.1.1 in order to check on near-rings. Suppose we define, for $|w| > 2$,

$$x \cdot w = x$$

Then, if $w_1, w_2 \in G$, and $|w_1| > 2$, $|w_2| > 2$, the relation

$$(x \cdot (w_1 \cdot w_2)) = ((x \cdot w_1) \cdot w_2)$$

will hold if we can show $|w_1 \cdot w_2| > 2$. I gave the full list of possibilities for w_1 and w_2 earlier, and one can see, for instance, that if $w_1 = (x_1 + y)$ and $w_2 = (x + 2y)$, then

$$\begin{aligned} (w_1 \cdot w_2) &= (-y + x + y) \cdot (x + 2y) + (y \cdot (x_1 + 2y) = \\ &(x_2 + y + x + x + 2y + x_1 + 2y) = \\ &(y + x - y + y + x + x + 2y - y + x + y + 2y) = \\ &(y + x + y + x) \end{aligned}$$

and $|2(y + x)| = 3$. If necessary, by plugging in all possibilities, but, more simply, by seeing the patterns, I claim that the associativity identity is true and that, consequently, the group does host near rings with identity.

4.1.5 Groups with Order 27

We need to investigate two non-isomorphic groups. One has exponent 9 and the other exponent 3. These two groups are members of two infinite classes of groups dealt with in Sect. 6.7. Both host near-rings with identity.

4.1.6 Coda

There are 44 non-isomorphic, non-abelian, groups with order 32. They have exponents (in GAP order)

$$\{4, 8, 8, 4, 8, 8, 8, 8, 8, 8, 8, 8, 8, 16, 16, 16, 16, 4, 4,$$
$$4, 4, 4, 4, 4, 4, 4, 4, 4, 4, 4, 4, 8, 8, 8, 8, 8, 8, 8, 8, 4, 4, 4, 4, 4\}$$

The reader will realise that it is relatively easy to fix the status of some of these groups, such as D_{16} and $(D_4 \oplus C_4)$, and that we also have knowledge of some of the groups with even higher order. There are also ten non-abelian groups with order 36, and it is possible they are within reach of some of these techniques. From there, one could easily get to groups with order 40 and, perhaps, beyond. However, this seems a reasonable point to terminate this investigation.

These are immensely detailed calculations, and the possibilities for errors are large, so I cannot guarantee their validity in all cases. I have also not really considered the isomorphism classes of near-rings with identity that appear, and that is an important question which should be attended to.

Chapter 5
Near-Rings on Some Families of Groups

I continue to look for near-rings hosted by groups—here concentrating on unital, zero-symmetric, near-rings hosted by the groups in various group families, although I also look at more general near-rings too.

5.1 Finite Symmetric Groups

Suppose $(\mathrm{S}_N, +, \cdot)$ is the symmetric group of degree n. Select a and b to be distinct integers in $\{1, 2, 3, \ldots, n\}$, so (a, b) is one fixed transposition. We may define a product on permutations as

$$\sigma \cdot \tau = \begin{cases} (a, b) & \text{if both permutations are odd} \\ 0 & \text{otherwise} \end{cases}$$

This extends to a bi-distributive near-ring hosted by $(\mathrm{S}_N, +, \cdot)$ (see Definition 87). Its zeroiser ideal (Definition 55) is $\mathcal{Z}_{(\mathrm{S}_N, +, \cdot)} = \mathrm{A}_n$. This is also its annular ideal (see Definition 86).

5.2 Finite Simple Non-abelian Groups

Assume $(S, +)$ is simple and non-abelian. Suppose 1 is the multiplicative left identity of a pre-near-ring $(S, +, \cdot)$. The additive order of 1 must be equal to the exponent of $(S, +)$ which is even (Feit and Thompson) and not two. Call that exponent (2m). Then $(1 + 1) \cdot (m1) = 0$, which means

$$\binom{(m1)}{0}_{\mathcal{L}} = S$$

R. Lockhart, *The Theory of Near-Rings*, Lecture Notes in Mathematics 2295,
https://doi.org/10.1007/978-3-030-81755-8_5

which in turn would give us the contradiction that $0 \neq (1 \cdot (m1)) = 0$.

Lemma 202 *Finite non-abelian simple groups cannot host pre-near-rings with left identities.*

If $(S, +, \cdot)$ is a near-ring and $x \cdot y = 0$ for non-zero, $x, y \in S$, the same argument shows that

$$\begin{pmatrix} y \\ 0 \end{pmatrix}_{\mathcal{L}} = S$$

In non-trivial cases, we can find some y such that $x \cdot y \neq 0$, and that implies that the mapping (see Notation 93)

$$\mathcal{R}_y : N \longrightarrow N$$

is an automorphism of the simple group $(S, +)$. The elements of S partition into disjoint sets, N and A, defined as

$$N = \{y \in S \; : \; S \cdot y = \{0\}\}$$

$$A = \{y \in S \; : \; S \cdot y = S\}$$

I shall assume the near-ring is non-trivial. In that case, neither A nor N is empty and the zeroiser ideal (Definition 55) must be trivial. We have that $S \cdot N = \{0\}$ and $N \cdot S \subseteq N$, so N is a two-sided sideal (Sect. 1.8). More, if $a \in A$, we have that $N = N \cdot a$ and $A = A \cdot a$. This suggests a possible near-ring product defined as follows:

(i) Select any $a \in S^*$.
(ii) Define, for $x, y \in S$

$$x \cdot y = \begin{cases} x & \text{if } y = a \\ 0 & \text{otherwise} \end{cases}$$

This is a near-ring in which $A = \{a\}$, and $N = \{n \in N \; : \; n \neq a\}$.

Observations 203 Suppose a and b define two near-rings hosted by $(S, +)$ in this way. These two near-rings can only be isomorphic when a and b are linked by an automorphism of the simple group. So this prescription gives rise to a great many structurally similar, but non-isomorphic, near-rings.

5.2.1 *Isotopy*

The non-isomorphic near-rings occurring can be collated using the more general notion of *isotopy*. I shall illustrate this. Suppose that the near-ring defined by a is $(S, + \cdot)$ and that the near-ring defined by b is $(S, +, :)$. One can define a bijective mapping: $\phi : S \to S$ such that

$$(x : y) = (x \cdot \phi(y))$$

The bijective mapping ϕ is defined as

$$\phi(x) = \begin{cases} b & \text{if } x = a \\ a & \text{if } x = b \\ x & \text{otherwise} \end{cases}$$

There is some additional information about isotopy in Appendix A.

5.2.2 *A Class of Non-trivial Near-Rings Hosted by Any Group*

More generally, starting with *any* additive group $(S, +)$ and any subset of $U \subseteq S^*$, we may define a non-zero-symmetric near-ring product by

$$x \cdot y = \begin{cases} x & \text{if } y \in U \\ 0 & \text{otherwise} \end{cases} \tag{5.1}$$

Definition 204 I shall call the near-ring $(S, +, \cdot)$ defined by (5.1) the **near-ring supported by the subset U** and use the notation S^U for it, in this subsection.

The prescription may be used to define a non-trivial near-ring products on any non-trivial group. In such a near-ring, we have, for all elements x

$$x^2 = \begin{cases} x & \text{if } x \in U \\ 0 & \text{if } x \notin U \end{cases}$$

Observations 205

(i) When U is not empty, these near-rings have multiplicative right identities and $r^{(2)}(N) = N$ (see Notation 61).
(ii) Suppose $N \lhd (S, +)$ and take $U = \{S - N\}$. Then N is an ideal of S^U.
(iii) Any subset of N is a right sideal in S^U (see Sect. 1.8).
(iv) Suppose U and V are subsets of S and neither contains 0. Then S^U and S^V can only be isomorphic in situations in which there is an isomorphism of the additive group $(S, +)$ which maps U bijectively onto V.

If U and V are subsets of N not containing zero and there is a bijection $\theta : N \to N$ mapping U to V, and $S^U = (S, +, \cdot)$ and $S^V = (S, +, *)$, then, of course, for all $a, b \in N$

$$\theta(a \cdot b) = (\theta(a) * \theta(b)) \text{ and}$$
$$(a \cdot b) = (a * \theta(b))$$

although θ may well not be an isomorphism of the additive group. This is a further occurrence of obvious symmetries which are not isomorphisms (see Sect. 5.2.1).

5.3 Unital Near-Rings on S_n

Suppose $(S_N, +, \cdot)$ is a pre-near-ring in which $1 \in S_n$ is a left identity. Assume $n \geq 5$. The additive order of 1 is the exponent of S_n (I don't claim there is such an element) and this is larger than 3. S_n does have an element x with additive order 3 and so $(1 + 1 + 1) \cdot x = 0$ which implies that the left annihilator of x is A_n. Now, $(1 + 1)$ is an even permutation and $(1 + 1) \cdot x = 0$, which means x has additive order equal to 2.

Lemma 206 *When $n \geq 5$, the symmetric group with degree n hosts no pre-near-rings with left identity.*

The group S_4 has exponent 12 and is not I-appropriate (page 49). It hosts no p.n.r. with identity.

5.4 The Quaternion Group with Order 8

This group has the presentation

$$Q_8 = \langle a, b \, : \, 4a = 2a - 2b = a + b + a - b = 0 \rangle \tag{5.2}$$

Symmetry considerations allow us to assume that a is the identity of any near-ring with identity hosted by Q_8. The commutator subgroup is the two-element set $\{0, 2a\}$, and, of course, $(2a)^2 = 0$. Since $2a \cdot Q_8 \cdot 2a = 0$, it must be that $Q_8 \cdot 2a$ consists entirely of elements with order 2 or 1, and so it is $\{0, 2a\}$. By order arguments based on the isomorphism

$$Q_8 \cdot (2a) \cong \frac{Q_8}{\begin{pmatrix} 2a \\ 0 \end{pmatrix}_{\mathcal{L}}}$$

we deduce that $\begin{pmatrix} 2a \\ 0 \end{pmatrix}_{\mathcal{L}}$ is an additively cyclic subgroup generated by an element with order 4. Without loss of generality, we may assume this element to be b. The elements of Q_8 are $\{0, a, b, 2a, 3a, (a+b), (2a+b), (3a+b)\}$. Q_8 is a useful example and I refer to it in other parts of the book, using other notations. The reader wishing to easily translate the symbols given here with those of Table 10.2, or who prefers the scientific notation, given on 355, should use this dictionary.

$$\begin{array}{lll} 2 \leftrightarrow a & & \leftrightarrow i \\ 4 \leftrightarrow b & & \leftrightarrow j \\ 5 \leftrightarrow (a+b) & & \leftrightarrow k \\ 1 \leftrightarrow 2a & & \leftrightarrow -1 \\ 3 \leftrightarrow 3a & & \leftrightarrow -i \\ 6 \leftrightarrow (2a+b) & & \leftrightarrow -j \\ 7 \leftrightarrow (3a+b) & & \leftrightarrow -k \\ 0 \leftrightarrow 0 & & \leftrightarrow 1 \end{array} \tag{5.3}$$

Determining the near-ring is equivalent to fixing values for $b \cdot x$ for

$$x \in \{b, 3a, (a+b), (2a+b), (3a+b)\}$$

and all these values must lie in $\langle b \rangle = \{0, b, 2a, (2a+b)\}$ because $b \cdot x$ annihilates (2a). Because $(3a)^2 = 1$, we know that either $b \cdot (3a) = b$ or $b \cdot (3a) = (2a+b)$. We can also say that $(3a) \cdot x = 3x = (2a+x)$. So, if $b \cdot (3a) = b$, we have that $b^2 = b \cdot (2a+b)$ and $b \cdot (a+b) = b \cdot (3a+b)$, whereas, if $b \cdot (3a) = (2a+b)$, we have that $b^2 = (2a + b \cdot (2a+b))$ and $b \cdot (a+b) = (2a + b \cdot (3a+b))$.

First Case: $b \cdot (3a) = b$
Suppose $b^2 = (2a+b)$. We have $b^2 = b \cdot (2a+b)$, and multiply on the right by b to get $b = b \cdot (2a + (2a+b)) = b^2 = (2a+b)$. On the other hand, if $b^2 = 2a$, the same argument gives $2a = b \cdot (2a+2a) = 0$. We conclude that $b^2 \in \{0, b\}$. Since $(b \cdot (a+b)) \cdot b = b \cdot (b+b^2)$, we can deduce that $b \cdot (a+b) \in \{0, b\}$. Setting $b^2 = 0 = b \cdot (a+b)$ and distributing linearly on the right (so $(2a+b) \cdot b$ is set to $(2a) \cdot b + b^2 = (2a+0))$ gives us this candidate near-ring (Table 5.1).
The two additive generators a and b are both associative elements, that is, if x is set to either one of them, $x(y \cdot z) = (x \cdot y) \cdot z$, for all $y, z \in Q_8$. If this table represented a pre-near-ring, Observations 70 would permit us to conclude that the p.n.r was actually a near-ring, in that the associativity subgroup would contain additive generators for the whole group. It is not a near-ring because, for example,

$$2a = (2a+b)^2 \cdot (2a+b) \neq (2a+b) \cdot (2a+b)^2 = 0$$

The product in the table is not actually right-distributive. Were it to be so, then since $0 = (a+b+a-b)$, one would expect $0 = (a+b+a-b) \cdot b = (b+b) = 2a$.

Table 5.1 A possible near-ring

*	0	a	$2a$	$3a$	b	$a+b$	$a+b$	$3a+b$
0	0	0	0	0	0	0	0	0
a	0	a	$2a$	$3a$	b	$2a+b$	$a+b$	$3a+b$
$2a$	0	$2a$	0	$2a$	$2a$	$2a$	$2a$	$2a$
$3a$	0	$3a$	$2a$	a	$2a+b$	b	$3a+b$	$a+b$
b	0	b	0	b	0	0	0	0
$2a+b$	0	$2a+b$	0	$2a+b$	$2a$	$2a$	$2a$	$2a$
$a+b$	0	$a+b$	$2a$	$3a+b$	b	$2a+b$	$a+b$	$3a+b$
$3a+b$	0	$3a+b$	$2a$	$a+b$	$2a+b$	b	$3a+b$	$a+b$

The reason that distributing linearly from the right does not produce a pre-near-ring is that the intended product does not satisfy the **pre-near-ring construction conditions**—see Definition 2.2. It will be seen that we cannot produce a pre-near-ring in which $b^2 = 0$, in this case. The alternative value for b^2 is b, but

$$0 = (a+b+a-b) \cdot b = (b+b+b-b) = 2a$$

Consequently this value for b^2 would not fit the p.n.r construction conditions either.

Second Case: $b \cdot (3a) = (2a+b)$
We know $b^2 \in \{0, 2a, b, (2a+b)\}$ and have established both that $b^2 = (2a+b(2a+b))$ and $b \cdot (a+b) = 2a + b \cdot (3a+b)$. These relations and similar arguments to those used above allow us to establish $b \in \{b, 2a+b\}$, but these values mean that the p.n.r. construction requirement

$$0 = (a+b+a-b) = (a \cdot b + b^2 + a \cdot b - b^2)$$

cannot be satisfied.

Lemma 207 *The quaternion group with order 8 hosts no near-rings with identity.*

5.4.1 Unital d.g. p.n.r. Hosted by Q_8

Suppose one asks for p.n.r. hosted by Q_8 in which a is a multiplicative identity and b is in $C_{\mathcal{D}}(Q_8)$ (Definition 69). Working with the relations given by the presentation (5.2), we can deduce that

$$b \cdot (2a) = 2a$$
$$b \cdot (3a) = (2a+b)$$

Table 5.2 Possible d.g. p.n.r with identity on Q_8

b^2	$b \cdot (a+b)$	$b \cdot (2a+b)$	$b \cdot (3a+b)$
$3a$	$(a+b)$	a	$(3a+b)$
a	$(3a+b)$	$3a$	$(a+b)$
$(a+b)$	a	$(3a+b)$	$3a$
$(3a+b)$	$3a$	$(a+b)$	a

Table 5.2 gives possible values for $b \cdot x$ consonant with right distributivity ($x \in \{b, (a+b), (2a+b), (3a+b)\}$).
So, for instance, $b^2 = 0$ is impossible because it implies $b \cdot (a+b+a-b) = (b+b) = 2a \neq 0$. The first two possibilities do not represent *right* p.n.r. because in the first case

$$(a+b+a-b) \cdot (3a+b) = ((3a+b)+(3a+b)+(3a+b)+(a+b)) = 2a \neq 0$$

and, in the second

$$(a+b+a-b) \cdot (3a+b) = ((3a+b)+(a+b)+(3a+b)+(3a+b)) = 2a \neq 0$$

The other two are genuine d.g. p.n.r. with identity, and the only possibilities. Could they be isomorphic? Automorphism Φ of the additive group which fix the element a must map b to one of the four elements $\{b, a+b, 2a+b, 3a+b\}$. These structures will be isomorphic provided distinct products can be linked by a relation of the form

$$(x * y) = \Phi^{-1}(\Phi(x) \cdot \Phi(y))$$

for some group automorphism Φ. So, one needs to check whether these products can be linked in this way using each of the three automorphisms which actually change b. I find they are not (such symmetries map the p.n.r. to other d.g. p.n.r. in which b is not a distributive element).

Lemma 208 *There are two d.g. non-isomorphic p.n.r. with identity hosted by* Q_8.

The two d.g. p.n.r. identified are given by Tables 5.3 and 5.4.
The *cores* of the first p.n.r. (page 43) are

$$C_{\mathcal{D}}(Q_8) = \{0, a, b\}$$
$$C_{\mathcal{A}}(Q_8) = \{0, a, 2a, 3a\}$$

Table 5.3 First d.g. p.n.r hosted by Q_8

$*$	0	a	$2a$	$3a$	b	$2a+b$	$a+b$	$3a+b$
0	0	0	0	0	0	0	0	0
a	0	a	$2a$	$3a$	b	$2a+b$	$a+b$	$3a+b$
$2a$	0	$2a$	0	$2a$	$2a$	$2a$	$2a$	$2a$
$3a$	0	$3a$	$2a$	a	$2a+b$	b	$3a+b$	$a+b$
b	0	b	$2a$	2a + b	$a+b$	$3a+b$	a	$3a$
$2a+b$	0	$2a+b$	$2a$	b	$3a+b$	$a+b$	$3a$	a
$a+b$	0	$a+b$	0	$a+b$	a	a	b	b
$3a+b$	0	$3a+b$	0	$3a+b$	$3a$	$3a$	$2a+b$	$2a+b$

Table 5.4 Second d.g. p.n.r hosted by Q_8

$*$	0	a	$2a$	$3a$	b	$2a+b$	$a+b$	$3a+b$
0	0	0	0	0	0	0	0	0
a	0	a	$2a$	$3a$	b	$2a+b$	$a+b$	$3a+b$
$2a$	0	$2a$	0	$2a$	$2a$	$2a$	$2a$	$2a$
$3a$	0	$3a$	$2a$	a	$2a+b$	b	$3a+b$	$a+b$
b	0	b	$2a$	2a + b	$3a+b$	$a+b$	$3a$	a
$2a+b$	0	$2a+b$	$2a$	b	$a+b$	$3a+b$	a	$3a$
$a+b$	0	$a+b$	$2a$	$3a+b$	$3a$	$3a$	$2a+b$	$3a$
$3a+b$	0	$3a+b$	$2a$	$a+b$	$3a$	a	b	b

Notice that $(a+b) \notin C_{\mathcal{D}}(Q_8)$ because, although the multiplication table would suggest that (a + b) is distributive ($a = (a+b)\cdot(2a+b) = 0+a = (a+b)\cdot 2a + (a+b)\cdot b$, etc.), we have

$$(a+b)\cdot a + (a+b)\cdot b + (a+b)\cdot a + (a+b)\cdot(2a+b) =$$
$$(a+b) + a + (a+b) + a = 2a \neq 0$$

Notice that $C_{\mathcal{D}}(Q_8)$ is not multiplicatively closed (see Observations 70).
The *cores* of the second p.n.r. are

$$\begin{aligned} C_{\mathcal{D}}(Q_8) &= \{0, a, b\} \\ C_{\mathcal{A}}(Q_8) &= \{0, a, 2a, 3a\} \end{aligned}$$

Once again, $C_{\mathcal{D}}(Q_8)$ is not multiplicatively closed. Distributive generation seems to impose quite a restriction on the p.n.r. hosted by Q_8: it is claimed in [102] that the group hosts 4352 non-isomorphic p.n.r. with left identity.

5.5 Dihedral Groups

The dihedral group of degree n and order 2n has the presentation given in (2.19) on page 83.

5.5.1 *The Dihedral Group of Order 8*

The dihedral group D_n ($n \geq 3$) has exponent equal to n when n is even, and 2n when it is not. Since it is not cyclic, it may possibly host a pre-near-ring with left identity only in the even case, when there actually is an element with the same order as its exponent.

$$D_4 = \langle a, b : 4a, 2b, 2(a+b) \rangle \tag{5.4}$$

is the dihedral group of degree four. Its Cayley table is given as Table B.8 on page 501. Its commutator subgroup is $\{2a, 0\}$.

I consider near-rings with identity definable on this group, and of course the commutator subgroup will be a right ideal in all of them. Thus, $2a \cdot D_4 \subseteq \{0, 2a\}$. We know that the multiplicative identity must be an element with order equal to the group exponent, in this case, 4, and, without loss of generality, may assume it to be a. Consider the set $D_4 \cdot (2a) = \{x \cdot (2a) : x \in D_4\}$. This set contains the element 2a, is a subgroup of D_4, and possesses no elements with order 4. Its order is either 2 or 4. Since $(2a)^2 = 0$, we know that all of its elements left annihilate $(2a)$. That is,

$$D_4 \cdot (2a) \subseteq \begin{pmatrix} 2a \\ 0 \end{pmatrix}_{\mathcal{L}} \tag{5.5}$$

We have

$$D_4 \cdot (2a) \cong \frac{D_4}{\begin{pmatrix} 2a \\ 0 \end{pmatrix}_{\mathcal{L}}}$$

and deduce that $D_4 \cdot (2a) = \{0, 2a\}$, the left annihilator of $(2a)$ has order 4 and the inclusion (5.5) is strict. This annihilator cannot contain a or 3a. It must be one of the two sets

$$\{0, 2a, b, 2a+b\} \quad \text{or} \quad \{0, 2a, a+b, 3a+b\}$$

We have so far placed no restrictions on b and may assume, without loss of generality, that

$$\begin{pmatrix} 2a \\ 0 \end{pmatrix}_{\mathcal{L}} = \{0, 2a, b, 2a + b\} \tag{5.6}$$

(a left ideal). So, I shall assume $b \cdot 2a = 0$ in what follows. Now, $D_4 \cdot 2a = \{0, 2a\}$, and it follows $\begin{pmatrix} 2a \\ 0 \end{pmatrix}_{\mathcal{L}}$ is a two-sided ideal, so D_4 hosts no simple near-rings with identity.

We habitually consider only zero-symmetric near-rings but might concede the possibility of D_4 hosting a non-zero-symmetric near-ring with identity. For such a near-ring, we would have $b \cdot 0 = ka + rb \neq 0$ for some $k \in \{0, 1, 2, 3\}, r \in \{0, 1\}$. Then, $2ka = (b \cdot 0) \cdot 2a = b \cdot 0 = ka + rb$, which is impossible. So D_4 hosts no non-zero-symmetric near-rings with identity. Our zero-symmetric near-ring would be determined provided that we can explain what the product $b \cdot x$ should be, for all possible choices of x.

- It remains to establish the value of $(b \cdot x)$ for $x \in \{3a, b, a + b, 2a + b, 3a + b\}$.
- And we know $b \cdot x \in \{0, 2a, b, 2a + b\}$.

Now, $b \cdot (a+b) = b \Rightarrow (a+b)^2 = a$ which is impossible, from order considerations and right distributivity. Consequently, $b \cdot (a + b) \in \{0, 2a, 2a + b\}$. For $b \cdot 3a$, we have that $b = b \cdot (3a)^2$ and deduce that $b \cdot 3a = 2ma + b$, where m is either 0 or 1.

D_4 First Case: $b \cdot (3a) = b$
We have that 3a acts as a right identity on $\{0, 2a, b, 2a+b\}$. We know $b \cdot (a+b) \cdot 3a = b \cdot (3a + b)$ and using this obtain Table 5.5 of possible values for $(b \cdot A)$.
Here, $x, y, z \in \{0, 2a, b, 2a + b\}$, and $y \neq b$. We know that $b \cdot (a + b) \cdot (a + b) = b \cdot (a + b + b \cdot (a + b))$. If it could happen that $b \cdot (a + b) = 2a$, we should then have that $0 = b \cdot (3a + b) = b \cdot (a + b) \cdot 3a = 2a$. We have to conclude that $y = b \cdot (a + b) \neq 2a$. Thus, in this first case, $y \in \{0, 2a + b\}$; and since $(2a + b) \cdot b = b^2$, we can also say that $b^3 = z \cdot b$.
Suppose $b \cdot (2a+b) = (2a+b)$. Left multiplication by b shows this is incompatible with $b^2 \in \{0, 2a\}$. That is, $z = (2a + b) \Rightarrow x = b^2 \in \{b, (2a + b)\}$.
We know that $y \in \{0, 2a + b\}$ and we have that $b \cdot (a + b) \cdot (2a + b) = b \cdot (2a + b + 2a + b) = 0$, deducing $y \neq (2a + b)$.
This gives our first subcase.

Subcase 1

$$\begin{aligned} z &= (2a + b) \\ y &= 0 \\ x &\in \{b, (2a + b)\} \end{aligned}$$

Table 5.5 First case: possible product values

A	$2a$	$3a$	b	$a+b$	$2a+b$	$3a+b$
$b \cdot A$	0	b	x	y	z	y

Subcase 2

We now consider the consequences of $z = b$.
If $y = b \cdot (a+b) = 2a+b$, then $b = b^2 \cdot (a+b)$ from which we deduce $b^2 = b$. But then, $b = (2a + b) \cdot b = b \cdot (a + b) \cdot b = b \cdot (b + b) = 0$.
It follows that $y = 0$.
We have that $b^3 = b^2$ here, so $x \neq 2a$.
On the other hand, if $x = 2a + b$, then $b^3 = b \cdot (2a + b) = z = b$, but $b^3 = z \cdot b = b^2 = 2a + b$.
We conclude $x \neq (2a + b)$.
The subcase is
$z = b$
$y = 0$
$x \in \{0, b\}$.

Subcase 3

Here, suppose $z = 2a$. Since $b \cdot 2a = b^2 \cdot (2a + b) = 0$, we know $x \notin \{b, 2a + b\}$. If $b \cdot (a + b) = (2a + b)$, then $0 = 2a \cdot (a + b) = b \cdot (2a + b) \cdot (a + b) = b \cdot (2a + b) = 2a$. Thus $y = 0$. The relation $b^3 = z \cdot b = 0$ indicates that $x \in \{0, 2a\}$.
The subcase is:
$z = 2a$
$y = 0$
$x \in \{0, 2a\}$.

Subcase 4

Finally, suppose $z = 0$
$0 = b \cdot (2a + b) \cdot b = b^3 \Rightarrow x = b^2 \notin \{b, 2a + b\}$.
The subcase is
$z = 0$
$x \in \{0, 2a\}$
$y \in \{0, 2a + b\}$.

The possibility that $x = 2a$ occurs in the third and fourth subcases.
In that event, we have $z = b \cdot (a + b) \cdot b$.
Thus, $y = 0 \Rightarrow z = 0$ and $y = (2a + b) \Rightarrow z = 2a$.
However, $y = (2a + b) \Rightarrow z = b \cdot (2a + b) = b^2 \cdot (a + b) = 2a \cdot (a + b) = 0$.
Thus, $x = 2a \Rightarrow y = 0$ and $z = 0$.

Table 5.6 summarises the situation as we presently have it. Its top and bottom rows come from Table 5.5.
It remains to be seen whether any of these eight pre-near-rings with identity are associative and whether or not they are isomorphic. Notice that associativity was implicit in many of the arguments so far advanced, so that we have no claim toward obtaining *all* pre-near-rings with identity element hosted by D_4.

D_4 Second Case: $b \cdot (3a) = (2a + b)$
We have

$$b \cdot (a + b) \cdot 3a = b \cdot (3a + 2a + b) = b \cdot (a + b)$$

Table 5.6 Candidate near-rings hosted by D_4—first case

	A	a	$2a$	$3a$	b	$(a+b)$	$(2a+b)$	$(3a+b)$
1	$b \cdot A$	b	0	b	b	0	$(2a+b)$	0
2	$b \cdot A$	b	0	b	$(2a+b)$	0	$(2a+b)$	0
3	$b \cdot A$	b	0	b	0	0	b	0
4	$b \cdot A$	b	0	b	b	0	b	0
5	$b \cdot A$	b	0	b	0	0	$2a$	0
6	$b \cdot A$	b	0	b	0	0	0	0
7	$b \cdot A$	b	0	b	0	$(2a+b)$	0	$(2a+b)$
8	$b \cdot A$	b	0	b	$2a$	0	0	0
	$b \cdot A$	b	0	b	x	y	z	y

and

$$b \cdot (3a+b) \cdot 3a = b \cdot (a+2a+b) = b \cdot (3a+b)$$

So, 3a acts as a right identity on the products $b \cdot (a+b)$ and $b \cdot (3a+b)$. These elements are in the set $\{0, 2a, b, 2a+b\}$; consequently they must actually be members of its subset, $\{0, 2a\}$. Moreover,

$$b^2 \cdot (3a) = b \cdot (2a+b)$$

Suppose $b \cdot (a+b) = 2a$. Then, $0 = b \cdot (3a+b+b \cdot (3a+b))$ and so $b \cdot (3a+b) \neq 2a$. Similarly, if $b \cdot (3a+b) = 2a$, then $0 = b \cdot (3a+b) \cdot (a+b) = b \cdot (a+b+b \cdot (a+b)) \Rightarrow b \cdot (a+b) = 0$. In general, either

$$b \cdot (a+b) = 2a + b \cdot (3a+b)$$

or

$$y = z = 0$$

If $b^2 = b$, and $b \cdot (a+b) = 2a$, then $0 = b \cdot 2a = b \cdot (a+b) = 2a$. Hence, $b^2 = b \Rightarrow y = 0$. Alternatively, if $b^2 = (2a+b)$, then $b \cdot (2a+b) = x \cdot 3a = b$. At the same time, $b \cdot (2a+b) = b^3 = (2a+b) \cdot b = b^2 = (2a+b)$. We conclude that the case $x = b^2 = (2a+b)$ does not occur. Finally, suppose $b^2 = 2a$. This means $b \cdot (2a+b) = b \cdot 3a = 2a$. If, now, $b \cdot (a+b) = 2a$, then $0 = b \cdot (b+b^2) = b \cdot (2a+b)$. We conclude that $b^2 = 2a \Rightarrow b \cdot (a+b) = 0$. Following the methods of the first case, we obtain Table 5.7.

The table represents seven pre-near-rings with identity which may also be near-rings. At this point we have a total of 15 candidate near-rings with identity on the dihedral group D_4. The next step is to divide them into isomorphism classes.

Table 5.7 Candidate near-rings hosted by D_4—second case

Number	A	2a	3a	b	a + b	2a + b	3a + b
9	$b \cdot A$	0	(2a + b)	0	0	0	0
10	$b \cdot A$	0	(2a + b)	0	0	0	2a
11	$b \cdot A$	0	(2a + b)	0	2a	0	0
12	$b \cdot A$	0	(2a + b)	2a	0	2a	0
13	$b \cdot$ A	0	(2a + b)	2a	0	2a	2a
14	$b \cdot$ A	0	(2a + b)	b	0	(2a + b)	0
15	$b \cdot$ A	0	(2a + b)	b	0	(2a + b)	2a

Isomorphisms

A potential isomorphism Φ of these structures would preserve a and send b to something of the form $ra + b$. We must have

$$0 = \Phi(b \cdot 2a) = (ra + b) \cdot (2a) = 2ra$$

so a non-trivial isomorphism would be

$$\Phi(a) = a$$
$$\Phi(b) = (2a + b)$$

One considers pre-near-rings isomorphic to our 12 candidates. The prescription defines a new product (*) as

$$(x * y) = \Phi^{-1} (\Phi(x) \cdot \Phi(y)) \tag{5.7}$$

Isomorphisms will respect our two subcases: $b \cdot (3a) = b$ and $b \cdot (3a) = (2a + b)$. I now consider whether any of these pre-near-rings might be isomorphic. Case 15 is represented by Table 5.8, and from it, one can derive Table 5.9.
For example, $(b * (a + b)) = \Phi^{-1} (b \cdot (3a + b)) = 2a$; and $(b * (3a + b)) = \Phi^{-1} ((2a + b) \cdot (a + b)) = \Phi^{-1}(0) = 0$. The problem is the derived structure given by Table 5.9 does not appear in Table 5.7. We deduce that case 15 cannot represent a near-ring and discount it. Supporting that analysis, it is clear that in case 13 we can argue $0 = b \cdot 2a = b^2 \cdot (3a + b) = b \cdot (3a + b) = 2a$, so we don't have a near-ring.

Table 5.10 gives the pre-near-rings produced by applying (5.7) to our 14 remaining candidates.

Table 5.8 Case 15

A	$2a$	$3a$	b	$a + b$	$2a + b$	$3a + b$
$b \cdot A$	0	$(2a + b)$	b	0	$(2a + b)$	$2a$

Table 5.9 Product derived from case 15

A	$2a$	$3a$	b	$(a + b)$	$(2a + b)$	$(3a + b)$
$b * A$	0	$(2a + b)$	b	$2a$	$(2a + b)$	0

Table 5.10 Isomorphisms of the 14 possibilities

No.	A	$2a$	$3a$	b	$(a+b)$	$(2a+b)$	$(3a+b)$
1	$b*A$	0	b	b	0	$(2a+b)$	0
2	$b*A$	0	b	b	0	b	0
3	$b*A$	0	b	$(2a+b)$	0	0	0
4	$b*A$	0	b	$(2a+b)$	0	$(2a+b)$	0
5	$b*A$	0	b	$2a$	0	0	0
6	$b*A$	0	b	0	0	0	0
7	$b*A$	0	b	0	b	0	b
8	$b*A$	0	b	0	0	$2a$	0
9	$b*A$	0	$(2a+b)$	0	0	0	0
10	$b*A$	0	$(2a+b)$	0	$2a$	0	0
11	$b*A$	0	$(2a+b)$	0	0	0	$2a$
12	$b*A$	0	$(2a+b)$	$2a$	0	$2a$	0
13	$b*A$	0	$(2a+b)$	$2a$	$2a$	$2a$	0
14	$b*A$	0	$(2a+b)$	b	0	$(2a+b)$	0

Table 5.11 Isomorphism classes

Case	1	2	5	6	9	10	12	14
Case	1	4	8	6	9	11	12	14

Table 5.12 Candidate near-rings

No.	A	$2a$	$3a$	b	$(a+b)$	$(2a+b)$	$(3a+b)$
1	$b*A$	0	b	b	0	$(2a+b)$	0
2	$b*A$	0	b	b	0	b	0
5	$b*A$	0	b	$2a$	0	0	0
6	$b*A$	0	b	0	0	0	0
9	$b*A$	0	$(2a+b)$	0	0	0	0
10	$b*A$	0	$(2a+b)$	0	$2a$	0	0
12	$b*A$	0	$(2a+b)$	$2a$	0	$2a$	0
14	$b*A$	0	$(2a+b)$	b	0	$(2a+b)$	0

Line 3 of Table 5.6 maps to line 3 of Table 5.10, and that is a product not described in the original Table 5.6 and therefore cannot be a near-ring. The same is true for lines 7 and 13. Isomorphisms occur in pairs; inspection reveals the ones given by Table 5.11.

Line 3 of Table 5.6 maps to line 3 of Table 5.10, and that is a product not described in the original Table 5.6 and therefore cannot be a near-ring. The same is true for lines 7 and 13. This offers the possibility of eight non-isomorphic near-rings. So, for example, case 2 of Table 5.6 matches case 4 of Table 5.10, and vice versa.

Table 5.12 summarises what we now have.

Case 12 says $0 = b \cdot (a+b) \cdot b = b \cdot (b+2a) = 2a$. So, this is not a near-ring.

The following result is described in [78] as having first been proved by J.R. Clay, using computers.

Table 5.13 Case 2

No.	A	$2a$	$3a$	b	$(a+b)$	$(2a+b)$	$(3a+b)$
10	$b * A$	0	b	b	0	b	0

Lemma 209 *The dihedral group of order 8 hosts seven non-isomorphic near-rings with multiplicative identities.*

Proof We need to verify that each of the seven pre-near-rings we have obtained is associative. We can do that by establishing that the element b is always in the associativity core of the pre-near-ring (see Definition 69). This involves verifying that $b \cdot (x \cdot y) = (b \cdot x) \cdot y$ for each of the 64 choices of x and y that might be made; however, we can reduce this to 36 choices, since we can suppose $x, y \notin \{0, a\}$ because those cases are obvious. The checking is unremarkable but tedious and I shall only complete it for case 2 (Table 5.13).

We only need to verify $b \cdot (x \cdot y) = (b \cdot x) \cdot y$, for $x, y \in \{2a, 3a, b, a+b, 2a+b, 3a+b\}$.

$x = 2a$

We have $b \cdot x = 0$, and the product $x \cdot y$ is non-zero only when $y = 3a$, when $0 = b \cdot 2a = b \cdot (2a \cdot 3a) = (b \cdot 2a) \cdot 3a = 0$.

$x = 3a$

We have $b \cdot x = b$ and need to verify $b \cdot y = b \cdot (3a \cdot y)$. Since $(3a \cdot y)$ actually is y, when y has order 2, we only need to consider $y = 3a$, and there the result is immediate.

$x = b$

We have $b \cdot x = b$ and must show $b \cdot (b \cdot y) = b \cdot y$. By inspection, $b \cdot (b \cdot y)) = b \cdot y$.

$x = a + b$

We have $b \cdot x = 0$ and must show $0 = b \cdot ((a+b) \cdot y) = b \cdot (y + b \cdot y)$. From Table 5.13 it is obvious $y + b \cdot y = 0$, so the result follows.

$x = 2a + b$

We have $b \cdot x = b$ and must verify $b \cdot y = b \cdot (2y + b \cdot y)$. Setting y in turn to be one of $\{2a, 3a, b, a+b, 2a+b, 3a+b\}$ gives $b \cdot y$ as

$$\{0, b, b, 0, b, 0\}$$

respectively, and makes $b \cdot (2y + b \cdot y)$ correspondingly

$$\{b \cdot 0, b \cdot (2a+b), b \cdot b, b \cdot 0, b \cdot b, b \cdot 0\}$$

which is

$$\{0, b, b, 0, b, 0\}$$

$x = 3a + b$

We have $b \cdot x = 0$ and must verify $0 = b \cdot (3y + b \cdot y)$. Using the same tactic, we set y in turn to one of $\{2a, 3a, b, a+b, 2a+b, 3a+b\}$ getting the products $\{b \cdot 2a, b \cdot (a+b), b \cdot 0, b \cdot (a+b), b \cdot 2a, b \cdot (3a+b)\}$, all of which are 0.

□

Appendix B offers the multiplication tables for these seven near-rings.

5.5.2 *Other Finite Dihedral Groups*

The remaining finite dihedral groups that may possibly host near-rings with identity have the form D_{2n} for $n > 2$. One might write $n = pm$, where p is some prime number; and, to preclude the case already considered, we may assume either that $p > 2$, or $m > 1$, or both. From order considerations

$$(b \cdot ma) = \begin{cases} 0 & \text{or} \\ pma & \text{or} \\ (ra + b) \end{cases}$$

and correspondingly we would have

$$((a + b) \cdot ma) = \begin{cases} ma & \text{or} \\ (ma + pma) & \text{or} \\ (ma + ra + b) \end{cases}$$

Each of these last three elements should have order 0 or 2. But the first has order 2p, as does the second. We conclude that

$$b \cdot ma = (ra + b)$$

A similar argument suggests that $b \cdot 2a = \lambda a + b$, giving $b \cdot 2ma = (2ra + \lambda a) + b$. This last relation implies that $b \cdot (pa) \in \langle a \rangle$ and implies p is an odd prime. Since $(a + b) \cdot pa$ has order 0 or 2, we can also conclude that $m = 1$.

Lemma 210 *If the dihedral group* D_{2n} *has order greater than eight and it hosts near-rings with multiplicative identities, then n is an odd prime.*

I now consider whether it is possible to define near-rings with identity on D_{2p} where p is some odd prime. The standard order argument tells us that

$$(b \cdot (pa)) = \begin{cases} 0 & \text{or} \\ pa \end{cases}$$

In the second case, $(a + b) \cdot pa = 0$, so simple re-labelling allows us to suppose $b \cdot (pa) = 0$ in all near-rings. From that it follows that $b \cdot (ra + b) = 0$ for all integers r. We can also conclude that $b \cdot (2a) = (2ka + b)$ for some integer k. Now select π as some primitive root modulo (2p). Each of the (p-1) units of the ring of integers C_{2p} is a power of π and $b \cdot \pi a = (\eta a + b)$ for some $0 \leq \eta < 2p$; but, of course, η has to be even. Any choice of $0 \leq \zeta < p$ and $0 \leq k < p$, together with the prescription

$$\begin{aligned} b \cdot (2a) &= (2ka + b) \\ b \cdot (pa) &= 0 \\ b \cdot (ra + b) &= 0 \\ b \cdot (\pi a) &= (2\zeta a + b) \end{aligned}$$

corresponds to a pre-near-ring with identity a. Expanding and equating $(b \cdot (2a)) \cdot (\pi a)$ and $(b \cdot (\pi a)) \cdot (2a)$ give us the identity

$$\zeta \equiv k(\pi - 1) \ \text{(modulo } p)$$

which fixes ζ once k is known. So, the pre-near-rings we are left with are fixed uniquely by the value of $b \cdot (2a)$. Suppose k_1 and k_2 are two such structures, and they are isomorphic. The isomorphism must have the form

$$\Phi(a) = a$$

$$\Phi(b) = (ta + b)$$

Referring to (5.7) leads us to require

$$0 = (b * pa) = \Phi^{-1}((ta + b) \cdot pa) = tpa$$

which forces $t = 2s$ to be even: we need only consider derived pre-near-rings in which $\Phi(b) = (2sa + b)$. Under this restriction, it is straightforward to check that all possible choices of k result in isomorphic structures. More than that, any choice is itself a near-ring.

Lemma 211

(i) The dihedral group D_{2p}, where p is an odd prime, hosts exactly one near-ring with identity, up to isomorphism.

(ii) In this near-ring

$$\langle 2a, b\rangle = \begin{pmatrix} pa \\ 0 \end{pmatrix}_{\mathcal{L}}$$

and so $\langle 2a, b\rangle$ is a non-trivial ideal with order 2p.

Marjorie Johnson seems to have been the first to tie up all possibilities for near-rings hosted by finite dihedral groups, though she herself mentions other authors [78]. The group D_6 is solvable but not nilpotent. It is the smallest such group hosting near-rings with identity. (S_3 is of course both solvable and non-nilpotent, but it does not host such near-rings).

5.5.3 Pre-Near-Rings

Interestingly, the dihedral group D_{2m} generally does host a pre-near-ring with identity a for all non-zero values of m, under the prescription

$$b \cdot x = \begin{cases} 0 & \text{if } x = 0 \\ b & \text{if } x \in \{\langle a \rangle - \{0\}\} \\ 0 & \text{otherwise} \end{cases}$$

because this prescription satisfies the corresponding p.n.r. construction conditions. The corresponding associativity core is equal to $\langle a \rangle$ in all cases.

$$0 = (b \cdot (2a \cdot ma)) \neq ((b \cdot 2a) \cdot ma) = b$$

The remaining case to consider is the infinite dihedral group.

5.5.4 The Infinite Dihedral Group

The infinite dihedral group has a presentation

$$D_\infty = \langle a, b : 2b = 0 = 2(a + b) \rangle \tag{5.8}$$

Suppose it hosts a near-ring with identity. The identity element must be in $\langle a \rangle$ and have the form $ra, r \in \mathbb{Z}$. Then, $b = (ra) \cdot b = r(a \cdot b) \Rightarrow r = \pm 1$. Without loss of generality, assume a to be the identity. Select any prime, p. Then using order arguments

$$(b \cdot (pa)) = \begin{cases} 0 & \text{or} \\ (\lambda a + b) & \end{cases}$$

The first case gives $(a + b) \cdot (pa) = pa$, which does not have additive order 2, so the second case must apply. We can write $(b \cdot (2a)) = ka + b$, $(b \cdot (pa)) = \lambda a + b$. Then expanding $b \cdot (2pa)$ two ways, we obtain

$$\lambda = k(p - 2)$$

which means

$$(b \cdot (pa)) = (k(p-1)a + b)$$

We also have that

$$(b \cdot (ra + b) = k(ra + b)$$

None of these relations fix $(b \cdot (-a)$, and this must have the form

$$b \cdot (-a) = sa + b \tag{5.9}$$

The value of $b \cdot (2a)$ controls much of the potential near-ring product. We have that

$$(\psi a + b) \cdot 2a = (2\psi + k)a + b$$

so replacing b with an appropriate choice for $(\psi a + b)$ enables us to assume

$$(b \cdot (2a)) = \begin{cases} (a+b) & \text{or} \\ b \end{cases}$$

depending on whether k was odd or even. In the first case, $(-a+b) \cdot 2a = (-a+b)$, so replacing b with $(-a+b)$ takes us to the second case, and we may generally assume for all integers m

$$\begin{aligned} b \cdot (ma) &= b \\ b \cdot (ma + b) &= 0 \end{aligned}$$

Then expanding $b \cdot (-3a)$ two ways gives $s = 0$ and

$$(b \cdot (-a)) = b$$

The product on D_∞ may be described as

$$\begin{aligned} b \cdot (ma) &= b \\ b \cdot (-a) &= b \\ b \cdot (ma + b) &= 0 \end{aligned}$$

Notice that this product is zero-symmetric.

Lemma 212 *The infinite dihedral group hosts exactly one near-ring with identity, up to isomorphism. It is zero-symmetric.*

Appendix B on page 497 gives some multiplication tables for small near-rings with identities.

5.6 Finite Groups from the Krimmel Class

The main reference in this section is to the paper [100], which is freely available online. That paper has something to say about the near-ring products which may be defined on a class of finite groups to which, in previous publications, I have referred to as the Krimmel class. My present interest is only on whether near-rings with identity may be defined at all, so I skip some of the details. I consider the class of groups with the presentation

$$\mathcal{K} = \langle e, \mu : m\mu = ie, ne = 0, (-\mu + x + \mu) = je\rangle \tag{5.10}$$

i and j are taken to be non-negative integers, with $0 < j < n$ and $0 \leq i < n$. m is prime and

$$j^m \equiv 1 \ (modulo\ n)$$
$$i(j-1) \equiv 0 \ (modulo\ n)$$

These groups have order mn and exponents either n or mn, so I assume that this exponent is actually n in what follows. I am only interested in non-abelian members of the class $\mathcal{K}$ here—that is, in non-abelian finite groups having a cyclic normal subgroup with prime index. With respect to non-zero-symmetric groups, we have Theorem 4.2, from [100]:

Theorem 213 *The group $H \in \mathcal{K}$ hosts non-zero-symmetric near-rings with identity e if and only if:*

(a) The exact sequence $\langle e\rangle \mapsto H \twoheadrightarrow \frac{H}{\langle e\rangle}$ splits.
(b) The commutator subgroup of H has order m.
(c) $m = 2 \Rightarrow n \equiv 8$ (modulo 8).

When this is true, the group hosts precisely one such near-ring, up to isomorphism.

This is a version of Theorem 4.3 from [100]

Theorem 214 *A non-nilpotent group $H \in \mathcal{K}$ hosts zero-symmetric near-rings with identity e if and only if:*

(a) The exact sequence $\langle e\rangle \mapsto H \twoheadrightarrow \frac{H}{\langle e\rangle}$ splits.
(a) The commutator subgroup of the group has prime order p distinct from m.

There is but one isomorphism class of such near-rings.

For nilpotent members of class $\mathcal{K}$, the Sylow decomposition of the nilpotent group H goes over to near-rings, and the problem reduces to m-groups in the class $\mathcal{K}$. These groups have a presentation of the form

$$H = \left\langle e, \mu : m^a e = 0, m\mu = 0, (-\mu + e + \mu - e) = m^{a-1}\right\rangle \tag{5.11}$$

where m and a are prime and $a \geq 2$. The case $(m + a) = 4$ corresponds to D_4, so I assume $(m + a) > 4$. Not all such groups host zero-symmetric near-rings with identity—this is Theorem 4.4 from [100].

Theorem 215 *A group H with presentation (5.11) hosts zero-symmetric near-rings with identity e if and only if:*

(a) The exact sequence $\langle e \rangle \mapsto H \twoheadrightarrow \frac{H}{\langle e \rangle}$ *splits.*
(b) The commutator subgroup of the group has prime order m.

Theorem 7.1 from [100] applies:

Theorem 216

(i) If $m = 2$ *and* $a = 3$*, then (5.11) hosts 32 non-isomorphic zero-symmetric near-rings in which e is the multiplicative identity.*
(ii) If $m = 2$ *and* $a > 3$*, that number is* 2^{a+3}.
(iii) If $m = 3$*, the number is* 3^{a-1}.
(iv) If $m > 3$*, the number is* m^{a-2}.

5.6.1 A Classification Theorem Reported in Gorenstein

Gorenstein [55] gives this classification theorem for non-abelian p-groups with order p^m containing a cyclic subgroup with order p^{m-1}. It is his Theorem 4.4. I use Notation 3 in what follows and should note that (his Theorem 4.3) no two of the groups $M_m(p)$, $D_{2^{m-1}}$, Q_{2^m}, S_m (all of which having order p^m) are isomorphic.

Theorem 217 *Suppose P is a non-abelian p-group with order* p^m *containing a cyclic subgroup with order* p^{m-1}.

(i) When p is odd, $P \cong M_m(p)$
(ii) When $p = 2$ *and* $m = 3$*, P is either* D_4 *or* Q_8.
(iii) When $p = 2$ *and* $m > 3$*, P is one of* $M_m(2)$, $D_{2^{m-1}}$, Q_{2^m} *or* S_m.

5.7 Generalised Quaternion Groups

The class of generalised quaternion groups involves groups with presentations (3). These are groups with order 2^n $(n > 2)$. They have cyclic normal subgroups with prime index 2, and their commutator subgroups are cyclic and have order 2^{n-2}. They have exactly one element with order 2. The zero-symmetric near-rings with identity definable on these groups are discussed in Theorem 215. Since the relevant exact sequence fails to split, there are none. The reader will see that there similarly are no non-zero-symmetric near-rings with identity.

5.8 Dicyclic Groups

Dicyclic groups have presentations of the form (4). They are non-abelian and so can only be I-appropriate (page 49) when the parameter n is even. The exponent then will be $2n$. These are groups with cyclic normal subgroups with order $2n$. Their commutator subgroups are cyclic and have order n. Since the sole involution in these groups is the element na, the key exact sequence mentioned in Theorem 215 fails to split, and these groups do not support near-rings with identity.

5.9 Finite Hamiltonian Groups

These are the finite non-abelian groups in which every subgroup is normal. According to [132], they have the form

$$H = (\mathrm{Q}_8 \oplus A) \tag{5.12}$$

where A is an abelian group containing no element with order 4.[1] When A is a cyclic group with odd order m, these groups have exponents $4m$. They possess a cyclic normal subgroup with prime index 2, and, according to the results in Sect. 5.6, cannot host near-rings with identity. A further one, G_7, occurs on page 175. It is important, because when A does not have odd order, it must be a direct sum: $A = (\mathrm{C}_2 \oplus B)$, where B is an abelian group with no element with order 4. This gives the decomposition:

$$H = (G_7 \oplus B) \tag{5.13}$$

If G_7 actually did host zero-symmetric, unital, near-rings, this would mean that H did too, using direct sums. It seemed to me that G_7 did not host such near-rings, but it would be interesting to know whether a group with the form

$$\left(\mathrm{Q}_8 \oplus \overbrace{(\mathrm{C}_2 \oplus \mathrm{C}_2 \oplus \ldots \mathrm{C}_2)}^{\text{(n times)}} \right) \tag{5.14}$$

could possibly host them, for some smallest value of n. This would go a long way toward determining the finite Hamiltonian groups that host zero-symmetric, unital, near-rings.

---

[1] I hope I don't exasperate the reader by remarking that the direct sum of two copies of Q_8 is not Hamiltonian. One feels it *ought* to be—try looking at j-conjugates of the subgroup generated by $(-i, j)$.

5.10 Semi-dihedral Groups

Semi-dihedral groups are non-abelian 2-groups with presentations given by (6). Their commutator subgroups do not have prime order, so, according to Theorem 215, they do not host near-rings with identity.

5.11 Gorenstein's Group $\mathbf{M}_m(p)$

Consider groups having presentations of the form (6). These are sometimes called the non-abelian modular p-groups.

The conditions of Theorem 215 are met in these groups. They do host near-rings with identity.

5.12 Central Products

The central product of two groups is a quotient of the direct product of those groups by a specific subgroup of the direct product of their centres. More formally, suppose A and B are finite additive groups with centres $Z(A)$, $Z(B)$. Suppose we have subgroups $\alpha < Z(A)$, $\beta < Z(B)$ and an isomorphism: $\theta \ : \ \alpha \mapsto \beta$. Now define

$$N \ = \ \{(x, y) \in \ (\alpha \otimes \beta) \ : \ (\theta(x) + y) = 0\}$$

Clearly, $N \triangleleft (A \times B)$.

The corresponding **central product** is defined as

$$\left(A \ \overset{Z}{\Pi_N} \ B\right) = \left(\frac{(A \oplus B)}{N}\right)$$

If it happens that $(A, \ +, \cdot)$ and $(B, \ +, \ \cdot)$ are near-rings, and that N is an ideal of their direct product, then, of course, the central product will be a near-ring.

5.13 Free Products

Free products correspond to categorial coproducts in the category of all groups [132]. Suppose $(R \ + \ \cdot)$ and $(S, \ +, \cdot)$ are right near-rings. The associated free product of the additive groups is $(R * S)$. Its elements are finite strings $(x_1x_2x_3 \dots x_n)$ such that each entry x_j is in exactly one of the groups R, S (which are to be regarded as disjoint), and no consecutive entries are in the same group. The strings are to be formally regarded as $(x_1 + x_2 + \ldots + x_n)$, and the addition of two such strings is as you would expect, a sort of concatenation resolving the chance abutment of two elements from the same group. The identity element is simply the empty string. Given such a string $(x_1 + x_2 + \ldots + x_n)$, I define two projection operators, π_R and π_S. The first simply replaces all string entries from the group S by the empty string. The second does the same but works with entries from the group R. I can define a product on strings with the formula

$$((x_1 + x_2 + \ldots + x_n) \cdot (y_1 + y_2 + \ldots + y_n)) = \sum_{i=1}^{n} (x_i \cdot \pi_{x_i} (y_1 + y_2 + \ldots + y_n)) \tag{5.15}$$

where

$$\pi_{x_i} = \begin{cases} \pi_R & \text{if } x_i \in R \\ \pi_S & x_i \in S \end{cases}$$

The resulting structure $((R * S),\ +,\ \cdot)$ is a right-distributive near-ring. If R and S have multiplicative right identity elements, 1_R and 1_S, then the string $(1_R + 1_S)$ will be a right identity of the new near-ring. The opposite is untrue of left identity elements, although if 1_R and 1_S are left identities in their near-rings, the element $(1_R + 1_S)$ is interesting: left multiplication by it has the effect of separating and collecting the string entries, producing a string with the form $(r + s)$, where $r \in R, s \in S$. That is,

$$(1_R + 1_S) \cdot (x_1 + x_2 + \ldots x_n) = (\pi_{1_R} (x_1 + x_2 + \ldots x_n) + \pi_{1_S} (x_1 + x_2 + \ldots x_n))$$

The structure just described is not a free product in the category of right near-rings.

5.14 Finite Non-solvable Groups

The transformation near-ring $T_0(A_5)$ is additively a direct sum of 59 copies of the group A_5 and so non-solvable. This is a huge[2] unital near-ring (see Chap. 7). Its order is 60^{59}. Certainly, then, some finite non-solvable groups host near-rings with identity, but one wonders whether there are smaller examples.

Such groups should have an element with the same additive order as their own exponent, and this is my initial criterion for narrowing the search. All groups smaller

[2] Somewhat larger than the "Eddington number", which is Arthur Eddington's estimate for the number of protons in the universe! [38].

than 60 are solvable, so simple combinatorial arguments will be problematic. If one were to obtain the *smallest* non-solvable group hosting unital, zero-symmetric near-rings, $(\Omega, +)$, then its near-rings are either simple or they only contain proper, non-trivial, ideals whose additive groups are non-solvable. In the simple case, the work of Chaps. 7 and 11 suggests that additively the group $(\Omega, +)$ should be a direct sum of isomorphic non-solvable subgroups (as happens with transformation near-rings, for instance).

In the non-simple case, the factor groups host zero-symmetric, unital, near-rings and must be solvable (Ω is, by hypothesis, minimal). Applying the Feit and Thompson theorem, to these ideals, one deduces that they have even order. The multiplicative identity element of the quotient generates a sub-near-ring which corresponds to a zero-symmetric, unital near-ring hosted by a non-solvable group and one which contains the ideal as a subgroup. In this minimal case, we would have an exact sequence

$$J \longmapsto \Omega \twoheadrightarrow \mathrm{C}_m \tag{5.16}$$

and so $(\Omega, +)$ should have a normal, non-solvable, subgroup whose corresponding factor group is cyclic. This markedly reduces the kind of groups one should search for. I look at the first few groups that are possible non-solvable hosts.

5.14.1 Groups with Order 360

The smallest non-solvable group with an element of the same order as its exponent is the direct sum:

$$G = (\mathrm{A}_5 \oplus \mathrm{C}_6) \tag{5.17}$$

This has exponent 30 and order 360. There are no other non-solvable groups with this order that are I-appropriate. Three of this group's proper normal subgroups contain A_5, but the maximum *anti-chain* of such subgroups has cardinality 2 (see Sect. 1.20). The exponent of A_5 is 30, and elements of G having order 30 have the form (g, x) where $g \in \mathrm{A}_5$ has order 5 and x is a generator of C_6. Suppose that $e = (g, x)$ is a multiplicative identity of a near-ring hosted by $(G +)$. We know that the annihilator $\binom{5e}{0}_{\mathcal{L}}$ includes $6e = (g, 0)$ but is a proper subset of G. We deduce that this normal subgroup contains the summand A_5, but since the annihilator cannot contain either $2e$ or $3e$, it must actually be that summand. Three distinct primes divide the exponent of this group: 2, 3 and 5. We deduce that

$$\binom{2e}{0}_{\mathcal{L}} = \langle 15e \rangle = \langle (0, 3x) \rangle$$

and

$$\begin{pmatrix} 3e \\ 0 \end{pmatrix}_{\mathcal{L}} = \langle 10e \rangle = \langle (0, 4x) \rangle$$

(see Lemma 60). Since annihilators are left ideals, $G \cdot (0, 3x) \in \langle (0, 3x) \rangle$. However, non-zero elements of A_5 may not annihilate 3x. So, if $y \in A_5$ and $y \neq 0$, we have that $(y, 0) \cdot (0, x) = (0, 3x)$, which has order 2. We can choose y to have additive order 3, but that would force the non-zero element $(y, 0) \cdot (0, x)$ to have the same order. I conclude that $(G, +)$ hosts no near-rings with identity.

5.14.2 Groups with Order 600

The next smallest non-solvable group containing an element with order equal to its exponent is

$$G = (A_5 \oplus C_{10}) \tag{5.18}$$

The group has order 600 and exponent 30. There are no other non-solvable groups with this order that are I-appropriate. A potential identity element would have the form $e = (g, x)$ where x has order 10 and g order 3. Almost the same argument as the one we just used shows that the group cannot host a near-ring with identity. The same three primes are used but now the argument is that

$$\begin{pmatrix} 3e \\ 0 \end{pmatrix}_{\mathcal{L}} = A_5$$

5.14.3 Groups with Order 720

Four of the 23 non-solvable groups with order 720 are I-appropriate. In two of these, the exponent is 60, and in the other two, it is 30.

Exponent 60: Case 1

My first candidate is the 12th non-solvable group with order 720; in GAP ordering, it is the GAP group (720, 419). It is $G = (A_5 \oplus C_{12})$. A_5 has elements with orders 2, 3 and 5, with 5 the maximal order occurring (page xix). There are 24 elements with this order and they are all conjugate. Any possible identity of G would have the form $e = (g + x)$, where $g \in A_5$ and $|g| = 5$, and $|x| = 12$. I think the same argument as we used above establishes the group does not host near-rings with identity.

There are three remaining possibilities for groups with order 720. I list these now, but this is as far as I have got because the remaining cases are less amenable to direct calculation. It would be nice to determine the smallest non-solvable host and say something of its structure. I know of non-solvable groups hosting zero-symmetric near-rings with identity which are smaller than $T_0(A_5)$; the details are given on page 261. Although these are vastly smaller than $T_0(S_3)$, they are still ridiculously large and smaller examples are needed.

5.14.4 Remaining Possibilities with Order 720

Exponent 60: Case 2
The second candidate has GAP number (720,420). Its commutator subgroup is $SL(2, 5)$ and there is an exact sequence:

$$SL(2,5) \rightarrowtail G \twoheadrightarrow C_6$$

Notice that the group is not $(SL(2, 5) \oplus C_6)$ because that is not I-appropriate (the element structure of $SL(2, 5)$ is reported on page xix).

Exponent 30: Case 1
The first candidate with this exponent has commutator subgroup equal to $A_5 \oplus C_3$ and exact sequence:

$$(A_5 \oplus C_3) \rightarrowtail G \twoheadrightarrow (C_2 \oplus C_2)$$

It is GAP number (720,772).

Exponent 30: Case 2
The second candidate with this exponent is $(A_5 \oplus C_3 \oplus C_2 \oplus C_2)$. This time, the commutator subgroup is A_5. It is GAP number (720,771).

5.14.5 Direct Sums of Simple Groups

One might ask whether, say, a group

$$\mathfrak{g}_r(A_5) = \overbrace{(A_5 \oplus A_5 \oplus \ldots A_5)}^{\text{(r times)}} \tag{5.19}$$

could host unital, zero-symmetric, near-rings, for some $r < (|A_5| - 1)$. More generally, one could look at groups $\mathfrak{g}_r(A)$ where A is *any* finite, non-abelian, simple group. Such a group is centreless and characteristically simple and there are rich decomposition theorems relating to groups of this sort [132]. Suzuki identifies these groups as **completely decomposable groups** [139]. In these groups, normal subgroups are direct factors, so quotients are also sums of copies of (in this case) A_5. If one supposes that r is the smallest positive integer such that $\mathfrak{g}_r(A_5)$ hosts zero-symmetric unital near-rings, then such near-rings must be simple, because otherwise the resulting quotient offers a smaller value to r.

From Theorem 457, the near-ring is 2-primitive and so is an additive direct sum of copies of some additive group $(\Gamma, +)$. From the theory of decomposable groups [132, 139], $\Gamma \cong A_5$, and from the theory of 2-primitive near-rings, $r = (|A_5| - 1)$. The analysis applies to all finite non-abelian simple groups and not just A_5.

Group Invariants

These observations suggest a series of invariants associated with groups involving the minimal number of direct summands that host various categories of near-ring—unital, zero-symmetric, near-rings, in this case.

Chapter 6
Near-Rings Hosted by p-Groups and Related Groups

In this chapter, p will always represent some finite prime, and many of the groups considered will be finite p-groups. All finite p-groups are I-appropriate but not all host near-rings with identity. The determination of the structure of groups with prime power order up to four was a nineteenth-century triumph associated with Hölder, and subsequent work has extended these results. The general story seems to be that the $p = 2$ case is exceptional. For instance, there are 14 non-abelian groups with order 16 but 15 with order p^4 $(p > 2)$. p-groups may well constitute a majority in the class of finite groups. Surprisingly, 99 percent of all groups with order less than 2000 actually have order 1024,[1] and a strongly held piece of group-theoretic folklore is that the fraction of groups that are 2-groups approaches one as one increases order (I don't think there is any proof of this, though there are asymptotic estimates for the number of groups with order p^n.).

Determining near-rings on general p-groups is likely to be quite impossible—perhaps even formally impossible. Fixing the near-rings with identity hosted by groups with small prime power order does seem more tractable; and I make a start on this here. The reader may suspect this to be a sort of mathematical stamp-collecting so I indulge in a little self-justification here.

Over the past 70 years, there has been a great deal of work on near-ring theory, and much interesting material has been found by many exceptional mathematicians. Some, perhaps the bulk, of this work could be characterised as an attempt to extend and generalise successful ideas from the theory of rings—the sort of thing considered in the fourth part of this book. Near-ring theory, in so far as it exists, is such a huge extension of the, already vast, theory of rings as to require an analysis based on subcases and, perhaps, ideas with no direct counterpart in ring theory. I suggest, then, a more inductive analysis of structure could be appropriate, one based on real case studies.

[1] I believe the exact number of groups with order 1024 may be 49,487,365,422.

R. Lockhart, *The Theory of Near-Rings*, Lecture Notes in Mathematics 2295,
https://doi.org/10.1007/978-3-030-81755-8_6

A recurring theme of the present work has been the application of group-theoretic constructs to near-rings; and one can imagine other lines of attack with no obvious ring-theoretic counterpart. One corollary to this is the need for specific accessible examples. Finite p-groups would be an obvious starting point. My own methods are rather ad hoc, but the reader will observe the possibilities for automation of their banalities, even for infinite families of groups.

6.1 Groups with Order p

Jim Clay determined the near-rings on p-groups for p smaller than 8 [22]. Groups with prime order p are, of course, cyclically generated by a single element of order p

$$G = C_p = \langle a : pa = 0 \rangle$$

They are the simple finite abelian group and, in common with cyclic groups of general prime power order, host the finite fields.

Suppose $(G, +, \cdot)$ is a near-ring hosted by such a group. If the near-ring has an identity, then we again have the finite field. We suppose that the near-ring does not have an identity. Elements of G fall into two disjoint sets. The first, N, contains the zero element and includes all elements $x \in G$ such that $G \cdot x = \{0\}$. The second, $M = G - N$, consists of all elements $x \in G$ for which $G \cdot x = G$. These are **right-cancellative** elements in the sense that if $x, y \in M$, and $g \in G$, and $g \cdot x = g \cdot y$, then it will be true that $x = y$. If the near-ring is not to be trivial, there must be an element $a \in G$ such that $a^2 \neq 0$, and without loss of generality, this is an additive generator of the whole group. I write

$$a^2 = ra \quad \text{for some } 0 < r < p$$

It is immediate that

$$a^{j+2} = a \cdot r^j a = r^{j+1} a \quad \text{j = 0,1,2...}$$

a relationship which decides $a \cdot x$ for $|r|_p$ distinct non-zero elements x (where $|r|_p$ represents the order of the integer r modulo p). I suppose that a has been chosen so that r has the maximal available order modulo p. This will be an integer k dividing $(p-1)$. There are $((p-1)-k)$ non-zero elements of G which are not of the form $r^j a$. We may choose some element as which is one of these.

The positive integer s is no power of r, and $|r|_p$ and $|s|_p$ are coprime, though both divide $(p-1)$. I write

$$a \cdot sa = ma \quad \text{where } 0 \leq m < p$$

If m is not zero, this gives us

$$a \cdot m^j sa = m^{j+1} a$$

6.2 The Klein Group

I take a detailed look at the near-rings with identity hosted by the Klein group in Sect. 16.1.6. Elementary abelian p-groups are considered in Chap. 16.

6.3 Groups with Order 2p ($p > 2$)

Groups with order $2p$ are cyclic or dihedral, so this case need not detain us [134, p. 82].

6.4 Groups with Order pq Where p and q Are Prime and ($p < q$)

Such groups are either cyclic or have a cyclic normal subgroup with order q. The latter is not I-appropriate.

6.5 Groups with Order $\mathbf{p^2}$

Groups with order p^2 are abelian.

6.6 Groups with Order $\mathbf{2p^2}$ ($p > 2$)

Of course there are non-I-appropriate groups with order $2p^2$. D_{p^2} is one. Groups with order $2p^2$ have a single, normal, Sylow p-subgroup. The exponent is either $2p$ or $2p^2$, so to be I-appropriate and non-abelian, we should require the $2p$ case and that the single Sylow p-subgroup has the form $S_p = (C_p \oplus C_p)$ (see Sect. 4.1.2). There are two possible presentations:

$$G = \langle x, y, z : px = py = 2z = 0, (x + y) = (y + x), \\ (z + y - z) = -y, (z + x - z) = x\rangle \tag{6.1}$$

$[G, G] = \langle y \rangle$.

$$H = \langle x, y, z \ : \ px = py = 2z = 0, (x + y) = (y + x), \\ (z + y - z) = -y, (z + x - z) = -x \rangle \tag{6.2}$$

$[H, H] = \langle y, x \rangle$. The number of Sylow 2-subgroups is a divisor of p^2, either 1, p or p^2. The single Sylow p-subgroup has order p^2, which means that the set of those elements having order either 2 or $2p$ must have cardinality p^2; and for the group to be I-appropriate, we would need to have fewer than p^2 elements with order 2. There cannot be a single Sylow 2-subgroup because it would have to be $\langle z \rangle$ and be normal. We deduce that there are exactly p elements with order 2 and, presumably, $(p^2 - p)$ elements with order $2p$. We know that

$$(-y + z + y) = (z + 2y) \text{ and, } (-my + z + my) = (z + 2my)$$

so the p order 2 elements are the elements $\{(z + my)\}_{m=0}^{(p-1)}$. None of the distinct elements $\{(z + mx)\}_{m=0}^{(p-1)}$ lies in S_p or has order 2, so these must be elements with order $2p$. But in the case of $(H, +)$, the argument just give implies that the same set of elements $\{(z + mx)\}_{m=0}^{(p-1)}$ should have order 2. We conclude that the group does not host near-rings with identity.

We are now restricted to the group $(G, +)$ of (6.1). It is fairly clear that

$$G = \left(\mathrm{D}_p \oplus \mathrm{C}_p\right)$$

where $\mathrm{D}_p = \langle z, y \rangle$ and $\mathrm{C}_p = \langle x \rangle$. The identity element of the posited near-ring may be taken as $e = (z + x)$. We define

$$(y \cdot w) = \begin{cases} (my) & \text{if } x = me = m(z + x) \\ (0) & \text{otherwise} \end{cases}$$

$$(z \cdot w) = (pw)$$

$$(x \cdot w) = ((p - 1)w)$$

This prescription fits the p.n.r. construction conditions for the group. We need to verify that for all $a, b, \in G$

$$(y \cdot (a \cdot b)) = ((y \cdot a) \cdot b) \tag{6.3}$$

and that is equivalent to proving that

$$(a \cdot b) \in \langle e \rangle_{(mult)} \Leftrightarrow a, b, \in \langle e \rangle_{(mult)}$$

(the remaining parts of the associativity core calculation being straightforward). Suppose $a = (my + sz + qx), b = (wy + uz + vx)$. We may rewrite this in the simpler form:

$$\begin{aligned} a &= (my + n(z + x)) \\ b &= (wy + r(z + x)) \end{aligned}$$

Then,

$$(a \cdot b) = (my \cdot (wy + r(z + x))) + n(wy + r(z + x))$$

Now,

$$(my \cdot (wy + r(z + x))) = \begin{cases} mry & \text{if } w = 0 \\ 0 & \text{otherwise} \end{cases}$$

and

$$(n(wy + r(z + x))) = \begin{cases} nre & \text{if } w = 0 \\ (n(wy + rz) + nrx) & \text{otherwise} \end{cases}$$

So,

$$(a \cdot b) = \begin{cases} (mry + nre) & \text{if } w = 0 \\ (n(wy + rz) + nrx) & \text{otherwise} \end{cases}$$

From this, it seems that

$$(a \cdot b) \in \langle e \rangle_{(mult)} \Leftrightarrow w = m = 0 \ \& \ nr \neq 0 \tag{6.4}$$

When this happens, $y \cdot (ab) = nry$. We also have that

$$(y \cdot a) \cdot b = (y \cdot ne) \cdot re = nry$$

I conclude that this class of groups does host near-rings with identity.

6.7 Groups with Order $\mathbf{p^3}$ ($p > 2$)

In this section, I consider non-abelian p-groups with order p^3 ($p > 2$). We know that in such groups, the centre and commutator subgroup coincide. Marshall Hall [62] offers these presentations for the two non-abelian groups with order p^3 (p an odd prime) in his classic text:

$$(G, +) = \left\langle a, b \ : \ p^2a = pb = 0, [b, -a] = pa \right\rangle \tag{6.5}$$

which has exponent p^2

$$(N, +) = \langle a, b, c : pa = pb = pc = 0, [a, b] = c, [a, c] = [b, c] = 0\rangle \quad (6.6)$$

which has exponent p.

Equation (6.5) is an example of a family of groups $\{\mathrm{Mod}_n(p)\}_{n=2,3,4...}$

$$\mathrm{Mod}_n(p) = \left\langle a, b : p^{n-1}a = pb = 0, [b, -a] = p^{n-2}a \right\rangle$$

which have been called **the modular p-groups** (see (6) and Sect. 5.11). They host near-rings with identity.

The exponent p-group is alternatively represented as

$$\mathbf{Heis}(C_p) = \left\{ \begin{bmatrix} 1 & a & b \\ 0 & 1 & c \\ 0 & 0 & 1 \end{bmatrix} : a, b, c \in C_p \right\} \quad (6.7)$$

I obtained this from an Internet paper by Keith Conrad called "Groups of order p^3", and the notation is his. This is a subgroup of SL(3, p) and is suggestive of further non-abelian groups of triangular matrices, lying in SL($(n+2)$, p) and with orders $\{p^{\frac{(n+1)(n+2)}{2}}\}$ $(n = 1, 2, \ldots)$. Matrices looking like these ones are sometimes called **upper unitriangular matrices**. They crop up in the theory of Lie groups.

Equation (6.6) is a group with order p^3 and exponent p. Its commutator subgroup is $\langle c\rangle_{(add)}$. This is also the group centre. These groups seem very amenable to calculation. Proper, non-trivial, additive subgroups have orders p or p^2 and are abelian. These identities are useful:

$$(-b + a + b) = (a + c) \quad (6.8)$$

$$(-a + b + a) = (b - c) \quad (6.9)$$

$$(-mb + na + mb) = (na + nmc) \quad (6.10)$$

$$(-ma + nb + ma) = (nb - nmc) \quad (6.11)$$

From them, one decides that additive normal subgroups with order p^2 have the form

$$\langle ma + nb, c\rangle_{(add)}$$

and, in fact, that every non-trivial normal subgroup contains $\langle c\rangle_{(add)}$. This additive subgroup must therefore left annihilate any element that has a non-trivial left annihilator. We know $\langle c\rangle.N \subseteq \langle c\rangle$, so the multiplicative identity, e, cannot lie in $\langle c \rangle$, and $c^2 = kc$ for some $k \in \mathbb{N}$, but then $(c - ke) \cdot c = 0$, so c has non-trivial annihilators and $\{N \cdot c\} \neq N$ ($k = 1$ is possible but does not affect the argument that c has non-trivial annihilators). One consequence of this is that $c^2 = 0$ and even that $\langle c \rangle \cdot \langle c \rangle = \{0\}$. Elements of the group are uniquely expressible in the form

$$(ma + nb + rc)$$

so it cannot be the case that $a \cdot c = b \cdot c = 0$. The additive group $\{N \cdot c\}$ has order p or p^2 and contains $\langle c \rangle$. If it has order p^2, then the annihilator of the element c has order p and is itself the additive group $\langle c \rangle$, which must be a (two-sided) ideal. But in that case, the group $\{N \cdot c\}$ also annihilates c, which is a contradiction. We deduce that $\{N \cdot c\} = \{\langle c \rangle\}$ and that the annihilator of c is an additive normal subgroup with order p^2. It's also clear that

$$\{\langle c \rangle\} = \{N \cdot c\} = \{N \cdot (rc)\} \quad (0 < r < p)$$

We do know that we can find $(ma + nb + rc)$ in this annihilator, implying $(ma \cdot c + nb \cdot c) = 0$. It cannot be the case that $(a \cdot c) = (b \cdot c) = 0$; else the near-ring has no multiplicative identity. Consequently, we may choose $a, b, \in N^*$ in such a way that

$$\begin{aligned}(a \cdot c) &= c \\ (b \cdot c) &= c \\ (c \cdot c) &= 0\end{aligned}$$

This would make the left annihilator of c equal to $\langle (a - b), c \rangle_{(add)}$. Replacing a by $(a - b)$ allows us to set

$$\begin{aligned}(a \cdot c) &= 0 \\ (b \cdot c) &= c \\ (c \cdot c) &= 0\end{aligned}$$

So, the near-ring identity element is either b or it is $(b + c)$; and the annihilator of c is $\langle a, c \rangle$. We may replace $(b + c)$ with b, if necessary, and assume the identity element is b itself. We have

$$\begin{aligned}((ma + b) + (ma + b)) \cdot w &= (2ma + 2b - mc) \cdot w \\ &= (2ma \cdot w + 2w - (mc \cdot w))\end{aligned}$$

which gives us

$$[-w, (ma \cdot w)] = (mc \cdot w)$$

The additive group $\{N \cdot a\}$ cannot have order p^3, since

$$\langle c \rangle = \{N \cdot c\} = \{N \cdot a\} \cdot c = \{N \cdot (a \cdot c)\} = \{0\}$$

is impossible. Alternatively, we know the left annihilator of c is $\langle a, c \rangle$, and if we have zero symmetry, this means

$$\langle a \rangle \supseteq \{N \cdot a\} \subseteq= \langle a, c \rangle$$

Thus, $\langle c \rangle \cdot a = \{0\}$. Either the group $\langle N \cdot a \rangle$ has order p^2 (first case) or it has order p (second case). In the first case, $\langle N \cdot a \rangle = \langle a, c \rangle$, and the annihilator of a is exactly $\langle c \rangle$, which is a two-sided ideal. In the second case, $\langle N \cdot a \rangle = \langle a \rangle$, and the annihilator is of the form $\langle (ma + nb), c \rangle$. We do know

$$(ma + nb) \cdot a = (ma^2 + na)$$

If $a^2 \in \langle a \rangle$, then we could construct an annihilator of a which was not in $\langle c \rangle$, so in the first case,

$$a^2 = (ra + sb + vc)$$

where $(s + v) > 0$. Applying (6.8) allows us to deduce $s = 0$, and so both $a^2 = (ra + vc)$, and $\{N \cdot a\} = \langle a, c \rangle$. In the second case, we know that for some m, n, $(ma+nb) \cdot a = 0$. So, $ma \cdot a = -na$, and we have that $a^2 = ka$ for some $0 < k < p$. Thus, generally,

$$a^2 = (ka + vc)$$

where $k \neq 0$ and, in the second case, $v = 0$. Select a primitive root modulo p, π. We know $\{N\} = \{N \cdot \pi b\}$, and we know $(c \cdot \pi b) = mc$ for some $1 \leq m < p$. This determines all products of the form $(c \cdot tb)$, because

$$(c \cdot \pi^j b) = (m^j c)$$

More, for any $w \in N^*$,

$$(c \cdot \pi^j w) = (c \cdot \pi^j b \cdot w) = (m^j c \cdot w)$$

Now, $a \cdot (nb) = (ma + rb + sc) \Rightarrow 0 = (ma \cdot c + rc + 0) = rc$. Thus,

$$(a \cdot nb) = (ma + rc)$$

but then,

$$[(-nb), (a \cdot nb)] = (c \cdot nb)$$

which means

$$(c \cdot nb) = nmc$$

As we observed, we can say a lot about the possible structure of hosted near-rings.

A Particular Case

I attempt to define a zero-symmetric near-ring in which b is the identity. Define

$$(a \cdot (ma + nb + rc)) = \begin{cases} 0 & \text{if } m \neq 0 \\ na & \text{otherwise} \end{cases}$$

and

$$(c \cdot (ma + nb + rc)) = \begin{cases} 0 & \text{if } m \neq 0 \\ nc & \text{otherwise} \end{cases}$$

These settings give us a right-distributive stem

$$(a \cdot (ma + nb + rc)) \cdot (la + tb + sc) = \begin{cases} 0 & \text{if } lm \neq 0 \\ nta & \text{otherwise} \end{cases}$$

and (using Kronecker's delta)

$$(ma + nb + rc) \cdot (la + tb + sc) = \big(\delta_{l,0} mta + n(la + tb + sc) + \delta_{l,0} rtc\big)$$

Consequently,

$$(a \cdot ((ma + nb + rc) \cdot (la + tb + sc))) = \begin{cases} 0 & \text{if } lm \neq 0 \\ nta & \text{otherwise} \end{cases}$$

So the group does host a near-ring with identity element.

6.8 Groups with Order $2p^3$ or Order $2p^4$ ($p > 2$)

In these groups, the Sylow p-subgroup is normal and cannot be cyclic if we are to have non-abelian I-appropriate groups. I believe there are 12 isomorphism classes for the first sort of groups and perhaps 50 for the second.

6.9 Groups with Order p^4 ($p > 2$)

There seem to be ten isomorphism classes to deal with. Presentations for these classes are well-known (indeed, I believe presentations for groups with order p^5 or with order p^6 are known).

I don't continue the calculations for groups with orders $2p^3$, p^4 and $2p^4$—something likely to try the patience of the most committed reader; but establishing the isomorphism classes of near-rings with identity hosted by all these small p-groups seems to me to be a worthwhile undertaking, something that might attract a postgraduate student, perhaps.

6.10 The Prüfer Groups

The Prüfer groups, $\mathbb{Z}(p^\infty)$, are the Sylow p-subgroups of the additive group $\left(\frac{\mathbb{Q}}{\mathbb{Z}}, +\right)$ (p is some fixed prime number). They are locally finite groups (all finite subsets generate finite subgroups), and they have elements with order equal to any finite power of p—that is, they have infinite exponent. Every proper subgroup of these groups is a finite cyclic group, and they are the only infinite abelian group all of whose proper subgroups are finite and cyclic. Kurosh [92] says that the corresponding problem for non-abelian groups—that is, determining those all of whose proper subgroups are finite—is unsolved; he was writing in the 1950s. I understand that examples of infinite non-abelian groups all of whose proper subgroups were cyclic were discovered in 1979 and that they are called **Tarski monster groups**.[2] These are finitely generated groups, and they have the property that each proper subgroup is isomorphic to C_p, (p fixed). They are simple groups, but they are I-appropriate. In 1979, Alexander Olshanskii proved that such groups exist for all primes $p > 10^{75}$. It would be interesting to explore the possible near-rings that these groups might host, and I wonder whether they might support unital near-rings.

The Prüfer groups are clearly not I-appropriate and so they host no unital near-rings. I am now interested in whether they might host more general near-rings. Prüfer groups are *divisible* groups. They are also direct limits (page 92) of sequences of cyclic p-groups: $C_p \subseteq C_{p^2} \subseteq C_{p^3} \subseteq \ldots$, [132]. They have presentations of the form

$$\langle x_1, x_2, \ldots : px_1 = 0, px_{j+1} = x_j, 1 \leq j < \infty \rangle$$

[2] Tarski is reported as having once described himself as the greatest living *sane* logician. He was a contemporary of Gödel.

Thus, they have a countable number of additive generators, one for each prime power order, and they are abelian. From right distributivity and order considerations, each subgroup $\langle a_k \rangle$ is a right ideal of any pre-near-ring. Indeed, if we have a product for which the two conditions

(i) Elements $a_j \cdot x$ are always in $\langle a_j \rangle$ (perhaps being trivial).
(ii) It is always true that $a_j \cdot x = p(a_{j+1} \cdot x)$.

Apply, then we can extend to a unique p.n.r. in the obvious way (these twin conditions corresponding to p.n.r. construction conditions). So in any p.n.r., the value of $a_k \cdot x$ is completely fixed once the value of $a_r \cdot x$ is known, for any fixed $r \geq k$ (in detail, putting $r = j + k$ gives $a_k \cdot x = p^j a_{k+j} \cdot x$). Suppose that we have a non-trivial pre-near-ring $(\mathbb{Z}(p^\infty) + \cdot)$ and that the pre-near-ring is bi-distributive (Definition 87). We may choose a smallest $i \in \mathbb{N}$ and, then, a smallest $j \in \mathbb{N}$ for which $a_i \cdot a_j \neq 0$ and then, of course, $a_i \cdot a_j = k_0 a_1$, for some fixed $0 < k_0 < p$. Right distributivity tells us that we may write $a_{i+r} \cdot a_j = k_r a_{r+1}$, where $0 < k_r < p^{r+1}$ and $k_0 \equiv k_r$ (modulo p). We know $a_{i+j+1} \cdot a_j = k_{j+1} a_{j+1}$ and $p^j a_j = 0$, whereas $p^j k_{j+1} a_{j+1} = k_0 a_1 \neq 0$, so left distributivity cannot occur and obtain the next result:

Lemma 218 *The additive group $(\mathbb{Z}(p^\infty), +)$ hosts no non-trivial bi-distributive pre-near-rings.*

Suppose $w \in \mathbb{Z}(p^\infty)$ and $(\mathbb{Z}(p^\infty) + .)$ are a p.n.r. The annihilator $\begin{pmatrix} w \\ 0 \end{pmatrix}_{\mathcal{L}}$ is, as usual, a left ideal of the p.n.r. and therefore is a two-sided ideal too. It must either be the case that $x_j \cdot w = 0$ for $j = 1, 2, 3, \ldots$ or that there is a smallest $k \geq 1$ such that $x_k \cdot w \neq 0$. In the first situation, the annihilator $\begin{pmatrix} w \\ 0 \end{pmatrix}_{\mathcal{L}}$ is equal to the whole group. In the second situation, it is $\langle x_{k-1} \rangle$, with the convention that $x_0 = 0$. Pursuing that second situation, it is clear that $x_k \cdot w \in \langle x_1 \rangle$, since otherwise k would not be minimal, and it follows that $\mathbb{Z}(p^\infty) = \mathbb{Z}(p^\infty) \cdot w$.

Lemma 219 *Suppose $(\mathbb{Z}(p^\infty), +, \cdot)$ is a p.n.r. and $w \in \mathbb{Z}(p^\infty)$. Then the product $\{\mathbb{Z}(p^\infty) \cdot w\}$ is equal either to $\mathbb{Z}(p^\infty)$ or to $\{0\}$.*

The following represents a non-trivial near-ring hosted by $(\mathbb{Z}(p^\infty) +)$:

- Select any fixed $k \in \mathbb{N}$.
- Set $x_1 \cdot x_k = \lambda_k x_1$ for any fixed $0 \leq \lambda_k < p$.
- Set $x_j \cdot m x_k = m \lambda_k x_j$, for all $0 \leq m < p$.
- Impose right distributivity.
- Repeat until all values for k are exhausted.

For example, take $p = 2$ and define

$$x_j \cdot x_k = \begin{cases} x_j & \text{if k is even} \\ 0 & \text{if k is odd} \end{cases}$$

This represents a near-ring in which each element x_{2m} is a right identity, and each element x_{2m+1} is a right annihilator of the whole near-ring.

Lemma 220 $\mathbb{Z}(p^\infty)$ *hosts an uncountable number of mutually non-isomorphic near-rings.*

Proof Select any subset S of $\mathbb{N}$, and define $(x_j) \cdot (x_k) = \chi_S(k)x_j$, where χ_S is the set characteristic function. That is,

$$x_j \cdot x_k = \begin{cases} x_j & \text{if } k \in S \\ 0 & \text{if } k \notin S \end{cases}$$

Then define $\lambda x_j \cdot \psi x_k = \lambda\psi(x_j \cdot x_k)$ for any $\lambda, \psi \in \mathbb{Z}$. This is easily seen to be a near-ring which I shall call $\mathbb{Z}(p^\infty)^S$. Suppose $\mathbb{Z}(p^\infty)^{S_j}$ and $j \in \{1, 2\}$ are two such near-rings and that they are isomorphic. Any isomorphism, Φ, between the two must map generators x_k to elements θx_k, where $(\theta, p) = 1$. If $S_1 \neq S_2$, then without loss of generality, we can find some smallest $k \in S_1$ such that $k \neq S_2$. Then, if our isomorphic near-rings are $\left(\mathbb{Z}(p^\infty)^{S_1} + \cdot\right)$ and $\left(\mathbb{Z}(p^\infty)^{S_2} + *\right)$, we have

$$\theta x_j = \Phi^{-1}\left(\Phi(x_j) \cdot \Phi(x_k)\right) = (x_j * x_k) = 0$$

Consequently, the two near-rings are not isomorphic. There are an uncountable number of ways in which we can select subsets $S \subset \mathbb{N}$, and each corresponds to a different near-ring. □

Remark 221 Suppose we select two equipotent subsets, $S_j \subseteq \mathbb{N}(j = 1, 2)$ and a bijection $\Phi : \mathbb{N} \to \mathbb{N}$ which maps S_1 bijectively onto S_2. The two near-ring products $\left(\mathbb{Z}(p^\infty)^{S_j}\right)$ and $j \in \{1, 2\}$ are *isotopic* (see (A.3) on page 480) with an isotopy involving bijections rather than group automorphisms and derived from

$$x_i * x_k = x_i \cdot x_{\Phi(k)}$$

6.11 A Research Suggestion

There are available estimates for the number of isomorphism classes of group with order p^n, and we have some knowledge of the near-rings with identity hosted by some p-groups. I only know of one useful construction that takes two near-rings with identity and delivers a third—the direct sum.[3] Perhaps others could be found, but the reader will discover that the obvious suspects (semi-direct products, central products ...) have alibis provided even by the material offered here.

---

[3] I don't see that ultra-products would help in this context.

It would be interesting to compare the number of cases that can be constructed from known subcases, using direct sums, with the number of non-isomorphic p-groups, for a given order. One might assume that all cases with order p^r, with $r < n$, were known. An asymptotic estimate of this fraction would be of some importance. Presumably the estimate tends to zero for ordinary groups; but if one instead considered isomorphism classes of near-rings, it may be that this is not the case, or perhaps one could obtain a non-zero asymptote for particular classes of near-ring. One might imagine a classification of algebraic products based on the asymptotes produced. Asymptotics for, say, the percentage of near-rings with identity as a fraction of the percentage of p.n.r. with identity, hosted by a given group, would be very interesting, though the problem may be intractable with current resources.

My own first foray into research involved determining the near-rings hosted by a particular class of groups [102]. At the time, I thought my work was rather dull and uninspired. Revisiting the same problem, at the other end of a lifetime, forced me to comprehend the possibilities for algebraic structure theorems in this sort of work. Having ignored these considerations for decades, I have come to realise that there is much interesting work to do here, some of it intensely algebraic. I hope someone might take up the challenge and not imagine, as I did, that in some way it was not what algebraists "do".

Part III
Representations and Cohomology

Chapter 7
Transformation Near-Rings

The next two chapters are on transformation near-rings. Much of my analysis follows that of Wielandt, Fröhlich, Betsch, Laxton and other authors and is reported in [119] or in [128]. Mappings are normally applied on the *left* of their arguments: $sin(x)$ rather than $(x)sin$. Addition is pointwise (where there is inherited group addition), and multiplication is by composition. Most of the vectors occurring in this chapter are actually row vectors, and I don't clutter things up with the transposition notation consistency might demand, but I do consistently refer to "tuples" rather than vectors. This chapter includes a very small amount of basic general topology; [68] and [87] are good sources. I also use a number of Gothic symbols here, to denote specific subgroups of transformation near-rings which I am not sure have been treated previously, and some other structures. These include $\mathfrak{H}$, $\mathfrak{T}$, $\mathfrak{A}$, $\mathfrak{M}$, $\mathfrak{Ph}$, $\mathfrak{i}$, $\mathfrak{B}$, $\mathfrak{V}$, $\mathfrak{J}$, $\mathfrak{s}$, and $\mathfrak{W}$.

7.1 Introduction

Transformation near-rings have a very basic role in near-ring theory, one that is at least as vital as the relationship between rings and matrices. This chapter contains a good deal of technical work describing the structure of these near-rings—their sideals, ideals, sub-near-rings, and automorphisms—and then goes on to consider particularly important examples of substructures including many that have not previously been described. Some of this work is important in the representation theory of near-rings, where transformation near-rings are omnipresent (Chap. 11). Chapter 8 then offers a further collection of generalisations, including the important stemhome near-ring (Sect. 8.7) which is central to my work on product theory in Chaps. 15 and 16.

R. Lockhart, *The Theory of Near-Rings*, Lecture Notes in Mathematics 2295,
https://doi.org/10.1007/978-3-030-81755-8_7

The small number of papers I could find on this subject is rather old and will have duplicated some, but not all, of the results in this chapter [65, 79, 80]. In particular, I believe that more than half of the sub-near-rings of transformation near-rings defined in Notation 227 are original to this book.

7.2 Preliminaries

Definition 222

(i) Suppose that M and N are sets; the set $\mathrm{T}(M, N)$ represents all mappings from M to N. Some authors also represent this as N^M.

(ii) When N is an additive group, $\mathrm{T}(M, N)$ acquires additive group structure in which one adds mappings by adding their images:

$$(\theta + \phi)(m) \;=\; (\theta(m) + \phi(m))$$

(iii) When M and N are additive groups, the subset $\mathrm{T}_0(\mathbf{M}, \mathbf{N}) \in \mathrm{T}(M, N)$ represents mappings which map the additive zero of M to the additive zero of N ("normalised" mappings—page 30).

(iv) I am particularly concerned with the situation in which M and N are the same additive group. I then write $\mathrm{T}(M)$ for $\mathrm{T}(M, M)$ and $\mathrm{T}_0(M)$ for $\mathrm{T}_0(M, M)$. These structures have an additional near-ring product defined by functional composition:

$$(\theta \cdot \phi)(m) \;=\; (\theta(\phi(m))) \;=\; (\theta \circ \phi) \tag{7.1}$$

They are called **transformation near-rings**.

(v) If M is a subgroup of N, then $\mathrm{T}(N, M)$ is a sub-near-ring of $\mathrm{T}(N)$.

(vi) I occasionally consider structures of the form $\mathrm{T}_0(M, H)$ where H is a general subset of an additive group. Since this always occurs within the context of an additive group, there is no ambiguity with respect to the identification of normalised mappings. The reader may assume that any such subset includes an additive zero in all cases, although this is not strictly essential.

(vii) I occasionally use the symbol $(\circ)$ in place of $(\cdot)$ when I wish to emphasise the compositional nature of the transformation near-ring product—$(\theta \circ \phi)$ for $(\theta \cdot \phi)$.

(viii) Later, I need to consider structures built from $\mathrm{T}_0(M)$ and need additional subscripts. Rather than writing barbarisms such as $(\mathrm{T}_0)_S(M)$ for these things, I employ the Gothic symbol $\mathfrak{M}$ to stand for T_0. This is a shorthand, and I employ $\mathfrak{M}$ in no other context in this book (see page 294).

(ix) I frequently represent the elements of a finite group $(N, +)$ as $\{n_0 = 0, n_1, \ldots, n_n\}$. This implies that the group has order $(n + 1)$. When dealing with non-zero-fixing mappings, I tend to number from n_1 rather than n_0.

Table 7.1 Symbols representing mappings

Number	Symbol	First used
(i)	$\tau_n(x) = (n + x)$	Page 262
(ii)	$\rho_n(x) = (x + n)$	Page 264
(iii)	$\mathbf{C_n(x)} = n$	Page 237
(iv)	$m_n(x) = [n, x]$	Page 266
(v)	$s_n(x) = n^x = -x + n + x$	Page 266
(vi)	$i_n(x) = x^n = -n + x + n$	Page 266

It should be noted that of the two binary operations in $\mathrm{T}(M)$, only addition depends on the group, multiplication being entirely set-theoretic (actually amounting to the multiplication of stochastic matrices in finite cases; see (7.54)). Although the near-ring $\mathrm{T_0}(M)$ is zero-symmetric, it is not negatively symmetric (defined on page 9).

7.2.1 Mapping Notation

For each $n \in N$, Table 7.1 describes six distinct mappings in $\mathrm{T}(N)$ which are important in what follows. These mappings apply to $\mathrm{T}(N)$, which is an additive group. This means that in finite cases, expressions such as $-(\tau_m)$ can be replaced by some sum of the mappings τ_m—a trivial point which is used and useful in some of the manipulations I do later.

Lemma 223 *There is no proper sub-near-ring of* $\mathrm{T}(N)$ *which properly contains* $\mathrm{T_0}(N)$.

Proof Selecting some $\phi \in \{\mathrm{T}(N) - \mathrm{T_0}(N)\}$, I know $\phi(0) \neq 0$, and so one can find $\theta \in \mathrm{T_0}(N)$ such that $(\theta \circ \phi)(0) = m$, for any fixed $m \in N$. But then we can find $\psi \in \mathrm{T_0}(N)$ such that $(\psi + (\theta \circ \phi)) = \mathrm{C}_m$. So, the sub-near-ring contains $\mathrm{T_0}(N)$ and all the mappings $\{\mathrm{C}_m\}_{m \in N}$. But that makes it equal to $\mathrm{T}(N)$. □

7.2.2 Ideals of $\mathbf{T}(N)$

I refer back to the second example in Sect. 1.8.7 both for notation and for results, here. Suppose J is an ideal of $\mathrm{T}(N)$. Is it possible that $(J \cap \mathrm{T_0}(N)) = \{0\}$? If $j \in J^*$, its fairly easy to see that whenever j has more than two elements of N in its image set, we may construct mappings $\theta, \phi \in \mathrm{T_0}(N)$ such that

$$\theta \cdot (\phi + j) - (\theta \cdot \phi)$$

is a non-trivial mapping in $\mathrm{T_0}(N)$ lying in J. Consequently, for the possibility to be realised, we should require that the ideal J must lie within the sideal $\mathfrak{k}$ (defined—

(1.22)) and that the group $(N, +)$ should be abelian. Given $\theta, \phi \in \mathrm{T}(N)$ and $\mathrm{C}_m \in J^*$, we have

$$(\theta \circ (\phi + \mathrm{C}_m)) - (\theta \circ \phi) = \mathrm{C}_n \in J \tag{7.2}$$

If now we insist $\theta, \phi \in \mathrm{T}_0(N)$, we obtain that $\theta(m) = n$. If we can arrange $\theta \in \mathrm{T}_0(N)$ which violates this property, we can conclude that J does not lie inside $\mathfrak{k}$.

Lemma 224 *When the group $(N, +)$ has at least three elements, any non-trivial ideal of* $\mathrm{T}(N)$ *has a non-trivial intersection with* $\mathrm{T}_0(N)$.

Proof Taking $\phi = Id$ in (7.2) gives us, for all $x \in N$,

$$\theta\,(x + m) = \theta(x) + \theta(m)$$

so, we require that $\mathrm{T}_0(N) = \mathrm{End}(N)$, that is, all zero-fixing mappings should be homomorphisms. □

Interestingly all elements of $\mathrm{T}_0(\mathrm{C}_2)$ are endomorphisms, and $\{\mathrm{C}_0, \mathrm{C}_1\}$ is a two-sided ideal in $\mathrm{T}(\mathrm{C}_2)$ which intersects $\mathrm{T}_0(\mathrm{C}_2)$ trivially.

7.2.3 Automorphisms of T(N)

Starting with a near-ring $(N, +, \cdot)$ and a bijection α of N, one may construct an isomorphic near-ring $(N, \overset{\alpha}{+}, \overset{\alpha}{\cdot})$ with the prescription

$$m \overset{\alpha}{+} n = \alpha^{-1}\,(\alpha(m) + \alpha(n)) \tag{7.3}$$

$$m \overset{\alpha}{\cdot} n = \alpha^{-1}\,(\alpha(m) \cdot \alpha(n)) \tag{7.4}$$

(see page 480). Letting I represent the identity mapping on the set N gives us this bijection of $\mathrm{T}(N)$

$$\alpha(\theta) = (\theta + I)\ \ \alpha : \mathrm{T}(N) \to \mathrm{T}(N) \tag{7.5}$$

and a situation familiar from the elementary algebra of fields, we obtain a near-ring

$$\left(\mathrm{T}(N), \overset{\alpha}{+}, \overset{\alpha}{\cdot}\right) \tag{7.6}$$

defined by

$$\left(\theta \overset{\alpha}{+} \phi\right) = (\theta + I + \phi) \tag{7.7}$$

$$\left(\theta \overset{\alpha}{\cdot} \phi\right) = (\theta \cdot (\phi + I) + \phi) \tag{7.8}$$

although, in field theory, left distributivity permits one to write

$$(\theta \cdot (\phi + I) + \phi) = (\theta \cdot \phi + \theta + \phi)$$

and the set $\{\{N\} - \{-I\}\}$ there becomes an abelian group under the operation $(\overset{\alpha}{\cdot})$. One may do this for any bijection of $\mathrm{T}(N)$, obtaining isomorphic near-rings. There are $(n^n)!$ bijections when $|N| = n$. On the other hand, there are a mere $n!$ bijections of the set N itself. Clearly, general bijections of $\mathrm{T}(N)$ will not preserve bijective mappings of the set N. Taking any general bijection α of $\mathrm{T}(N)$, the isomorphic near-ring $(\mathrm{T}(N), \overset{\alpha}{+}, \overset{\alpha}{\cdot})$ becomes of interest, particularly in relation to (7.4). The right-hand side of the equation involves a product of the images of two elements of T(N), and that product is just the standard composition of mappings. $\alpha^{-1}(0)$ is the zero of the new group, as $\alpha^{-1}(I)$ is the multiplicative identity; but that latter mapping may well not act bijectively on the original set N, because there are no such restrictions on the bijection α.

One would suppose that this new near-ring should recognisably be a transformation near-ring; but then the identity mapping would then bijectively isolate the set of elements acted upon—the new correspondence is abstract, something I find more than a little confusing. Specifically, the *left*-hand side of (7.4) should be identifiable with a function composition.

7.2.4 The Finite Topology

The set of all transformation from M to N, $\mathrm{T}(M, N) = N^M$, has the *finite topology*, discussed by Jacobson in [76] and used by Betsch, Laxton and others in their work on representations of near-rings [10]. This topology is defined by the collection of all sets of the form $\{\mathfrak{C}(\theta, m)\}_{\theta \in \mathrm{T}(M,N), m \in M}$, where

$$\mathfrak{C}(\theta, m) = \{f \in \mathrm{T}(M, N) : f(m) = \theta(m)\} \tag{7.9}$$

in which collection constitutes a topological sub-base.[1]

The sets $\{\mathfrak{C}(\theta, m)\}_{m \in M}$ form a **neighbourhood sub-base** for the element $\theta \in \mathrm{T}(M, N)$. If $\theta \in \mathrm{T}(M, N)$, then the sets

$$\mathfrak{C}(\theta, m_1, m_2, \ldots . m_k) = \{f \in \mathrm{T}(M, N) : f(m_j) = \theta(m_j), j = 1, 2 \ldots . k\} \tag{7.10}$$

[1] So, arbitrary unions of finite intersections generate all the open sets.

(where $m_1, m_2, \ldots m_k$ are a finite collection of elements of M) form a neighbourhood *basis* for the element θ.[2] In finite cases, in this topology, density is the same thing as equality. The topology is produced by first giving N the discrete topology, then forming the product topology on $\mathrm{T}(M, N)$, and finally taking the subspace topology if, say, $\mathrm{T}_0(M, N)$ were wanted. I do exactly this when discussing the finite topology on $\mathfrak{M}_S(\Gamma)$ later (Sect. 7.6.2).

7.2.5 Sub-near-Rings

In this chapter

M and N should be regarded as additive groups rather than general sets.

The non-zero-symmetric near-ring **T**(N) consists of the near-ring of all mappings from $(N, +)$ to itself. It contains $\mathrm{T}_0(N)$ as a sub-near-ring (it is a left ideal, of course—it is actually $(\mathrm{T}(N))_0$ (see Lemma 80)). Note that

$$\mathrm{T}(N) \cdot \mathrm{T}_0(N) \supsetneq \mathrm{T}_0(N) \tag{7.11}$$

$$\mathrm{T}_0(N) \cdot \mathrm{T}(N) \supsetneq \mathrm{T}_0(N) \tag{7.12}$$

So, the left-ideal condition is distinct from (7.11), something not obvious from ring theory. The near-ring $\mathrm{T}(N)$ is not zero-symmetric, though $\mathrm{T}_0(N)$ is. In finite situations, $\mathrm{T}_0(M, N)$ is, additively, a direct product of $(|M| - 1)$ copies of $(N, +)$; and $\mathrm{T}(M, N)$ is a direct product of $|M|$ copies. More formally and using the notation described on page xvi,

$$\mathrm{T}_0(M, N) \cong \prod_{m \neq 0} N \tag{7.13}$$

is an isomorphism of additive groups. Any near-ring, $(M, +, \cdot)$, embeds in a transformation near-ring, $\mathrm{T}(N)$: simply take a group N properly containing M, and define for each $m \in M$

$$\mathcal{E}_m(x) = \begin{cases} m \cdot x & \text{if } x \in M \\ m & \text{otherwise} \end{cases} \tag{7.14}$$

[2] So all open sets are unions of these open sets.

When $(M + \cdot)$ is zero-symmetric, the embedding is into $\mathrm{T}_0(N)$. The obvious special case is

$$N = M \oplus \mathrm{C}_2$$

and we then might write

$$\mathcal{E}_m(m, z) = \begin{cases} (m \cdot x, 0) & \text{if } z = 0 \\ (m, 0) & \text{if } z = 1 \end{cases}$$

—embedding $(N, +, \cdot)$ into $\mathrm{T}_0(N \oplus \mathrm{C}_2)$.

Lemma 225 *When A and $\{B_j\}_{j \in J}$ are additive groups,*

$$\mathrm{T}_0\left(A, \prod_j (B_j)\right) \cong \prod_j \mathrm{T}_0(A, B_j)$$

(the isomorphism being one of additive groups).

Remark 226 The near-ring $\mathrm{T}_0(N)$ acts on the additive group $(N, +)$ in the obvious way. If $n \in N - \{0\}$, then $\mathrm{T}_0(N) \cdot n = N$, so N possesses no proper, non-trivial, subsets which are closed under the action of $\mathrm{T}_0(N)$. Thus, N is a minimal left—$\mathrm{T}_0(N)$—group (see Definition 20).

7.2.6 E(N), I(N), A(N), 𝔅(N), *and* Phom(N)

The following extensive set of notations and comments describes some of the sub-near-rings of $\mathrm{T}(N)$ and of $\mathrm{T}_0(N)$ that are of interest to me. In this section and much of the rest of the chapter, I need the idea of a **sideal**—defined in Sect. 1.8. Also, I need the notation in Sect. 1.15.

Notation 227

(i) The near-ring $\mathrm{T}_{(C)}(N)$ is the set of all constant mappings from the group $(N, +)$ to itself. It is a sub-near-ring of $\mathrm{T}(N)$. Constant mappings have the form C_g where g is some fixed member of N and

$$\mathrm{C}_g(n) = g \quad \forall n \in N \tag{7.15}$$

and $\mathrm{T}_{(C)}(N)$ is a constant near-ring in the sense of Definition 78 (i.e. $x \cdot y = x$, for all x). Additively

$$\mathrm{T}_{(C)}(N) \cong (N, +)$$

The corresponding near-constant sub-near-ring $T_{(\mathcal{C},0)}(N)$ can be defined as the mappings $C_{(g,0)}(n) = \begin{cases} g & \text{if } n \neq 0 \\ 0 & \text{if } n = 0 \end{cases}$ and that embeds in $T_0(N)$.

Of course

$$\left(T_{(\mathcal{C},0)}(N) +\right) \cong (N, +)$$

More, since, for any $\theta \in T(N)$, we have

$$\left(C_g \circ \theta\right)(m) = g = C_g(m)$$

we have that

$$T_{(\mathcal{C})}(N) \cdot T(N) = T_{(\mathcal{C})}(N)$$

although

$$T_{(\mathcal{C},0)}(N) \cdot T_0(N) \not\subset T_{(\mathcal{C},0)}(N)$$

Thus, $T_{(\mathcal{C})}(N)$ is right sideal of $T(N)$, but $T_{(\mathcal{C},0)}(N)$ is not a right sideal of $T_0(N)$.

(ii) Taking any subset $S \subseteq N$ which contains the group identity element, we may form

$$T_{(S\to 0)}(N) = \{\theta \in T_0(N) : \theta(S) = \{0\}\} \tag{7.16}$$

This is a sub-near-ring of $T_0(N)$, and we have the obvious additive isomorphism

$$T_{(S\to 0)}(N) = \prod_{S^c} (N, +) \tag{7.17}$$

(set complement notation page xvii). More

$$S \subsetneqq V \Rightarrow T_{(V\to 0)}(N) \subsetneqq T_{(S\to 0)}(N)$$

bequeathing us a rich collection of chains of near-rings within the near-ring $T_0(N)$. Naturally (see Notation 48),

$$T_{(S\to 0)}(N) = \binom{S}{0}_{\mathcal{L}}^{N} \tag{7.18}$$

so the sub-near-ring is actually a left ideal of $T_0(N)$. More, when the group is infinite, we may construct infinitely ascending and descending chains of left ideals. Sub-near-rings of $T_0(N)$ not properly contained in one of these chains are those mappings taking each non-zero element of $(N, +)$ to something other than zero, that is, sub-near-rings whose kernel sets (defined below) are actually equal to $\{0\}$.

(iii) (7.18) describes what left annihilators of subsets of N look like. I am also interested in left and right annihilators of subsets of $T_0(N)$. Suppose Θ is a subset of $T_0(N)$. Define

$$\mathcal{I}(\Theta) = \{\theta(n) : \theta \in \Theta \text{ and } n \in N\} \tag{7.19}$$

This is the **image set** of the mappings in Θ. It is a subset of N, containing 0. Then,

$$\begin{pmatrix} \Theta \\ 0 \end{pmatrix}_{\mathcal{L}}^{T_0(N)} = \begin{pmatrix} \mathcal{I}(\Theta) \\ 0 \end{pmatrix}_{\mathcal{L}}^{N} \tag{7.20}$$

so left annihilators of subsets of $T_0(N)$ are the same as left annihilators of their image sets. I reinforce that point that (7.20) shows that subsets of $T_0(N)$ have non-trivial annihilators precisely when their image sets do not encompass the whole of N. In finite cases, relation (7.17) permits us to count the left annihilator ideals of the transformation near-ring. Turning to right annihilators and again fixing a subset $\Theta \subseteq T_0(N)$ define

$$\mathcal{K}(\boldsymbol{\Theta}) = \{n \in N : \theta(n) = 0, \text{ for all } \theta \in \Theta\} \tag{7.21}$$

This is the **kernel set** of the mappings in Θ. It is a subset of N containing 0. Don't confuse this with (7.16), which is a subset of $T_0(N)$. Then,

$$\begin{pmatrix} \Theta \\ 0 \end{pmatrix}_{\mathcal{R}}^{T_0(N)} = T_0(N, \mathcal{K}(\Theta)) = \begin{pmatrix} N \\ \mathcal{K}(\Theta) \end{pmatrix}_{\mathcal{L}}^{T_0(N)} \tag{7.22}$$

Right annihilators are right sideals of the form $T_0(N, S)$, where S is a subset of N. Additively, they are of the form

$$\prod_{N^*} (S)$$

and, again, they may be counted in finite cases.

(iv) $\mathfrak{B}(\mathbf{N})$ is the *group* generated additively by the bijective mappings in $T(N)$. This has the subgroup $\mathfrak{B}_0(N)$ consisting of its zero-fixing members. Neither group is likely to consist entirely of mappings which are themselves bijective; but notice that the bijective mappings in $T_0(N)$ do additively generate a subgroup of $\mathfrak{B}_0(N)$. The bijective elements of $T(N)$ are precisely the multiplicatively invertible elements. These elements have additive order equal to the exponent of the group $(N, +)$, and we may generalise again to consider groups $\mathcal{C}(\mathbf{N})$ and $\mathcal{C}_0(\mathbf{N})$ by enlarging the generating set to include all elements with that additive order. Such larger groups will be characteristic subgroups of their respective transformation near-rings.

(v) The near-ring $\mathbf{I}(\mathbf{N})$ is the distributively generated near-ring (see Definition 75) generated additively from the inner automorphisms of the additive group $(N, +)$.

(vi) The near-ring $\mathrm{A}(\mathbf{N})$ is the distributively generated near-ring generated additively from the automorphisms of the additive group $(N, +)$. We have

$$\mathrm{I}(N) \subseteq \mathrm{A}(N) \subseteq \mathfrak{B}_0(N) \subseteq \mathcal{C}_0(N) \tag{7.23}$$

(vii) The near-ring $\mathrm{E}(\mathbf{N}) = \langle \mathrm{End}(N, +) \rangle$ is the distributively generated near-ring generated additively from the endomorphisms of the additive group $(N, +)$. It is usually called **endomorphism near-ring** of the group $(N, +)$, even though it is only generated by the additive endomorphisms and is not usually a group of them; so, its members are sums of additive endomorphisms rather than endomorphisms per se. When $(N, +)$ is a is abelian, $\mathrm{End}(N) = \mathrm{E}(N)$ is another ring. There is a theorem associated with Baer and Kaplansky which says that for torsion abelian groups isomorphic endomorphism rings imply isomorphic groups [90]. Of course (page 43)

$$\mathrm{C}_{\mathcal{D}}(\mathrm{T}_0(N)) = \mathrm{End}(N) \tag{7.24}$$

This means that (from the observation about multiplicatively invertible elements, above, and from (1.39))

Lemma 228 *Inner automorphisms of* $\mathrm{T}_0(N)$ *are generated by* $\mathrm{Aut}(N)$.

It is possible for a group $(N, +)$ to be non-abelian and yet one of $\mathrm{I}(N)$ or $\mathrm{A}(N)$ or $\mathrm{E}(N)$ to still be a ring. There are several papers on this topic. Pilz refers to a 1932 paper of Fitting investigating automorphisms of non-abelian groups which sum to further automorphisms [128]; and it is shown in [20], for instance, that $\mathrm{I}(N)$ is a ring exactly when $(N, +)$ has the property that conjugate elements always commute. Such a group is nilpotent of class no more than three, and this class of groups is generally known as the class of **L-groups**. The nilpotent groups of class two are exactly the non-abelian groups for which $\mathrm{I}(N)$ is a *commutative* ring.

There are non-abelian groups for which $\mathrm{E}(N)$ is a ring. I think it is true that all known finite examples are nilpotent of class two. People speak of **E-groups** in this context and **A-groups** when discussing automorphisms, though these terms are not quite so standard. This paper [111] gives an example of a non-abelian E-group. Of course, $\mathrm{I}(N)$ is only a near-ring consisting entirely of endomorphisms when $(N, +)$ is actually abelian, and it is then a ring. Sect. 8.5 describes near-rings in which endomorphisms are additively closed with a rather different sort of addition employed (see (8.20)).

(viii) The near-ring Phom(N) consists of the subset of all mappings in $T_0(N)$ which are also *phomomorphisms* (see Definition 47). We have

$$A(N) \subseteq E(N) \subseteq \text{Phom}(N) \subseteq T_0(N) \tag{7.25}$$

I repeatedly use Lemma 354 in calculations of the size of Phom(N) in this and other chapters. As defined here, phomomorphisms are already zero-fixing mappings (normalised). If one waives that requirement, one obtains a larger near-ring, $\mathcal{P}(\mathbf{N}) \subseteq T(N)$, of pseudo-homomorphic mappings (defined in (1.25)). Such mappings have the form

$$\alpha = (\theta + C_\delta) \tag{7.26}$$

where θ is a *normalised* phomomorphism and C_δ is a constant mapping (Table 7.1) with $\delta \in [N, N]$ and $\theta(0) = -\delta$. The sub-near-ring $T_{(\mathcal{C})([\mathbf{N},\mathbf{N}])} = \{C_\delta : \delta \in [N, N]\} \subseteq T(N)$ combines with Phom(N) to produce this larger near-ring of not-necessarily normalised phomomorphisms. Phom(N) is a normal subgroup of the larger structure (which additively is a semi-direct sum of Phom(N) with $T_{(\mathcal{C})}([N, N])$). That is, additively

$$(\mathcal{P}(N), +) = \big(\text{Phom}(N) \rtimes T_{(\mathcal{C})}([N, N])\big) \tag{7.27}$$

(compared with Lemma 80). Of course, $T_{(\mathcal{C})}([N, N])$ is, additively, isomorphic to $([N, N], +)$.

(ix) Another near-ring that seems relevant is the Neumann near-ring, $(\mathfrak{N}, \frac{2}{+} \cdot)$. In this near-ring, all mappings are endomorphisms, and $(N, +)$ is a reduced free group (see page 323). Each mapping in $\mathfrak{N}$ lies in $T_0(N)$, but the near-ring is not itself a sub-near-ring of it, because its own addition is incompatible with that of $(T_0(N), +)$, although the multiplication is the same.

(x) The near-ring $\mathcal{F}T_0(\mathbf{N})$ is the sub-near-ring of $T_0(N)$ consisting of mappings which are non-zero on only a finite number of elements of N—that is, mappings having **finite support** (see Sect. 1.8.7). Thus,

$$\mathcal{F}T_0(N) = \{\theta \in T_0(N) : \theta(n) = 0 \text{ p.p. } n \in N\} \tag{7.28}$$

and we have, additively,

$$\mathcal{F}T_0(N) = \bigoplus_{n \neq 0} N \tag{7.29}$$

We may extend this idea to consider

$$\mathcal{F}T(\mathbf{M}, \mathbf{N}) = \{\theta \in T(M, N) : \theta(m) = 0 \text{ p.p. } m \in M\} \tag{7.30}$$

where M and N are additive groups.

Lemma 229 *The near-ring* $\mathcal{F}\mathrm{T}_0(N)$ *is dense in the near-ring* $\mathrm{T}_0(N)$ *(referring to the finite topology—page 235—and the definition on page 6).*

Proof Select $\theta \in \mathrm{T}_0(N)$. A neighbourhood of θ has the form $\mathfrak{C}(\theta, n_1, n_2, \ldots . n_k)$. Clearly there are elements of $\mathrm{T}_0(N)$ in this neighbourhood. □

When R is a commutative ring, we have the group ring, **R(G)**, and the identification of sets:

$$R(G) \equiv \mathcal{F}\mathrm{T}(G, R) \tag{7.31}$$

(the reader may be more familiar with group rings in which the group is actually finite; [122] discusses the infinite case).

Jacobson noticed this result and reports it in [76].

Lemma 230 *When* $(G, +)$ *and* $(H, +)$ *are additive groups,* $\mathrm{Hom}(G, H)$ *is a dense subset of* $\mathrm{T}_0(G, H)$ *in the finite topology.*

Proof Jacobson selects any $\theta \in \mathrm{T}_0(G, H)$ belonging to the closure of $\mathrm{Hom}(G, H)$. Any neighbourhood of θ intersects with some $f \in \mathrm{Hom}(G, H)$. He selects any $g_1, g_2 \in G$ and considers the neighbourhood:

$$\mathfrak{C}(\theta, g_1, g_2, (g_1 + g_2))$$

By hypothesis, we can find some $f \in \mathrm{Hom}(G, H)$ such that

$$\begin{aligned} f(g_1) &= \theta(g_1) \\ f(g_2) &= \theta(g_2) \\ f(g_1 + g_2) &= \theta(g_1 + g_2) \end{aligned}$$

Not only that, but we can do this for whichever g_1, g_2 we choose. This forces θ into being a homomorphism and proves the lemma. □

Corollary 231 $\mathrm{Phom}(G, H)$ *is a dense subset of* $\mathrm{T}_0(G, H)$.

A Research Suggestion

It would be an interesting problem to determine presentations for $\mathrm{E}(N)$, $\mathrm{I}(N)$, $\mathrm{A}(N)$ and perhaps even $\mathrm{Phom}(N)$. Such presentations would be based on a pre-existing presentation for $(N, +)$. $\mathrm{T}_0(N)$ is, of course, straightforward.

7.3 Multiplicative Structure

The near-ring product in $\mathrm{T}(N)$ is simply a composition of mappings and is entirely set-theoretic in nature, not depending on the additive structure of the group. In this section, I look chiefly at this monoid of mappings and multiplicative structure, turning to the additive side of things in the next. Most recent authors write $\mathrm{T}(N)$ as $\mathrm{M}(N)$ and $\mathrm{T}_0(N)$ as $\mathrm{M}_0(N)$ —generally using M in place of T. My notation was used in early work on near-rings but is now non-standard. As a genuflection to current practice, I do use the single symbol $\mathfrak{M}$ for T_0. In later work, I replace $\mathrm{T}_0(Q, G)$ by $C^1(Q, G)$ (see 9.2), in order to be consistent with early researchers in group cohomology. Very occasionally, I shall abbreviate $\mathrm{Phom}(N)$ to $\mathrm{P}_0(N)$; this occurs mostly within this chapter or Chap. 9.

Notation 232 Suppose $(\Gamma, +)$ is an additive group and $a, b, \in \Gamma$ $(b \neq 0)$. I define $\Delta^{(a,b)} \in \mathrm{T}_0(\Gamma)$ by

$$\boldsymbol{\Delta}^{(a,b)}(x) = \delta_{b,x}a$$

That is to say,

$$\Delta^{(a,b)}(x) = \begin{cases} a & \text{if } x = b \\ 0 & \text{otherwise} \end{cases} \quad (b \neq 0) \tag{7.32}$$

This does not necessarily describe a unique member of $\mathrm{T}_0(N)$, $\Delta^{(0,x)} = \Delta^{(0,y)}$ for all x, y. It is also convenient to allow $b = 0$ on occasion, but then one must ensure only $\Delta^{(0,b)}$ is allowed.

When $(N, +)$ is finite, any $\theta \in \mathrm{T}_0(N)$ has the form

$$\theta = \sum_{n \in N, n \neq 0} \Delta^{(\theta(n), n)} \tag{7.33}$$

This will also be true for mappings with finite support (7.28). The following, rather, pyramidical, result is of some technical importance:

Lemma 233 *Fix some $b \in N^*$.*

(i) The subgroup

$$\left\langle \{\Delta^{(a,b)}\}_{a \in N} \right\rangle = \{\Delta^{(a,b)}\}_{a \in N} \tag{7.34}$$

is additively isomorphic to $(N, +)$.

(ii)

(a) That subgroup is normal in $(\mathrm{T}_0(N), +)$, being a direct summand of it.

(b) If $\theta \in \mathrm{T}_0(N)$, then (using additive notation for conjugates)

$$\left(\Delta^{(a,b)}\right)^{\theta} = \Delta^{(a^{\theta(b)}, b)} \tag{7.35}$$

(iii)

(a)

$$\left(\Delta^{(a,b)} + \Delta^{(d,b)}\right) = \Delta^{(a+d,b)} \tag{7.36}$$

(b)

$$-\left(\Delta^{(a,b)}\right) = \Delta^{(-a,b)} \tag{7.37}$$

(c) If $b \neq d$, then

$$\left(\Delta^{(a,b)} + \Delta^{(c,d)}\right) = \left(\Delta^{(c,d)} + \Delta^{(a,b)}\right) \tag{7.38}$$

(d)

$$\Delta^{(a,b)} \cdot \Delta^{(c,d)} = \delta_{b,c}\, \Delta^{(a,d)} \quad (d \neq 0). \tag{7.39}$$

(e) If $a, b, c, d \in N$, $d \neq 0$, $b \neq 0$, $d \neq b$, then for all $\theta \in \mathrm{T}_0(N)$

$$\left(\theta \circ \left(\Delta^{(a,b)} + \Delta^{(c,d)}\right)\right) = \left(\left(\theta \circ \Delta^{(a,b)}\right) + \left(\theta \circ \Delta^{(c,d)}\right)\right) \tag{7.40}$$

(f) If $a, b, c, d \in N^*$ and $f, g, h, k \in N$ and $c \neq d$, then

$$\left(\Delta^{(f,a)} + \Delta^{(g,b)}\right) \circ \left(\Delta^{(h,c)} + \Delta^{(k,d)}\right) = \left(\Delta^{(f,a)} \circ \Delta^{(h,c)}\right) + \left(\Delta^{(f,a)} \circ \Delta^{(k,d)}\right) + \left(\Delta^{(g,b)} \circ \Delta^{(h,c)}\right) + \left(\Delta^{(g,b)} \circ \Delta^{(k,d)}\right) \tag{7.41}$$

(iv) *$\{\Delta^{(a,c)} : a, c \in N, c \neq 0\}$ additively generates $\mathcal{F}\mathrm{T}_0(N)$.*
In fact, any non-zero $\theta \in \mathcal{F}\mathrm{T}_0(N)$ has the form

$$\theta = \left(\Delta^{(a_1,b_1)} + \Delta^{(a_2,b_2)} + \ldots + \Delta^{(a_k,b_k)}\right) \tag{7.42}$$

where $b_1, b_2, \ldots, b_k$ are some finite number of distinct non-zero members of N and $a_1, a_2, \ldots, a_k$ are members of N (not necessarily distinct).

(v) *If $\theta \in \mathrm{T}_0(N)$ then*

$$\left(\theta \circ \Delta^{(a,b)}\right) = \Delta^{(\theta(a),b)} \tag{7.43}$$

In particular, the set $\{\Delta^{(a,b)} : a \in N\}$ is a left sideal of $\mathrm{T}_0(N)$ (b is fixed and non-zero).

(vi) *The set* $\{\Delta^{(a,b)} : a \in N\}$ *is a left ideal of* $\mathrm{T}_0(N)$.
In fact, if $\theta, \phi \in \mathrm{T}_0(N)$,

$$\theta \circ \left(\phi + \Delta^{(a,b)}\right) - (\theta \circ \phi) = \Delta^{((\theta(\phi(b))+a-\theta(\phi(b))),b)}$$

(vii) *If* $\theta \in \mathrm{T}_0(N)$, *suppose that*

$$\{c : \theta(c) = b\} = \{c_1, c_2, \ldots, c_k\}$$

for some $k < \infty$, *then*

$$\left(\Delta^{(a,b)} \circ \theta\right) = \left(\Delta^{(a,c_1)} + \Delta^{(a,c_2)} + \ldots + \Delta^{(a,c_k)}\right) \tag{7.44}$$

If we can find no $c \in N$ *for which* $\theta(c) = b$, *then, of course,* $\left(\Delta^{(a,b)} \circ \theta\right) = 0$.

(viii) *If* $\phi \in \mathrm{T}_0(N)$ *and* $a_1, a_2, \ldots, a_k, c_1, c_2, \ldots, c_k \in N$, *with all* $c_1, c_2, \ldots, c_k$ *both distinct and non-zero, then*

$$\phi \circ \left(\Delta^{(a_1,c_1)} + \Delta^{(a_2,c_2)} + \ldots + \Delta^{(a_k,c_k)}\right) =$$

$$\left(\Delta^{(\phi(a_1),c_1)} + \Delta^{(\phi(a_2),c_2)} + \ldots + \Delta^{(\phi(a_k),c_k)}\right) \tag{7.45}$$

(ix) *Using the notation defined above, the collection of all expressions of the form*

$$\left\{\left(\Delta^{(\phi(a_1),c_1)} + \Delta^{(\phi(a_2),c_2)} + \ldots + \Delta^{(\phi(a_k),c_k)}\right) : \phi \in \mathrm{T}_0(N)\right\} \tag{7.46}$$

will be a left sideal of $\mathrm{T}_0(N)$.

(x) *If* $\{a_0, a_1, \ldots, a_n\}$ *is some set of additive generators for* $(N, +)$, *then any* $\theta \in \mathrm{T}(N)$ *is a sum of terms from the set* $\{\Delta^{(a_j,x)} : 0 \le j \le n, x \in N\}$.

7.3.1 Sideals and Cleiks

I shall often assume that the additive group $(N, +)$ is finite in this subsection, though not always, and indeed much of the material generalises. Subsequent work makes some of my observations otiose, in finite cases. I gave a simple example of two-sided sideals in transformation near-rings on page 21.

Lemma 234 *Suppose* $S \subseteq \mathrm{T}_0(N)$ *is non-trivial and both a two-sided sideal and an additive subgroup. Then*

$$\mathcal{F}\mathrm{T}_0(N) \subseteq S \tag{7.47}$$

Proof There is some $s \in S$ such that for some $a \in N$, $s \cdot a = b \neq 0$. Then, for any $c, d \in N$, with $d \neq 0$

$$\Delta^{(c,d)} = \Delta^{(c \cdot b)} \cdot s \cdot \Delta^{(a,d)} = \Delta^{(c,d)} \in S$$

□

Cleiks and Super-Cleiks

Definition 235 Suppose $(N, +)$ is a finite group and $a_1, a_2, \ldots, a_k \in N$, and $c_1, c_2, \ldots, c_k$ are mutually distinct elements of N. The expression

$$\left(\Delta^{(a_1,c_1)} + \Delta^{(a_2,c_2)} + \ldots + \Delta^{(a_k,c_k)}\right) = \sum_{i=1}^{k}\left(\Delta^{(a_i,c_i)}\right) \tag{7.48}$$

is called a **cleik**[3] fixed by $\{a_j\}_{j=1}^k$ and $\{c_j\}_{j=1}^k$ or the cleik

$$\{a_1/c_1, a_2/c_2, \ldots, a_k/c_k\} = \{a_i/c_i\}_{i=1}^k$$

If $\{c_1, c_2, \ldots c_k\} = N$, the cleik is called a **full cleik**. The set of expressions

$$\left\{\left(\Delta^{(\phi(a_1),c_1)} + \Delta^{(\phi(a_2),c_2)} + \ldots + \Delta^{(\phi(a_k),c_k)}\right) : \phi \in \mathrm{T}(N)\right\} \tag{7.49}$$

is the **associated super-cleik**.

The addition used in Definition (7.48) is formally related to that of the underlying group but is actually independent of it. In finite cases, it turns out that the addition is actually closer to the standard addition of $(0-1)$ matrices (see 7.54). The definition was given for transformation near-rings $T(N)$. When dealing with $\mathrm{T}_0(N)$, one must insist that

$$\Delta^{(a,0)} \textit{ with } a \neq 0$$

is **banned** and make obvious minor modifications to the terms used. Cleiks are structures that are entirely related to the multiplicative part of the transformation

[3] On the authority of Mr Oliver and Mrs Maureen McLauchlan, I can say that in Scottish dialect, a "cleik", is a thin piece of bent iron used to extract lobsters and crabs from their hiding places by exploiting their aggression—inducing them to catch hold and then pulling them out. I had intended to christen these structures "cliques" and the two words are pronounced identically, but that term is used in [54] in structures relating to the spectrum of prime ideals (page 413) and in [17] relating to graph theory—a subset of vertices such that any two are connected—a concept that may well be connected to my one!

near-ring, and we can even weaken the additive properties of $(N, +)$ without affecting them (see Sect. 8.6.4). Using the definition given in Table 7.1, we can say

$$\tau_m = \left(\Delta_{(m,n_0)} + \Delta_{(m,n_1)} + \ldots + \Delta_{(m,n_n)}\right) \tag{7.50}$$

so τ_m is a full cleik.

Lemma 236 *If* $\sum_{i=1}^{k}\left(\Delta^{(a_i,c_i)}\right)$ *is a cleik and* $b, d \in N^*$*, then*

$$\Delta^{(b,d)} \cdot \left(\sum_{i=1}^{k}\left(\Delta^{(a_i,c_i)}\right)\right) = \left(\sum_{i=1}^{k}\left(\Delta^{(b,d)} \cdot \Delta^{(a_i,c_i)}\right)\right) \tag{7.51}$$

and this is itself a cleik (see (7.39)).

Suppose we have a cleik $\{a_i/c_i\}_{i=1}^{2}$ and $\theta \in \mathrm{T}_0(N)$. From Lemma 233, we can say

$$\left(\{a_i/c_i\}_{i=1}^{2}\right)^{\theta} = \left(\Delta^{(a_1^{\theta(c_1)},c_1)} + \Delta^{(a_2^{\theta(c_2)},c_2)}\right)$$

and might suppose that this *additive* conjugation keeps us within the associated super-cleik. That would require us to imagine the conjugation as relating to a mapping $\phi \in \mathrm{T}_0(N)$, where

$$\phi(a_1) = a_1^{\theta(c_1)} \; ; \; \phi(a_2) = a_2^{\theta(c_2)}$$

and it will be seen that this cannot generally be arranged (for instance, suppose $\theta(c_2) = 0$ and $\theta(c_1) \neq 0$ and $a_1 = a_2$). So, cleiks and super-cleiks cannot be claimed as normal subsets. $\mathrm{T}_0(N)$ can be thought of as the Cartesian product of $|N| - 1$ isomorphic copies of the group $(N, +)$, so we may identify a mapping $\theta \in \mathrm{T}_0(N)$ with the tuple of elements of $(N, +)$ defined by its image on the non-zero group elements. To set this up less informally, we should really define an ordering of the elements of the group which would then index the tuple elements. In infinite cases, these tuples would have to have infinitely many non-zero elements although in $\mathcal{F}\mathrm{T}_0(N)$ this will not be the case. I return to these ideas in Definition 242.

Suppose S is a left sideal of $\mathrm{T}_0(N)$, so $\mathrm{T}_0(N) \cdot S \subseteq S$. For each integer $m \in \mathbb{Z}$, right distributivity gives us that for all $\theta \in S$, we have that $m\theta \in S$. Thus, the set S contains the additive identity mapping and all positive and negative sums of its elements.

Right Sideals

When S is a right sideal of $\mathrm{T}_0(N)$, it will certainly contain the additive identity mapping (the zero). Moreover, the subgroup $\langle S\rangle_{(add)}$ will also be a right sideal—that is, a *right* near-ring group (Definition 20). The singleton $\{0\}$ is an ideal of $\mathrm{T}(N)$ and a right sideal, but not a left sideal. It is, of course, a left sideal of $\mathrm{T}_0(N)$.

Suppose S is a non-trivial right sideal of $\mathrm{T}_0(N)$. Lemma 233 tells us that for any $\theta \in S$, S contains each of the elements $\Delta^{(\theta(a),b)}$ (where $a, b \in N$, $b \neq 0$ and $\theta \in S$). It also contains the zero mapping.

We already know $0 \in \mathcal{I}(S)$. Choosing any proper non-trivial subgroup, $0 < H < N$, permits us to exhibit a right sideal which itself is both proper and non-trivial, namely, $S = \mathrm{T}_0(N, H)$. Here $\mathcal{I}(S) = H$. Of course, this is also a right near-ring group; and when $H \lhd N$, it is a right ideal.

Lemma 237 *Suppose N is finite and S is a non-trivial right sideal of* $\mathrm{T}_0(N)$. *Then,*

$$\langle S\rangle_{(add)} = \langle \mathrm{T}_0(N, \mathcal{I}(S))\rangle_{(add)} = \mathrm{T}_0(N, \langle \mathcal{I}(S)\rangle_{(add)})$$

Proof It is clear that

$$\langle S\rangle_{(add)} \subseteq \langle \mathrm{T}_0(N, \mathcal{I}(S))\rangle_{(add)}$$

If $a \in (\mathcal{I}(S))^*$, then $s \cdot x = a$; for some $x \in N^*$. This means $\Delta^{(a,x)} = \left(s \circ \Delta^{(x,x)}\right) \in S$, and it means $\Delta^{(a,y)} = \left(\Delta^{(a,x)} \circ \Delta^{(x,y)}\right) \in S$, for any $y \in N^*$. Elements of $\langle \mathrm{T}_0(N, \mathcal{I}(S))\rangle_{(add)}$ are finite sums of positive and negative copies of elements $\Delta^{(a,b)}$, where $a \in \mathcal{I}(S)$, $b \in N$, and so lie in $\langle S\rangle_{(add)}$. To show $\langle \mathrm{T}_0\,(N, \mathcal{I}(S))\rangle_{(add)} = \mathrm{T}_0(N, \langle \mathcal{I}(S)\rangle_{(add)})$, first notice that

$$\langle \mathrm{T}_0(N, \mathcal{I}(S))\rangle_{(add)} \subseteq \mathrm{T}_0\left(N, \langle \mathcal{I}(S)\rangle_{(add)}\right)$$

since both expressions represent additive subgroups. An element $\theta \in \mathrm{T}_0(N, \langle \mathcal{I}(S)\rangle)$ maps any $b \in N$ to an expression of the form

$$\theta(b) = (\pm a_1 \pm a_2 \pm a_3 \ldots \pm a_t)$$

where $a_j \in \mathcal{I}(S)$. This is modelled, on b, by the expression

$$\left(\pm\Delta^{(a_1,b)} \pm \Delta^{(a_2,b)} \pm \Delta^{(a_3,b)} \ldots \pm \Delta^{(a_t,b)}\right)$$

and we can form similar sums for every other element in N. The complete set of these sums is equivalent to θ. □

Shortly, I shall describe a matrix formulation which makes the symmetries easier to visualise in finite cases. It is obvious that any right sideal containing a bijective mapping of $(N, +)$ must be identical to $\mathrm{T}_0(N)$.

Left Sideals

Lemma 238 *Suppose S is a left sideal of* $\mathrm{T}_0(N)$. *For each non-zero* $s \in S$, *we have the group isomorphism*

$$\mathrm{T}_0(N) \cdot s \cong \frac{\mathrm{T}_0(N)}{\binom{s}{0}_{\mathcal{L}}}$$

and

$$\bigcap_{s \neq 0} \binom{s}{0}_{\mathcal{L}}$$

is a trivial ideal of $\mathrm{T}_0(N)$.

Proof That the intersection is a right ideal is a consequence of S being a left sideal. This intersection would represent all elements of $\mathrm{T}_0(N)$ left annihilating every element of S. $\mathrm{T}_0(N)$ is simple (see page 311), so this ideal is trivial. □

Corollary 239 *Thus, no, non-zero, member of* $\mathrm{T}_0(N)$ *left annihilates every member of S. In particular, when b is non-zero,* $\Delta^{(a,b)}$ *cannot annihilate S, so there is some* $c \in N$ *and some* $\theta \in S$ *such that* $\theta(c) = b$.

Using the image set notation, we might say

Corollary 240

$$\mathcal{I}(S) = N \tag{7.52}$$

More, S contains all the expressions

$$\left(\Delta^{(\phi(a),c_1)} + \Delta^{(\phi(a),c_2)} + \ldots + \Delta^{(\phi(a),c_k)}\right)$$

where $\{c_1, c_2, \ldots, c_k\}$ *represents the non-zero elements of N mapped to b under* θ *and* $\phi \in \mathrm{T}_0(N)$. *This is a super-cleik (page 246).*

It's easy to see from Lemma 233 that all elements of S give rise to super-cleiks that are in S.

Lemma 241 *Suppose N is finite and S is a non-trivial left sideal of* $\mathrm{T}_0(N)$. *Then S is a union of some collection of super-cleiks. Conversely, any union of a collection of super-cleiks is a left sideal of* $\mathrm{T}_0(N)$.

Two-Sided Sideals

Suppose now that S is a non-trivial, two-sided sideal of $\mathrm{T}_0(N)$. Using the notation we developed above and Lemma 233, we can say that S contains the set $\{\Delta^{(a,b)} : a, b \in N, b \neq 0\}$, that $\mathcal{I}(S) = N$ and that S additively generates $\mathrm{T}_0(N)$. More than this, S must contain all possible expressions of the form

$$\left(\Delta^{(a,c_1)} + \Delta^{(a,c_2)} + \ldots + \Delta^{(a,c_k)}\right) \tag{7.53}$$

(with $c_j \neq 0$ and distinct). It will be seen that this set of expressions is actually a sideal and that, in non-trivial cases, it is not equal to $\mathrm{T}_0(N)$. This phenomenon is illustrated in Example 246.

7.3.2 A-Matrices

> For the remainder of Sect. 7.3, assume all groups are finite.

I shall now describe a representation of transformation near-ring products that is similar to the standard regular representation of group theory. I shall be dealing with **column stochastic matrices**. These will always be special in that they will be $(0-1)$ matrices whose column sums total 1. Similarly, the reader should understand "doubly stochastic matrices" to be $(0-1)$ matrices which are both column and row stochastic.

In finite cases, we enumerate N as $\{n_0, n_1, \ldots, n_m\}$ and fix the element n_0 as 0 (the additive identity).

Definition 242 One associates any $\theta \in \mathrm{T}(N)$ with a $(m+1) \times (m+1)$ column stochastic matrix $(\theta_{i,j})$, where if $\theta(n_j) = n_k$, then

$$\theta_{i,j} = \delta_{i,k}$$

(and number rows and columns: $0, 1, 2, \ldots, m$).

Trivially, if $\theta, \phi \in \mathrm{T}(N)$, then

$$(\theta \circ \phi)_{i,j} = \big((\theta_{i,j}) \cdot (\phi_{i,j})\big)_{i,j} \tag{7.54}$$

—that is, this notation respects standard matrix multiplication. I tend to blur the distinction between mappings θ and their representing matrix $\theta_{i,j}$ in what follows. Of course, this presupposes a fixed enumeration of the elements of $(N, +)$.

The product (7.54) takes two column stochastic matrices and delivers a third. Such a product has the effect of rearranging the columns of the left-hand matrix but can involve deletions and repetitions. An alternative view is that the product represents a rearrangement of the *rows* of the *right-hand* matrix but now deletions cannot occur; instead, one may replace a row with a sum of rows—"merge rows". I illustrate all this with an example.

Suppose we have the additive group $(N, +) = (\mathrm{C}_4, +)$ and number the elements $\{0, 1, 2, 3\}$. Suppose $\theta \in \mathrm{T}(\mathrm{C}_4)$ is: $\{0 \mapsto 3, 1 \mapsto 3, 2 \mapsto 0, 3 \mapsto 1\}$, while ϕ is $\{0 \mapsto 1, 1 \mapsto 0, 2 \mapsto 2, 3 \mapsto 0\}$. The composition $(\theta \circ \phi)$ would be expressed as

$$\begin{bmatrix} 0&0&1&0 \\ 0&0&0&1 \\ 0&0&0&0 \\ 1&1&0&0 \end{bmatrix} \cdot \begin{bmatrix} 0&1&0&1 \\ 1&0&0&0 \\ 0&0&1&0 \\ 0&0&0&0 \end{bmatrix} = \begin{bmatrix} 0&0&1&0 \\ 0&0&0&0 \\ 0&0&0&0 \\ 1&1&0&1 \end{bmatrix} \tag{7.55}$$

The Effect of Matrix Operations

(i) Regarding the left-hand product matrix (θ), as operated on by the right-hand one (ϕ) ("right action"), the effect is to preserve the first three columns in situ and to move one of the first two columns to the fourth position. The original fourth column simply disappears.
(ii) Regarding the right-hand product matrix (ϕ), as operated on by the left-hand one (θ) ("left action"), the effect is to move the fourth row to the second and third positions, to move the third row to the first position and to replace the fourth row with the "sum" of rows one and two.
(iii) The "sum" mentioned in (ii) is purely formal and has no connection with the additive group involved. It is a sum of 0s and 1s, and multiple 1s cannot possibly be added together, because they don't occur in the same positions in these column stochastic matrices. Perhaps it would be better to talk of "column merging" in this context.
(iv) In this example, the addition of rows is caused by multiple occurrences of 1s in row 4 of θ. The single occurrence of 1 in the third position of row one moves the third row of ϕ to the first, and the single occurrence of 1 in row two, last position, moves the last row of ϕ to the second row.
(v) In general, left action (θ acting on ϕ) is more complex than right. Positions of 1s indicate where rows are to be moved, and additions correspond to multiple occurrences of 1s in a row and imply some rows being completely replaced (though they must appear as constituents of sums).
(vi) Later, I look at right and left actions when the acting matrix is doubly stochastic. Such actions neither remove columns nor merge them (page 268).

These observations serve to isolate the structure of left or right sideals in finite transformation near-rings. The left case is significantly more complicated to determine. The matrix construction is a graphic illustration that the various sorts of sideal in a transformation near-ring are unconnected to the structure of the additive group.

The mapping $\Delta^{(a,b)}$ is represented as a matrix whose top row consists entirely of 1s, with the exception of the column relating to the element b, which has a 1 in the row relating to the element a. This is a rank two matrix (when $a \neq 0$). Any two such rank two matrices are *equivalent* matrices.

Multiplicative Centre of $\mathbf{T}_0(N)$

It seems obvious that the multiplicative centre of $\mathrm{T}_0(N)$ must consist of scalar matrices and consequently is $\{0, Id\}$. Then from (1.40) and from (7.24), one gets, of course, Lemma 228.

Column Stochastic Matrices and (T(N), +)

The $\mathcal{A}$-matrix notation associates any $(m+1) \times (m+1)$ column stochastic matrix with some unique member of $\mathrm{T}(N)$. Each column of one of these matrices represents a cleik. The set of columns corresponds to a full cleik, so there is a one-to-one correspondence between column stochastic matrices and full cleiks. Bijective members of $\mathrm{T}(N)$ correspond exactly to doubly stochastic matrices.

The use of this notation is restricted to products. The addition to be applied is based on that of the underlying group and is distinct from the usual matrix addition, although aspects of it have some useful similarities. To continue with the above example, using $(N, +) = C_4$, I calculate $(\theta + \phi)$ as

$$\begin{bmatrix} 0010 \\ 0001 \\ 0000 \\ 1100 \end{bmatrix} + \begin{bmatrix} 0101 \\ 1000 \\ 0010 \\ 0000 \end{bmatrix} = \begin{bmatrix} 1000 \\ 0001 \\ 0010 \\ 0100 \end{bmatrix} \tag{7.56}$$

Elements of $T_0(N)$ fix zeros and are characterised by the constraint: $\theta_{i,0} = \delta_{i,0}$—that is, they have 1 in the top left position. There must be some interplay between matrices and near-rings here because the notation suggests structures related to matrices which would have near-ring analogues—Kronecker products, for example. I start looking at this in Chap. 11.

People have invented other matrix products that might be considered, for example, **Hadamard products**, in which two matrices of the same shape are combined into another matrix of the same shape by multiplying elements in the same position.

If one did this for the column stochastic matrices we have been looking at and made the obvious modifications to cater for columns consisting entirely of 0s (simply putting a 1 on the top row), one obtains a commutative and associative product operation on transformation near-rings in which all elements are idempotent and the product of two transformations produces the transformation which maps to 0 all elements upon which the two transformations do not agree. The product does not generally distribute over the addition, corresponding to the original group action, of the constituent matrices, so the structure is not a Boolean near-ring (Sect. 2.5.3). I revisit products like this in the context of transformation near-rings in Sect. 8.3.

7.3.3 *Operating on* $\Delta^{(\mathbf{a},\mathbf{b})}$

Restricting ourselves to $T_0(N)$, the $\mathcal{A}$-matrix corresponding to $\Delta^{(a,b)}$ has 1*s* in its top row apart from the entry corresponding to b, where the 1 is in the row corresponding to a. We may move a row to any row other than the top one simply by operating on the right with the appropriate doubly stochastic matrix. For example,

$$\begin{bmatrix} 1000 \\ 0001 \\ 0100 \\ 0010 \end{bmatrix} \cdot \begin{bmatrix} 1011 \\ 0000 \\ 0000 \\ 0100 \end{bmatrix} = \begin{bmatrix} 1011 \\ 0100 \\ 0000 \\ 0000 \end{bmatrix}$$

might be said to change $\Delta^{(3,1)}$ into $\Delta^{(1,1)}$. In the same way, we may transform any $\Delta^{(a,b)}$ into any $\Delta^{(a,c)}$ (i.e. swap columns b and c), by operating on the *right* with a suitably configured doubly stochastic matrix (here we assume $b, c \neq 0$). Notice these activities always preserve top rows.

7.3.4 Left and Right Sideals

If S is a right sideal of the transformation near-ring $\mathrm{T}_0(N)$, we know it as a collection of full cleiks. Any matrix in S gives birth to a set of distinct cleiks—its columns. One may build new column stochastic matrices from those columns by selecting any non-trivial subset of them and disbursing them at random as the new columns of a row stochastic matrix. Thus, from

$$\begin{bmatrix} 0\,0\,1\,0 \\ 0\,0\,0\,1 \\ 1\,0\,0\,0 \\ 0\,1\,0\,0 \end{bmatrix}$$

we might build

$$\begin{bmatrix} 1\,0\,0\,1 \\ 0\,1\,0\,0 \\ 0\,0\,0\,0 \\ 0\,0\,1\,0 \end{bmatrix}$$

(which is obtained from the first matrix by multiplying on the right by)

$$\begin{bmatrix} 0\,0\,0\,1 \\ 0\,0\,1\,0 \\ 1\,0\,0\,0 \\ 0\,1\,0\,0 \end{bmatrix}$$

and we may build many other matrices. If $\theta \in \mathrm{T}(N)$ is associated with a matrix which is not invertible, that is, the mapping is not bijective, then the set

$$\{(\theta \circ \phi) \,:\, \phi \in \mathrm{T}(N)\} \tag{7.57}$$

must consist entirely of matrices having zero determinants—non-bijective mappings. The set is a right sideal, so every non-zero, non-bijective mapping in $\mathrm{T}(N)$ does lie in a proper, non-trivial, right sideal. The right sideal in (7.57) is clearly the minimal right sideal containing θ (I am not claiming it is a minimal right sideal—simple arguments based on matrix rank show that this cannot generally be the case).

The dual observation applies. Each non-bijective mapping lies in a smallest proper left sideal too. The set of all matrices in $\mathrm{T}(N)$ which are non-invertible (and so have rank smaller than $|N|$) must be a (two-sided) sideal of $\mathrm{T}(N)$. Clearly, it is a maximal sideal. Interestingly, it is also maximal both as a left and as a right sideal.

7.3.5 Nilpotence

Given $\theta \in \mathrm{T}_0(N)$, one can define descending chains of image sets

$$\mathcal{I}(\{\theta\}) \supseteq \mathcal{I}(\{\theta^2\}) \supseteq \mathcal{I}(\{\theta^3\}) \supseteq \ldots \mathcal{I}(\{\theta^k\}) \supseteq \ldots \tag{7.58}$$

The mapping θ is *nilpotent* (see Sect. 1.22) if, for some finite integer k, $\mathcal{I}(\{\theta^k\}) = 0$. For this to happen, we should require $\mathcal{I}(\{\theta\})$ to consist of a set of elements with no fixed points other than 0; and more than that, there must be no "loops" of elements within $\mathcal{I}(\{\theta\})^*$ which are closed under repeated applications of θ. All this comes down to the requirement that, given any $m \in N^*$, the succession of elements

$$\theta(m), \theta^2(m), \theta^3(m) \ldots \tag{7.59}$$

are all mutually distinct until the first occurrence of 0 (whereupon the sequence is thereafter fixed on 0).

Lemma 243 *If $\theta \in \mathrm{T}_0(N)$ is (multiplicatively) nilpotent, then the smallest integer t such that $\theta^t = 0$ is no larger than $|N|$.*

General Mappings in Finite Cases
The sequence (7.58) can be infinitely descending in infinite cases, but when groups are finite, it must stabilise; and this is true for *all* mappings θ, not merely the nilpotent ones. I call the set of elements defined by this means the **stabilised image**—$\mathcal{SI}(\theta)$. θ acts upon the stabilised image bijectively. If the original mapping θ were surjective, the stabilisation point is immediate—the whole group N. In nilpotent cases, the stabilised image is $\{0\}$. The point at which the stabilised image is reached depends on its size. I fix that size at $(k+1)$ and assume there are $(m-k)$ elements *not* in the stabilised image. Then the value of t at which $\theta^t : N \mapsto \mathcal{SI}(\theta)$ is $t = (m-k)$ (we can construct a mapping that simply sequences through elements not in the stabilised image). Nilpotent members of $\mathrm{T}_0(N)$, of course, correspond to nilpotent matrices. Their eigenvalues are equal to zero. The matrix notation from Sect. 7.3.2 can be used here, and the notation I have already introduced was chosen to conform to it. If we make elements $\{n_0 = 0, n_1, n_2, \ldots, n_k\}$ correspond to the elements of the stabilised image—$\mathcal{SI}(\theta) = \mathcal{I}(\theta^t)$—then the $\mathcal{A}$-matrix corresponding to θ can have the block form:

$$\begin{bmatrix} A & B \\ C & D \end{bmatrix} \tag{7.60}$$

where A is the $((k+1)\times(k+1))$ matrix of the mapping of the stabilised image; B is a $((k+1)\times(m-k))$ matrix which, together with the $(m-k)\times(m-k))$, D, is column stochastic (and represents the effect of θ on elements not in the stabilised image); and, finally, the matrix C is a $((m-k)\times(k+1))$ matrix consisting entirely of zeros. A represents a bijective mapping and is an invertible matrix. The tth. power of the matrix of θ has the same format as (7.60) except that the matrix D now consists entirely of 0s (A and B are column stochastic).

When the original mapping is nilpotent, the matrix A has a single element—the top left entry 1—and a sufficiently large power of that matrix results in a rank one matrix with 1s along its top row.

Note that (7.60) cannot be achieved simultaneously, for all matrices. It is just that we can rig the generators so that any mapping would look like it, while other mappings, even of nilpotent elements, would not.

7.3.6 Idempotence

Related to nilpotence, one can discuss mappings $\theta \in \mathrm{T}_0(N)$ for which $\theta^2 = \theta$. These are the **idempotent** mappings, and they are precisely the mappings which act as identity mappings on their own image sets. So, their image sets are precisely the fixed points of the original mapping. Following the analysis of Sect. 7.3.5, we can say that it is possible to describe these mappings as $\mathcal{A}$-matrices with the form

$$\mathrm{i} = \begin{bmatrix} A & B \\ C & D \end{bmatrix} \tag{7.61}$$

where, this time, A is a square identity matrix, C is zero, D is zero and B is a column stochastic matrix with the same number of rows as A. Obviously, the matrices i are idempotent. Any idempotent mapping, θ, partitions the elements of N^* into distinct subsets $(b_1, b_2, \ldots b_r\}$ and $\{c_1, c_2, \ldots c_{n-r}\}$ such that θ is a sum of two cleiks

$$\theta = \left(\Delta^{(b_1,b_1)} + \Delta^{(b_2,b_2)} + \ldots \Delta^{(b_r,b_r)}\right) + \left((\Delta^{(x_1,c_1)} + \Delta^{(x_2,c_2)} + \ldots \Delta^{(x_{n-r},c_{n-r})}\right) \tag{7.62}$$

where $\{x_j\}_{j=1}^{n-r} \subset \{0, b_1, b_2, b_3, \ldots b_r\}$. From these observations, one can easily calculate the number of idempotents that the near-ring has.

7.3.7 $\mathrm{T_0(N)}$ *Generalised*

One can generalise from $\mathrm{T}_0(N)$ and consider column stochastic matrices having 1s at other specified fixed point on the leading diagonal. These will be multiplicatively closed. Multiplicatively, they are monoids. They are only *additively* closed for $\mathrm{T}_0(N)$.

Definition 244 The monoid $T_j(N)$ is defined for each fixed $j \in \{0, 1, 2, \dots n\}$ as the set of column stochastic matrices θ in which

$$\theta_{i,j} = \delta_{i,j}(\mathit{fixed})$$

Of course, $T_j(N) \cap T_k(N)$ represents column stochastic matrices with 1s on the ith. and jth. diagonal entries. These correspond to mappings of N which fix both n_i and n_j. I write

$$\mathrm{T}_{i,j} = T_j(N) \cap T_k(N) \; \mathit{and} \; \mathrm{T}_{i,j,k} = \left(T_i \cap T_j \cap T_k\right) \tag{7.63}$$

and so on: $\mathrm{T}_{i,j,k,l,m\dots}$. Each $n_j \in N$ generates an additive subgroup $\langle n_j \rangle_{(add)}$. There will be a corresponding additive subgroup of $(\mathrm{T}(N), +)$

$$\mathcal{W}_j = \left\{\theta \in \mathrm{T}(N) \; : \; \theta(n_j) \in \langle n_j \rangle_{(add)}\right\} \tag{7.64}$$

Notice that this subgroup is quite distinct from the near-ring $T\left(N, \langle n_j \rangle_{(add)}\right)$, which is contained in it.

Lemma 245 *Additively,*

$$\mathcal{W}_j \cong \left(\left(\oplus_{1 \le k < j}(N, +)\right) \oplus \langle n_j \rangle_{(add)} \left(\oplus_{j < k \le n}(N, +)\right)\right) \tag{7.65}$$

—a direct sum of $(n-1)$ copies of $(N, +)$ and one copy of $\langle n_j \rangle_{(add)}$.

Example 246 I look at the cleiks defined in (7.53) in the near-ring $\mathrm{T_0(C_3)}$ and represent $\mathrm{C_3}$ as $\{0, 1, 2\}$. These cleiks include the individual expressions

$$\left\{\Delta^{(1,1)}, \Delta^{(1,2)}, \Delta^{(2,1)}, \Delta^{(2,2)}\right\},$$

together with the zero mapping, $\Delta^{(0,1)}$, and the two expressions

$$\left(\Delta^{(1,1)} + \Delta^{(1,2)}\right),$$
$$\left(\Delta^{(2,1)} + \Delta^{(2,2)}\right),$$

The corresponding column stochastic matrices are (in the same order) The four matrices

$$\begin{bmatrix}1&0&1\\0&1&0\\0&0&0\end{bmatrix}, \begin{bmatrix}1&1&0\\0&0&1\\0&0&0\end{bmatrix}, \begin{bmatrix}1&0&1\\0&0&0\\0&1&0\end{bmatrix}, \begin{bmatrix}1&1&0\\0&0&0\\0&0&1\end{bmatrix}$$

The zero mapping

$$\begin{bmatrix}1&1&1\\0&0&0\\0&0&0\end{bmatrix}$$

Plus the two matrices

$$\begin{bmatrix}1&0&0\\0&1&1\\0&0&0\end{bmatrix} \begin{bmatrix}1&0&0\\0&0&0\\0&1&1\end{bmatrix}$$

The reader will be able to verify that this system of matrices is closed under premultiplication and postmultiplication by a (0,1) column stochastic matrix of the form

$$\begin{bmatrix}1&A&B\\0&C&D\\0&E&F\end{bmatrix}$$

The system represents a non-trivial sideal properly contained in $T_0(C_3)$. If one examines the cleik

$$\left\{\left(\Delta^{(\phi(1),1)}+\Delta^{(\phi(2),2)}\right)\ :\ \phi\in T_0(C_3)\right\}$$

(which does not occur in the previous system), one sees at once that it contains the mapping

$$\begin{bmatrix}1&0&0\\0&1&0\\0&0&1\end{bmatrix}$$

and that it is equal to $T_0(C_3)$.

The set of all bijective mappings of N is neither a left nor a right sideal since it does not contain the zero mapping. It's easy to construct examples in which the set of bijective mappings fixing zero is not a normal subset of $T_0(N)$. This happens when $N = D_4$, for instance.

It would be appropriate to attempt to relate the bijective and zero-fixing mappings to the structure of $T_0(N, +)$. The matrices in question are part of a larger set of $(0-1)$ column stochastic matrices in $T_0(N)$ which represent elements with additive order equal to the exponent of the group $(N, +)$. The subgroup generated by this larger set must be a characteristic subgroup of the additive group $(T_0(N), +)$.

It was shown by Fröhlich [41] that when $(N, +)$ is simple and non-abelian, $I(N) = T_0(N)$, which would force the characteristic subgroup to be the whole group in that situation.

The bijective mappings in $T_0(N)$ are $(0-1)$ column stochastic matrices with rank equal to the order of the group $(N, +)$. Since near-ring products in $T_0(N)$ are simply matrix products, we observe that the column stochastic matrices with rank less than this (i.e. the non-singular, column stochastic $(0-1)$ matrices) form a proper and non-trivial sideal of $T_0(N)$ (we should exclude $N = C_2$ here).

7.3.8 A Sub-near-Ring of $T_0(S_3)$

I consider the symmetric group S_3 with presentation

$$\langle a, b \; : \; 3a = 2b = 2(a+b) = 0 \rangle$$

and the additive subgroup of $T_0(S_3)$

$$\{\theta \in T_0(S_3) \; : \; \theta(x) = 0, \forall x \in \{b, (a+b), (2a+b)\}\} \tag{7.66}$$

This structure is a left sideal of $T_0(S_3)$. It is also a left ideal of $T_0(S_3)$. It can be identified as the set of mappings

$$A = \left\{ \left(\Delta^{(\phi(a), a)} + \Delta^{(\phi(2a), 2a)} \right) \; : \; \phi \in T_0(S_3) \right\}$$

In $\mathcal{A}$-matrix terms, there are 6×6 column stochastic matrices in which the last three columns have 1s along the top three entries and 0s elsewhere (ordering the group elements: $0, a, 2a, b, (a+b), (2a+b)$). This is a super-cleik. I unite this set with the eight-element set

$$B = \left\{ \left(\Delta^{(m_1 b, b)} + \Delta^{(m_2(a+b), (a+b))} + \Delta^{(m_3(2a+b), (2a+b))} \right) \right\}$$

where the numbers m_j range over $\{0, 1\}$. In matrix terms, this is a set of 6×6 column stochastic matrices in which the first three columns have 1 in the top row and either the next three have entries in which the 1 is on the top row or they have entries in which the 1 is on the leading diagonal. B forms an additive subgroup of $T_0(S_3)$ but is neither a left nor a right sideal. I use $(A \cup B)$ to generate a subgroup of $\mathrm{T}_0(S_3)$. The subgroup consists of column stochastic (0-1) matrices of the form

$$C = \begin{bmatrix} 1 & X & X & X & X & X \\ 0 & X & X & 0 & 0 & 0 \\ 0 & X & X & 0 & 0 & 0 \\ 0 & X & X & X & 0 & 0 \\ 0 & X & X & 0 & X & 0 \\ 0 & X & X & 0 & 0 & X \end{bmatrix}$$

where each column has exactly one non-zero value of X (so the symbol X should not be interpreted as representing a fixed value). There are 288 such matrices, and they form a multiplicatively closed subset of $\mathrm{T}_0(S_3)$. This is a near-ring with identity containing S_3 as a normal subgroup and left ideal.

Indeed, the subgroup is additively of the form

$$C = (S_3 \oplus S_3 \oplus C_2 \oplus C_2 \oplus C_2)$$

and each direct summand is a minimal left ideal of the sub-near-ring (not always of the near-ring). Note that S_3 is not a right sideal here but that $(S_3 \oplus S_3)$ is actually an ideal of the sub-near-ring. This can be verified by matrix multiplication, but a more direct proof would be to isolate the second and third columns of these matrices—the ones corresponding to images of a and of 2a—these have only zeros as entries in the second and third rows for entries ouside these columns, so the subgroup is a sideal of the near-ring, though not of $\mathrm{T}_0(S_3)$, where the zero condition does not occur. What we have is an extension of S_3 by a group of order 48, an extension which hosts near-rings with identity. Alternatively, we have a near-ring exact sequence

$$(S_3 \oplus S_3) \longmapsto C \twoheadrightarrow (C_2 \oplus C_2 \oplus C_2)$$

I should say that we know of smaller extensions of S_3 which host near-rings with identity (Sect. 4.1.3). Something like this process is applicable more generally; but it is very dependent on the additive structure of the group and the element singled out. In this case, things were simpler because the subgroup $\langle a \rangle$ is *isolated*, in the sense that no member of S_3 outside the subgroup generates members of it. Groups all of whose elements have prime order, such as A_5, have isolated subgroups. The key idea is to select an element whose range set is the whole group. The columns corresponding to that element can then have non-zero entries (i.e. 1s) at any row position. This will, additively, represent a subgroup isomorphic to $(N, +)$. One must then arrange that for all elements b of the group, the transformation $\Delta^{(b,b)}$ is one of our mappings. This gives 1s on the leading diagonals and ensures that the full cleik $\{\Delta^{(x,x)} : x \in N^*\}$ is one of our mappings and our set must contain the identity matrix. But seeking near-rings, we require additive closure. This means that there will be matrices with 1s around leading diagonals in positions corresponding to sums of the element defining any particular column. This business of *isolation* simply restricts the 1s in the columns of elements not mapping to the whole group to places relating to sums of group elements—so $2a + 2a \notin \{b, (a+b), (2a+b)\}$, for instance. One could modify the construction just given, producing 144 matrices with the form

$$C' = \begin{bmatrix} 1 & X & X & X & X & X \\ 0 & X & X & 0 & 0 & 0 \\ 0 & X & X & 0 & 0 & 0 \\ 0 & X & 0 & X & 0 & 0 \\ 0 & X & 0 & 0 & X & 0 \\ 0 & X & 0 & 0 & 0 & X \end{bmatrix}$$

The exact sequence is now

$$S_3 \longmapsto C' \twoheadrightarrow (C_3 \oplus C_2 \oplus C_2 \oplus C_2)$$

S_3 is an ideal and this is an exact sequence of near-rings. One might repeat this construction using the element b rather than a. Since the element has additive order 2, we have a single cleik plus the union of diagonal elements to handle. The collection of matrices arrived at is, essentially (it depends on ordering), of the form

$$D = \begin{bmatrix} 1 & X & X & X & X & X \\ 0 & X & 0 & 0 & 0 & 0 \\ 0 & X & X & X & 0 & 0 \\ 0 & X & X & X & 0 & 0 \\ 0 & X & 0 & 0 & X & 0 \\ 0 & X & 0 & 0 & 0 & X \end{bmatrix}$$

and represents a near-ring D with identity with order 216 containing S_3 as an ideal. The second column here represents b, third and fourth are a and $2a$, and fifth and sixth are $(a+b)$ and $(2a+b)$. D is additively

$$D = (S_3 \oplus C_3 \oplus C_2 \oplus C_2)$$

The exact sequence is now

$$S_3 \longmapsto D \twoheadrightarrow (C_3 \oplus C_2 \oplus C_2)$$

This construction embeds finite groups as subgroups of finite zero-symmetric near-rings with identity.

A_5
The subgroups generated by elements of A_5 are all isolated (page xix gives element details), and one can build a near-ring from $T_0(A_5)$ using this technique. A_5 is an additive subgroup of this near-ring, so additively, it is non-solvable. The very smallest near-ring I can find in this way has order $60 \times 5^{23} \times 3^{20} \times 2^{15} = (2^{15} \times 3^{20} \times 5^{23}) \approx 8.2 \times 10^{31}$. Sadly, this is the smallest non-solvable group that I know to host zero-symmetric, unital, near-rings. Although this is vastly smaller in size than $T_0(A_5)$, it is still huge and illustrates why the search for the smallest non-solvable group hosting zero-symmetric, unital, near-rings might be thought important.

7.4 $\mathfrak{T}$(N), $\mathfrak{H}$(N) and $\mathfrak{B}$(N)

This section makes extensive use of the notation in Table 7.1.

Throughout this section, the single symbol I represents the identity mapping (on N).

In Sect. 7.4, all groups are finite.

7.4.1 The Structure of $\mathfrak{H}$(N)

Recall the definition of the holomorph given in Notation 3, the definition of $\mathfrak{B}(N)$ on page 239; and suppose $(N, +)$ is any finite additive group,

$$N = \{n_0 = 0, n_1, \ldots, n_n\}$$

We have that

$$(\mathrm{T}(N), +) > \mathfrak{B}(N) \supseteq \mathrm{Hol}(N) \supseteq \mathrm{Aut}(N, +) \tag{7.67}$$

The holomorph is a multiplicative subgroup of the *additive* group $\mathfrak{B}(N)$ and

$$\mathrm{Hol}(N) \subset \mathfrak{B}(N) < \mathrm{T}(N) \tag{7.68}$$

It is well-known that the automorphism of $(N, +)$ is just the zero-fixing mappings of the holomorph.

The *additive* group the holomorph generates, $\mathfrak{H}(N)$, is an *additive* subgroup of $\mathfrak{B}(N)$ and one which contains $\mathrm{A}(N)$. Thus,

Definition 247

$$\mathfrak{H}(N) \;=\; \langle \mathrm{Hol}(N)\rangle_{(add)} \tag{7.69}$$

and record that

$$\mathrm{A}(N) \;\subseteq\; \mathfrak{H}(N) \;\subseteq \mathfrak{B}(N) \;\subseteq\; \mathrm{T}(N) \tag{7.70}$$

One of the two constituents of the holomorph is the multiplicative group of left translations—that is,

$$\mathcal{T}(N) \;=\; \{\tau_n \;:\; n \in N\} \;\subset \mathrm{T}(N) \tag{7.71}$$

obviously, $\{\mathcal{T}(N) \cap \mathrm{T}_0(N)\} \;=\; \{\tau_0\} \;=\; \{0\}$. The other constituent is the multiplicative group of automorphisms, $\mathrm{Aut}(N) \subset \mathrm{T}_0(N)$. Notice that

$$(\mathcal{T}(N) \cap \mathrm{A}(N)) \;=\; \{\tau_0\} \tag{7.72}$$

(automorphisms are zero-fixing).

Lemma 248 *Suppose* $\theta, \phi \in \mathrm{T}(N)$ *and* $m \in N$. *Then,*

$$(\tau_m \circ (\theta + \phi)) \;=\; ((\tau_m \circ \phi) + \phi) \tag{7.73}$$

Lemma 249 *The multiplicative group* $(\mathcal{T}(N), \;\cdot)$ *is isomorphic to the additive group* $(N, \;+)$.

Proof Suppose $m, n \in N$. Then,

$$(\tau_m \circ \tau_n) \;=\; \tau_{(m+n)} \tag{7.74}$$

□

Lemma 250 $\mathfrak{H}(N)$ *is a sub-near-ring of* $\mathrm{T}(N)$.

Proof The proof relies on (7.73) and (7.74). Elements of $\mathfrak{H}(N)$ are sums of elements of the form: $(\theta \cdot \tau_m)$, where $\theta \in \mathrm{Aut}(N)$.

$$\begin{aligned}
\left(\theta_1 \cdot \tau_{m_1}\right) \circ \left(\theta_2 \cdot \tau_{m_2} + \theta_3 \cdot \tau_{m_3}\right) \;&=\\
\theta_1 \circ \left(\tau_{m_1} \cdot \theta_2 \cdot \tau_{m_2} + \theta_3 \cdot \tau_3\right) \;&=\\
\left(\left(\theta_1 \cdot \tau_{m_1} \cdot \theta_2 \cdot \tau_{m_2}\right) \;+\; \left(\theta_1 \cdot \theta_3 \cdot \tau_3\right)\right)&
\end{aligned}$$

And

$$(\theta_1 \cdot \tau_{m_1} \cdot \theta_2 \cdot \tau_{m_2}) = \left(\theta_1 \cdot \theta_2 \cdot \theta_2^{-1} \cdot \tau_{m_1} \cdot \theta_2 \cdot \tau_{m_2}\right)$$
$$= \left(\theta_1 \cdot \theta_2 \cdot \tau_{(\theta^{-1}(m))} \cdot \tau_{m_2}\right)$$

□

Lemma 251

(i)

$$A(N) \ \triangleleft \ \mathfrak{H}(N)$$

(ii)

$$\left(\frac{\mathfrak{H}(N)}{A(N)}\right) \cong (N, +) \tag{7.75}$$

Proof

(i) Suppose that $\theta \in \text{Aut}(N)$. We know

$$(\theta)^{\tau_a} = (-I - C_a + \theta + C_a + I)$$

and it is easy to verify that the mapping $(-C_a + \theta + C_a)$ is an automorphism. This means that $(\theta)^{\tau_a} \in \text{Aut}(N)$.

(ii) The cosets $\tau_b \cdot A(N)$ and $\tau_d \cdot A(N)$ are equal precisely when

$$\tau_a = \tau_d \cdot \theta,$$

for some $\theta \in A(N)$ (not necessarily an automorphism). But then because $A(N)$ is zero-fixing,

$$a = b$$

Thus, the cosets $\{\tau_{n_j} \cdot A(N)\}_{j=0}^{n}$ are mutually distinct.

□

$A(N)$ is *never* a right ideal of $\mathfrak{H}(N)$.

7.4.2 The Structure of $\mathfrak{T}$(N)

I am interested in the *additive* group generated by $\mathcal{T}(N)$ and put

Definition 252

$$\mathfrak{T}(N) = \langle \mathcal{T}(N) \rangle_{(add)} = \langle \{\tau_n : n \in N\} \rangle_{(add)} \tag{7.76}$$

In finite cases, elements of $\mathfrak{T}(N)$ are actually sums of elements τ_m (see page 233). The additive group $\mathfrak{T}(N)$ is exactly the group additively generated by the interior rows and columns of a *Cayley table* (page xvii) for $(N, +)$ considered as vectors and as such may be of purely group-theoretical interest. I mean by this that its generating elements can be thought of as group element vectors—tuples with as many coordinates as there are elements in the group. All such generating elements have additive order equal to the group exponent.

My convention on Cayley tables is that the sum (x + y) is to be the element in the row indexed by x and the column indexed by y. With that convention, interior Cayley table rows correspond to the *left* translations $\{\tau_n\}_{n \in \Gamma}$, whereas interior table columns correspond to *right* translations $\{\rho_n\}_{n \in \Gamma}$.

Notation 253 I now introduce notation that has been used before by other authors (see [112]). The distinct group elements of $(N, +)$ are enumerated

$$\{n_0, n_1, \ldots, n_n\}$$

(I usually make the group identity the first element, $n_0 = 0$). The mapping $\theta \in \mathrm{T}(N)$ is represented as a tuple:

$$(\theta_0, \ldots, \theta_n)$$

where

$$\theta(n_i) = \theta_i \in \{n_0, n_1, \ldots, n_n\}$$

When confined to $\mathrm{T}_0(N)$, it is to be understood that $\theta_0 = 0$.

I utilise this notation throughout this subsection and, in particular, only consider finite groups. Confusingly, my standard group $(N, +)$ is deemed to contain $(n + 1)$ elements. The notation permits us to view elements of $\mathrm{T}(N)$ simply as sequences of elements of $(N, +)$. These sequences would be infinite in infinite cases but in finite cases have as many entries as the order of the group. The sequences are row vectors, called here "tuples". The notation is reminiscent of the sequence spaces occurring in functional analysis, particularly in infinite cases.

The Finite Topology
The sets $\{\mathfrak{C}(\theta, m)\}$ described in (7.9) now have an immediate interpretation as all sequences in which the entry corresponding to the element m is $\theta(m)$—that is, sequences having the same value in the mth. position as θ has. Similarly, the finite topology (page 235) has a basis consisting of entire collections of tuples all having the same values at a finite fixed set of coordinates (different sets of coordinates for different collections). If we take two distinct transforms $\theta, \phi \in \mathrm{T}(N)$, we can be sure that for some $a \in N$, $\theta(a) \neq \phi(a)$. This means the open sets $\mathfrak{C}(\theta, a)$, $\mathfrak{C}(\phi, a)$ are disjoint and the topological space is Hausdorff.

Taking the set $\mathfrak{C}(\theta, a_1, a_2, \ldots, a_k)$ as part of the basis for the topology described in Sect. 7.9 and taking $\phi \notin \mathfrak{C}(\theta, a_1, a_2, \ldots, a_k)$, it must be the case that ϕ and θ disagree about where to send some a_j and so $\mathfrak{C}(\phi, a_1, a_2, \ldots, a_k)$ and $\mathfrak{C}(\theta, a_1, a_2, \ldots, a_k)$ are disjoint. This means that our original open set $\mathfrak{C}(\theta, a_1, a_2, \ldots, a_k)$ is closed—it is a *clopen set.*

To summarise,

(i) The finite topology on $T(N)$ has a basis of clopen sets.
(ii) That topology is Hausdorff.

In finite cases, the finite topology is of course **compact**, and our topological space is then a **Boolean space**—that is, a compact, Hausdorff space which is totally disconnected [9].

7.4.3 More on the Representation

It is often more convenient to use the actual group elements in this notation, so, for example, in the case of a dihedral group with order 8, D_4, we might discuss the mapping

$$(0, a, 2a, 3a, (a+b), (2a+b), (3a+b), b)$$

(which is an automorphism in which $a \mapsto a$ and $b \mapsto (a+b)$). We have $I = \tau_0$. This is the single normalised mapping in $\mathcal{T}(N)$.

Then,

$$\rho_m = ((I) + \tau_m + (-I)) \tag{7.77}$$

which means that all the mappings $\{\rho_m\}_{m \in N}$ are in $\mathfrak{T}(N)$. We know

$$\mathrm{C}_m = (\tau_m - I) \tag{7.78}$$

so all of the constant mappings lie within $\mathfrak{T}(N)$. More,

$$i_m = (\mathrm{C}_{-m} + \rho_m) \tag{7.79}$$

and so, $\mathrm{I}(N) \subseteq \mathfrak{T}(N)$. We also have that

$$(\tau_m + \tau_n)(x) = m + x + n + x = m + x + 0 + x + n^x \quad (7.80)$$
$$= (\tau_m + \tau_0 + s_n)(x)$$

concluding that, for all $n \in N$, both $s_n \in \mathfrak{T}(N)$ and $m_n \in \mathfrak{T}(N)$.

Lemma 254 *All of the transformations given in Notation 7.2.1 lie in* $\mathfrak{T}(N)$.

Lemma 255 $\mathfrak{T}(N)$ *is a sub-near-ring of* $\mathrm{T}(N)$.

Proof Elements of $\mathfrak{T}(N)$ have the form

$$\sum_{j}^{k} \mp\tau_{n_j} \quad (n_j \in N) \quad (7.81)$$

Products of such elements require us to look at expressions of the form

$$\tau_m \cdot \left(\pm\tau_{n_1} \pm \tau_{n_2} \ldots \pm \tau_{n_k}\right)$$

Such expressions would map an element x to

$$(m \pm (n_1 + x) \pm (n_2 + x) \ldots \pm (n_k + x))$$

and this means the function involved is

$$\mathrm{C}_m \pm \tau_{n_1} \pm \tau_{n_2} \ldots \pm \tau_{n_k}$$

which is in $\mathfrak{T}(N)$. □

Lemma 256

(i) Suppose $m, n \in N$. *Then,*

$$\tau_{(m+n)} = (-i_{-m} + \tau_m + \tau_n) \quad (7.82)$$

(ii) Suppose $\{a_j\}_{j=1}^n \in N$. *Then,*

$$\sum_j \tau_{a_j} = \left(\sum_{j=1}^{n-1} i_{s_j}\right) + \tau_{(a_1+a_2+\ldots+a_n)} \quad (7.83)$$

where $s_j = -(a_1 + a_2 + \ldots + a_k)$. *So, all mappings in* $\mathfrak{T}(N)$ *can be written as a sum* $(x + y)$, *where* $x \in \mathrm{I}(N)$, *and if* $y \neq 0$, *then* $y = \tau_m$ *for some* $m \in N$. *In particular, any normalised mapping in* $\mathfrak{T}(N)$ *lies in* $\mathrm{I}(N)$.

(iii) Suppose $w(x_1, x_2, \ldots, x_k)$ is some word in the indeterminates $\{x_j\}_{j=1}^n$. Then,

$$w\left(\tau_{a_1}, \tau_{a_2}, \ldots \tau_{a_k}\right) = i + \tau_{w(a_1, a_2, \ldots a_k)}$$

where i here represents some element of the near-ring $\mathrm{I}(N)$.

This means that in considering $\mathfrak{T}(N)$, we need only to analyse the group generated by $\mathrm{I}(N)$ plus the set of mappings $\{\tau_{g_s}\}_{s\in S}\}$, where $\{g_s \; : \; s \in S\}$ is any additive generating set for $(N, +)$.

More than that, we have, for any $a, b \in N$

$$(-\tau_a + i_b + \tau_a) = \left(-I + i_{(b+a)} + I\right) \in \mathrm{I}(A) \tag{7.84}$$

(note, $I \in \mathrm{I}(A)$). Consequently,

$$\mathrm{I}(N) \lhd \mathfrak{T}(N) \tag{7.85}$$

I(N) will be a proper subgroup in all but trivial cases (remember that $\mathfrak{T}(N)$ involves non-zero-fixing mappings).

Lemma 257 *When $(N, +)$ is a finite group,*

$$\left(\frac{\mathfrak{T}(N)}{I(N)} +\right) \cong (N, +) \tag{7.86}$$

Proof Using the tuple notation of (page 264), we can distinguish terms τ_a by the fact that unless a is zero (and they represent the identity mapping), they involve tuples whose first entry is non-zero (in fact, a). Suppose $\{g_j\}_{j=1}^k$ is some set of additive generators for $(N, +)$, and suppose $w(g_1, g_2, \ldots, g_k)$ is some word representing a relation satisfied by these generators. Then, $w\left(\tau_{g_1}, \tau_{g_2}, \ldots, \tau_{g_k}\right)$ will be a tuple with first element equal to zero, that is, some mapping in $\mathrm{I}(N)$. It follows that because the additive coset generators $\left\{\left\{\tau_{g_j} + \mathrm{I}(N)\right\}\right\}_{j=1}^k$ satisfy precisely the same relations as the additive generators $\{g_j\}_{j=1}^k$, the two groups are isomorphic.

□

Corollary 258 *A necessary and sufficient condition that $\mathfrak{H}(N)$ should be contained in $\mathfrak{T}(N)$ is that* $\mathrm{A}(N)$ *should be equal to* $\mathrm{I}(N)$*. When this is not the case, $\mathfrak{T}(N)$ contains no outer automorphisms of $(N, +)$.*

Corollary 259 *When $(N, +)$ is a finite simple non-abelian group,*

$$\mathfrak{T}(N) = \mathrm{T}(N)$$

$\mathrm{I}(N)$ is *never* a right ideal of $\mathfrak{T}(N)$. The additive generators of $\mathfrak{T}(N)$ do not constitute a distributively generating set, but they have the interesting distributivity property given by (7.73). Elements of $\mathfrak{T}(N)$ have the form given by (7.81).

7.4.4 Permutations and Additive Isomorphisms

Notation 253 suggests further symmetries of $\mathrm{T}(N)$, which correspond to interesting symmetries of the $\mathcal{A}$-matrices described in Sect. 7.3.2. Select $\sigma \in S_{n+1}$. This can be represented as a doubly stochastic matrix, $\mathcal{S}$, in which the $(i - j)$th. element is 1 if and only if $\sigma(i) = j$. It may thus also be identified with an element of $\mathfrak{B}$. There are two new tuples, ${}^{\sigma}\theta$ and θ^{σ}, formed from θ using σ. If $\theta \in \mathrm{T}(N)$, I write

$$\theta = (\theta_0, \theta_1, \ldots, \theta_n)$$

and define tuples (further transformations), ${}^{\sigma}\theta$, and θ^{σ} with the formulae

$$\left(\theta^{\sigma}\right)_j = \theta_{\sigma(j)} = (\theta \circ \sigma)(j) \tag{7.87}$$

$$\left({}^{\sigma}\theta\right)_j = \sigma_{\theta j} = (\sigma \circ \theta)(j) \tag{7.88}$$

The inverse of the permutation matrix $\mathcal{S}$, is, of course, its transpose, $\mathcal{S}^t$, since permutation matrices are **orthogonal**. The new column stochastic matrices relating to these new transformations are simply left and right products of the transformation matrix $(\theta_{i,j})$ and its inverse. Specifically,

$$\left(\theta^{\sigma}\right)_{i,j} = \left((\theta)_{i,j} \cdot \mathcal{S}^t\right) \tag{7.89}$$

$$\left({}^{\sigma}\theta\right)_{i,j} = \left(\mathcal{S} \cdot (\theta)_{i,j}\right) \tag{7.90}$$

$\mathcal{S}$ is *doubly stochastic*. The first product, above, simply rearranges the columns of the matrix $\theta_{i,j}$, in a way dictated by the permutation σ^{-1}. The second product rearranges the rows of that matrix as prescribed by the permutation σ. Since $\mathcal{S}$ has maximal rank, the column merging, described on page 250 and involving left products, does not occur; neither does the column deletion and repetition, which is a feature of more general right products. These processes leave the matrix rank of $\theta_{i,j}$ quite unchanged. As such, they preserve bijective mappings, which is as one would expect. The processes correspond to a right and a left action of S_{n+1} on $\mathrm{T}(N)$; and these actions must preserve left and right sideals (though the bijective mappings, which are also preserved, are neither, and $\mathrm{T}_0(N)$ is not itself preserved). The fact that right actions simply shuffle columns gives us this result

Lemma 260 *The action* $\mathrm{T}(N) \times \mathcal{S} \longmapsto \mathrm{T}(N)$ *is an additive homomorphism of the additive group* $\mathrm{T}(N)$.

It is obvious that row shuffling—left action—does not constitute an additive homomorphism, at least in non-abelian cases.

Definition 261 Subsets of $\mathrm{T}(N)$ which are preserved under all right actions of doubly stochastic matrices will be called **right permutation closed** subsets. **Left permutation closed** subsets are defined correspondingly.

7.4.5 *Automorphisms of* $T_0(N)$

Suppose α is an automorphism of $T_0(N)$. α certainly maps doubly stochastic (i.e. *invertible*) $\mathcal{A}$-matrices to other doubly stochastic matrices. The key to determining α is to fix what it does to mappings of the form $\Delta^{(a,b)}$ because general elements of $T_0(N)$ are simply sums of such mappings.

Remember, for the moment, I am assuming all groups are finite. These observations, plus the observations in Sect. 7.3.3, bring us closer to pinning down what α does to these mappings. Doubly stochastic, that is, *invertible*, matrices are sums of the form

$$\Theta = \left(\Delta^{(a_1,n_1)} + \Delta^{(a_2,n_2)} + \Delta^{(a_3,n_3)} + \ldots + \Delta^{(a_n,n_n)}\right)$$

(this enumerates the group N as $\{n_0 = 0, n_1, \ldots n_n\}$). The elements $\{a_1, a_2, \ldots a_n\}$ must be *distinct* members of N^*. Since we know $\alpha(\Theta)$ is doubly stochastic (i.e. invertible) and that

$$\alpha(\Theta) = \left(\alpha\left(\Delta^{(a_1,n_1)}\right) + \alpha\left(\Delta^{(a_2,n_2)}\right) + \ldots + \alpha\left(\Delta^{(a_n,n_n)}\right)\right)$$

we might suspect that α should act *bijectively* on the set $\{\Delta^{(a_j,n_j)}\}_{j\in\{1,2,\ldots,n\}}$. If that were the case, α would be nothing more than a rearrangement of the rows and columns of the $\mathcal{A}$-matrices, and α would simply be a transformation on $\mathcal{A}$-matrices of the form

$$A \mapsto (S \cdot A \cdot T) \tag{7.91}$$

where S and T are doubly stochastic $(0-1)$ matrices, each with 1 in the top left corner. That is to say, A is transformed into an *equivalent matrix*. Here, S rearranges rows, and T rearranges columns. From Lemma 260, the right action by T actually is an additive automorphism of $T_0(N)$. Clearly, we would need further restrictions on S and T in order to get a near-ring automorphism—that is, to handle the multiplicative part.

Inner Automorphisms

Any automorphism of θ of $(N\ +)$ gives rise to a corresponding doubly stochastic $\mathcal{A}$-matrix, β. Since $\theta(x+y) = (\theta(x)+\theta(y))$ for all $x, y \in N$, it follows that for all $\mathcal{A}$-matrices, a, b, we have that $\beta \cdot (a+b) = (\beta(a) + \beta(b))$, and so, taking $\beta = S$ and $\beta^{-1} = T$ in (7.91), we obtain an inner automorphism of $T_0(N)$ (see Lemma 228). It is apparent that if α maps any $\Delta^{(a,b)}$ to another $\Delta^{(c,d)}$, then it performs this service for all of them and so does indeed act bijectively on the set of such mappings. Assuming that α does act *bijectively* on the set $\{\Delta^{(a_j,n_j)}\}_{j\in\{1,2,\ldots,n\}}$, then the transformation

$$\Delta^{(a,b)} \mapsto \Delta^{(c,d)}$$

means that α has the form given by (7.91). Consequently,

$$S \cdot (\theta \circ \phi) \cdot T = ((S \cdot \theta \cdot T) \circ (S \cdot \phi \cdot T)) \Rightarrow (\theta \circ \phi) = (\theta \cdot T \cdot S \cdot \phi)$$

and since θ and ϕ could themselves be invertible, one deduces

$$(T \cdot S) = I \tag{7.92}$$

It is straightforward to see that T must be an additive automorphism of $(N, +)$, so bijection implies an automorphism which is inner.

I now use some notation and results from Lemma 233, again assuming α is an automorphism. For any $a, b \neq 0$, we know $\alpha\left(\Delta^{(a,b)}\right) \neq 0$. The mapping will have finite support consisting of some non-empty subset of N^*. If $c \in N^*$, then

$$\Delta^{(a,b)} \circ \Delta^{(c,d)} = \delta_{b,c} \Delta^{(a,d)} \tag{7.93}$$

so

$$\alpha(\Delta^{(a,b)}) \circ \alpha(\Delta^{(c,d)}) = \delta_{b,c} \alpha(\Delta^{(a,d)}) \tag{7.94}$$

This means that when $b \neq c$,

$$\left(\mathcal{I}(\alpha(\Delta^{(c,d)})) \cap Support(\alpha(\Delta^{(a,b)}))\right) = \phi \tag{7.95}$$

and, when $b = c$:

$$Support(\alpha(\Delta^{(a,d)})) \subset Support(\alpha(\Delta^{(b,d)})) \tag{7.96}$$

which in turn implies

$$Support(\alpha(\Delta^{(a,d)})) = Support(\alpha(\Delta^{(b,d)})) \tag{7.97}$$

and

$$\mathcal{I}(\alpha(\Delta^{(a,d)})) = \mathcal{I}(\alpha(\Delta^{(b,d)})) \tag{7.98}$$

We now know that if $b \neq a$

$$\left(\mathcal{I}(\alpha(\Delta^{(a,a)})) \cap Support(\alpha(\Delta^{(b,b)}))\right) = \phi \tag{7.99}$$

If we put

$$I(a) = \mathcal{I}(\alpha(\Delta^{(a,a)})) \cap Support(\alpha(\Delta^{(a,a)})) \neq \phi \tag{7.100}$$

then $I(a)$ and $I(b)$ are disjoint. By the Dirichlet pigeon-hole principle, they must be singleton sets.

But *Support*$(\alpha(\Delta^{(a,a)}))$ must also be a singleton unless we violate (7.95). This means that for each a, there is some x such that

$$\alpha(\Delta^{(a,a)}) = \Delta^{(x,x)} \tag{7.101}$$

Putting $\theta(a) = x$ defines an additive automorphism of $(N, +)$ and since

$$\Delta^{(a,a)} \circ \Delta^{(a,b)} \circ \Delta^{(b,b)} = \Delta^{(a,b)} \tag{7.102}$$

it must be that

$$\alpha(\Delta^{(a,b)}) = \Delta^{(\theta(a),\theta(b))} \tag{7.103}$$

This means that α is an inner automorphism.

There is a well-known theorem in ring theory associated with Emmy Noether and Thoralf Skolem which says that every automorphism of a finite dimensional central simple algebra is inner [66]. I believe they each discovered this independently, although Skolem was marginally first. This is true for finite transformation near-rings — extending Lemma 228. There may be scope for further investigation here, perhaps involving sub-near-rings of transformation near-rings and automorphism towers.

7.4.6 *The Structure of* $\mathfrak{B}$(N)

I want to investigate the additive group $\mathfrak{B}(N)$. Recall the notation developed in Notation 227 and that $\mathfrak{B}(N)$ is defined there as a *group* rather than a near-ring. I use the notation for mappings introduced in Notation 253, so $I = (n_0, n_1, \ldots n_n)$. The reader should continue to regard the additive group $(N, +)$ as finite. Of course

$$\mathfrak{H}(N) \subset \mathfrak{B}(N) \tag{7.104}$$

Although the additive generators of $\mathfrak{B}(N)$ comprise the fixed set of bijective mappings of N, the group itself depends on the additive structure of $(N, +)$. That is, the additions involved are not the customary additions of matrices but depend on the additive structure of the group $(N, +)$. There are nice theorems associated with Birkhoff and von Neumann which describe additions of doubly stochastic matrices [95]. These won't apply here, because the matrix addition is defined by the group involved, but one might hope, and perhaps look, for something similar.

A typical bijection has the form

$$\left(n_{\sigma(0)}, n_{\sigma(1)}, \ldots, n_{\sigma(n)}\right)$$

where $\sigma \in S_{(n+1)}$. The group $\mathfrak{B}(N)$ is additively generated by the bijective mappings

$$(b_0, b_1, \ldots, b_{(n)})$$

which, notationally, are simply permutations of the group elements ($b_j = n_{\sigma(j)}$). The elements $\{b_j\}_{j=0}^{n}$ are the distinct elements of the group $(N, +)$. I am *not* using the cycle notation so common in permutation theory:

$$\boxed{(b_0, b_1, \ldots, b_n) \text{ means } n_j \mapsto b_j.} \text{ (page 264).}$$

The operation (7.90) permutes the terms of the tuples representing mappings in $\mathfrak{B}(N)$, so that all re-orderings of any tuple occurring must occur—$\mathfrak{B}(N)$ is right permutation closed. This will be assumed in what follows.

Suppose $g \in N$ has additive order in excess of 2. Then, there will be two elements of $\mathfrak{B}(N)$ having the forms

(i) $\big(0, g, -g, a_0, a_1, \ldots a_{(n-3)}\big)$.
(ii) $\big(g, -g, 0, -a_0, -a_1, \ldots - a_{(n-3)}\big)$.

Here, $\{a_j\}_{j=0}^{(n-3)}$ are distinct elements of the set $(N - \{0, g, -g\})$. Adding, we conclude that $(g, 0, -g, 0, 0, \ldots 0) \in \mathfrak{B}(N)$. The same can be proved when $|g| = 2$, using the tuples $\big(g, a_0, 0, a_1, a_2, \ldots a_{(n-3)}\big)$ and $\big(0, -a_0, g, -a_1, -a_2, \ldots - a_{(n-3)}\big)$. This next result is true for both finite and infinite groups.

Lemma 262 *Suppose* $(N, +)$ *is a group. Then,*

(i) All elements with the form $(0, 0, \ldots, 0, g, 0, 0 \ldots, 0, -g, 0, 0 \ldots)$ *lie in* $\mathfrak{B}(N)$ *(by this, I mean that the two non-zero entries may occur anywhere, and may not necessarily be distinct).*

(ii)

$$\mathfrak{B}(N) \supseteq [\mathrm{T}(N), \mathrm{T}(N)]$$

and so $\mathfrak{B}(N) \lhd \mathrm{T}(N)$.

(iii) All elements $(g, g, \ldots, g)$ $(g \in N)$ *lie in* $\mathfrak{B}(N)$.

Proof

(i) Taking any non-zero $g, h \in N$, we recognise the tuple

$$(0, 0, \ldots, 0, g, 0, 0, ..0, -g, 0, 0, \ldots 0)$$

as in $\mathfrak{B}(N)$, as is

$$(0, 0, \ldots, 0, (g + h), 0, 0, ..0, (-h - g), 0, 0, \ldots 0)$$

By addition

$$(0, 0, \ldots, 0, (g+h), 0, 0, ..0, (-g-h), 0, 0, \ldots 0)$$

is in $\mathfrak{B}(N)$. So, by subtraction,

$$(0, 0, \ldots, 0, 0, 0, 0, ..0, [h, g], 0, 0, \ldots 0)$$

is in $\mathfrak{B}(N)$. Now apply (8.1).

(ii) This is already known, since $\mathfrak{T}(N) \subseteq \mathfrak{B}(N)$. Alternatively,

$$(g, g, \ldots, g) = (g+n_0, g+n_1, \ldots, g+n_n) - (n_0, n_1, \ldots, n_n)$$

and both the mappings on the right are bijections.

□

Corollary 263 *For any positive integer n smaller than the order of N*

$$\left(0, 0, \ldots, \overbrace{g, g, \ldots, g}^{\text{n times}}, 0, 0, \ldots, -ng, 0, 0, \ldots, 0\right) \in \mathfrak{B}(N)$$

and, of course, the gs may be distributed as one wishes, although, shown as clumped, they don't have to be.

Corollary 264 *For any positive integer n smaller than the order of N and any integer partition of n of the form* $(a_1 + a_2 + \ldots a_r) = n$

$$(0, 0, \ldots, a_1 g, a_2 g, \ldots, a_r g, 0, 0, \ldots, -ng, 0, 0, \ldots, 0) \in \mathfrak{B}(N)$$

and, of course, the $a_j gs$ *may be distributed as one wishes, although, shown as clumped, they don't have to be*

The mappings described in Lemma 262 (i) are of interest.

We might regard N as a sort of near-ring group under the action of T(N), although, of course, the zero symmetric condition does not apply. With that proviso, these mappings can be thought of as being the cleik

$$(0, 0 \ldots, g, 0, \ldots, 0, -g, 0, \ldots 0) = \left(\Delta^{(g,x)} + \Delta^{(-g,y)}\right) \tag{7.105}$$

When $(N, +)$ has odd order, these cleiks actually generate $\mathfrak{B}(N)$ additively—they generate the additive generators of that group. This is alluringly similar to the multiplicative situation—of transpositions generating symmetric groups. If $(N, +)$ has even order, there will be elements g with additive order 2—in point of fact, an odd number of such elements; and there will be an even number of non-trivial elements with order larger than 2. The *bijections* then involve just a single occurrence of g since $g = -g$; but the set of cleiks from the lemma has to include $\left(\Delta^{(g,x)} + \Delta^{(g,y)}\right)$—two occurrences of g. We can say that if $\mathrm{T}(N) \neq \mathfrak{B}(N)$, there will be mappings $\Delta^{(a,x)} \notin \mathfrak{B}(N)$. Right permutation closure says that if for some $x \in N$ we have $\Delta^{(a,x)} \notin \mathfrak{B}(N)$, then for any $y \in N$, we must have $\Delta^{(a,y)} \notin \mathfrak{B}(N)$.

Define

$$\mathcal{H}(N) = \{a \in N \ : \ \Delta^{(a,x)} \notin \mathfrak{B}(N) \textit{ for some } x \in N\} \tag{7.106}$$

Notice that from Lemma 262,

$$(\mathcal{H}(N) \cap \mathfrak{B}(N)) = \{\phi\} \tag{7.107}$$

Right permutation closure gives us the following result:

Lemma 265 *Suppose $a, b, \in \mathcal{H}(N)$. Then the following conditions are equivalent:*

(i) $(a + b) \notin \mathcal{H}(N)$.
(ii) For some $x, y \in N$, $\left(\Delta^{(a,x)} + \Delta^{(b,y)}\right) \in \mathfrak{B}(N)$.
(iii) For any $x, y \in N$, $\left(\Delta^{(a,x)} + \Delta^{(b,y)}\right) \in \mathfrak{B}(N)$.

The situation $a = g, b = -g$, described in Lemma 262, is simply a special case of this. Define

$$\mathfrak{J}(N) = \{\mathcal{H}(N)\}^c = \left\{a \in N \ : \ \Delta^{(a,x)} \in \mathfrak{B}(N)\right\} \tag{7.108}$$

$\mathfrak{J}(N)$ is a normal subgroup of $(N, +)$, and it contains the commutator subgroup $[(N, +), (N, +)]$. In fact,

$$\mathrm{T}(N, \mathfrak{J}(N)) \subseteq \mathfrak{B}(N) \tag{7.109}$$

Any subgroup of $(N, +)$ for which $\mathfrak{J}(N) \subseteq A \subseteq N$ must include elements $a \in A$ such that $\Delta^{(a,x)} \notin \mathfrak{B}(N)$. Consequently, $\mathfrak{J}(N)$ is the maximal subgroup of N with property (7.109). We know that if $\mathfrak{B}(B) \neq \mathrm{T}(N)$, then $\mathfrak{J}(N)$ is a proper subgroup of $(N, +)$. Elements of $\mathrm{T}(N, \mathfrak{J}(N))$ consist entirely of tuples from $\mathfrak{J}(N)$. Choosing $a \in \mathcal{H}(N)$, we know $\left(\Delta^{(a,x)} + \Delta^{(-a,y)}\right) \in (\mathfrak{B}(N) - \mathrm{T}(N, \mathfrak{J}(N)))$. In this situation, at least, $\mathrm{T}(N, \mathfrak{J}(N)) \subsetneq \mathfrak{B}(N)$.

Lemma 266

$$\mathfrak{B}(N) = \mathrm{T}(N) \ \Leftrightarrow \ \mathfrak{J}(N) = N \ \Leftrightarrow \ \mathcal{H}(N) = \phi.$$

Suppose $(a_0, a_1, \ldots, a_n) \in \mathrm{T}(N)$, with $(N, +)$ finite. Then, $(a_0 + a_1, 0, a_2, \ldots a_n) = (a_0, a_1, \ldots, a_n) + (a_1, -a_1, 0, 0, \ldots 0)$ and, continuing in this way,

$$(a_0 + a_1 + \ldots + a_n, 0, 0, \ldots 0)$$

may be formed by adding a single element of $\mathfrak{B}(N)$ to the original element.

Thus,

Lemma 267 *Suppose* $(N, +)$ *is a finite group with order* $(n+1)$ *and* $(a_0, a_1, \ldots, a_n) \in \mathrm{T}(N)$.
Suppose $(a_0 + a_1 + a_2 + \ldots a_n) \in \mathfrak{J}(N)$.
Then,

(i) $(a_0, a_1, \ldots, a_n) \in \mathfrak{B}(N)$.
(ii) When σ *is any permutation of the numbers* $\{0, 1, 2, \ldots, n\}$,

(i) $\left(a_{\sigma(0)}, a_{\sigma(1)}, \ldots, a_{\sigma(n)}\right) \in \mathfrak{B}(N)$.
(ii) $\left(a_{\sigma(0)} + a_{\sigma(1)} + a_{\sigma(2)} + \ldots + a_{\sigma(n)}\right) \in \mathfrak{J}(N)$.

The permutation symmetries above may be less interesting than they appear at first sight. They are obvious, given the fact that $\mathfrak{B}(N)$ contains the commutator subgroup of $\mathrm{T}(N)$ and is right permutation closed.

Corollary 268 *Suppose* $(a_0, a_1, a_2, \ldots, a_n) \in \mathfrak{B}(N)$. *Then,* $(a_0 + a_1 + a_2 + \ldots a_n) \in \mathfrak{J}(N)$.

So, the condition that the coordinates add up to something in $\mathfrak{J}(N)$ actually characterises the mappings in $\mathfrak{B}(N)$. Symmetry considerations require that additive automorphisms of $(N, +)$ preserve $\mathcal{H}(N)$ and $\mathfrak{J}(N)$. They are both **characteristic subsets** of the additive group; so $\mathfrak{J}(N)$ is a characteristic subgroup of $(N, +)$, and not merely normal.

7.4.7 Further Investigation

I have repeatedly remarked that the $\mathcal{A}$-matrix notation describes multiplicative properties which are purely set-theoretic. Taking the ensemble of $((n+1) \times (n+1))$ column stochastic matrices, one *could* investigate the possible additions that would result in a transformation near-ring or something weaker. This would be an exercise in a reasonable generalisation of the additive host problem (see page 329) and seems an interesting thing to investigate, although I am not suggesting a complete solution would be possible!

7.5 Some Examples

7.5.1 The Cyclic Group $\mathbf{C_3}$

The cyclic group $C_3 = \{0, 1, 2\}$ has $\mathrm{T}(C_3) \cong (C_3 \oplus C_3 \oplus C_3)$, additively. The elements τ_x form the set

$$\mathcal{T}(C_3) = \{(1, 2, 0), (2, 0, 1), (0, 1, 2)\}$$

The group $\mathfrak{T}(C_3)$ (page 264) has elements

$$\{(0,0,0),(1,1,1),(2,2,2),(0,1,2),(0,2,1),(1,0,2),\\ (1,2,0),(2,0,1),(2,1,0)\} \quad (7.110)$$

This is additively isomorphic to $(C_3 \oplus C_3)$. We have

$$A(C_3) = \langle(0,1,2),(0,2,1)\rangle = \{(0,0,0),(0,1,2),(0,2,1)\} \cong C_3 \cong E(C_3)$$

and, of course, $I(C_3) = \langle(0,1,2)\rangle = A(C_3)$. We have

$$\mathfrak{B}(C_3) = \mathfrak{H}(C_3) = \mathfrak{T}(C_3) \quad (7.111)$$

The subgroup $\mathfrak{K} = \{(0,0,0),(0,0,1),(0,0,2)\}$ has trivial intersection with the groups given in (7.111). This subgroup lies wholly within $T_0(C_3)$ and, of course, additively

$$T(C_3) = (\mathfrak{T}(C_3) \oplus \mathfrak{K}) \quad (7.112)$$

This is an additive direct sum of structures which are sub-near-rings of $T(C_3)$.

7.5.2 Finite Dihedral Groups

Consider the family of dihedral groups D_n with degree n and order 2n, and stipulate $n \geq 3$. I use the presentation given by (2.19) (page 83). The family divides into two subclasses with distinct additive properties, depending on whether n is odd or even (more details are given in Appendix B, page 497). The multiplicative group of inner automorphisms is isomorphic to D_n when n is odd (the centreless case) and isomorphic to $D_{\frac{n}{2}}$ when n is even. We can fix endomorphisms by specifying what they do to the two generators, a and b. Inner automorphisms generated by elements $(-ra)$ map

$$a \mapsto a; \quad b \mapsto (2ra + b)$$

Inner automorphisms generated by elements (ra + b) map

$$a \mapsto -a; \quad b \mapsto (2ra + b)$$

Adopting Notation 253, I enumerate the group with the code

$$\{0 \leftrightarrow 0, a \leftrightarrow 1, \dots (n-1)a \leftrightarrow (n-1), b \leftrightarrow n, \dots, ((n-1)a + b \leftrightarrow (2n-1)\} \quad (7.113)$$

then the first n elements of tuples representing inner automorphisms involve only the code numbers $\{0, 1, 2, \ldots, (n-1)\}$, and the last n elements involve only the remaining code numbers.

Exclamation Mark Notation
It is occasionally useful to indicate the point at which images of $\langle a \rangle$ stop and images of its complement start; I do this using an exclamation mark. The identity mapping would then be

$$(0, a, 2a, \ldots, (n-1)a, \ !\, b, (a+b), \ldots, ((n-1)a+b))$$

or, possibly,

$$(0, 1, \ldots, (n-1), \ !\, n, (n+1), \ldots, (2n-1))$$

I use this bracketed notation a lot in what follows, sometimes with group elements, sometimes with code numbers; and I use it on other groups too.

7.5.3 $\mathbf{D_n}$ *when n Is Odd*

When n is an odd number, element order considerations mean we may treat the first n entries of automorphisms and the second n entries, separately, because n and 2 are coprime.

Inner Automorphisms
The first n entries are either $(0, 1, 2, 3, \ldots, (n-1))$ or $(0, (n-1), (n-2), \ldots, 1)$ (corresponding to images +a and -a, respectively). These additively generate the cyclic group C_n. The second n entries represent tuples with either the general form

$$((2ma+b), (2ma+a+b), (2ma+2a+b), \ldots, (2ma+(n-1)a+b)) \tag{7.114}$$

or the general form

$$((2ma+b), (2ma-a+b), (2ma-2a+b), \ldots, (2ma-(n-1)a+b)) \tag{7.115}$$

$(m \in \{0, 1, \ldots, (n-1)\})$. The first case arises when a is mapped to itself, the second when it is mapped to its own additive inverse. Taking $m = 0$ and adding the resulting two tuples give us

$$(0, 2a, 4a, \ldots, 2(n-1)a)$$

and, hence,

$$(0, a, 2a, 3a, \ldots, (n-1)a)$$

(n is odd). From this and the first tuple, we get

$$(b, b, \ldots, b)$$

and then,

$$(2ma, 2ma, \ldots, 2ma)$$

from which we get

$$(a, a, \ldots, a)$$

In summary, the second n entries involve us in investigating the group generated by

$$\{(a, a, \ldots, a), (b, b, \ldots, b), (0, a, 2a, \ldots (n-1)a)\}$$

or, alternatively,

$$\{(1, 1, \ldots, 1), (n, n, \ldots, n), (0, 1, 2, \ldots (n-1))\} \tag{7.116}$$

Such expressions are isomorphic to the group $(\mathrm{D}_n \oplus \mathrm{C}_n)$.

Hence, when n is an odd number

$$\mathrm{I}(\mathrm{D}_n) \cong (\mathrm{C}_n \oplus \mathrm{D}_n \oplus \mathrm{C}_n) \tag{7.117}$$

The subgroup defined by (7.116) is a near-ring but it has no identity.

Full Automorphisms

When $n > 3$, there are outer automorphisms. These have the form

$$a \mapsto ra, \quad b \mapsto (sa + b) \tag{7.118}$$

where (r,n) = 1 (coprime) and $1 \lneqq r \lneqq (n-1)$. and actually present us with nothing new, both being subsumed by situations we have already covered. Thus, when n is odd

$$\mathrm{A}(\mathrm{D}_n) \cong \mathrm{I}(\mathrm{D}_n) \cong (\mathrm{C}_n \oplus \mathrm{D}_n \oplus \mathrm{C}_n) \tag{7.119}$$

The near-ring $\mathrm{A}(\mathrm{D}_n)$ is distributively generated, and so its commutator subgroup is an ideal (page 49). This subgroup is additively isomorphic to C_n, and the quotient group is a *ring* (page 55) with identity, hosted by $(\mathrm{C}_n \oplus \mathrm{C}_2 \oplus \mathrm{C}_n)$. It has order $2n^2$.

E(D$_n$), Phom(D$_n$)

Endomorphisms of D_n have to map b either to zero or to something of the form $(sa+b)$, and they map a to something of the form ra, $(0 \leq r < n)$. Each such mapping is already catered for within $A(D_n)$, so, when n is an odd number, additively

$$E(D_n) \cong A(D_n) \cong (C_n \oplus D_n \oplus C_n) \tag{7.120}$$

The order of Phom(D_n) (when n is odd) will be $n^{(n-1)} \times 2$. The group must contain the commutator subgroup of the corresponding transformation near-ring, which is

$$\oplus_{(g \neq 0, g \in D_n)} C_n$$

and the group must contain $E(D_n)$; this means that, additively

$$\text{Phom}(D_n) \cong \left(\left(\bigoplus_{j=1}^{(n-2)} C_n \right) \oplus D_n \right) \tag{7.121}$$

$\mathfrak{T}$(D$_n$) & $\mathfrak{H}$(D$_n$)

Turning to $\mathfrak{T}(D_n)$, we know that (using (7.113))

$$\tau_a = (1, 2, 3, \ldots, (n-1), 0, \; ! \; (n+1), (n+2), \ldots, (2n-1), n)$$

and that

$$\tau_b = (n, (2n-1), (2n-2), \ldots, (n+1), \; ! \; 0, (n-1), (n-2), \ldots, 1)$$

By Lemma 257, we need only to consider the group generated additively by the above two mappings and $I(D_n)$. Simple order considerations mean that we may treat the first n entries of τ_a and τ_b and the last n, separately. For the first n entries, our generating set (including $I(D_n)$) is

$$\{(0, 1, 2, \ldots, (n-1)), (1, 2, \ldots, (n-1), 0), (n, (2n-1), (2n-2), \ldots (n+1))\}$$

This is equivalent to the generating set

$$\{(0, 1, 2, \ldots, (n-1)), (1, 1, \ldots, 1), (n, n, \ldots n)\}$$

which generates a group additively isomorphic to $(C_n \oplus D_n)$. For the second n entries (see (7.116)), it is

$$\{(1, 1, \ldots, 1), (n, n, \ldots, n), (0, 1, 2, \ldots (n-1)),$$
$$((n+1), (n+2), \ldots, (2n-1), n), (0, (n-1), (n-2), \ldots, 1)\}$$

This is equivalent to

$$\{(1, 1, \ldots, 1), (n, n, \ldots, n), (0, 1, 2, \ldots (n-1)), \\ ((n+1), (n+2), \ldots, (2n-1), n), \}$$

which is equivalent to

$$\{(1, 1, \ldots, 1),\ \ (n, n, \ldots, n),\ \ (0, 1, 2, \ldots (n-1)),\ \ (1, 2, \ldots, (n-1), 0)\}$$

which is equivalent to

$$\{(1, 1, \ldots, 1),\ \ (n, n, \ldots, n),\ \ (0, 1, 2, \ldots (n-1))\}$$

This produces a group additively isomorphic to $(\mathrm{C}_n \oplus \mathrm{D}_n)$. That is, additively, when n is odd,

$$\mathfrak{T}(\mathrm{D}_n) \cong ((\mathrm{C}_n \oplus \mathrm{D}_n) \oplus (\mathrm{C}_n \oplus \mathrm{D}_n)) \cong \mathfrak{H}(\mathrm{D}_n) \tag{7.122}$$

$\mathfrak{B}(\mathbf{D_n})$

Turning to $\mathfrak{B}(\mathrm{D}_n)$ and using Lemma 262, we know that

$$((0, a, 2a, \ldots, (n-1)a,\ !\,b, (a+b), \ldots, ((n-1)a+b)) \\ -\ (0, 0, \ldots, 0,\ !\,0, b, b, b, \ldots, b, b)) \in \mathfrak{B}(\mathrm{D}_n)$$

and taking an appropriate odd number of additions of the resulting term, we can claim terms such as

$$(0, 0, \ldots, 0, b, 0, 0, \ldots 0) \tag{7.123}$$

(a single occurrence of b, somewhere) as being in $\mathfrak{B}(N)$. From this,

$$((0, a, (2a+b), \ldots, ((n-1)a+b),\ !\,b, (a+b), \ldots, ((n-1)a+b)) \in \mathfrak{B}(\mathrm{D}_n)$$

(crucially, the second entry does not have even order), and so, we can claim terms such as

$$(0, 0, \ldots, 0, a, 0, 0, \ldots, 0) \tag{7.124}$$

(a single occurrence of a, somewhere) as being in $\mathfrak{B}(N)$. From (7.123) and (7.124), this must mean that when n is odd

$$\mathfrak{B}(\mathrm{D}_n) = T(\mathrm{D}_n) \tag{7.125}$$

The near-rings $I(D_n)$, $A(D_n)$ and $E(D_n)$ (n odd) were first determined by Malone and Lyons [113]. They also handled the even case, in a subsequent paper. Meldrum [119] covers these near-rings and gives rather more information about them. The orders I give for $I(D_n)$, $A(D_n)$ and $E(D_n)$, when n is odd, match the orders he reports for these near-rings. From that, one might take some, fragile, confidence that the analysis is correct, if only with respect to these first three structures (I don't think anyone has looked at $\mathfrak{B}(N)$ before, for instance).

7.5.4 D_n *when n Is Even*

Now suppose n is an even number larger than three.

Inner Automorphisms

I have covered the structure of inner automorphisms on page 276. The first sort of inner automorphism generates mappings with the form

$$(0, a, 2a, \dots, (n-1)a,\ !\,(2ra+b), (2ra+a+b), \dots, (2ra+(n-1)a+b))$$

and the second sort generates mappings with the form

$$(0, -a, -2a, \dots, -(n-1)a,\ !\,(2ra+b), ((2r-1)a+b), ((2r-2)a+b), \dots, ((2r-(n-1))a+b))$$

Additively, $I(D_n) =$

$$\langle\, ((0, 0, \dots, 0,\ !\,2a, 2a, , \dots, 2a)\,, (0, a, 2a, \dots, (n-1)a,\ !\,b, (a+b), \dots, ((n-1)a+b))\,, (0, 0, \dots, 0,\ !\,0, 2a, \dots, 2(n-1)a)\,\rangle \tag{7.126}$$

Elements of $I(D_n)$ have the form $(x + y + x)$, where

$$x \in \langle((0, 0, \dots, 0,\ !\,2a, 2a, , \dots, 2a)\rangle$$

$$y \in \langle(0, 0, \dots, 0,\ !\,0, 2a, \dots, 2(n-1)a)\rangle$$

and

$$z \in \langle(0, a, 2a, \dots, (n-1)a,\ !\,b, (a+b), \dots, ((n-1)a+b))\rangle$$

Structurally, when n is even and bigger than four

$$I(D_n) = \left(C_n \rtimes \left(C_{\frac{n}{2}} \oplus C_{\frac{n}{2}}\right)\right) \tag{7.127}$$

This group has order $\left(\frac{n}{2} \times \frac{n}{2} \times n\right) = \frac{n^3}{4}$. It is abelian exactly when n is equal to 4 and is then a direct sum of the three cyclic groups

$$\mathrm{I(D_4)} \cong (\mathrm{C_4} \oplus \mathrm{C_2} \oplus \mathrm{C_2}) \tag{7.128}$$

In other cases, the group is non-abelian, with commutator subgroup

$$\langle(0, 0, \ldots, 0 \,!\, 4a, 4a, \ldots, 4a)\rangle \oplus \langle(0, 0, \ldots, \,!\, 0, 4a, 8a, \ldots, 4(n-1)a)\rangle \tag{7.129}$$

Outer Automorphisms

Outer automorphisms of D_n take a to an element of the form ra, where $(r, n) = 1$, and they take b to elements with the form $(sa + b)$, where $0 \leq s < n$. They are distinguished from *inner* automorphisms either by s being odd or by r lying *strictly* between 1 and $(n - 1)$. Subject to those restrictions, they have the form

$$(0, ra, \ldots, (n-1)ra, \,!\, (sa+b), ((r+s)a+b), \ldots, (((n-1)r+s)a+b)) \tag{7.130}$$

and while each such mapping, ϕ, has additive order n, it is always true that $2\phi \in \mathrm{I(D}_n)$. Values $r = 1, s = 0$ and $r = 1, s = 1$ lie in $\mathrm{A(D}_n)$, at least, and a subtraction implies that so does

$$\underline{w} = (0, 0, \ldots, 0, \,!\, a, \ldots, a) \tag{7.131}$$

The group therefore contains

$$(0, ra, 2ra, \ldots, (n-1)ra, \,!\, b, (ra+b), \ldots, ((n-1)ra+b))$$

for *any* r coprime to n, and so it also contains

$$(0, a, 2a, \ldots, (n-1)a, \,!\, b, (ra+b), \ldots, ((n-1)ra+b))$$

(r being invertible modulo n). Writing $r = (2.\lambda + 1)$, then

$$\begin{aligned}(0, a, 2a, \ldots, (n-1)a, \,!\, b, (ra+b), \ldots, ((n-1)ra+b)) = \\ (\lambda \times (0, 0, \ldots, 0, \,!\, 0, 2a, \ldots, 2(n-1)a) + \\ (0, a, 2a, .., (n-1)a, b, (a+b), \ldots, ((n-1)a+b)))\end{aligned}$$

We conclude that

$$\mathrm{A(D}_n) = \langle \mathrm{I(D}_n), \underline{w} \rangle \tag{7.132}$$

and that the non-abelian group $A(D_n)$ has order $(2 \times \frac{n^3}{4}) = \frac{n^3}{2}$. Elements of $A(D_n)$ have the form $(\lambda \underline{w} + \mu \underline{y} + \xi \underline{x})$, where

$$\underline{w} = (0, 0, \ldots, 0, \ ! \, a, a, a, \ldots, a) \tag{7.133}$$

$$\underline{y} = (0, 0, \ldots, 0, \ ! \, 0, 2a, 4a, 6a, \ldots, 2(n-1)a) \tag{7.134}$$

and

$$\underline{z} = (0, a, 2a, \ldots, (n-1)a, \ ! \, b, (a+b), \ldots, ((n-1)a+b)) \tag{7.135}$$

$\underline{z}$ is simply the multiplicative identity of the transformation near-ring; and I write

$$\underline{A} = \left\{\underline{w}, \underline{y}, \underline{z}\right\}$$

The group $A(D_n)$ has the form

$$A(D_n) \cong \left(C_n \rtimes \left(C_n \oplus C_{\frac{n}{2}}\right)\right) \tag{7.136}$$

$\mathfrak{T}(D_n)$ & $\mathfrak{H}(D_n)$

We now know the order of $\mathfrak{T}(D_n)$ when n is even; it will be $\frac{n^4}{2}$. Similarly, the order of $\mathfrak{H}(D_n)$ is n^4 (see (7.75) and (7.86)).

$E(D_n)$

Endomorphisms of D_n are determined by the images of a and b. They can be represented as

$$a \mapsto (ma + sb) \tag{7.137}$$

$$b \mapsto (ka + rb) \tag{7.138}$$

Here, $0 \leq s, r \leq 1$ and $0 \leq m, k \leq (n-1)$.

The image of b should be either 0 or have order 2. This gives us three distinct cases:

(A) $k = r = 0$.
(B) $r = 1$.
(C) $k = \frac{n}{2}$, & $r = 0$.

The condition

$$(b + a + b) = -a$$

imposes further conditions on the images of a and b in (7.137) and (7.138). It must be that

$$((ka + rb) + (ma + sb) + (ka + rb)) = -(ma + sb)$$

I analysed these three cases and produced the following:

(A)

$$[a \mapsto (ma + sb), \ b \mapsto 0].$$

(0) $a \mapsto 0, \ b \mapsto 0.$
(1) $a \mapsto \frac{n}{2}a, \ b \mapsto 0.$
(2) $a \mapsto (ma + b), \ b \mapsto 0.$

(B)

$$[a \mapsto (ma + sb), \ b \mapsto (ka + b)].$$

(3) $a \mapsto ma, \ b \mapsto (ka + b).$
(4) $a \mapsto b, \ b \mapsto b.$
(5) $a \mapsto b, \ b \mapsto (\frac{n}{2}a + b).$
(6) $a \mapsto (\frac{n}{2}a + b), \ b \mapsto b.$
(7) $a \mapsto (ma + b) \ b \mapsto (ma + b).$
(8) $a \mapsto (ma + b) \ b \mapsto ((\frac{n}{2} + m)\, a + b).$
(9) $a \mapsto ((k + \frac{n}{2})\, a + b) \ b \mapsto (ka + b).$

(C)

$$[a \mapsto (ma + sb), \ b \mapsto \frac{n}{2}a].$$

(10) $a \mapsto 0, \ b \mapsto \frac{n}{2}a.$
(11) $a \mapsto \frac{n}{2}a, \ b \mapsto \frac{n}{2}a.$
(12) $a \mapsto (ma + b), \ b \mapsto \frac{n}{2}a.$

Table 7.2 lists the forms of the possible endomorphisms of D_n when n is even. We can assume $\underline{A}$, the additive generators of $A(D_n)$, is present too; and the reader will see how these mappings can be constructed from the mappings listed in Table 7.2, although they are not all endomorphisms.

Additive Generation of $E(D_n)$

Consider the mappings needed to augment $A(D_n)$ and thereby generate $E(D_n)$. I attempt to construct a smaller subset of mappings that additively generate the near-ring and automatically include $\underline{A}$. From θ_3, we may construct the mapping

$$\beta_1 = (0, 0, \ldots, 0, \ !\, b, b, b, b, \ldots, b, b)$$

Table 7.2 Endomorphisms of D_n

	a	b
θ_0	0	0
θ_1	$\frac{n}{2}a$	0
θ_2	$(ma+b)$	0
θ_3	ma	$(ka+b)$
θ_4	b	b
θ_5	b	$(\frac{n}{2}a+b)$
θ_6	$(\frac{n}{2}a+b)$	b
θ_7	$(ma+b)$	$(ma+b)$
θ_8	$(ma+b)$	$((m+\frac{n}{2})a+b)$
θ_9	$((k+\frac{n}{2})a+b)$	$(ka+b)$
θ_{10}	0	$\frac{n}{2}a$
θ_{11}	$\frac{n}{2}a$	$\frac{n}{2}a$
$\theta 12$	$(ma+b)$	$\frac{n}{2}a$

and from θ_3 and β_1 obtain the mapping

$$(0, ma, 2ma, 3ma, \ldots, (n-1)ma \,!\, ka, (ma+ka), (2ma+ka), (3ma+ka), \ldots, ((n-1)ma+ka))$$

which gives us the two mappings:

$$(0, ma, 2ma, 3ma, \ldots, (n-1)ma, \,!\, 0, ma, 2ma, 3ma, \ldots, (n-1)ma)$$

and

$$(0, 0, 0, \ldots, 0, \,!\, ka, ka, ka, \ldots, ka) = k\underline{w}$$

We know

$$\theta_4 = (0, b, 0, b, \ldots 0, b, \,!\, b, 0, b, 0, \ldots, b, 0)$$

so from β_1 and θ_4 obtain

$$\beta_2 = (0, b, 0, b, \ldots 0, b, \,!\, 0, b, 0, b, 0, \ldots, b, 0, b)$$

I analyse whether the seven mappings of Table 7.3 additively generate $E(D_n)$. They do lie inside it. From β_1 and β_2, we can generate

$$\beta_3 = (0, b, 0, b, \ldots, 0, b, \,!\, b, 0, b, 0, \ldots, b, 0, b, 0)$$

and armed with these three, we may remove all references to b from the transformations given in Table 7.2, concentrating solely on the a terms and only worrying

Table 7.3 Posited generators of $E(D_n)$

Name	Effect
β_1	$(0, 0, 0, 0, \ldots, 0, 0, \; ! \, b, b, b, b, b, \ldots, b, b, b)$
β_2	$(0, b, 0, b, \ldots, 0, b, \; ! \, 0, b, 0, b, 0, \ldots, b, 0, b)$
θ_4	$(0, b, 0, b, \ldots, 0, b, \; ! \, b, 0, b, 0, b, \ldots, 0, b, 0)$
$\underline{w}$	$(0, 0, 0, 0, \ldots, 0, 0, \; ! \, a, a, a, a, \ldots, a, a, a)$
$\underline{y}$	$(0, 0, \ldots, 0, \; ! \, 0, 2a, 4a, 6a, \ldots, 2(n-1)a)$
$\underline{\alpha}$	$(0, a, 2a, 3a, \ldots, (n-1)a, \; ! \, 0, a, 2a, 3a, \ldots, (n-1)a)$
$\underline{\gamma}$	$(0, a, 0, a, \ldots, a, 0, a, \; ! \, 0, a, 0, a, 0, \ldots 0, a)$

about the final four transformations of Table 7.3, because $\underline{\gamma}$ is obtained from θ_2 and β_2.

From $\underline{w}$ and $\underline{\gamma}$, one can obtain

$$\underline{q} = (0, a, 0, a, \ldots, a, 0, a, \; ! \, a, 0, a, 0, a, \ldots, 0, a, 0)$$

and so θ_{11}. It's actually easier to obtain the endomorphisms of Table 7.2 for oneself than be told how to do it so I leave the proofs to the reader. The key to it is the ability to ignore b terms, but one needs to be circumspect in handling mappings such as θ_8, because the b term ensures that image points must have order 2 and if this is removed incautiously, there would be problems. From the mappings in Table 7.3, I can generate all the mappings in Table 7.2.

Exploring These Mappings

The generating mappings are in three subsets which generate disjoint subgroups:

(i) $\{\underline{w}, \underline{\alpha}\}$
(ii) $\{\underline{\gamma}, \underline{y}\}$
(iii) $\{\beta_1, \beta_2, \theta_4\}$

The first two subsets generate a subgroup $\mathfrak{V}$, defined as the set of mappings in $E(D_n)$ in which no terms include occurrences from $\{D_n - \langle a \rangle\}$. This will be a normal subgroup of $E(D_n)$.

The third set of generators builds a subgroup:

$$\underline{\mathfrak{W}} = (\langle \beta_2 \rangle \oplus \langle \theta_4 \rangle) \cong (C_2 \oplus C_2) \tag{7.139}$$

The first generating set gives a subgroup with the form $\left(C_n \oplus C_{\frac{n}{2}}\right)$. The second set does the same, and these two subgroups are disjoint.

From this, I claim an exact sequence

$$\left(\left(C_n \oplus C_{\frac{n}{2}}\right) \oplus \left(C_n \oplus C_{\frac{n}{2}}\right)\right) \longmapsto E(D_n) \twoheadrightarrow (C_2 \oplus C_2) \tag{7.140}$$

which would give the order of $E(D_4)$ to be $64 \times 4 = 256$.

Later, I become interested in the subset of E(D_4) relating to mappings with image sets within the commutator subgroup $\langle 2a \rangle$. The group of these mappings is generated by $\{2\underline{w}, \underline{y}, 2\underline{\gamma}\}$ and has order 8.

The first work attempting to determine endomorphism near-rings was due to Malone and Lyons [114]. Meldrum [119] lists their results without proof. The cited values match my ones for the n odd cases and agree for $A(D_n)$ and $I(D_n)$ in even cases. He gives these values for the order of E(D_n) when n is even:

$$|\mathrm{E}(\mathrm{D}_n)| = \begin{cases} \left(\frac{n^7}{64}\right) & \text{if 4 divides n but 8 does not} \\ \left(\frac{2n^7}{(n,4)^4}\right) & \text{otherwise} \end{cases} \tag{7.141}$$

The reader will see that my value for the order of E(D_4) matches the value above but the extension of my methods to more general cases would not. For that reason, I do not make that, obvious, extension, here.

$\mathfrak{B}$(D$_4$)

Turning to $\mathfrak{B}$(D$_4$), we know from Lemma 262

$$\mathfrak{B}(\mathrm{D}_4) \supsetneq \oplus_{j=1}^{8} \langle 2a \rangle = [T(\mathrm{D}_4), T(\mathrm{D}_4)]$$

so its order exceeds $(2)^8 = 256$. Using the notation on page 274, the subgroup $\mathfrak{J}$(D$_4$) must be one of the characteristic subgroups:

$$\{\langle 2a \rangle, \quad \langle a \rangle, \quad \langle a, b \rangle\}$$

The mappings $\{\underline{w}, \underline{y}, \underline{z}\}$ (page 283), though not all surjective, do lie in $\mathfrak{B}$(D$_n$) and A(D$_4$) $\subseteq$ $\mathfrak{B}$(D$_4$). From Lemma 267, we know that transformations of the form $\left(\Delta^{(g,x)} + \Delta^{(g,y)}\right)$ always lie in $\mathfrak{B}$(D$_4$), for any $g, x, y, \in$ D$_4$. Elements of T(D$_4$) are n-dimensional row vectors in the elements of D$_4$. Some of these tuples contain elements which involve the element b of the group. All bijective mappings have an even number of such elements. If we add two tuples and both originally contained an even number of terms involving the element b, the resulting tuple must have an even number of terms involving the element b (thought that even number may, of course, be zero). That means that the tuples in $\mathfrak{B}$(D$_4$) must each contain only an even number of terms involving b, and it means that $\mathfrak{B}$(D$_4$) is a proper subgroup of T(D$_4$) because $b \in \mathcal{H}$(D$_4$). It also means $\mathfrak{J}$(D$_4$) is either $\langle 2a \rangle$ or $\langle a \rangle$. Almost the same argument might be applied to a. The identity mapping is

$$\underline{z} = (0, a, 2a, 3a, b, (a+b), (2a+b), (3a+b))$$

But in analysing the generation of $\mathfrak{B}$(D$_4$), we may discount b terms and terms involving $2a$ and therefore only need to consider additions of permutations of the tuples with the form (this being derived from $\underline{z}$, as a specific example)

$$(0, a, 0, a, \; !\, 0, a, 0, a)$$

with the convention that $(a + a) \equiv 0$. Again, we always obtain an even number of a terms here (perhaps zero such terms). This implies that transformations with the form $\Delta^{(a,x)}$ do not lie within $\mathfrak{B}(D_4)$ and so $\mathfrak{J}(D_4)$ is actually $\langle 2a \rangle$. We now know $a, 3a, b, (a+b), (2a+b), (3a+b), \in \mathcal{H}(D_4)$ and that—with some looseness of language—elements of $\mathfrak{B}(D_4)$ only involve even numbers of a and b terms, with all such tuples being in $\mathfrak{B}(D_4)$. This essentially determines $\mathfrak{B}(D_4)$ and permits us to tell, at a glance, whether a given tuple is in $\mathfrak{B}(D_4)$.

7.5.5 D_∞ and $\mathfrak{A}(D_\infty)$

I now look at the group D_∞ with presentation (5.8) on page 204. The group has no centre and its commutator subgroup is $\langle 2a \rangle$. The situation for this group is very like that of D_{2n}. Inner automorphisms map $a \mapsto \pm a$ and $b \mapsto (2sa + b)$, with $s \in \mathbb{Z}$.

A modified form of Notation 253 can be used, one which now refers to infinite tuples and which again uses the ! notation. We would represent the identity mapping as

$$I = (0, a, -a, 2a, -2a, \ldots, \; ! \, b, (a+b), (-a+b), (2a+b), \ldots) \tag{7.142}$$

Our possible inner automorphisms have one of the two forms:

$$(0, a, -a, 2a, \ldots, \; ! \, (2sa+b), (2sa+a+b), (2sa-a+b), \\ (2sa+2a+b), \ldots, \ldots,) \tag{7.143}$$

or

$$(0, -a, a, -2a, \ldots, \; ! \, (2sa+b), (2sa-a+b), (2sa+a+b), \ldots,) \tag{7.144}$$

The group of inner automorphisms is

$$\mathrm{I}(D_\infty) \cong \left\langle \underline{x}, \underline{y}, \underline{z} \right\rangle \tag{7.145}$$

where

$$\underline{x} = (0, 0, 0, \ldots, 0, \; ! \, 2a, 2a, 2a, 2a, \ldots)$$

$$\underline{y} = (0, \ldots, 0, \; ! \, 0, 2a, -2a, 4a, -4a, 6a, -6a, \ldots)$$

and

$$\underline{z} = (0, a, -a, 2a, -2a, \ldots, \; ! \, b, (a+b), (-a+b), (2a+b), (-2a+b), \ldots,) \\ = I$$

Then,

$$\mathrm{I}(\mathrm{D}_\infty) = (\mathbb{Z} \rtimes (\mathbb{Z} \oplus \mathbb{Z})) \quad (7.146)$$

Compare this with (7.126). Automorphisms take a to $\pm a$ and b to $(sa + b)$, $s \in \mathbb{Z}$. Again, the situation follows $\mathrm{A}(\mathrm{D}_{2n})$, automorphisms having the form ($\underline{w}$ + $\underline{y}$ + $\underline{z}$), where

$$\underline{w} \in \langle((0, 0, \ldots, 0, \ ! \, a, a, , \ldots, a)\rangle$$

Interestingly, $\mathrm{A}(\mathrm{D}_\infty)$ has the same structure as $\mathrm{I}(\mathrm{D}_\infty)$—one simply replaces the generator $\underline{x}$ by the generator $\underline{w}$, but at the same time

$$\mathrm{I}(\mathrm{D}_\infty) \subsetneq \mathrm{A}(\mathrm{D}_\infty) \quad (7.147)$$

$\mathfrak{A}(\mathrm{D}_\infty)$

I don't much concern myself with infinite groups in this book, but the above observation is suggestive of further structure in infinite cases. The infinite group D_∞ possesses injective, but non-surjective, endomorphisms. These map a to ra and b to (sa + b), where $r, s \in \mathbb{Z}$. The set of all injective endomorphisms of the group generates a *supergroup* $\mathfrak{A}(\mathrm{D}_\infty) \supsetneq \mathrm{A}(\mathrm{D}_\infty)$, where

$$\begin{aligned}\mathfrak{A}(\mathrm{D}_\infty) = \langle &(0, 0, \ldots, \ ! \, a, a, ..), (0, 0, \ldots, \ ! \, 0, 2a, -2a, 4a, -4a, \ldots), \\ &(0, 0, \ldots, \ ! \, b, b, ..), (0, a, -a, 2a, -2a, \ldots, \ ! \, 0, a, -a, 2a, -2a, \ldots)\rangle\end{aligned} \quad (7.148)$$

and, additively,

$$\mathfrak{A}(\mathrm{D}_\infty) = (\mathrm{D}_\infty \rtimes (\mathbb{Z} \oplus \mathbb{Z})) \quad (7.149)$$

Either non-injective endomorphisms of D_∞ map $a \mapsto b, b \mapsto \{0, b\}$ or they map $a \mapsto 0, b \mapsto (sa+b)$. The second class of endomorphisms involves endomorphisms of the form

$$(0, 0, \ldots, \ ! \, (sa + b), (sa + b), \ldots)$$

and such mappings are already present in $\mathfrak{A}(\mathrm{D}_\infty)$. The other class adds the generators

$$(0, b, b, 0, 0, b, \ldots ! \, b, 0, 0, b, b, 0, 0, \ldots) \text{ and } (0, b, b, 0, 0, b, \ldots . ! \, 0, b, b, 0, 0, b, b, \ldots)$$

to the mix, but the second can be constructed from the first and the mapping $(0, 0, \ldots !b, b, \ldots)$ already present. Thus,

$$\begin{aligned}
E(D_\infty) &= \langle (0, 0, \ldots, \, !\, a, a, ..), (0, 0, \ldots, \, !\, 0, 2a, -2a, 4a, -4a, , ..), \\
&(0, 0, \ldots, \, !\, b, b, ..), (0, a, -a, 2a, -2a, \ldots, \, !\, 0, a, -a, 2a, -2a, \ldots), \qquad (7.150) \\
&(0, b, b, 0, 0, b, \ldots !\, b, 0, 0, b, b, 0, 0 \ldots)\rangle
\end{aligned}$$

The commutator subgroup of $T(D_\infty)$ has coordinates in $\langle 2a\rangle$. It will be of uncountable order. Phomomorphisms of D_∞ are precisely the zero-fixing mappings θ for which, for any $x \in D_\infty, k \in \mathbb{Z}$

$$\theta(kx) \equiv k\theta(x) \pmod{\langle 2a >} \qquad (7.151)$$

(this is directly to do with the form of the commutator subgroup). The phomomorphism group is

$$\text{Phom}(D_\infty) = \left(\prod_{x \neq 0} \langle 2a\rangle\right) \rtimes (D_\infty \oplus D_\infty) \qquad (7.152)$$

(see this by mapping a and b independently to elements of D_∞ and then representing the congruences, coordinate by coordinate). D_∞ is an infinite group of countable order which is finitely presented. Such groups should have automorphism near-rings which are countable, but what of $\mathfrak{B}(D_\infty)$?

When a group $(N, +)$ has finite order n, one might think that Stirling's approximation to the factorial would suggest that the set of bijective mappings of N becomes small compared to the set of all mappings, as n becomes large, and that the equality $\mathfrak{B}(N) = T(N)$ is unlikely. Countering that, when N has countable order, the set of bijective mappings is itself *uncountable*, as is T(N). This is the situation for $\mathfrak{B}(D_\infty)$. It would be interesting to find a mapping in $T(D_\infty)$ that is not in $\mathfrak{B}(D_\infty)$. I have not yet done this.

7.5.6 Q_8

I choose to represent the additive structure of $(Q_8, +)$ using the Hamiltonian quaternions (page 113). The identity is the transformation

$$I = (1, -1, i, -i, j, -j, k, -k)$$

and the reader is warned of the strong possibility of confusion as I wilfully mix additive and multiplicative notation. The "zero" mapping is now

$$0 = (1, 1, 1, 1, 1, 1, 1, 1)$$

and the mapping $(I + I)$ is

$$(1, 1, -1, -1, -1, -1, -1, -1)$$

This notation groups coordinates into four, two-element, "boxes". Box one is the first two coordinates of the tuple; box three is the fifth and sixth coordinates. The set $\{i, j\}$ is a set of "additive" generators for Q_8. One identifies endomorphisms by stipulating their effect on the first coordinate of boxes two and three. Including the identity mapping, there are four distinct inner automorphisms of $(Q_8, +)$. The other three are

$$\begin{aligned}
\theta_1 &= (1, -1, i, -i, -j, j, -k, k) \ \ (x :\mapsto x^i) \\
\theta_2 &= (1, -1, -i, i, j, -j, -k, k) \ \ (x :\mapsto x^j) \\
\theta_3 &= (1, -1, -i, i, -j, j, k, -k) \ \ (x :\mapsto x^k)
\end{aligned}$$

The various "additions" of these tuples result in tuples whose first boxes have the form $[1, 1]$ and whose other boxes are any assortment of the boxes $[-1, -1]$ or $[1, 1]$. There are eight such possibilities for this, and these eight mappings (which, of course, do not include the original mappings $\{\theta_j\}_{j=1,2,3}$) constitute an additive subgroup H of $\mathrm{I}(Q_8)$ which is both normal and central in $(\mathrm{T}_0(Q_8), +)$. For example,

$$\begin{aligned}
(\theta_1 + \theta_2) &= (1, 1, 1, 1, 1, 1, -1, -1) \\
(\theta_2 + \theta_3) &= (1, 1, -1, -1, 1, 1, 1, 1)
\end{aligned}$$

We can say

$$H \cong (\mathrm{C}_2 \oplus \mathrm{C}_2 \oplus \mathrm{C}_2) \tag{7.153}$$

θ_1 plus elements of H generate $\mathrm{I}(Q_8)$. There is a group exact sequence

$$(\mathrm{C}_2 \oplus \mathrm{C}_2 \oplus \mathrm{C}_2) \longmapsto \mathrm{I}(Q_8) \twoheadrightarrow \mathrm{C}_2$$

and

$$\mathrm{I}(Q_8) \cong \langle H, \theta_1 \rangle \cong (\mathrm{C}_2 \oplus \mathrm{C}_2 \oplus \mathrm{C}_4) \tag{7.154}$$

This means the order of $\mathrm{I}(Q_8)$ is 16 and it is abelian.

General automorphisms of $(Q_8, +)$ have the form

$$(1, -1, x, -x, y, -y, z, -z) \tag{7.155}$$

and that includes the inner automorphisms just discussed. In this notation, x represents any one of the elements of Q_8 which have order 4. y is another such but distinct from $\pm x$, and z is an order 4 element distinct from $\{\pm x, \pm y\}$. Note that I use the term "distinct" loosely in what follows, not distinguishing elements which simply have different sign. Distinction relates directly to the bijective nature of automorphisms. The mapping

$$(1, -1, j, -j, -i, i, k, -k)$$

is an automorphism. The tuples involved are fixed by the first value of the second box and the first value of the third—their third and fifth coordinates. These must be distinct, and they determine all other values, the first value of the fourth box is fixed from them—in the above case, $k = j + (-i)$ (think, $j \times (-i)$: I said it was confusing!). There are exactly $6 \times 4 = 24$ possible automorphisms.

Automorphisms are very nearly characterised by the pattern of having distinct box entries and having alternating signs. However, tuples such as

$$(1, -1, k, -k, i, -i, -j, j) \tag{7.156}$$

are not automorphisms ($k + i \neq -j$). However, the related tuple

$$(1, -1, k, -k, i, -i, j, -j)$$

is an automorphisms—and they may be obtained from (7.156) by operating on it by $(\theta_1 + \theta_2)$ (explicitly calculated earlier). All we ever need to do to turn one of these pattern-matching mappings into a proper automorphism is to switch the signs for the elements in the last box; and that can be done by the addition of $(\theta_1 + \theta_2)$. The upshot of all this is that $\mathrm{A}(Q_8)$ should contain all tuples which have distinct elements between distinct boxes and which have alternating signs within boxes; but there are other tuples within that group.

For example,

$$(1, -1, i, -i, j, -j, k, -k) + (1, -1, -i, i, k, -k, -j, j) = (1, 1, 1, 1, i, i, i, i)$$

and

$$(1, 1, i, i, 1, 1, i, i) + (1, 1, j, j, j, j, 1, 1) = (1, 1, k, k, j, j, i, i)$$

and

$$(1, -1, i, -i, j, -j, k, -k) + (1, 1, i, i, j, j, k, k) = (1, -1, -1, 1, -1, 1, -1, 1)$$

which suggests we should augment H, forming a new group K, by including the mapping

$$(1, -1, 1, -1, 1, -1, 1, -1)$$

So,

$$K = (H \oplus \langle (1, -1, 1, -1, 1, -1, 1, -1) \rangle) \cong (C_2 \oplus C_2 \oplus C_2 \oplus C_2)$$

We have the following observations (and I am sometimes Cavalier with signs because in the context of the above result they are irrelevant):

$$(1, -1, i, i, j, j, k, k) + (1, -1, j, j, i, i, k, k) = (1, 1, k, k, -k, -k, -1, -1)$$

From this, I conclude that all tuples with the form

$$(1, 1, x, x, y, y, z, z)$$

where exactly two of $\{x, y, z\}$ are identical and in $\{i, j, k\}$ and the other one is 1, are in $A(Q_8)$. If we now consider the subset of these, consisting of tuples in which the two identical terms are either i or j, we obtain tuples such as

$$(1, 1, j, j, 1, 1, j, j)$$

From these, we may generate all other tuples with the prescribed form and all of the automorphisms of $(Q_8, +)$—I said I was Cavalier with signs! There are six mappings of this sort. They are additively independent, and they lead me to claim that there is an exact sequence

$$(C_2 \oplus C_2 \oplus C_2 \oplus C_2) \longmapsto A(Q_8) \twoheadrightarrow (C_2 \oplus C_2 \oplus C_2 \oplus C_2 \oplus C_2 \oplus C_2) \tag{7.157}$$

which would give $|A(Q_8)| = 2^{10}$.

For non-surjective endomorphisms, one cannot have, say, $|\theta(i)| = 4$ and $|\theta(j)| = 2$ because, then,

$$\theta(i^2) = -1 = \theta(-1) = \theta(j^2) = (\theta(j))^2 = 1$$

Similar arguments show that for non-surjective endomorphisms, one cannot have an element with order 4 in an image set; and so

$$\theta(i), \theta(j) \in \{\pm 1\}$$

It turns out that there are just two non-trivial non-surjective endomorphisms. They are

$$(1, 1, -1, -1, 1, 1, -1, -1) \tag{7.158}$$
$$(1, 1, -1, -1, -1, -1, 1, 1)$$

Both tuples already lie within H. I conclude $\mathrm{A}(Q_8) = \mathrm{E}(Q_8)$.

Meldrum [119] cites work by other authors stating that for generalised quaternion groups Q_{2^n}, $\mathrm{I}(Q_{2^n})$ has order $2^{(3n-5)}$, and $\mathrm{A}(Q_{2^n})$ has order $2^{(3n-4)}$. He does agree that $\mathrm{E}(Q_{2^n}) = \mathrm{A}(Q_{2^n})$. My own results are different from these for Q_8, agreeing only on $\mathrm{I}(Q_8)$.

The reader may find a flaw in my reasoning, at least in the $\mathrm{A}(Q_8)$ case, and because I could not match these results, I do not offer an analysis of generalised quaternion groups here, issuing instead this health warning.

7.6 Additive Structure

I now look more closely at the group-theoretic side of transformation near-rings. I repeatedly need the idea of an NR-group in this section (page 16).

I am only concerned with zero-fixing transformations in this section.

Some sets may be infinite.

7.6.1 $\mathfrak{M}$(N)

In what follows, it is convenient to affix a suffix to my notation for zero-fixing transformations; but because such notation such as

$$(\mathrm{T}_0)_S(\Gamma)$$

becomes pointlessly unwieldy, I use the single symbol $\mathfrak{M}$ to represent T_0 and then work with symbols such as $\mathfrak{M}_S(\Gamma)$ in the way one would expect. This seems doubly appropriate because current authors use notation such as $\mathrm{M}_0(N)$ for what I would call $\mathrm{T}_0(N)$. I do this throughout the rest of this chapter but be warned: in this chapter, there are sections where $\mathfrak{M}_S(\Gamma)$ is a multiplicative semi-group, but not a near-ring (page 296).

7.6.2 Centraliser Near-Rings

Suppose that S is some semi-group of endomorphisms of an additive group $(\Gamma, +)$. I define a near-ring

Notation 269

$$\mathfrak{M}_S(\Gamma) = \{\theta \in \mathrm{T}_0(\Gamma) : \theta(\gamma \cdot s) = (\theta\gamma) \cdot s, \ \forall s \in S, \ \forall \gamma \in \Gamma\}$$

Definition 270 $\mathfrak{M}_S(\Gamma)$ is called a **centraliser near-ring.**

Elements of $\mathfrak{M}_S(\Gamma)$ are transformations acting on Γ. I use functional notation $\theta(\gamma)$ and product notation $\theta \cdot \gamma$ interchangeably in what follows in this chapter.

The finite topology on $\mathrm{T}_0(\Gamma)$ bequeaths a subspace topology on $\mathfrak{M}_S(\Gamma)$ which I also refer to as "the finite topology".

We may map the centraliser set S to a subset $\bar{S} \in \mathrm{T}_0(\Gamma)$. I map $s \in S$ to $\bar{s} \in \mathrm{T}_0(\Gamma)$ with the prescription

$$\bar{s}(w) = (w \cdot s), \quad w \in \Gamma \tag{7.159}$$

S and $\bar{S}$ are isomorphic semi-groups. If $\theta \in \mathfrak{M}_S(\Gamma)$, then

$$\theta(\bar{s} \cdot w) = \theta(w \cdot s) = \theta(w) \cdot s = \bar{s} \cdot \theta(w) \Rightarrow (\theta \circ \bar{s}) = (\bar{s} \circ \theta)$$

But then,

Lemma 271

$$\mathfrak{M}_S(\Gamma) = \{\theta \in \mathrm{T}_0(\Gamma) : (\theta \circ \bar{s}) = (\bar{s} \circ \theta)\}$$

Note that $\mathfrak{M}_S(\Gamma)$ probably does not contain $\bar{S}$. Suppose $(N, +, \cdot)$ is a zero-symmetric near-ring with identity. For each $n \in N$, we may define an endomorphism s_n of $(N, +)$, which is to be applied on the *right*

$$(x)s_n = (x \cdot n) \tag{7.160}$$

The set $S = \{s_n : n \in N\}$ is a semi-group of endomorphisms of the group $(N, +)$. We may form the near-ring $\mathfrak{M}_S(N)$. There is an obvious map, $\Phi : N \longmapsto \mathfrak{M}_S(N)$, defined by

$$\Phi(m) = \phi_m, \ \ where\ \phi_m(x) = (m \cdot x) \tag{7.161}$$

This is an isomorphism of near-rings. Only the surjectivity part is not so obvious.

Suppose, then, $\theta \in \mathfrak{M}_S(N)$. Then,

$$\Phi(\theta(1))(x) = \phi_{\theta(1)}(x) = (\theta(1) \cdot x) = (\theta(1) \cdot s_x) = (\theta(1 \cdot s_x)) = \theta(x)$$

Lemma 272 *Any zero-symmetric near-ring with identity is a centraliser near-ring.*

Generalisations
One can generalise these structures, considering quite general multiplicative semi-groups $(\bar{S}\,.)$ in $T_0(\Gamma)$, insisting $I, 0 \in \bar{S}$ and defining the centraliser

$$\mathfrak{M}_{\bar{S}}(\Gamma) = \left\{\theta \in T_0(\Gamma) : (\theta \circ \bar{s}) = (\bar{s} \circ \theta) \textit{ for all } \bar{s} \in \bar{S}\right\} \tag{7.162}$$

The distributivity core of $T_0(\Gamma)$ is $\mathrm{End}(\Gamma)$ (see page 240). The multiplicative semi-group $\mathfrak{M}_{\overline{S}}(\Gamma)$ might not necessarily be a near-ring itself, so this extends my notation. It will certainly be a near-ring when $\bar{S} \subseteq \mathrm{End}(\Gamma)$. The full condition for obtaining a near-ring is
$\overline{S} \subseteq$

$$\{\theta \in T_0(\Gamma) : (\theta \cdot (\alpha + \beta)) = ((\theta \cdot \alpha) + (\theta \cdot \beta)) \;\forall\, \alpha, \beta \in \mathfrak{M}_S(\Gamma)\} \tag{7.163}$$

(7.162) suggests the existence of some semi-group S (acting on the *right*) of Γ, which might perhaps be achieved by writing

$$(\gamma \cdot s) = (\bar{s} \cdot \gamma), \;\; \textit{for } \bar{s} \in \bar{S}. \tag{7.164}$$

I am less and less concerned with right actions here and propose, indicatively, to replace the notation $\mathfrak{M}_S(\Gamma)$ in (7.162) by $\mathfrak{M}_{\bar{S}}(\Gamma)$, but this applies only in this subsection.

Pseudo-Centralisers
Extending this definition a little, consider semi-groups $\bar{S}$ containing I and 0 which are *phomomorphisms* of $(\Gamma, +)$ (Chap. 9 and Lemma 306). We would then consider the set $\mathcal{PZ}_{\bar{S}}(\Gamma) =$

$$\{\theta \in \mathrm{Phom}(\Gamma) : (\theta \circ \bar{s}) \equiv (\bar{s} \circ \theta) \; (\textit{modulo}[\Gamma, \Gamma]) \textit{ for all } \bar{s} \in \bar{S}\} \tag{7.165}$$

This, again, is a sub-near-ring of $\mathrm{Phom}(\Gamma)$ and so, of course, of $T_0(\Gamma)$. It might be described as a **pseudo-centraliser** (compare Sect. 14.1).

Using Notation 96, I construct another generalisation of Definition 270 as the near-ring

$$\mathfrak{Ph}_S(\Gamma) = \{\theta \in \textit{Phom}\,(\Gamma) : ((\theta(\gamma)) \cdot s) = (\theta(\gamma \cdot s)) \;\forall\, s \in S\} \tag{7.166}$$

Then, we have this corollary to Lemma 272.

Corollary 273 *Any (zero-symmetric) F-near-ring with identity is a sub-near-ring of some* $\mathfrak{Ph}_S(\Gamma)$.

7.6.3 A Duality of Semi-Groups

One can apply (7.162) to any subset $S_0 \subseteq \mathrm{T}_0(\Gamma)$. Then, writing $\bar{S}_1 = \mathfrak{M}_{S_0}(\Gamma), \ldots \bar{S}_{n+1} = \mathfrak{M}_{\bar{S}_n}(\Gamma), \ldots$, we obtain a sequence of semi-groups

$$\bar{S}_1, \bar{S}_2, \bar{S}_3, \ldots \tag{7.167}$$

I don't include S_0 in this sequence since it may not be a semi-group. Notice that I am not claiming these structures as near-rings. Each of these semi-groups includes the identity and zero mappings. I continue to **overline** the semi-groups which interest me—$\bar{S}$, not S—both to fit in with what went before and to stress their special nature—they are supposed to contain the identity and zero mappings; and I continue to write $\bar{S}_{j+1} = \mathfrak{M}_{\bar{S}_j}(\Gamma)$.

Semi-groups should be supposed to contain the zero and identity mappings.

An Example: $\mathfrak{L}_0$(N)

One may modify the near-ring $\mathfrak{L}(N)$ defined on page 335 by using nearly the same definition as that of (8.34) except that only zero-fixing transforms are considered. This gives a near-ring $\mathfrak{L}_0(N)$ contained in $\mathrm{T}_0(N)$. Its $\mathcal{A}$-matrices have 1 in the top left position but are otherwise the same as those of $\mathfrak{L}(N)$. These are rank two matrices, with the single exception of the matrix corresponding to zero, which is the sole element in $(\mathfrak{L}(N) \cap \mathfrak{L}_0(N))$.

One can consider the centraliser of $\mathfrak{L}_0(N)$ in $\mathrm{T}_0(N)$. This is simply the semi-group $\{0, 1\}$. Notice that the centraliser of *that* semi-group is actually $\mathrm{T}_0(N)$. This means that if $S_0 = \mathfrak{L}_0(N)$, then $\bar{S}_1 = \{0, 1\}$ and $\bar{S}_2 = \mathrm{T}_0(N)$. Notice that the multiplicative centre of the near-ring $\mathrm{T}_0(\Gamma)$ is $\{0, 1\}$.

The Duality

Now, $S_0 \subseteq \bar{S}_2$, and, quite generally,

$$\bar{S}_j \subseteq \bar{S}_{j+2}, \quad j = 1, 2, 3, \ldots \tag{7.168}$$

If $A and B$ are two subsets of $\mathrm{T}_0(\Gamma)$ and $A \subseteq B$, then

$$\mathfrak{M}_A(\Gamma) \supseteq \mathfrak{M}_B(\Gamma) \tag{7.169}$$

The sets can be identical here, if, for example, A does not contain the identity mapping but B does. So, we have

$$\mathfrak{M}_{\bar{S}_{j+2}} \subseteq \mathfrak{M}_{\bar{S}_j}, \quad j = 0, 1, 2, \ldots$$

which implies

$$\bar{S}_{j+2} \subseteq \bar{S}_j \quad j = 1, 2, 3, \ldots \tag{7.170}$$

The conclusion is that $\bar{S}_1 = \bar{S}_3 = \bar{S}_5 = \ldots$ and $\bar{S}_2 = \bar{S}_4 = \bar{S}_6 = \ldots$ Thus, any subset S of $\mathrm{T}_0(\Gamma)$ gives rise to two semi-groups, $\bar{S}_1 = \mathfrak{M}_S(\Gamma)$ and $\bar{S}_2 = \mathfrak{M}_{\bar{S}_1}(\Gamma)$, which are dual, in the sense

$$\mathfrak{M}_{\bar{S}_a}(\Gamma) = \bar{S}_b \; \textit{for } a, b, \in \{1, 2\}, (a \neq b) \tag{7.171}$$

Note strongly that I am not asserting that S is one or the other of these two semi-groups and it frequently will not be.

Lemma 274

(i) S is contained in the multiplicative centraliser of $\mathfrak{M}_S(\Gamma)$ in $\mathrm{T}_0(\Gamma)$.
(ii) If we denote this multiplicative centraliser by $\bar{S}_2$, then

$$\mathfrak{M}_S(\Gamma) = \mathfrak{M}_{\bar{S}_2}(\Gamma)$$

Lemma 275 *Suppose S is some multiplicative semi-group contained in $(\mathrm{T}_0(\Gamma)\ \cdot)$ and containing the identity and zero mappings. Then,*

$$\bar{S}_2 = \{\theta \in \mathrm{T}_0(\Gamma) : (\theta \circ \phi) = (\phi \circ \theta) \; \textit{for all } \phi \in \mathfrak{M}_S(\Gamma)\} \tag{7.172}$$

I described earlier that from a semi-group S, one could obtain two new semi-groups, $\bar{S}_1$ and $\bar{S}_2$. I now temporarily adopt the convenient shorthand: $S^* = \mathfrak{M}_S(\Gamma)$. Then, of course, $\bar{S}_1^* = \bar{S}_2$ and $\left(\bar{S}_1^*\right)^* = \bar{S}_2^* = \bar{S}_1$.

The two semi-groups S and S^* commute. The smallest semi-group containing both of them is simply their product

$$\left(S \cdot S^*\right) = \left\{(x \cdot y) : x \in S, \; y \in S^*\right\} \tag{7.173}$$

Then, $(S \cdot S^*)^*$ is the set of elements commuting simultaneously with everything in S and everything in S^*.

Lemma 276 $(S \cdot S^*)^* = \left(S^* \cap (S^*)^*\right)$

More generally, when S is merely a *subset* of $\mathrm{T}_0(\Gamma)$ containing the zero and identity mappings, the set of all finite products of elements from S

$$\mathfrak{X}(S) = \left\{(s_1 \cdot s_2 \cdot s_3 \cdot \ldots s_n) \; : \; s_j \in S \; \forall \, j = 1, 2, \ldots n\right\} \tag{7.174}$$

is the smallest semi-group containing S (see page 335); and elements of $\mathrm{T}_0(\Gamma)$ commute with $\mathfrak{X}(S)$ if and only if they commute with its generators (the elements of S). That is to say,

$$S^* = (\mathfrak{X}(S))^* \tag{7.175}$$

If we take any two semi-groups containing the zero and identity mappings, S and T, the smallest semi-group containing them both is simply the set Z of all elements with the form

$$(a_1 \cdot b_2 \cdot a_3 \cdot b_3 \cdot \ldots a_k \cdot b_k) \tag{7.176}$$

where $a_j \in S$ and $b_j \in T$. Then,

$$Z^* = \left(S^* \cap T^*\right) \tag{7.177}$$

Lemma 277 *Suppose S and T are as above, and suppose $(S^*)^* = S$ and $(T^*)^* = T$. Then,*

$$\left(Z^*\right)^* = (S \cap T) \tag{7.178}$$

7.6.4 Density

I now return to the original definition of $\mathfrak{M}_S(\Gamma)$ (Definition 270).

Density is a standard idea in topology, but it is so important in this book that I define it on page 6. This Lemma appears in Gerhard Betsch's paper [10]. It is little more than an application of (7.10).

Lemma 278 *Suppose N is a sub-near-ring of $\mathfrak{M}_S(\Gamma)$. Then, N is dense in the finite topology on $\mathfrak{M}_S(\Gamma)$ if and only if, for any $m \in \mathfrak{M}_S(\Gamma)$ and any finite collection of elements in Γ, such as $\{\gamma_1, \gamma_2, \ldots, \gamma_t\}$, we can find some $n \in N$ such that $n\gamma_j = m \cdot \gamma_j$ for all $j = 1, 2, \ldots, t$.*

7.7 $\mathfrak{M}_S(\Gamma)$ when S Is Fixed-Point-Free

An important special case occurs when S is a fixed-point-free automorphism group of an additive group $(\Gamma, +)$, which we act on the *right* of the group. That is, S is a group of automorphisms of Γ, and each $s \neq 1 \in S$ has the property that $(\gamma)s \neq \gamma$ when $\gamma \neq 0$. In this situation, S partitions Γ into disjoint *orbits* (page xxii), each of the form $(\gamma)S$ for some $\gamma \in \Gamma$ and each with the same cardinality. When S is trivial, that is, the identity automorphism, the orbits are exactly the members of Γ, and $\mathfrak{M}_S(\Gamma)$ becomes $T_0(\Gamma)$. A theorem of Thompson says that when Γ is finite, it cannot have non-trivial fixed-point-free automorphisms of prime order unless it is additive nilpotent [139]. That is, in such cases S is trivial, and $\mathfrak{M}_S(\Gamma) = T_0(\Gamma)$ is automatic. When S is an automorphism group but not fixed-point-free, orbits can have different sizes, as happens, for instance, when S is the group of inner automorphisms of $(\Gamma, +)$ and the orbits are conjugacy classes. In such situations, we do not have the automatic ability to link any two distinct orbit representatives by a transformation $\theta \in \mathfrak{M}_S(\Gamma)$, and this is a key technique of my analysis. I shall consider the near-ring $\mathfrak{M}_S(\Gamma)$ in some detail and

S is fixed-point-free automorphism group throughout this section.

When S is non-trivial, it has orbits with more than one element in them. Mappings are completely determined by their effect on any single element of each orbit. More than that and most importantly,

We may actually *define* a mapping simply by prescribing a value to any set of orbit representatives and extending in the obvious way.

Lemma 278 now implies:

Lemma 279 *When the automorphism group S has only a finite number of orbits in Γ and N is a sub-near-ring of $\mathfrak{M}_S(\Gamma)$ which is dense in the finite topology, then*

$$N = \mathfrak{M}_S(\Gamma) \tag{7.179}$$

and the near-ring $(N, +, \cdot)$ is unital.

A Dense Sub-near-Ring Which Is Not Unital

If we consider the additive group of real numbers $(\mathbb{R}, +)$ and the transformation near-ring $T_0(\mathbb{R})$, we can construct a sub-near-ring $\Sigma_0(\mathbb{R})$ consisting of all mappings in $T_0(\mathbb{R})$ with *finite support* (meaning non-zero on finite subsets only, see page 21). This will be an uncountable near-ring (a consequence of there being uncountably many finite subsets of $\mathbb{R}$). It is not unital. However, from Lemma 278, we can identify it as a dense sub-near-ring of $T_0(\mathbb{R})$. The conclusion is that dense sub-near-rings of $\mathfrak{M}_S(\Gamma)$ are not necessarily unital.

Key Observations

It's worth keeping in mind that elements of $\mathfrak{M}_S(\Gamma)$ annihilate either all elements of an orbit or none of them and that every non-zero element of $\mathfrak{M}_S(\Gamma)$ must map at least *one* of the orbit representatives to a non-zero member of Γ. Another useful fact is that if γ_i, γ_j are representatives of two orbits, there will be some $\theta \in \mathcal{M}_S(\Gamma)$ for which $\theta(\gamma_i) = \gamma_j$. Furthermore, members of a given orbit have exactly the same left annihilator sets in $\mathfrak{M}_S(\Gamma)$, that is,

Lemma 280

$$\begin{pmatrix} \gamma_j \\ 0 \end{pmatrix}_{\mathcal{L}} = \begin{pmatrix} \{\gamma_j \cdot S\} \\ 0 \end{pmatrix}_{\mathcal{L}} = \begin{pmatrix} \Gamma_j \\ 0 \end{pmatrix}_{\mathcal{L}} \tag{7.180}$$

And so, two elements of Γ^* will have the same left annihilators if and only if they are in the same orbit of S.

(a result that appears trivial contextually but which is useful in Chap. 11)

It's also worth pointing out that if M is *any* non-trivial left sideal of $\mathfrak{M}_S(\Gamma)$, then

$$M \cdot \Gamma = \Gamma$$

because $M \cdot \Gamma \neq \{0\}$ and we may operate on the left by any element of $\mathfrak{M}_S(\Gamma)$. In terms of (7.19), this means

$$\mathcal{I}(M) = \Gamma$$

These observations underlay many of the following deductions.

Some Important Notation

Some of the notation I am about to define is important for much of the rest of this section. The partition of Γ into orbits under S can be written: $\Gamma = \{0\} \bigcup \{\Gamma_i\}_{i \in I}$. Here I is some indexing set, which will be amended to include 0, so that $\Gamma_0 = \{0\}$ (I shall refer to I without further comment throughout this chapter). Choose one orbit representative from each orbit—obtaining the set $\{\gamma_i\}_{i \in I}$, where $\gamma_i \in \Gamma_i$; consequently, $\Gamma_i = \gamma_i \cdot S$. I shall set $\gamma_0 = 0$, which gives us Γ_0 as an anomalous, singleton, orbit. Lemma 278 now becomes

Lemma 281 *The near-ring N is dense in the finite topology on $\mathfrak{M}_S(\Gamma)$ if and only if, given any $m \in \mathfrak{M}_S(\Gamma)$ and any finite subset of I, such as $\{\gamma_1, \gamma_2, \ldots \gamma_n\}$, we can find $n \in N$ such that $(n \cdot \gamma_j) = (m \cdot \gamma_j)$ for $1 \leq j \leq n$.*

Suppose $a \in \mathfrak{M}_S(\Gamma)^*$. For each orbit representative, γ_j. I write

$$a \cdot \gamma_j = w_j$$

a is characterised as a transformation by the (perhaps infinite) tuple (w_j). Notice that w_j may or may not be in the orbit Γ_j; and notice that w_i, w_j may well be in the same orbit. Clearly, the near-ring $\mathfrak{M}_S(\Gamma)$ is additively isomorphic to a direct product of $|I^*|$ copies of $(\Gamma, +)$. We know (see Notation 48)

$$\begin{pmatrix} a \\ 0 \end{pmatrix}_{\mathcal{L}} \subseteq \bigcap_j \begin{pmatrix} w_j \\ 0 \end{pmatrix}_{\mathcal{L}}$$

Conversely, if $w \in \bigcap_j \begin{pmatrix} w_j \\ 0 \end{pmatrix}_{\mathcal{L}}$ and $w \cdot a \neq 0$, then for some orbit representative, γ_k, we must have $w \cdot a \cdot \gamma_k = w \cdot w_k \neq 0$. Since $w \in \begin{pmatrix} w_k \\ 0 \end{pmatrix}_{\mathcal{L}}$, this is impossible, giving us

Lemma 282 *With the notation just established,*

$$\begin{pmatrix} a \\ 0 \end{pmatrix}_{\mathcal{L}} = \bigcap_j \begin{pmatrix} w_j \\ 0 \end{pmatrix}_{\mathcal{L}}$$

We know that $\Gamma_j = \gamma_j \cdot S$ for each orbit Γ_j and that

$$\begin{pmatrix} \Gamma_j \\ 0 \end{pmatrix}_{\mathcal{L}} = \begin{pmatrix} \gamma_j \\ 0 \end{pmatrix}_{\mathcal{L}} \tag{7.181}$$

Taking any non-zero orbit representative, γ_j (or, indeed, any element $\gamma_j \in \Gamma$), we can say that

$$\mathfrak{M}_S(\Gamma) \cdot \gamma_j = \Gamma \tag{7.182}$$

This in turn implies that Γ and $\left(\frac{\mathfrak{M}_S(\Gamma)}{\begin{pmatrix} \gamma_j \\ 0 \end{pmatrix}_{\mathcal{L}}} \right)$ are isomorphic as near-ring groups (see Definition 20). A consequence of that is that $\begin{pmatrix} \gamma_j \\ 0 \end{pmatrix}_{\mathcal{L}}$ is both a maximal proper near-ring group and a maximal left ideal of $\mathfrak{M}_S(\Gamma)$. The annihilators of distinct orbits are themselves distinct (this can be arranged by *fiat*), from which intersections of such annihilators cannot be maximal left ideals. As defined above, w_j may or may not be a non-zero member of Γ. If it is non-zero, it lies within an orbit Γ_k and, of course,

$$\begin{pmatrix} w_j \\ 0 \end{pmatrix}_{\mathcal{L}} = \begin{pmatrix} \Gamma_k \\ 0 \end{pmatrix}_{\mathcal{L}} = \begin{pmatrix} \gamma_k \\ 0 \end{pmatrix}_{\mathcal{L}}$$

It also follows that $\begin{pmatrix} a \\ 0 \end{pmatrix}_{\mathcal{L}}$ is maximal, precisely when, for all orbits, Γ_k, we have some fixed j such that $a \cdot \gamma_k = \delta_{j,k} w_j$. That is, the element $a \in \mathfrak{M}_S(\Gamma)$ maps precisely one orbit, Γ_j, to non-zero elements. I use this notation in the next lemma.

Lemma 283 *Choosing an element $a \in \mathfrak{M}_S(\Gamma)^*$ which annihilates all orbits but one, Γ_j, we have*

$$\mathfrak{M}_S(\Gamma) \cdot a = \{t \in \mathfrak{M}_S(\Gamma) : t \cdot \Gamma_k = 0, \forall\, k \neq j\}$$

$$= \{t \in \mathfrak{M}_S(\Gamma) \;:\; t \cdot \gamma_k = 0, \forall\, k \neq j\}$$

Proof Our notation gives $a \cdot \gamma_j = w_j$. Select $\theta \in \{t \in \mathfrak{M}_S(\Gamma)^* : t \cdot \Gamma_k = 0, \forall\, k \neq j\}$, and put $\theta \cdot \gamma_j = w'_j \neq 0$. Although w_j and w'_j may be in distinct orbits, we can find a mapping $\beta \in \mathfrak{M}_S(\Gamma)$ such that $\beta \cdot w_j = w'_j$. This means

$$\theta \cdot \gamma_j \;=\; w'_j \;=\; \beta \cdot w_j \;= \beta \cdot a \cdot \gamma_j$$

and it means that the two mappings, θ and $(\beta \cdot a)$, have exactly the same effect on Γ—that is, they are the same mapping. Thus, $\theta \;=\; (\beta \circ a) \in \mathfrak{M}_S(\Gamma) \cdot a$. □

Keeping the definition of a from the theorem, the set $\mathfrak{M}_S(\Gamma) \cdot a$ is those elements of $\mathfrak{M}_S(\Gamma)$ which annihilate all orbit representatives except, possibly, γ_j. In fact these mappings map γ_j to *all* elements of Γ. In this chapter, I reserve the notation L_j for this set $\mathfrak{M}_S(\Gamma) \cdot a$.

Definition 284

$$L_j = \{t \in \mathfrak{M}_S(\Gamma) \;:\; t \cdot \Gamma_k = 0,\;\; \forall\, k\, \neq j\} \;=\; \bigcap_{k \neq j} \begin{pmatrix} \Gamma_k \\ 0 \end{pmatrix}_{\mathcal{L}} \tag{7.183}$$

Distinct elements of L_j map γ_j to distinct elements of Γ. This means that additively

$$\left(L_j,\, +\right) \;\cong\; T\left(\{\gamma_j\}, \Gamma\right) \;\cong\; (\Gamma,\, +) \tag{7.184}$$

and with the same definition of a

$$L_j \cdot a \;=\; L_j \tag{7.185}$$

Lemma 285 *Suppose $\gamma_a, \gamma_b \in \Gamma$ with $\gamma_a \neq 0$. Then, there is some (unique) $n \in \mathfrak{M}_S(\Gamma)$ such that*

$$(n \cdot \gamma_a) \;=\; \gamma_b$$

and if $\gamma \in \Gamma_j$, we can say that $n \in L_j$.

More Notation
I define $e_{i,j} \in L_j$ with the formula

$$e_{i,j} \cdot \gamma_k \;=\; \delta_{j,k}\gamma_i \tag{7.186}$$

These are exactly the mappings that take γ_j to each of the other coset representatives and take other coset representatives to zero. $e_{i,j}$ is only non-zero on Γ_j which it transforms into Γ_i. If $i \neq j$, $L_j \cdot e_{i,k} = \{0\}$. $e_{j,j}$ is a right identity for the set L_j.

Thus,

$$L_r \cdot e_{s,t} = \delta_{s,r} L_t \tag{7.187}$$

and

$$L_i \cdot L_j = L_j \tag{7.188}$$

Lemma 286

(i) *L_j is a normal subgroup of $\mathfrak{M}_S(\Gamma)$.*
(ii) *L_j is a left ideal of $\mathfrak{M}_S(\Gamma)$.*
(iii) *When $j \neq k$, $L_j \bigcap L_k = \{0\}$.*
(iv) *Suppose $t \in L_j$ and $t \neq 0$. Then, $\mathfrak{M}_S(\Gamma) \cdot t = L_j$.*
(v) *If J is a left sideal of $\mathfrak{M}_S(\Gamma)$ and $J \cdot L_j \neq \{0\}$, then $L_j = J \cdot L_j$.*
(vi) *Each L_j is a minimal left sideal of $\mathfrak{M}_S(\Gamma)$.*
(vii) *Each L_j is a minimal left ideal of $\mathfrak{M}_S(\Gamma)$.*
(viii) *Each L_j is a near-ring group under the action of $\mathfrak{M}_S(\Gamma)$, one that is $\mathfrak{M}_S(\Gamma)$-linearly isomorphic to Γ.*

Proof

(ix) Choose a as before, something that annihilates all elements of Γ not in Γ_j. From (iv), we know

$$\mathfrak{M}_S(\Gamma) \cdot a = L_j$$

We have

$$\mathfrak{M}_S(\Gamma) \cdot a \cdot \gamma_j = \Gamma$$

and

$$(\mathfrak{M}_S(\Gamma) \cdot a) \bigcap \begin{pmatrix} \gamma_j \\ 0 \end{pmatrix} = \{0\}$$

The mapping in question is $\Theta : L_j \longmapsto \Gamma, \theta(t \cdot a) = t \cdot a \cdot \gamma_j$. Then,

$$\theta(t \cdot a) = (t \cdot a) \cdot \gamma_j = t \cdot (a \cdot \gamma_j) = t \cdot \theta(a)$$

□

Lemma 287

$$(\mathfrak{M}_S(\Gamma), +) \cong \prod_{i \in I^*} (L_i, +)$$

Proof A non-zero member of $\mathfrak{M}_S(\Gamma)$ maps each Γ_i either to zero or to some Γ_j. For each $i \in I$, we may associate $x \in \mathfrak{M}_S(\Gamma)$ with an element $x_i \in L_i$ such that $x(\gamma_i) = x_i(\gamma_i)$ and, consequently, x and x_i have exactly the same effect on Γ_i. This gives us a bijective mapping $x \mapsto (x_1, x_2, \dots x_n \dots)$ and the obvious isomorphism. □

The isomorphism is actually $\mathfrak{M}_S(\Gamma)$-linear, though it is not an isomorphism of near-rings per se. Finally, I observe that for any $j \in I^*$,

$$\begin{pmatrix} L_j \\ 0 \end{pmatrix}_{\mathcal{L}}^{\mathfrak{M}_S(\Gamma)} = \{0\} \tag{7.189}$$

(this follows since $L_j \cdot \Gamma = \Gamma$). If $\theta \in \mathfrak{M}_S(\Gamma)$, then θ is a tuple $\left(\theta_j\right)$ of elements from the L_j; and

$$\theta_j = \left(\theta \circ e_{j,j}\right) \tag{7.190}$$

I find this interesting because it applies when the tuples have an infinite number of elements.

There are conceptual problems with direct products in this context. The implicit identity

$$(x_1, x_2, \dots, x_n, x_{n+1}, \dots) \cdot \gamma =$$

$$(x_1 \cdot \gamma + x_2 \cdot \gamma + \dots, +x_n \cdot \gamma + x_{n+1} \cdot \gamma + \dots)$$

makes one uneasy because it appears to offer the prospect of an infinite number of additions occurring in $(\Gamma, +)$. The apparent problem arises from a conflict of notations. In this and other contexts where I use it, there is but a single coordinate, x_k, mapping any particular orbit representative and γ, to something which might be non-zero. Only one of the coordinates offered could map an element of Γ to a non-zero element, and the purported infinite addition is specious, both for zero and non-zero elements. The coordinate notation is not intended as representing a full cleik.

7.7.1 The Structure of Minimal Left Ideals

Suppose that J is a minimal left ideal of $\mathfrak{M}_S(\Gamma)$ throughout this subsection.

We know that $J \cdot \gamma_j \neq 0$ for some γ_j (J cannot map all elements of Γ to zero) so there is some γ_j for which $\Gamma = J \cdot \gamma_j$. Indeed, this is true of all orbit representatives that J does not annihilate. Taking $x, y \in J$, and writing $x \cdot \gamma_j = y \cdot \gamma_j$, gives us that

$(x - y) \in \begin{pmatrix} \gamma_j \\ 0 \end{pmatrix}_{\mathcal{L}}$. So, if $x \cdot \gamma_j \neq 0$, we deduce, by minimality in J, that x and y are identical and obtain

Lemma 288

(i) Given any orbit representative, $\gamma_j \in \Gamma$, either $J \cdot \gamma_j = 0$ or all non-zero elements of J map γ_j to distinct, non-trivial elements of Γ or $J \cdot \gamma_j = \Gamma$.
(ii) No two distinct elements of J map any orbit representative to the same, non-zero, element of Γ.
(iii) J contains an element a which acts as a multiplicative right identity on J.
(iv) $J = \mathfrak{M}_S(\Gamma) \cdot a$, where a is the right identity defined above.
(v) If $j \in J$, and $j \neq 0$, then

1. $\mathfrak{M}_S(\Gamma) \cdot j = J$.
2. Either $J \cdot j = \{0\}$ or $J \cdot j = J$, and both situations can occur within the same minimal left ideal.

(vi) Minimal left ideals of $\mathfrak{M}_S(\Gamma)$ are minimal left near-ring groups, too.

Proof

(iii) If we put $a \cdot \gamma_j = \gamma_j$, for some (unique) $a \in J$, then for all $j \in J$, we have that $j \cdot \gamma_j = ja \cdot \gamma_j$, which means $(j - ja)$ annihilates γ_j and is an element of J, which, being minimal, argues that $j = ja$ and that a is a right identity for elements of J.

(iv) $J = J \cdot a \subseteq \mathfrak{M}_S(\Gamma) \cdot a \subseteq J$, so $\mathfrak{M}_S(\Gamma) \cdot a = J \cong \mathfrak{M}_S(\Gamma) / \begin{pmatrix} a \\ 0 \end{pmatrix}_{\mathcal{L}}$, and all minimal left ideals have the form $\mathfrak{M}_S(\Gamma) \cdot a$ for some idempotent element a.

(v)

1. There is some orbit representative γ_k for which $j \cdot \gamma_k \neq 0$. This means we can find some $x \in \mathfrak{M}_S(\Gamma)$ such that $x \cdot j \cdot \gamma_k = w$, for any $w \in \Gamma$. We know $J \cdot \gamma_k = \Gamma$, and since we have a minimal left ideal, this means that the single element j' of J for which $j' \cdot \gamma_k = w$ must be equal to $x \cdot j$.
2. If $J \cdot j \neq \{0\}$, then for some γ_k and some $j' \in J$, we have $j' \cdot j \cdot \gamma_k \neq 0$, which means we may find some $j'' \in J$ such that $j'' \cdot j \cdot \gamma_k = \gamma_k$. This in turn means that $j'' \cdot j$ acts as a right identity on elements of J, so $J = J \cdot j'' \cdot j \subseteq J \cdot j$.

Examples of both situations occurring can be found in L_j.

□

Using the notation just developed (with $a \cdot \gamma_j = \gamma_j$),

$$J = \mathfrak{M}_S(\Gamma) \cdot a \cong \frac{\mathfrak{M}_S(\Gamma)}{\begin{pmatrix} a \\ 0 \end{pmatrix}_{\mathcal{L}}}$$

Clearly, $L_j \cdot a = J = \mathfrak{M}_S(\Gamma) \cdot a$, and since J is minimal as a near-ring group, the annihilator of a must be maximal, as a left ideal. This means that the element a maps only one orbit representative to a non-zero element (in this case, γ_j), and consequently

Lemma 289 *Any minimal left ideal of $\mathfrak{M}_S(\Gamma)$ has the form L_j.*

7.7.2 Right Near-Ring Groups

In this subsection, suppose that J is a subgroup of $\mathfrak{M}_S(\Gamma)$ and $J \cdot \mathfrak{M}_S(\Gamma) \subseteq J$. If J is non-zero, we expect to find some γ_j such that $J \cdot \gamma_j \neq 0$. Of course, $J \cdot \gamma_j$ is a subgroup of Γ.

Lemma 290 *Suppose J is a non-zero, right near-ring group.*

(i) For any $w \in \Gamma^*$,

$$J \cdot w \neq \{0\} \tag{7.191}$$

(ii) For any L_j

$$J \cdot L_j \neq \{0\} \tag{7.192}$$

(iii) For any L_j

$$J \cdot L_j = \left(J \cap L_j\right) \neq \{0\} \tag{7.193}$$

Proof

(i) Select any $w \in \{\Gamma - \{0\}\}$. We can find $t \in \mathfrak{M}_S(\Gamma)$ such that $t \cdot w = \gamma_j$. J being a right near-ring group, we conclude that for all non-zero $w \in \Gamma$,

$$J \cdot w \neq \{0\}$$

(ii) This is immediate from the previous part.
(iii) That $J \cdot L_j \subseteq \left(J \cap L_j\right)$ is obvious, but L_j does have the right identity $e_{j,j}$.

□

If $t \in \mathfrak{M}_S(\Gamma)$, $t \neq 0$, we know that there must be some $w \in \Gamma$ such that $t \cdot w \neq 0$. From this, it follows that $J \cdot t \neq \{0\}$ and so

$$\binom{J}{0}_{\mathcal{R}}^{\mathfrak{M}_S(\Gamma)} = \{0\}$$

Notation 291 I set

$$J_k = J \cdot L_k \tag{7.194}$$

and note that if $r \neq k$,

$$(J_k \cap J_r) = \phi \tag{7.195}$$

If γ_i and γ_j are distinct orbit representatives and $x = j \cdot \gamma_i \neq 0\,(j \in J)$, we can find $t \in \mathfrak{M}_S(\Gamma)$ such that $t \cdot \gamma_j = \gamma_i$. That means that $x = j \cdot t \cdot \gamma_j \in J.\gamma_j$ and shows

Lemma 292 *For any two orbit representatives γ_i, γ_j*

(i) $J \cdot \gamma_i = J \cdot \gamma_j$
(ii) $J \cdot \Gamma = J \cdot \gamma_j = J_j \cdot \gamma_j$

Everything we have said so far applies when J is merely a right sideal of $\mathfrak{M}_S(\Gamma)$, but one consequence of these results is that when J actually is a subgroup, $J \cdot \Gamma$ will be a subgroup of Γ. Of particular interest is the case in which $J = [\mathfrak{M}_S(\Gamma), \mathfrak{M}_S(\Gamma)]$ because then $J \cdot \Gamma$ is $[\Gamma, \Gamma]$ (see Lemma 306).

Lemma 293

(i)

$$J \subseteq \prod_k J_k$$

(ii) When S has only finitely many orbits,

$$J = \prod_k J_k = \oplus_k J_k$$

Proof

(i) Using the notation (7.186),

$$J \cdot \gamma_k = J \cdot e_{k,k} \cdot \gamma_k \subseteq (J \cdot L_j) \cdot \gamma_k \subseteq J \cdot \gamma_k$$

and, so,

$$J \cdot \gamma_k = J \cdot L_k \cdot \gamma_k = J_k \cdot \gamma_k$$

Of course, $j \cdot e_{k,k} \in L_k$. Appealing to Lemma 287, one deduces that (in coordinate notation)

$$j = \big(\ldots, (j \cdot e_{k,k}) \ldots, (j \cdot e_{t,t}) \ldots\big)$$

(ii) Finite cases are covered by ordinary sums of elements.

□

When there are an infinite number of orbits, the above inclusion can be strict. For example, suppose J is the set of mappings of Γ which take it to only finitely many orbits. J is a sub-near-ring of $\mathfrak{M}_S(\Gamma)$ and a right -$\mathfrak{M}_S(\Gamma)$ group, but

$$J \subsetneq \prod_k J_k$$

J will be a subdirect product of the components $\{J_k\}$ (Notation 1). We can define a topology on $\mathfrak{M}_S(\Gamma)$ (see (7.9), page 235) by taking as a sub-base all sets of the form

$$S\left(\theta; \gamma_{i_1}, \gamma_{i_2}, \ldots \gamma_{i_k}\right) =$$

$$\{\phi \in \mathfrak{M}_S(\Gamma) \text{ such that } \phi(\gamma_{i_j}) = \theta(\gamma_{i_j}),\ \text{for } j = 1, 2, \ldots, k\}$$

where θ is a fixed element of $\mathfrak{M}_S(\Gamma)$ and $\{\gamma_{i_j}\}_{j=1}^{k}$ is a finite subset of the orbit representatives. With this topology, J is dense in $\prod_k J_k$. I define

$$\mathfrak{M}_S(\Gamma, J \cdot \Gamma) = \{\theta \in \mathfrak{M}_S(\Gamma) \ : \ \theta(\Gamma) \subseteq J \cdot \Gamma\} \tag{7.196}$$

This set is a subgroup of $(\mathfrak{M}_S(\Gamma), +)$ and also a right near-ring group. Recall that a topology is discrete if the point sets are open. Also, two mappings in $\mathfrak{M}_S(\Gamma)$ having the same values on any complete set of orbit representatives must be identical.

Lemma 294 *A necessary and sufficient condition that this topology be discrete is that the set of orbit representatives should be finite.*

The following results are related to Lemma 237.

Lemma 295 *Suppose J is a right near-ring group and S has only finitely many orbits. Then $J = \mathfrak{M}_S(\Gamma, J \cdot \Gamma)$.*

Proof Suppose $t \in \mathfrak{M}_S(\Gamma, J \cdot \Gamma)$. Using the decomposition in Lemma 287, we may write $t = \left(t_j\right)$ where $t_j \in L_j$ and the brackets indicate coordinates. We know $J \cdot \Gamma = J_j \cdot \gamma_j$ and can therefore choose $x_j \in J_j$ such that $t_j \cdot \gamma_j = x_j \cdot \gamma_j$. In fact our knowledge of L_j shows that that must mean $t_j = x_j$. From Lemma 293, since all coordinates of t lie in respective J_k, we have $x \in \prod_k J_k$. □

Notice that the previous lemma classifies all right $\mathfrak{M}_S(\Gamma)$ subgroups of $\mathfrak{M}_S(\Gamma)$ in terms of subgroups of Γ—at least, when there are only a finite number of orbits. The multiplicative structure of transformation near-rings is determined set-theoretically and is wholly innocent of group action. Consider, then, a set $h \subseteq \mathfrak{M}_S(\Gamma)$ with the property $h \cdot \mathfrak{M}_S(\Gamma) \subseteq h$—that is, a right sideal of this transformation near-ring. The subset $h \cdot \Gamma \subseteq \Gamma$ may or may not be a subgroup. In the case $S = \{1\}$, for instance, h could be the singleton set $\{\theta\}$ where, with $w \in \Gamma$ fixed ($w \neq 0$), and θ is defined as

$$\theta(x) = \begin{cases} w & \text{if } x \neq 0 \\ 0 & \text{otherwise} \end{cases}$$

then, of course, $h \cdot \Gamma = \{w, 0\}$. We may generate the subgroups

$$(\Omega, +) = \langle h \cdot \Gamma >_{(add)} \subseteq (\Gamma, +)$$

and

$$H = \langle h >_{(add)} \subseteq (\mathfrak{M}_S(\Gamma), +)$$

H is a right sideal of $\mathfrak{M}_S(\Gamma)$ and a right near-ring group.

Lemma 296 *When there are only a finite number of orbits,*

$$H = \mathfrak{M}_S(\Gamma, \Omega)$$

7.7.3 Annihilators

The notation I use here is given in Sect. 1.15.

$$\begin{pmatrix} \Gamma_j \\ 0 \end{pmatrix}_{\mathcal{L}} = \begin{pmatrix} \gamma_j \\ 0 \end{pmatrix}_{\mathcal{L}} = \{x \in \mathfrak{M}_S(\Gamma) \,:\, x \cdot \gamma_j = 0\} = \bigcup_{k \neq j} L_k \tag{7.197}$$

Elements of Γ have the same left annihilator ideal if and only if they are in the same orbit of S. The structure described in (7.197) is a left ideal. On page 302, I pointed out that it was also a *maximal* left ideal (see also Lemma 303).

7.7.4 Chains of Left Ideals

If we select a sequence of distinct orbit representatives, $\{\gamma_1, \gamma_2, \gamma_3 \ldots\}$, there is a corresponding collection of maximal left ideals: $\{J_1, J_2, \ldots\}$, where

$$J_k = \begin{pmatrix} \Gamma_k \\ 0 \end{pmatrix}_{\mathcal{L}}$$

and a corresponding descending chain of left ideals

$$\mathcal{J}_1 \supset \mathcal{J}_2 \supset \mathcal{J}_3 \supset \ldots \tag{7.198}$$

where

$$\mathcal{J}_{k+1} = \left(\mathcal{J}_k \bigcap J_{k+1}\right) \quad k = 0, 1, 2, \ldots \tag{7.199}$$

(7.198) represents a strictly descending chain of left ideals. This gives us a result relating to what one might describe as **Artinian near-rings** (see [7], page 97).

Lemma 297 *Suppose the near-ring $\mathfrak{M}_S(\Gamma)$ does not have an infinite strictly descending chain of left ideals. Then the number of orbits of S is finite.*

7.7.5 Simple Near-Rings

Until chapter end, Γ will have a finite number of orbits—$|I| < \infty$.

Suppose that J is a non-trivial ideal of $\mathfrak{M}_S(\Gamma)$. In so far as it is a left ideal, we know $J \cdot \Gamma = \Gamma$. In so far as it is a right ideal, we know $J = \mathfrak{M}_S(\Gamma, J \cdot \Gamma)$. We conclude that $J = \mathfrak{M}_S(\Gamma)$.

Lemma 298 $\mathfrak{M}_S(\Gamma)$ *is a simple as a near-ring.*

Suppose, more generally, that $V \neq \{0\}$ is a subgroup of $\mathfrak{M}_S(\Gamma)$ such that both $\mathfrak{M}_S(\Gamma).V \subseteq V$ and $V \cdot \mathfrak{M}_S(\Gamma) \subseteq V$—a subgroup and two-sided *sideal.* $\alpha \in \mathfrak{M}_S(\Gamma)$ can be written

$$\alpha = \alpha_1 + \alpha_2 + \ldots + \alpha_n$$

where $\alpha_j \in L_j$. There is some $v(\neq 0) \in V$ for which

$$v \cdot \gamma = \gamma_t \neq 0$$

for some $\gamma \in \Gamma$ and some fixed orbit representative $\gamma_t, \in \Gamma$. I define mappings $\{\phi_j\}_{j \in I^*}$ with the prescription

$$\phi_j(\gamma_m) = \delta_{m,j}\gamma$$

and note that $\phi_j \in L_j$ and

$$\left(v \circ \phi_j\right)(\gamma_m) = \delta_{m,j}\gamma_t$$

Finally, I define mappings $\{\theta_j\}_{j \in I^*}$ by

$$\theta_j(\gamma_m) = \delta_{t,j}\alpha_j \cdot \gamma_t$$

(where t is as defined above). Note that $\theta_j \in L_t$. Sideal structure ensures $v_j = (\theta_j \circ v \circ \phi_j) \in L_j$. We also have

$$(\theta_j \circ v \circ \phi_j(\gamma_m)) = \alpha_j \cdot \gamma_m$$

So, $v_j = \alpha_j$ and

$$\alpha = v_1 + v_2 + \dots v_n$$

from which we conclude that $\alpha \in V$.

Lemma 299 *The near-ring $\mathfrak{M}_S(\Gamma)$ contains no proper, non-trivial, subgroups which are two-sided sideals.*

Lemma 300

(i)

$$(\mathfrak{M}_S(\Gamma),\ +)\ \cong\ \oplus_{i\in I^*}(L_i,\ +) \tag{7.200}$$

(ii) $\oplus_{i\in I^*}(L_i,\ +)$ *is a simple near-ring.*

Proof

(i) comes from Lemma 287.

□

Similar methods show that the non-zero-symmetric near-ring $T(\Gamma)$ is simple when Γ is any group, finite or not, other than the cyclic group of order 2.

Observations 301 Lemma 300, together with the Jordan-Hölder theorem and Sect. 11.4.7, means that when we have a finite 2-primitive near-ring, type 2 near-ring group upon which it acts faithfully will be unique, up to isomorphism.

Lemma 302 *Suppose $m \in \mathfrak{M}_S(\Gamma)$ and $a_i \in L_i$,, $j = 1, 2, \dots, k$. Then,*

$$(m \cdot (a_1 + a_2 + \dots + a_k))\ =\ ((m \cdot a_1) + (m \cdot (a_2) + \dots + (m \cdot a_k)) \tag{7.201}$$

Proof The two expressions have identical effects on the orbit representatives of S.

□

7.7.6 Left Ideals

In this subsection, suppose J to be a proper, non-trivial left ideal of $\mathfrak{M}_S(\Gamma)$. Our finiteness restriction ensures

$$\mathfrak{M}_S(\Gamma) \;=\; \oplus_{j \in I^*} L_j \tag{7.202}$$

This is a simple near-ring. Left annihilators of left ideals are actually ideals (for zero-symmetric near-rings) and so, in this case, trivial. This means both

$$J \cdot L_j \;\neq\; \{0\}$$

and

$$L_j \cdot J \;\neq\; \{0\}$$

for all $j \in I$. The ideals $\{L_j\}_{j \in I}$ are minimal, so either are contained in J or have trivial intersection with it. If all of them lie within J, then J would be improper. We may therefore define a proper subset $K \subsetneq I$ which is maximal with respect to the property $J \bigcap \oplus_K L_k = \{0\}$ by a standard application of Zorn's lemma. The addition of any new L_j to $\oplus_K L_k$ results in a direct sum which contains elements of J. If this direct sum were not the whole of $\oplus_I L_j$, we have to conclude the existence of some L_t which has been missed out, but that would contradict the maximality of K.

Lemma 303

(i) *Any left ideal of $\oplus_I L_j$ is a direct summand of it and has the form $\oplus_M L_m$, where M is some subset of I^*.*
(ii) *Any non-trivial ideal of $\oplus_{I^*} L_j$ contains a minimal left ideal of the form L_j.*
(iii) *Maximal left ideals of $\oplus_{I^*} L_j$ have the form*

$$M_k \;=\; \oplus_{\{j \,\in\, I^* \,:\, j \neq k\}} L_j \;=\; \begin{pmatrix} \Gamma_j \\ 0 \end{pmatrix}_{\mathcal{L}} \tag{7.203}$$

(iv) *No proper, non-trivial, left sideal S can properly contain a maximal left idea—they are maximal left sideals too.*

Proof I only prove (iv). Suppose S is such a proper, non-trivial left sideal. S has elements that are in L_j. $\left(S \cap L_j\right)$ is a left near-ring group. L_j is a minimial left ideal, so from Lemma 288, S contains L_j. This means that S contains $\oplus_{I^*} L_j$. □

Returning to "Artinian near-rings",

Corollary 304 *Under the restrictions of Sect. 7.7.5, $\mathfrak{M}_S(\Gamma)$ does not possess an infinite, strictly descending, chain of left ideals.*

And if one were to consider "Noetherian near-rings" (see [7]), then,

Lemma 305 *Under the restrictions of Sect. 7.7.5, $\mathfrak{M}_S(\Gamma)$ does not possess an infinite, strictly ascending, chain of left ideals.*

There has been a lot of work on chain conditions in near-rings [119, 128]. The story rapidly becomes complicated because one can consider distinct conditions relating to distinct substructures: ideals, sideals, mideals, etc.

7.7.7 Modular Left Ideals

The notations and definitions I need here are given in Sect. 1.29.1, and I assume the decomposition (7.202), which represents the near-ring as a direct sum of $s < \infty$ isomorphic left ideals. I select a subset of these ideals, which, for convenience, I number $\{L_j\}_{j=1}^{k}$. The multiplicative identity element may be written in the form

$$1 = (e_1 + e_2 + \ldots + e_k) + \left(e_{(k+1)} + e_{(k+2)} + \ldots + e_s\right) \tag{7.204}$$

Of course, I intend $e_j \in L_j$ here, and at this point, I need Lemma 17 to observe that e_j is a *left identity* for L_j. I define

$$e_{\hat{k}} = \left(e_{(k+1)} + e_{(k+2)} + \ldots + e_s\right) \tag{7.205}$$

This acts as a left identity for the left ideal

$$\left(L_{(k+1)} \oplus L_{(k+2)} \oplus \ldots \oplus L_s\right) \tag{7.206}$$

Selecting any $\theta \in \mathfrak{M}_S(\Gamma)$, the element $\left(\theta - \left(\theta \cdot e_{\hat{k}}\right)\right)$ is in the left ideal

$$L_{\hat{k}} = (L_1 \oplus L_2 \oplus \ldots \oplus L_k) \tag{7.207}$$

This means $L_{\hat{k}}$ is actually a modular left ideal with modular right identity $e_{\hat{k}}$. It is apparent that all modular left ideals of the near-ring have this form and that there are $(2^s - 2)$ proper non-trivial ones and s maximal modular left ideals; but that coincides exactly with all possible proper non-trivial left ideals.

Chapter 8
Generalisations and Sub-near-Rings of Transformation Near-Rings

This chapter continues the ideas of the last one and uses the same notation. I also introduce the Gothic symbol $\mathfrak{S}$ in (8.18), $\mathfrak{L}$ on page 335, $\mathfrak{X}$ and $\mathfrak{Y}$ on page 335 and $\mathfrak{G}$ on page 336.

The chapter is about near-rings arising from standard transformation near-rings. I think this is a rich area and that I have barely skimmed the surface of it. I had intended to do rather more here, and I hope others will find the material worth extending.

8.1 Commutators

In this section, Q, G and Γ are additive groups. Consider $\theta \in \mathrm{T}_0(Q, [G, G])$ and $q \in Q$. We know that $\theta(q) = \mathrm{C}_1 + \ldots \mathrm{C}_n$, where each $\mathrm{C}_j \in [G, G]$ and a simple commutator. So, any one of these C_j is of the form $[g_1, g_2]$, where $g_1, g_2 \in G$. If $\alpha, \beta \in \mathrm{T}_0(Q, G)$ have the properties $\alpha(q) = g_1$ and $\beta(q) = g_2$, then $[\alpha, \beta](q) = \mathrm{C}_j$, and so members of $[\mathrm{T}_0(Q, G), \mathrm{T}_0(Q, G)]$ can be found to mimic the action of θ on the element q; and this can be done for each element $q \in Q$.

Lemma 306

$$\mathrm{T}_0(Q, [G, G]) = [\mathrm{T}_0(Q, G), \mathrm{T}_0(Q, G)]$$

An alternative way to view this is to note that additively $\mathrm{T}_0(Q, G)$ is a direct product of copies of $(G, +)$ and so the commutator subgroup is a similar direct product of copies of $[G, G]$. In the same way,

$$[\mathrm{T}(G), \mathrm{T}(G)] = \mathrm{T}(G, [G, G]) \tag{8.1}$$

The next observation is simply an application of Lemma 306 to itself.

R. Lockhart, *The Theory of Near-Rings*, Lecture Notes in Mathematics 2295,
https://doi.org/10.1007/978-3-030-81755-8_8

Corollary 307

$$\mathrm{T_0}((Q, [[G, G], [G, G]]) = [\mathrm{T_0}((Q, [G, G]), \mathrm{T_0}((Q, [G, G])]$$

I now use the notation of (1).

Lemma 308

(i) Suppose H and K are non-empty subsets of G. Then,

$$\mathrm{T_0}(Q, [H, K]) = [\mathrm{T_0}(Q, H), \mathrm{T_0}(Q, K)]$$

(ii)

$$\mathrm{T_0}((Q, [G, [G, G]]) = [\mathrm{T_0}(Q, G), [\mathrm{T_0}(Q, G), \mathrm{T_0}(Q, G)]]$$

Lemma 309 *Suppose $(Q, +)$ is an additive group.*

(i) $\mathrm{Phom}(Q)$ *is a near-ring which contains* $\mathrm{E}(Q)$ *(see Notation 227) as a sub-near-ring.*

(ii) There is an epimorphism of near-rings:

$$\Phi : \mathrm{Phom}(Q) \to \mathrm{E}\left(\frac{Q}{[Q, Q]}\right)$$

defined by $\Phi(\theta) \mapsto \bar{\theta}$, where $\bar{\theta}\{q + [Q, Q]\} = \{\theta(q) + [Q, Q]\}$.

(iii) The above epimorphism gives an exact sequence of near-rings:

$$K \mapsto \mathrm{Phom}(Q) \overset{\Phi}{\twoheadrightarrow} \mathrm{E}\left(\frac{Q}{[Q, Q]}\right)$$

where $K = kernel(\Phi) = [\mathrm{T_0}(Q), \mathrm{T_0}(Q)] = \mathrm{T_0}(Q, [Q, Q])$.

An alternative expression of this is

$$K \mapsto \mathrm{Phom}(Q) \overset{\Phi}{\twoheadrightarrow} \mathrm{Hom}\left(\frac{Q}{[Q, Q]}, \frac{Q}{[Q, Q]}\right) \tag{8.2}$$

Thus, $\mathrm{Phom}(Q)$ *has a non-trivial homomorphic image which is a ring, whenever $(Q, +)$ is not a perfect group.*

(iv) $\mathrm{Phom}(Q) \lhd \mathrm{T_0}(Q)$.

(v) $\frac{\mathrm{T_0}(Q)}{\mathrm{Phom}(Q)}$ *is an abelian group.*

(vi)

$$\mathrm{Phom}(Q) = \mathrm{T_0}(Q) \Leftrightarrow (Q, +) \text{ is additively perfect.} \tag{8.3}$$

One proves (vi) by noticing that phomomorphisms preserve commutators but ordinary mappings do not.

Part (iii) of the lemma gives useful information about Phom(Q). For example, we know that the order of $E\left(\frac{Q_8}{[Q_8,Q_8]}\right)$ is 2^4 and that the order of $[\mathrm{T}_0(Q_8), \mathrm{T}_0(Q_8)]$ is 2^7, so Phom(Q_8) must have order 2^{11} (see Lemma 389). Interestingly, Phom(D_4) will have exactly the same order.

I return to the notation developed in Sect. 7.6 with additive group $(\Gamma, +)$ and fixed-point-free automorphism group S and assume our transformation near-ring is a direct sum of the ideals L_j. I consider phomomorphisms contained in $\mathrm{T}_S(\Gamma)$, obtaining the near-ring $\mathrm{Phom}_S(\Gamma) = \mathrm{P}_S(\Gamma)$. We have

$$\mathrm{P}_S(\Gamma) \subseteq \oplus L_j = \mathrm{T}_S(\Gamma)$$

The commutator subgroup of $(\Gamma, +)$ involves an integral number of orbits of S, say $\{\gamma_j\}_{j=1}^{c}$. The subset $\mathrm{T}_S(\Gamma, [\Gamma, \Gamma])$ is contained in $\mathrm{P}_S(\Gamma)$; and since $e_{i,j} \in L_j$, for $i \leq c$, in non-abelian cases, L_j and $\mathrm{P}_S(\Gamma)$ have non-trivial intersection. So, when $(\Gamma, +)$ is non-abelian, for all k,

$$H_k = \mathrm{P}_S(\Gamma) \bigcap L_k \neq \{0\}$$

Each H_k is a left ideal of $\mathrm{P}_S(\Gamma)$, which therefore contains $\oplus H_k$ as a left ideal. I am not claiming that this structure is a left ideal of $\mathrm{T}_S(\Gamma)$. We actually have that

$$[L_k, L_k] \subseteq H_k$$

Elements of $\mathrm{P}_S(\Gamma)$ are finite sums of elements of the left ideals L_j. Right multiplication by $e_{j,j}$ is equivalent to projection of the jth. coordinate. In non-abelian cases,

$$L_j \supseteq \mathrm{P}_j = \mathrm{P}_S(\Gamma) \cdot e_{j,j} \supseteq H_j = \left(\mathrm{P}_S(\Gamma) \bigcap L_j\right) \neq \{0\}$$

Moreover, since $\mathrm{P}_S(\Gamma) \lhd \mathrm{T}_S(\Gamma)$ and $e_{j,j}$ is a right identity for L_j, it follows that

$$\mathrm{P}_j \lhd L_j$$

Actually, each P_j is a normal subgroup of $\mathrm{T}_S(\Gamma, +)$ itself (only the jth coordinates need be checked because others are mapped to zero). I do not assert that $\mathrm{P}_j \subseteq \mathrm{P}_S(\Gamma)$.

Lemma 310 $\mathrm{T}_S(\Gamma, [\Gamma, \Gamma]) =$

$$[\mathrm{T}_S(\Gamma), \mathrm{T}_S(\Gamma)] \subseteq \oplus_j H_j \subseteq \mathrm{P}_S(\Gamma) \subseteq \oplus_j \mathrm{P}_j \subseteq \oplus_j L_j$$

$$= \mathrm{T}_S(\Gamma)$$

8.2 More Sub-near-Rings

Phom(Γ) is a sub-near-ring of $\mathrm{T_0}(\Gamma)$. There are related sub-near-rings (see Notation 227), and I look at some of them here. The reader may wish to compare the following definition with (1.23).

Notation 311 Whenever $\Omega \lhd \Gamma$, we can form the two near-rings
(i)

$$\mathrm{T_0}({}^{\Omega}(\Gamma)) = \{\theta \in \mathrm{T_0}(\Gamma) : \theta(x+w) \equiv \theta(x) \text{ (Mod } \Omega), \ \forall\, x \in \Gamma, \text{ and } \forall\, w \in \Omega\}$$

(ii)

$$\mathrm{T_0}(\Gamma, \Omega) = \{\theta \in \mathrm{T_0}(\Gamma) : \theta(x) \in \Omega, \forall x \in \Gamma\}$$

Lemma 312

(i) $\mathrm{T_0}({}^{\Omega}(\Gamma))$ *and* $\mathrm{T_0}(\Gamma, \Omega)$ *are sub-near-rings of* $\mathrm{T_0}(\Gamma)$.
(ii) $\mathrm{T_0}(\Gamma, \Omega)$ *is a right ideal of* $\mathrm{T_0}(\Gamma)$.
(iii) $\mathrm{T_0}(\Gamma, \Omega)$ *is an ideal of* $\mathrm{T_0}({}^{\Omega}(\Gamma))$.
(iv) $\mathrm{T_0}(\Gamma, \Omega)$ *is, additively, a direct product of* $(|\Gamma| - 1)$ *copies of* Ω.

Now there is an exact sequence of near-rings:

$$\mathrm{T_0}(\Gamma, \Omega) \rightarrowtail \mathrm{T_0}({}^{\Omega}(\Gamma)) \overset{\Phi}{\twoheadrightarrow} \mathrm{T_0}\left(\frac{\Gamma}{\Omega}\right) \tag{8.4}$$

where $\Phi(\theta) = \bar{\theta}$ and $\bar{\theta}(\gamma + \Omega) = (\theta(\gamma) + \Omega)$.

8.2.1 Special Cases

Notation 313 In this subsection, if $(\Gamma, +)$ is any additive group, then

$$\mathcal{Z}(\Gamma) = \{w \in \Gamma \ : \ w + x = x + w, \forall x \in \Gamma\} \tag{8.5}$$

is the usual group centre (see Notation 2). I shall usually write the centre as $\mathcal{Z}$ when the group involved is clear.

A Generalisation of Phomomorphisms
Phomomorphisms may be generalised in many ways. Here I outline a generalisation suggested by class $\mathcal{F}$ near-rings (Sect. 1.24.2). One considers two groups, G and W,

and normal subgroups: $D \lhd G$, $V \lhd W$. I then look at the collection of normalised (page 30) mappings from G to W satisfying the property

$$\theta\,(g+d) \;\equiv\; (\theta(g))\;(\textit{modulo}\; V) \tag{8.6}$$

for all $g \in G$, $d \in D$. This collection has obvious group structure based on the standard addition of functions and essentially consists of the mappings between G and W which map D-cosets to V-cosets. I call it $S_0\,(G_D, W_V)$. We may then recognise the exact sequence of groups:

$$\mathrm{T_0}(G, V) \rightarrowtail S_0\,(G_D, W_V) \twoheadrightarrow \mathrm{T_0}\left(\frac{G}{D}, \frac{W}{V}\right) \tag{8.7}$$

I don't pursue this idea further here; but I do think a systematic account of the possible generalisations, and their properties would be worthwhile and easy enough to accomplish.

Sub-near-Rings Associated with the Group Centre

Observations 314

(i) $\mathrm{T_0}({}^{\mathcal{Z}}(\Gamma))$ contains all the automorphisms of the additive group $(\Gamma, +)$, but does not necessarily contain all its endomorphisms (the centre not, in general, being fully invariant).
(ii) $\mathrm{T_0}(\Gamma, \mathcal{Z}) = \mathcal{Z}(\mathrm{T_0}(\Gamma))$.

Automorphisms of $(\Gamma, +)$ fix $\mathcal{Z}$. The near ring, $\mathrm{I}(\Gamma)$, generated by the inner automorphisms of $(\Gamma, +)$ is a d.g. sub-near-ring of $\mathrm{T_0}({}^{\mathcal{Z}}(\Gamma)$. It contains the identity mapping. The exact sequence (8.4) is now

$$\mathcal{Z}(\mathrm{T_0}(\Gamma)) \rightarrowtail \mathrm{T_0}\left({}^{\mathcal{Z}}(\Gamma)\right) \twoheadrightarrow \mathrm{T_0}\left(\frac{\Gamma}{\mathcal{Z}(\Gamma)}\right)$$

Sub-near-Rings Associated with the Commutator Subgroup

Observations 315 $\mathrm{T_0}({}^{[\Gamma,\Gamma]}(\Gamma))$ contains all the phomomorphisms of the additive group, in particular, it contains $A(\Gamma)$, $\mathrm{E}(\Gamma)$ and $\mathrm{P_0}(\Gamma)$ (see Notation 227).

The exact sequence (8.4) is now

$$[\mathrm{T_0}(\Gamma), \mathrm{T_0}(\Gamma)] \rightarrowtail \mathrm{T_0}\left({}^{[\Gamma,\Gamma]}(\Gamma)\right) \twoheadrightarrow \mathrm{T_0}\left(\frac{\Gamma}{[\Gamma, \Gamma]}\right)$$

This should be compared with the exact sequence (8.2).

8.3 Hadamard Products

It may happen that one considers $\mathrm{T}_0(A)$ when $(A,\ +,\ \cdot)$ is a unital, zero-symmetric, right-distributive, near ring. One may then form another product operation on $\mathrm{T}_0(A)$ which I call the **Hadamard product**. The product would be defined by the formula

$$(\theta \circledcirc \phi)(a) \;=\; (\theta(a) \cdot \phi(a)) \tag{8.8}$$

$(\mathrm{T}_0(A),\ +,\ \circledcirc)$ is then a zero-symmetric, unital, near-ring, and the identity element is the mapping C_1 (page 233). When I is the identity mapping on A

$$(I \circledcirc \theta)(a) \;=\; (a \cdot \theta(a)) \tag{8.9}$$
$$(\theta \circledcirc I)(a) \;=\; (\theta(a) \cdot a)$$

Hadamard products do not preserve phomomorphisms; when investigating that possibility, one gets interesting cross-term obstructions of the form

$$(\theta(a) \circ \phi(b)) \;+\; (\theta(b) \circ \phi(a)) \tag{8.10}$$

which may be worth analysing in their own right.

I now use some notation and theory first presented on page 243 and restrict to situations in which the near-ring $(A,\ +,\ \cdot)$ is finite.

Suppose one asks for an additive group homomorphism

$$\partial \;:\; \mathrm{T}_0(A) \longmapsto \mathrm{T}_0(A) \tag{8.11}$$

which satisfies the *chain rule*

$$\partial(\theta \circ \phi) \;=\; (\partial(\theta) \circ \phi) \circledcirc (\partial(\phi)) \tag{8.12}$$

If this mapping is to be non-trivial, there must be $a, b \in A^*$ such that $\partial(\Delta^{(a,b)}) \neq 0$. Taking $c \in A^*$, we know $\left(\Delta^{(a,b)}\right) = \left(\Delta^{(a,c)} \circ \Delta^{(c,b)}\right)$, and therefore

$$\partial\left(\Delta^{(a,b)}\right) = \left(\left(\partial\left(\Delta^{(a,c)}\right) \circ \Delta^{(c,b)}\right) \circledcirc \left(\partial \Delta^{(c,b)}\right)\right) \tag{8.13}$$

It follows from Lemma 233 that for any $c, d \in A^*$, we would then have $\partial(\Delta^{(c,d)}) \neq 0$ and that the mapping $\partial(\Delta^{(x,b)})$ (with b fixed, x varying) acts bijectively on A^* and is an additive automorphism of the group $(A,\ +)$. It must also be true that $\partial\Delta^{(a,b)} = \Delta^{(c,b)}$ for some $c \in A^*$. Now, $\left(\Delta^{(a,b)} \circ \Delta^{(b,b)}\right) = \Delta^{(a,b)} = \left(\Delta^{(a,a)} \circ \Delta^{(a,b)}\right)$, from which it follows that for any $a \in A^*$, $\partial\Delta^{(a,a)} = \Delta^{(1,a)}$.

Given $a, b \in A^*$, we know

$$\Delta^{(a,a)} \;=\; \left(\Delta^{(a,b)} \circ \Delta^{(b,a)}\right) \tag{8.14}$$

From which,

$$\Delta^{(1,a)} = \left(\partial\Delta^{(a,b)} \circ \Delta^{(b,a)}\right) \circledcirc \partial\Delta^{(b,a)} \tag{8.15}$$

If we put $\partial\Delta^{(a,b)} = \Delta^{(h,b)}$ and $\partial\Delta^{(b,a)} = \Delta^{(k,a)}$, we obtain

$$\Delta^{(1,a)} = \left(\Delta^{(h,a)} \circledcirc \Delta^{(k,a)}\right) = \Delta^{((h\cdot k),a)} \tag{8.16}$$

This means that A^* is a multiplicative group and that $(A, +, \cdot)$ is a near-field.
At best, this analysis is cursory, and it may even be vacuous, since I don't even offer an example of the phenomenon actually occurring. If further work would be profitable, one might well need to consider subsets of functions and infinite groups—as happens in traditional real-valued calculus.

8.4 Endomorphism Near-Rings

I remark, again, that an endomorphism near-ring is not usually a near-ring of endomorphisms. In this section, I say what I can about $\mathrm{I}(\Gamma)$, $A(\Gamma)$ and $\mathrm{E}(\Gamma)$ using Notation 227 and the symbol defined in (1.30) on page 35. I further require that $(\Gamma, +)$ should be finite, with exponent n. The symbol γ usually means any element of Γ. The identity mapping is an endomorphism, and it has additive order equal to the exponent of the original group. This means that both $(\Gamma, +)$ and $\mathrm{E}(\Gamma)$ have elements with the precisely the same prime orders. If $\theta \in \mathrm{T}(\Gamma)$, then

$$(\theta + \theta + \ldots + \theta)(\gamma) = (\theta(\gamma) + \theta(\gamma) + \ldots + \theta(\gamma))$$

Consequently, $|\theta(\gamma)|$ divides $|\theta|$. An immediate consequence of this is that for all $\gamma \in \Gamma$:

$$\mathcal{S}_p^{\mathrm{T}_0(\Gamma)}(\gamma) \subseteq \mathcal{S}_p^{\Gamma} \tag{8.17}$$

When θ is actually an endomorphism of $(\Gamma, +)$, we additionally have

$$(\theta + \theta + \ldots + \theta)(\gamma) = (\theta(\gamma) + \theta(\gamma) + \ldots + \theta(\gamma)) = \theta(\gamma + \gamma + \ldots + \gamma)$$

which means $|\gamma| = k \Rightarrow k\theta(\gamma) = 0$. The elements of $\mathrm{End}(\Gamma)$ (pure endomorphisms) with orders some power of p generate the subgroup

$$\mathfrak{S}_p^{\mathrm{E}(\Gamma)} = \left\langle \mathcal{S}_p^{\mathrm{End}(\Gamma)} \right\rangle \tag{8.18}$$

(note that this subgroup may well contain elements with prime orders other than p).

Lemma 316

(i) $\mathfrak{S}_p^{\mathrm{E}(\Gamma)}$ *is a d.g. near-ring contained in* $\mathrm{E}(\Gamma)$.
(ii) When p and q are distinct primes,

$$\left(\mathfrak{S}_p^{\mathrm{E}(\Gamma)} \cdot \mathcal{S}_q^{\Gamma}\right) = \{0\}$$

8.4.1 Related Sub-near-Rings

Endomorphisms of the group $(\Gamma, +)$ map conjugacy classes to conjugacy classes. Let the class number of $(\Gamma, +)$ be $(n + 1)$ (i.e. assume that the group has (n+1) distinct conjugacy classes). Consider the sub-near-ring of $\mathrm{T}_0(\Gamma)$ generated by those mappings which map conjugacy classes to conjugacy classes. This is a sub-near-ring, $(\Psi, +, \cdot) \subseteq \mathrm{T}_0(\Gamma)$, and it contains $\mathrm{I}(\Gamma)$, $A(\Gamma)$ and $\mathrm{E}(\Gamma)$. The near-ring of mappings which take conjugate elements to the same image is a sub-near-ring of $(\Psi, +, \cdot)$. Additively, this is a direct product of n copies of $(\Gamma, +)$.

8.4.2 Sequences of Endomorphism Near-Rings

Suppose that $(G, +)$ is the finitely presented group given by (2.1). I write the verbal subgroup generated by the relations in this presentation as $V = V(G)$ (see Notation 3). This, of course, is simply the subgroup generated by the entire set of images of the relations, regarded as multivalued functions, and it is fully invariant. I define a mapping $\Phi : (\mathrm{E}(G), +) \longrightarrow (\mathrm{E}(\frac{G}{V}), +)$, $\Phi(\alpha) = \bar{\alpha}$, where

$$\bar{\alpha}(g + V) = (\alpha(g) + V) \tag{8.19}$$

Lemma 317 Φ *is a homomorphism of near-rings.*

The reader may wish to compare this result with Sect. 8.2 and, in particular, the exact sequence (8.4). The kernel of Φ, above, consists of members of $\mathrm{E}(\Gamma)$ mapping into V. Onc might wonder whether Φ is always surjective. When Γ is D_4 (using presentation (2.19)), the verbal subgroup V is $\langle 2a \rangle$, which is isomorphic to C_2, and $\mathrm{E}(\frac{D_4}{V(D_4)})$ has order 16. On page 286, I give the value for the order of $\mathrm{E}(D_4)$ as 256, so, if Φ were surjective, its kernel should have order 16. According to page 287, its order is 8.

8.5 Other Kinds of Endomorphism Near-Ring

This section requires the definition of reduced free groups on page 74. The standard addition and product of mappings give us transformation near-rings and endomorphism near-rings. Endomorphism near-rings are d.g., but historically another sort of d.g. endomorphism near-ring occurs very early in the research literature, in work associated with Hanna Neumann, in the 1950s, and, later, Albrecht Fröhlich, in the early 1960s.

Pilz gives an account of this in [128]. The idea is to start with a reduced free group with finite generating set $\{e_j\}_{j=1}^{n}$. Every mapping of the generating set to elements of the group extends uniquely to an endomorphism of that group (rather in the way of arbitrary maps of generating sets of free groups extending uniquely to homomorphisms). Faced with two endomorphisms of our reduced free group, one may define an addition endomorphism by a formula of the form

$$(\theta \overset{2}{+} \phi)(e_j) = (\theta(e_j) + \phi(e_j)) \tag{8.20}$$

(which defines the required endomorphism). This sort of addition takes two endomorphisms and delivers a third and is called by Fröhlich **addition of the second type**. The product endomorphism is, of course, the extension of

$$(\theta \circ \phi)(e_j)$$

One obtains a d.g. near-ring of endomorphisms of our original group; and it will be distinct from the endomorphism near-ring of the group, when the group is non-abelian. I call it $(\mathfrak{N}, \overset{2}{+}, .)$. The d.g. near-ring is distributively generated by the set of endomorphisms defined by

$$\phi_{i,j}(e_k) = \delta_{i,k} e_j$$

This sort of addition was important to Fröhlich in his "non-abelian homological algebra" papers; and it was discussed in relation to left and right representation theory in a thesis by Alan Machin [107].[1] The near-ring you get depends on the reduced free basis that you choose. Meldrum says more about these near-rings in [119] and calls them **Neumann near-rings**. The same idea can be applied to infinite generating sets, but one does not get a d.g. near-ring in such cases. I had intended to look at this idea in the context of transformation near-rings but could not find time to do this.

[1] Sadly, I have no access to this thesis. I read it more than 40 years ago and wish it were better known.

8.6 Change of Groups

One feature of transformation near-rings is that the near-ring product is a composition of set mappings unconditioned by the underlying group. This means that we may, in some part, utilise the structural decompositions that I have just reported in reverse and consider going from a multiplicative semi-group to an additive group making the whole structure a near-ring. It turns out that by significantly weakening additive structure, one can obtain entities which themselves have additive properties.

8.6.1 Near-Loops

People have looked at a range of binary structures, and two particularly well-known ones are given on page 482. These next definitions are similar to those ones but non-standard.

Definition 318 Suppose $(S, +)$ is a set together with a binary operation such that

(i) There is an element 0 such that $(x + 0) = x$, for all elements $x \in S$.
(ii) For all, $a, b \in S$, there is a **unique** element $x \in S$, such that $(x + a) = b$ Then, I call $(S, +)$ a **near-loop** and I refer to the element 0 as its **zero**. A near-loop $(S, +, 0)$ in which $(0 + x) = x$ for all $x \in S$ is called a **zero-symmetric** near-loop. A subset of a near-loop which is itself a near-loop, with the same operations and zero, is called a **sub-near-loop**.

Observe that additive associativity is not demanded; nor do I insist on uniqueness for the zero, although in our applications, or when the near-loop is zero-symmetric, it actually is unique. Table 8.1 represents a non-associative near-loop hosted by the three-element set $\{A, B, C\}$. It has both A and B acting as zero elements. Of course, given a, b, c in a near-loop $(S, +)$, we know that for some unique $x \in S$, $((a + b) + c) = (x + (b + c))$, we simply do not generally have that x and a will be the same (as the above table illustrates). This structure is neither a loop nor a quasi-group (see Definition 511).
Near-loops are characterised as having Cayley tables in which the interior columns consist of all the distinct elements of the underlying set. I believe that people working in combinatorics may call them **column Latin squares** [86].

Table 8.1 Near-loop with order 3

+	A	B	C
A	A	A	C
B	B	B	A
C	C	C	B

Notation 319 When dealing with these structures, it is occasionally useful to distinguish the zero element of the near-loop in our notation. I then talk of near-loops $(S,\ +,\ 0)$.

Lemma 320 *The collection of all near-loops hosted by a given set S and having some fixed element* $0 \in S$ *as zero itself forms a near-loop. This near-loop is zero-symmetric.*

Proof Suppose $(S\ +,\ 0)$ and $(S\ \overset{+}{+}\ \ 0)$ are near-loops. Define a new near-loop, $(S \oplus\ 0)$, which is to be the sum of the previous two near-loops, using the formula

$$a \oplus b = ((a+b) \overset{+}{+} b) \tag{8.21}$$

Clearly, $a \oplus 0 = ((a+0) \overset{+}{+} 0) = a$, and given $a, b \in S$, we can find a unique $y \in S$ such that $y \overset{+}{+} a = b$ and a unique $x \in S$ such that $x + a = y$, from which $x \oplus a = ((x+a) \overset{+}{+} a) = b$ and x is unique. The binary operation

$$a * b = a \tag{8.22}$$

defined for all $a, b \in S$, gives us a near-loop (compare this with Definition 78). Further, using the notation described in (8.21), $a(+\oplus*)b = (a+b)$ for all $a, b, \in S$, so we can say that $(S\ *)$ represents a zero element for the continuum of near-loops. Next, starting with two arbitrary near-loops $(S,\ +,\ 0)$ and $(S,\ \vee,\ 0)$, we should prove there is a unique near-loop $(S,\ \wedge,\ 0)$ such that $(S,\ \wedge,\ 0) \oplus (S,\ +,\ 0) = (S,\ \vee,\ 0)$. Taking $a, b \in S$, we determine the unique $x \in S$ for which $x+b = a \vee b$ and set $a \wedge b = x$. It is clear that no other product could work and remains to be shown that this is indeed a near-loop. When b is 0, x is equal to a and so $a \wedge 0 = a$. Given $a, b \in S$, one must show there is a unique $x \in S$ such that $x \wedge a = b$. One can find x such that $x \vee a = b + a$ and it is unique. Then our definition of $\wedge$ gives that $x \wedge a = b$ and x is unique. □

Notation 321 We now have an addition of near-loops and might write $(+\oplus \overset{+}{+})$ to represent the resulting near-loop operation produced by the addition of $(S,\ +,\ 0)$ and $(S,\ \overset{+}{+},\ 0)$. In this subsection, I reserve the symbol * to represent the product given in (8.22).

Lemma 322 *The addition* $\oplus$ *of near-loops defined above is actually a group operation.*

Proof The addition is easily seen to be associative. Suppose $(S,\ +,\ 0)$ is any near-loop. I need to construct another near-loop, $(S,\ \wedge,\ 0)$, such that for all a, b in S, $a(+\oplus\wedge)b = a * b = a$ (right inverse). Set $a \wedge b = c$, where c is the unique element for which $c + b = a$. Clearly, $a \wedge 0 = a$. Take any $a, b \in S$ and set $x = b + a$. Then, $x \wedge a = c$ means that $c + a = x$ consequently, $c = b$, and $x \wedge a = b$. Thus,

$(S \wedge 0)$ is a near-loop. Now, $(a+b) \wedge b = c$ means that c is the unique element for which $c + b = a + b$; thus $c = a$, and $((a+b) \wedge b = a = a * b$. Consequently, $(+ \oplus \wedge) = *$. □

Notation 323 If S is a set with one distinguished element $0 \in S$, then $(\Sigma(S) \oplus)$ represents the group of near-loops $(S, +, 0)$.

Suppose $(S, +, 0)$ and $(S, \wedge, 0)$ are near-loops and S is a finite set with order $(n+1) > 3$. I investigate whether $(+ \oplus \wedge)$ and $(\wedge \oplus +)$ can be the same. Taking, $a, b \in S$, this requires

$$((a+b) \wedge b)) = ((a \wedge b) + b) \tag{8.23}$$

With $(S, \wedge, 0)$ given, we can take $(S, +, 0)$ as an additive cyclic group, choosing a and b in (8.23) to be any elements of the group we please. In particular, taking a to be (n) and b to be 1, (8.23) becomes

$$(0 \wedge 1)) = ((n \wedge 1) + 1)$$

If we may designate some other element, s of S, as (n), as we may, since $n > 2$, and if we keep the same element of the set as 1, we would obtain (for a different Cayley table for C_3—that is, an isomorphic but distinct group)

$$(0 \wedge 1)) = ((s \wedge 1) + 1)$$

the two values, $(s \wedge 1)$ and $((n) \wedge 1)$, are distinct, so this is impossible. Thus,

Lemma 324 *The group* $(\Sigma(S), \oplus)$ *has trivial additive centre.*

In the case of our finite set S, which I enumerate as $\{s_0 = 0, s_1, s_2 \ldots s_n\}$, one may define a near-loop product $\wedge$ with the recipe:

$$s_m \wedge s_t = \begin{cases} s_m & \text{if t} = 0 \\ s_{n-m} & \text{if t} \neq 0 \end{cases}$$

$(S \wedge 0)$ is a near-loop; moreover, for all $a, b \in S$

$$((a \wedge b) \wedge b) = a$$

that is, $(S \wedge \oplus \wedge 0) = (S * 0)$. From this one adduces that the group $(\Sigma(S) \oplus)$ is of even order in finite cases. When S has finite order (n+1), we know that one member of Σ is the cyclic group with order n+1. This has the same order, n+1, when viewed as a member of the group $(\Sigma(S) \oplus)$, from which we deduce that (n+1) divides the order of that group.

8.6.2 *Homomorphisms and Normal Sub-Loops*

Definition 325 A **homomorphism** between two near-loops is a mapping

$$\theta : (G,\ +,\ 0) \to (H,\ +,\ 0) \tag{8.24}$$

for which

$$\theta(a+b) = \theta(a) + \theta(b) \text{ for all } a, b \in G$$

and

$$\theta(0) = 0.$$

The **kernel** of such a homomorphism is simply

$$\{x \in G\ :\ \theta(x) = 0\}$$

Notation 326 If $(G,\ +,\ 0)$ is a near-loop and $a \in G$, I define $(-a)$ to be the unique member of G for which $(-a) + a = 0$.

The same element of the near-loop may be variously and legitimately described as $(-b)$, $(-c)$, $(-d)$, etc. for distinct, $a, b, c \in G$; and we have no knowledge of whether $a + (-a)$ is generally equal to 0 and cannot claim $(-(-a)) = a$. This notation only really seems appropriate in situations in which the near-loop operation is denoted by "+"; but it is nevertheless useful.

Until the end of this section, all near-loops are zero-symmetric.

Definition 327 If $(K,\ +,\ 0)$ is a sub-near-loop of the near-loop $(G,\ +,\ 0)$, I define, for $a \in G$

$$K + a = \{(k+a)\ :\ k \in K\}$$

and call this a **right coset** of K in G. I might then write the collection of right cosets as $\frac{(G,\ +,\ 0)}{K}$.

Lemma 328 *Suppose (8.24) is homomorphism of (zero-symmetric) near-loops, with kernel, K. Suppose that* $a, b \in G$ *and* $k \in K$*. Then,*

(i) $\theta(-a) = -(\theta(a))$
(ii)

$$((-a+k)+a) \in K$$

and

$$(-a + (k + a) \in K$$

(iii) $(K, +, 0)$ *is a near-loop.*
(iv)

$$(K + a) + (K + b) = (K + (a + b))$$

(v) *If* $\theta(a) = \theta(b)$, *then* $a = (k + b)$ *for some* $k \in K$.
(vi)

$$(K + a) + (K + b) = (K + (a + b))$$

Proof

(i)

$$\theta(-a + (k + a)) = \theta(-a) + (0 + \theta(a))$$

now apply zero symmetry.

(ii) Write $a = (x + b)$, deducing $\theta(a) = \theta(x) + \theta(b)$.
Zero symmetry gives us that $x \in K$.

(iii) Taking $k, k' \in K$, we have that

$$(k + a) + (k' + b) = (x + (a + b))$$

and applying θ to this equality, plus part (iv), gives us that $x \in K$ and so

$$(K + a) + (K + b) \subseteq (K + (a + b))$$

The opposite inclusion is proved in the same way. □

Definition 329 When $(G, +, 0)$ is a near-loop and $(K, +, 0)$ is a sub-near-loop, K is said to be **normal** provided that for all $a, b \in G$

$$((K + a) + (K + b)) = (K + (a + b))$$

It will be seen that kernels of near-loop homomorphisms are normal sub-near-loops. We now have the standard addition of right cosets of normal sub-near-loops.

$$(K + a) + (K + b) = (K + (a + b)) \quad a, b \in G. \tag{8.25}$$

One corollary to the last result is that $(K + k) = K$ for all $k \in K$. More, if $(K + a)$ and $(K + b)$ have elements in common, they are equal, and any two right cosets have the same cardinality.

Lemma 330 *Right cosets of K partition G and form a near-loop* $\left(\frac{(G,\,+,\,0)}{K},\ +\right)$ *under the addition defined in* (8.25). *The coset K forms the zero for this near-loop, and the zero is two-sided in the sense*

$$(K + (K + a)) \;=\; (K + a)$$

In fact, it is not hard to show that

$$(K + a) \;=\; \{(a + k)\ :\ k\ \in K\}$$

We might describe $\{(a + k)\ :\ k\ \in\ K\}$ as $(a + K)$, obtaining a more typical view of normal substructures.

Lemma 331 *When K is a normal sub-near-loop of the near-loop* $(G,\ +,\ 0)$*, the collection of right cosets of K in G is a near-loop in the obvious way. There is surjective homomorphism of near-loops:*

$$\pi : G \twoheadrightarrow H = \frac{(G,\ +,\ 0)}{K}$$

in which $\pi(a) \;=\; (K + a)$. *K is the kernel of that homomorphism and the zero of the quotient near-loop.*

It is likely that near-loops have been discussed elsewhere, although at the time of writing I am unaware of previous work on them. Their point, in my context, relates to the *additive host problem*, treated next.

8.6.3 The Host Problem

Remark 332 The **(multiplicative) host problem** for near-rings starts with a fixed additive group $(A,\ +)$ and attempts the construction of a semi-group product, $(A\ \cdot)$, in such a way that $(A,\ +, \cdot)$ becomes a near-ring. One seeks to determine the isomorphism classes definable, with their cardinality referred to as the **multiplicative host number** for $(A,\ +)$. The multiplicative host problem considers the possible semi-group products that may be made to coexist with a fixed additive group structure.

An alternative is to start with a semi-group $(A^*,\ \cdot)$ and ask for the additive groups $(A,\ +)$ that one might define to make $(A,\ +,\ \cdot)$ into a near-ring. One would describe this as the **additive host problem** for near-rings. The additive host problem considers the possible group additions that may be made to coexist with a fixed semi-group structure. Variants of this idea occur in the theory of near-fields, where one considers the semi-group to be a subset of the elements of the near-ring (see Sect. 3.3.1).

In exploring the additive host problem, it becomes convenient to weaken the additive structures that concern us and to consider a class of structures rather wider than that of standard near-rings. Our viewpoint is that we start with a multiplicative semi-group $(S, \cdot)$, one that has an **absorbing element 0** (i.e. $x \cdot 0 = 0 = 0 \cdot x$). In what follows, all semi-groups will have a fixed absorbing element, 0, and such a thing, of course, is unique.

Definition 333 A (right-distributive) **lachs** is a structure $(A, +\cdot)$ for which

- $(A, +, 0)$ is a near-loop (see Definition 318).
- $(A, \cdot)$ is a semi-group.
- For all, $x \in A$, $0 \cdot x = 0 = x \cdot 0$ (*absorption*—i.e. bi-zero symmetry).
- For all, $a, b, c \in A$, $(a + b) \cdot c = (a \cdot c) + (b \cdot c)$.

The next result uses notation introduced on page 325.

Lemma 334 *Suppose* $(A, +, \cdot)$ *and* $(A, \overset{+}{+}, \cdot)$ *are two right-distributive lachs both hosted by the semi-group* $(A, \cdot)$. *Then* $\left(A, (+\oplus \overset{+}{+}), \cdot\right)$ *is also a right-distributive lachs hosted by the same semi-group.*

Proof Suppose $a, b, c \in A$, then

$$\left(a\left(+\oplus \overset{+}{+}\right)b\right)\cdot c = \left((a+b)\overset{+}{+}b\right)\cdot c = \left((a+b)\cdot c \overset{+}{+} b\cdot c\right) =$$
$$\left(((a\cdot c)+(b\cdot c))\overset{+}{+} b\cdot c\right) = \left((a\cdot c)(+\oplus \overset{+}{+})(b\cdot c)\right)$$

□

Notation 335 If $(S, \cdot, 0)$ is a semi-group with zero $0 \in S$, then $(\Sigma(S, \cdot, 0), \oplus)$ represents the group of all lachs hosted by $(S, \cdot)$ with addition $\oplus$ defined in (8.21).

8.6.4 *Transformations on Near-Loops*

Starting with a near-loop $(G, +, 0)$, one can form the collection of zero-fixing mappings from G to itself. I shall write this $\mathrm{T}_0(G)$.

Lemma 336 *With the standard operations,* $\mathrm{T}_0(G, +)$ *becomes a lachs.*

Proof The zero mapping plays the role of zero in this lachs. Suppose $\alpha, \beta \in \mathrm{T}_0(G)$. For each $g \in G$, we know we can find $x \in G$ such that $x + \alpha(g) = \beta(g)$. Define $x = \gamma(g)$. Then, $\gamma \in \mathrm{T}_0(G)$ and $\gamma + \alpha = \beta$, and γ is unique. □

Definition 337 When $(G, +, 0)$ is a near-loop, I shall call $\mathrm{T}_0(G, +, 0)$ its **associated transformation lachs**.

Observations 338 It seems reasonable to investigate the continuum of transformation lachs obtained by varying the near-loops definable on some fixed set (keeping the zero element fixed). We have a natural addition, formed by adding the host near-loops using the addition $\oplus$ defined in (8.21), and this results in an additive group of transformation lachs, whose own associated transformation near-ring now becomes of interest.

8.6.5 Transformations on Sets

Take a finite set, S, with order n+1. Take n isomorphic copies of S: $S_1, S_2, \ldots S_n$. Ordering the elements of S as $\{s_0, s_1, \ldots s_n\}$ induces a corresponding listing in each of the copies. So, one may represent the elements of S_j as $\{s_{k,j} : k = 0, 1, 2, \ldots n\}$. Distinguish a specific element of S that will be called zero, and adopt the convention that $s_0 = 0$ and that $s_{0,j} = 0$, for all $j = 1, 2 \ldots n$. Now regard each of the elements $\{s_{i,j}\}$ as mappings from S to itself, with the policy

$$s_{i,j}(s_k) = \delta_{j,k} s_i \tag{8.26}$$

There is then a natural product on $\bar{S} = \bigcup_{k=1}^{n} S_k$ given by the formula

$$s_{i,j} \cdot s_{k,r} \;=\; \delta_{j,k} s_{i,r}$$

This product, being based on the composition of mappings, implies that $(\bar{S}\ \cdot)$ is a semi-group. Next, we form ordered n-tuples, $(x_1, x_2, \ldots x_n)$, where $x_k \in S_k$, and call the set of such tuples T. There is a natural product

$$T \times \bar{S} \rightarrow T$$

in which $(x_1, x_2, \ldots x_n) \cdot s_{i,j} \;=\; (y_1, y_2, \ldots y_n)$, where $y_k = \delta_{k,j}(x_i \cdot s_{i,j})$. One may interpret T as the set of all zero-fixing mappings from S to itself:

$$(s_{i,j})(s_k) \;=\; s_r \tag{8.27}$$

Here, one supposes the kth coordinate of $(s_{i,j})$ to be $s_{r,k}$. This, in turn, implies a composition product:

$$T \times T \rightarrow T$$

In which, (using an abbreviated notation to represent tuples)

$$(s'_{i,j}) \cdot (s_{k,t}) \;=\; (x_{r,t})$$

where $(x_{r,t})$ is a tuple, with general, tth element $x_{r,t} = \left((s'_{i,j}) \cdot s_{k,t}\right)$.

The reader will observe that we have gone a considerable way into the structure of a transformation near-ring without recourse to any underlying additive structure and that what we have described represents a commonality shared by a considerable number of distinct transformation near rings. One could regard the tuples $(s_{i,j})$ as formal sums $\sum_j s_{i,j}$. Recalling the interpretation given in (8.26) and (8.27), we obtain a formal left distributivity law:

$$(\underline{x}) \cdot (s_{k,t}) = (\underline{x} \cdot s_{k,t})$$

(this time representing the left-hand tuple in vector form). This form of distributivity is based on no more than the fact that the kth coordinate of each tuple is in S_k and is simply a direct consequence of our notation. Right distributivity is a more complicated issue because elements of the form $s'_{i,j} \cdot (\sum_t s_r, t)$ are not inevitably to be found in S_j.

8.7 The Stemhome Near-Ring

Suppose $(B, +)$ is a group, then the near-ring $\mathrm{T}_0(B)$ is a left ideal of the near-ring $\mathrm{T}(B)$, and we have an exact sequence of groups

$$\mathrm{T}_0(B) \rightarrowtail \mathrm{T}(B) \overset{\pi}{\twoheadrightarrow} B \tag{8.28}$$

In finite cases, $\mathrm{T}(B)$ is, additively, a direct product of $|B|$ copies of $(B, +)$, while $\mathrm{T}_0(B)$ is a direct product of one fewer copies. One can think of the epimorphism π as a projection and represent elements of $\mathrm{T}(B)$ in the form (b, θ), where $b \in B$ and $\theta \in \mathrm{T}_0(B)$. The first coordinate of this representation can be viewed as the image of 0 under the member of $\mathrm{T}(B)$ that concerns us. Of course, T(B) is not a zero-symmetric near-ring. The addition on $\mathrm{T}(B)$ is

$$(b_1, \theta) \oplus (b_2, \phi) = (b_1 + b_2, \theta + \phi) \tag{8.29}$$

and one can define a product ($\otimes$) using the recipe[2]

$$(b_1, \theta) \otimes (b_2, \phi) = (\theta(b_2), (\theta \circ \phi)) \tag{8.30}$$

Lemma 339 $(\mathrm{T}(B), \oplus, \otimes)$ *is a zero-symmetric and right-distributive near-ring.*

Definition 340 I call $(\mathrm{T}(B), \oplus, \otimes)$ the **Stemhome** near-ring associated with $(B, +)$ and write this $\mathbf{ST}(B)$, or $(\mathbf{ST}(B), +, \cdot)$, only retaining the $\otimes$ notation for occasional emphasis.

[2] This is absolutely not the tensor product!

8.7.1 The Stemhome Functor

Lemma 341

(i) **ST** *is a full and faithful functor from the category of groups and automorphisms to the category of near-rings and automorphisms.*
(ii) Automorphisms of the near-ring $\mathbf{ST}(B)$ *fix* $\mathrm{T}_0(B)$.

Proof

(i) The automorphism α of $(B, +)$ is mapped to

$$\mathbf{ST}(\alpha) : (b, \theta) \mapsto \left(\alpha(b), \left(\alpha \circ \theta \circ \alpha^{-1}\right)\right) \tag{8.31}$$

and that is an automorphism of $\mathbf{ST}(B)$. When ξ is an automorphism of the near-ring ST(B), I can write $\xi(b, \theta) = (\beta(b), \gamma(\theta))$, thereby obtaining two new automorphisms:

$$(\beta(\theta(b)), \gamma(\theta\phi)) = \xi\,(\theta(b), \theta\phi) = \xi\,((a, \theta) \otimes (b, \phi)) =$$
$$(\gamma(\theta)(\beta(b), \gamma(\theta\phi))$$

This implies that $\beta \circ \theta = \gamma(\theta) \circ \beta$ and consequently that $\gamma(\theta) = \beta \circ \theta \circ \beta^{-1}$. Consequently, $\xi = ST(\beta)$.

(ii) Suppose $\triangledown$ is an automorphism of the near-ring and that $\theta \in \mathrm{T}_0(B)$. For suitable choices of symbols, I can write

$$\triangledown\left(\theta'(c), \theta' \circ \psi\right) = \triangledown\left(\left(b', \theta'\right) \otimes (c, \psi)\right) = (\triangledown\,(b, \theta) \otimes \triangledown\,(0, 1)) =$$
$$\triangledown\,((b, \theta) \otimes (0, 1)) = \triangledown\,(0, \theta)$$

Thus, $\triangledown\,(0, \theta) = \left(\theta'(c), \theta' \circ \psi\right)$ and so $\theta'(c) = 0$ and $\theta' \circ \psi = \theta$. There are no restrictions on our choice of b or θ in this argument, and I conclude $\triangledown\,(0, 1) = (0, 1)$. Writing $\triangledown\,(b, \theta) = \left(b', \theta'\right)$, as before, one gets

$$\triangledown\,(0, \theta) = \triangledown\,((b, \theta) \otimes (0, 1)) = \left(\left(b', \theta'\right) \otimes (0, 1)\right) = \left(0, \theta'\right)$$

□

Definition 342 I call a normalised mapping $j : B \to \mathrm{T}(B)$ such that $\pi \circ j = id_B$ (the identity mapping on B) a **stem-lifting**.

(compare this with (15.6) on page 427), although in that chapter I normally refer to "liftings" rather than "stem-liftings". Given a stem-lifting, there is an associated stem product ($\star$) hosted by the additive group $(B, +)$, given by

$$a \star b = \pi\,(j(a) \otimes j(b)) \tag{8.32}$$

(see Definition 472). I revisit these ideas in Chap. 15.

8.8 The Wurzel

Definition 343 Suppose $(A, +)$ is an additive group. The **Wurzel** group of $(A, +)$ is the structure

$$\mathbf{W}(A) = \mathrm{T}_0((A, \mathrm{T}_0(A))$$

I normally talk of 'the Wurzel' rather than "the Wurzel group". In finite cases, it is a direct sum of $(|A| - 1)^2$ copies of $(A, +)$. A generalisation of this would be the group

$$\mathrm{T}_0((A, \mathrm{T}_0((B, C))$$

where A, B and C are additive groups. The next definition is motivated by the corresponding construction in topology [116].

Definition 344 The **smash product** of the two groups $(A, +)$ and $(B, +)$ is the set of pairs of elements of the groups in which zero only occurs as a double entry:

$$A \wedge B = \{(a, b) : a \in A, b \in B, a = 0 \Leftrightarrow b = 0\}$$

In the resulting algebra, one makes the identification

$$(a, 0) = (0, a) = (0, 0)$$

and stipulates

$$(a, b) \overset{+}{+} (c, d) = (a + c, b + d)$$

and then $(A \wedge B \overset{+}{+})$ is a loop. Taking the zero of $A \wedge B$ to be $(0, 0)$, the definition of $\mathrm{T}_0((A \wedge B, C)$ is obvious, as is that it is an additive group isomorphic to a product of copies of $(C, +)$. Then,

Lemma 345

$$\mathrm{T}_0((A \wedge B, C) \cong \mathrm{T}_0((B, \mathrm{T}_0(A, C)) \cong \mathrm{T}_0((A, \mathrm{T}_0(B, C)) \tag{8.33}$$

The inverses of a typical non-zero element (a, b) form a set:

$$\{(c, d) : \text{ either } c = -a, \text{ or } d = -b, \text{ or both}\}$$

The associativity law

$$(x + (y + z)) = ((x + y) + z)$$

does hold whenever both $(y+z)$ and $(x+y)$ are non-zero, but not in other situations. The Wurzel is of interest in Chap. 15.

8.9 Elementary Closure Procedures

Suppose $\mathrm{T}(N)$ is a transformation near-ring. There is always a sub-near-ring of it, formed by this subset of the "rank one" elements:

$$\mathfrak{L}(N) = \{\theta \in \mathrm{T}(N) \; : \; \theta(m) = \theta(n), \; \forall \, m, n \in N\} \tag{8.34}$$

This near-ring is a two-sided sideal of $\mathrm{T}(N)$. The $\mathcal{A}$-matrices (pages 250, 485) of the transformations in $\mathcal{L}(N)$ are simply column stochastic matrices each having a single row consisting entirely of 1s. Of course, these transformations are additively closed. The near-ring is, in some sense, the dual of the set of permutation matrices (which are the matrices with maximal rank), and I use it as a source of examples in this section.

8.9.1 *Additive and Multiplicative Closures*

Suppose S is a subset of a transformation near-ring, $\mathrm{T}_0(N)$, which may, or may not, be infinite. Two closure operations are immediate:

(i) $\mathfrak{X}(S)$—which is the smallest semi-group of $(\mathrm{T}_0(N)\ \cdot)$ containing S
(ii) $\mathfrak{Y}(S)$—which is the smallest subgroup of $(\mathrm{T}_0(N),\ +)$ containing S

One may introduce *sequences* of these symbols:

(i) $\mathfrak{X}_0(S) \; = \; S.$
(ii) $\mathfrak{X}_1(S) \; = \; \mathfrak{X}(S).$
(iii) $\mathfrak{X}_{(m+1)}(S) \; = \mathfrak{X}\,(\mathfrak{Y}\,(\mathfrak{X}_m(S))) \; = \; \mathfrak{X}\,(\mathfrak{Y}_m(S)).$
(iv) $\mathfrak{Y}_m(S) \; = \; \mathfrak{Y}\,(\mathfrak{X}_m(S)).$

$\mathfrak{X}_m(S)$ is always a multiplicative semi-group; $\mathfrak{Y}_m(S)$ is always an additive subgroup. If S is the set of bijective mappings, then $\mathfrak{B}(N) \; = \; \mathfrak{Y}(S)$ (defined on page 239). There are subgroup chains

$$\mathfrak{Y}_0(S) \subseteq \mathfrak{Y}_1(S) \subseteq \mathfrak{Y}_2(S) \subseteq \mathfrak{Y}_3(S) \subseteq \dots \tag{8.35}$$

and semi-group chains

$$\mathfrak{X}_0(S) \subseteq \mathfrak{X}_1(S) \subseteq \mathfrak{X}_2(S) \subseteq \mathfrak{X}_3(S) \subseteq \dots \tag{8.36}$$

and each chain terminates, in the sense of ceasing to ascend, after a finite number of steps, in finite cases. When one chain terminates, so does the other, and the result is the smallest sub-near-ring containing S.

Definition 346 For any set $S \subseteq T_0(N)$, I define $\mathfrak{S}(S)$ as the smallest sub-near-ring of $T_0(N)$ containing S.

The closure operation $\mathfrak{X}$ is entirely set-theoretic and based on the composition of set functions. The other closure operation, $\mathfrak{Y}$, depends on the additive group structure of $(N, +)$.

8.9.2 A Topological Closure

I need some of the ideas in Sect. 7.4.4. One can define an alternative closure operation using the structure of T(N). As in Chap. 7, one regards mappings $\theta \in$ T(N) as (possibly infinite) tuples:

$$\theta = (a_0, a_1, a_2, \ldots, a_n, a_{n+1} \ldots) \tag{8.37}$$

(I assume the identity to be represented as a unique tuple $(n_0, n_1, \ldots,)$, with $\theta(n_j) = a_j$). There is a symmetric group S_N which permutes the N elements of a typical tuple (the group may be infinite). These permutations simply rearrange the coordinates of the tuple in the way defined by the permutation. Formerly, my representation of permutations was chosen to fit in with the $\mathcal{A}$-matrices I used to represent elements of T(N), because I wanted the permutations to appear as transformations (for that reason, you see transpositions occurring when permutations operate on the right—Sect. 7.4.4). I don't do that here, so there is a bit of inconsistency in my notation, and these permutations are the transposes of the previous ones. For example, in $(C_3, +)$, the mapping $\theta \in T(C_3)$ defined by

$$(2, 0, 1)$$

and acted upon by the permutation σ which in cycle notation is written $(0, 1, 2)$, results in a mapping

$$(1, 2, 0) = (2, 0, 1) \cdot \begin{bmatrix} 0\,1\,0 \\ 0\,0\,1 \\ 1\,0\,0 \end{bmatrix}$$

(I've written this trivia out explicitly to avoid confusion with what I did earlier).

Definition 347 Suppose S is a subset of T(N). I define

$$\mathfrak{c}(S) = \bigcup_{\sigma \in S_N} (S^{\sigma}) \tag{8.38}$$

The easiest way to understand what's going on here is to think of transformations as tuples, the closure operation operates on tuples generating all possible permutations of their entries.

Lemma 348 *The mapping $S \mapsto \mathfrak{c}(S)$ satisfies the Kuratowski closure axioms (page 6); so the collection of sets $\{\mathfrak{c}(S) \ : \ S \subseteq \mathrm{T}(N)\}$ constitute the closed sets of a topological space $\mathfrak{t}(N)$ defined on $\mathrm{T}(N)$.*

The topological space $\mathfrak{t}(N)$ is entirely set-theoretic and not dependent on the group structure of $\mathrm{T}(N)$. As an illustration, $\mathfrak{c}(\mathfrak{L}(N)) = \mathfrak{L}(N)$. The complement of the closure of a set—$C = (\mathfrak{c}(S))^c$ is itself closed—$\mathfrak{c}(C) = C$. This means the every open set is closed and every closed set opens in this topology—these sets are **clopen**, and the space is **disconnected** [87]. The space is not, however, Hausdorff. Mappings which are permutations of each other, in the obvious sense, will each be inside any open set containing the other. Nor, in infinite cases, is the space **compact**. I now ask the reader to recall the notation defined on page 239.

Lemma 349 *Suppose $(N, +)$ is an additive group.*

(i) $\mathfrak{c}(\mathfrak{B}(N)) = \mathfrak{B}(N)$.
(ii) $\mathfrak{B}(N)$ *is a closed set in the topological space* $\mathfrak{t}(N)$.

Proof From Lemma 260, $\sigma \in S_N$ acts homomorphically upon $\mathfrak{B}(N)$. σ also preserves bijective mappings, which means that $\mathfrak{B}(N)$ is right permutation closed and gives the result. □

Corollary 350

(i) If a subset $S \subseteq \mathrm{T}(N)$ contains no bijective mappings, then the subset $\mathfrak{c}(S)$ contains none too.
(ii) If a subset $S \subseteq \mathrm{T}(N)$ contains any rank one $\mathcal{A}$-matrices, then the subset $\mathfrak{c}(S)$ contains $\mathfrak{L}(N)$.

Suppose $G(N)$ is an additive subgroup of $\mathrm{T}(N)$ and a closed set under the topology just defined. Representing the identity mapping as $(n_0, n_1, n_2, \ldots)$, one can form sets

$$\mathcal{P}_j = \{\theta(n_j) \ : \ \theta \in G(N)\} \tag{8.39}$$

$\mathcal{P}_j$ is an additive subgroup of $(N, +)$. In fact, for any indices, i, j, closure symmetry gives us that $\mathcal{P}_i = \mathcal{P}_j$, and I represent this subgroup by the un-indexed symbol $\mathcal{P}$. To illustrate, if $G(N) = \mathfrak{L}(N)$, then $\mathcal{P} = N$. Given any subgroup $(H, +) < (N, +)$, one knows that the group $\mathrm{T}(N, H)$ is right permutation closed and a closed subgroup of $\mathrm{T}(N)$ under the topology just described. We have closed subgroups

$$G(N) < \mathrm{T}(N, \mathcal{P}) < \mathrm{T}(N) \tag{8.40}$$

And

$$\mathfrak{L}(N) \lneq \mathrm{T}(N, \mathcal{P}) = \mathrm{T}(N) \tag{8.41}$$

8.10 Polynomials

I'd like to have a brief look at polynomials in a near-ring setting. Originally, mathematicians considered polynomials with coefficients in fields because that is what had been happening by default. Later, this was extended to rings and then commutative rings. People working in universal algebra discussed the idea of polynomials in more general, and more abstract, settings. The interested reader can find more in the book by Lausch and Nöbauer [97]. Start with a standard **ring** with identity, $(R, +, \cdot)$. Form the ring of polynomials in one indeterminate $(R[x], +, \cdot)$. If $f(x), g(x) \in R[x]$, we have an evaluation mapping $ev : (R[x] \times R[x]) \mapsto R[x]$, as

$$ev\,(f(x), g(x)) \longmapsto f(g(x)) \tag{8.42}$$

Here, I suppose one should regard constant terms, elements of R, as polynomials of the form ax^0 and define

$$ev\left(ax^0, g(x)\right) = ag(x) \tag{8.43}$$

and, of course, $ag(x)$ should be interpreted as is standard in the polynomial ring. But how to interpret $ev(x^2, g(x))$? The obvious answer is as the element $((g(x))^2)$ in the ring $R[x]$. That is, the composition in (8.42) should take place in $R[x]$. The evaluation map can be regarded as a binary operator on $R[x]$,

$$(f(x) \boxtimes g(x)) = ev\,(f(x), g(x)) \tag{8.44}$$

Evaluation maps, or compositions, as one might now call them, were first studied a century ago by Joseph Ritt and others. They are important in numerical analysis and computer science theory where they crop up in algorithms used to evaluate polynomial functions. Compositions involving polynomials occur widely in mathematics. They are used in the construction of the **Nottingham group** [81]. The Nottingham group is a finitely generated topological group which is an inverse limit of finite p-groups [88].

Lemma 351 *The structure* $(R[x], +, \boxtimes)$ *just defined is an abelian, zero-symmetric, right near-ring with identity.*

The algebra $(R[x], +, \boxtimes)$ is not usually left-distributive, so not a ring, then. There will be a symmetric, left distributive near-ring with the opposite evaluation map: $(f(x) \boxtimes g(x)) = g(f(x))$. This is actually an opposite near-ring. Meldrum [119] says that a number of people have considered these near-rings in various papers, and he lists some of them. He cites [97] for further information.

8.10.1 Near-Rings

One imagines that one could apply the same sort of ideas when $(R, +, \cdot)$ is a right near-ring, but of course there are problems with multiplication.

8.10.2 Skew Polynomial Near-Rings

One can consider polynomial-like structures of the form

$$\left(a_1x^{n_1} + a_2x^{n_2} + \ldots + a_mx^{n_m}\right) \tag{8.45}$$

where $a_j \in R, m < \infty, n_j \in \mathbb{Z}$, and $0 \leq n_j < \infty$. One can add such structures and insist that contiguous terms of the form $(a_jx^k + a_{j+1}x^k)$ should always be resolved into $(a_j + a_{j+1})x^k$ in the way beloved by schoolchildren ("apples and oranges"). A general polynomial can always be written

$$\left(a_0 + a_1x + a_2x^2 + a_3x^3 + \ldots + a_mx^m\right) = \sum_{j=0}^{m} a_jx^j \tag{8.46}$$

Here, many of the coefficients could be zero, and hence, discardable, but $a_m \neq 0$, unless the polynomial actually is zero. One should also insist that in these structures $n_1 < n_2 < \ldots < n_m$ and that after all algebraic operations, the symbols are rearranged correctly and respect that. So, for example,

$$(ax^3 + bx^4) + (cx^2 + dx^3) = (cx^2 + (a+d)x^3 + bx^4) \tag{8.47}$$

In this addition, terms involving distinct powers of x commute, whereas terms involving the same power of x do not necessarily commute. The result is an additive group $(R[x], +)$ which contains the additive group $(R, +)$. Suppose, now, that $\sigma : R \mapsto R$ is an endomorphism of the near ring $(R, +)$. I define a multiplication of polynomials

$$\begin{aligned}\left(\sum_{j=0}^{m} a_jx^j\right) ¥ \left(\sum_{j=0}^{k} b_jx^j\right) &= \left(\sum_{j=0}^{m} a_jx^j ¥ \left(\sum_{t=0}^{k} b_tx^t\right)\right)\\ &= \left(\sum_{j=0}^{m} a_j \left(\sum_{t=0}^{k} \sigma^j(b_t)x^{(j+t)}\right)\right)\end{aligned} \tag{8.48}$$

except that it should be understood that the final term here is subject to re-ordering so that the indices are in ascending order. So, for example,

$$
\begin{aligned}
(ax^3 + bx^4)¥(cx^2 + dx^3) &= \\
\left(ax^3¥(cx^2 + dx^3)\right) + \left(bx^4¥(cx^2 + dx^3)\right) &= \\
\left(a\sigma^3(c)x^5 + a\sigma^3(d)x^6 + b\sigma^4(c)x^6 + b\sigma^4(d)x^7\right) &= \\
\left(a\sigma^3(c)x^5 + \left(a\sigma^3(d) + b\sigma^4(c)\right)x^6 + b\sigma^4(d)x^7\right)
\end{aligned}
$$

Engagingly, Lam refers to this technique as "Hilbert's twist" [94].

Lemma 352 *Under these prescribed operations, the polynomials just defined form a right-distributive, zero-symmetric, near-ring with identity.*

This construction is modelled directly on the important construction of skew polynomials in general ring theory [7]. I have not had time to do any more than simply describe it.

I believe a number of mathematicians have investigated polynomials in near-rings and would have a better tale to tell than mine. The interested reader can find references in [128] and in the online bibliography maintained by the University of Linz [101].

Chapter 9
Phomomorphisms

Although I do use some of the notation for group transformations from Chaps. 7 and 8 in what follows, I also occasionally use some slightly different notation (see Sect. 9.2). I occasionally mention modules, although I actually define them later, in Chap. 11. This chapter, and the next one, makes extensive use of the δ-operator defined in Definition 45. I shall also prefer to use the symbol $\mathbb{Z}_n$ for the cyclic group of order n, rather than C_n, which I employ everywhere else. The Gothic symbol $\mathfrak{a}$ appears on page 350 and the Gothic symbol $\mathfrak{PC}$ on page 235. The work in this chapter is related to some of the earliest papers on near-rings. There is more about that in Sect. 12.4.

Some of the motivation for this work is from homological algebra. The classic papers of Eilenberg and Maclane are appropriate background to these ideas [37], as are the by now venerable books [18, 77, 109], as well as the papers by Albrecht Frölich cited in the bibliography, although I hope the mathematics is intelligible in a stand-alone way and there is no immediate need to access this information, which rather provides a historical pathway to these ideas.

Recall a *phomomorphism* is a zero-fixing mapping between two groups which is a homomorphism modulo the commutator subgroup of the co-domain (see Definition 47 and Sect. 1.14.2). The phomomorphic status of a given mapping will depend upon what we identify as the co-domain, in non-abelian cases. When the co-domain is abelian, phomomorphisms are the standard group homomorphisms.

Notation 353 Suppose $(A, +)$ and $(B, +)$ are two additive groups. There is the canonical group epimorphism:

$$\Pi : B \twoheadrightarrow \frac{B}{[B, B]} \qquad b \mapsto \{b + [B, B]\}$$

Suppose

$$\theta : A \to B$$

R. Lockhart, *The Theory of Near-Rings*, Lecture Notes in Mathematics 2295,
https://doi.org/10.1007/978-3-030-81755-8_9

is a phomomorphism. Then there is an induced group *homomorphism*

$$\bar{\theta} : A \to \frac{B}{[B, B]} \qquad \bar{\theta}(a) = \Pi(\theta(a))$$

Lemma 354

(i) The mapping

$$\Phi : \text{Phom}(A, B) \twoheadrightarrow \text{Hom}\left(A, \frac{B}{[B, B]}\right) \qquad \Phi(\theta) = \bar{\theta} \tag{9.1}$$

is an epimorphism of additive groups.

(ii) Its kernel is the group $T_0(A, [B, B])$.

(iii) In finite cases, $|\text{Phom}(A, B)| = \left|\text{Hom}\left(A, \frac{B}{[B,B]}\right)\right| . |[B, B]|^{|A|-1}$

This lemma is repeatedly used in calculations in Chap. 7 and elsewhere.

The Pseudo-Finite Topology

This mapping Φ may be used to define a topology on Phom(A, B). First, give Hom$\left(A, \frac{B}{[B,B]}\right)$ the subspace topology induced from the finite topology, and then form the coarsest topology on Phom(A, B) such that the mapping Φ is continuous (see page 235).

In this topology,

$$\mathfrak{PC}(\theta, a) = \{f \in \text{T}_0(A, B) \ : \ f(a) \equiv \theta(a)(\textit{modulo}\, [B, B])\} \tag{9.2}$$

constitutes a neighbourhood sub-base for a topology on $\text{T}_0(A, B)$ which induces that coarsest topology on Phom(A, B). I might call it the **pseudo-finite** topology.

I now need some notation that was introduced in Chap. 7.

Lemma 355 *When* $(G, +)$ *is an additive group, there is a sequence of groups and homomorphisms*

$$K \longmapsto \text{E}(G) \xrightarrow{\Phi} \text{E}\left(\frac{G}{[G, G]}\right) \tag{9.3}$$

where $K = \left(\text{T}_0\left(G, [G, G] \bigcap \text{E}(G)\right)\right)$. *In finite cases, the mapping is surjective exactly when* E(G) *is dense in* P(G) *in the pseudo-finite topology.*

Proof Density is the only non-straightforward part. When the mapping is surjective, any image of a phomomorphism identifies (as an inverse image) a mapping in the endomorphism near-ring which establishes density. When density is true, one may find a finite collection of neighbourhoods identify a finite collection of endomorphism near-ring mappings, each, matching the image of our original phomomorphism at a single point. The intersection of this finite collection of neighbourhoods in turn gives a neighbourhood that allows one to identify a single endomorphism mapping (and finiteness of intersections is the key point). □

It would perhaps be worthwhile to further analyse the sequence (9.3) and perhaps to consider dense and closed subsets of phomomorphism near-rings more generally.

9.1 General Theory

Suppose A and B are additive groups. We know from Lemma 306

$$
\begin{aligned}
[\mathrm{Phom}(A, B), \mathrm{Phom}(A, B)] &\subseteq [\mathrm{T}_0(A, B), \mathrm{T}_0(A, B)] \\
&\subseteq \mathrm{Phom}(A, B) \subseteq \mathrm{T}_0(A, B)
\end{aligned}
$$

This means $[\mathrm{Phom}(A, B), \mathrm{Phom}(A, B)] \trianglelefteq [\mathrm{T}_0(A, B), \mathrm{T}_0(A, B)]$ and that

$$\frac{[\mathrm{T}_0(A, B), \mathrm{T}_0(A, B)]}{[\mathrm{Phom}(A, B), \mathrm{Phom}(A, B)]}$$

is abelian. That is,

$$[[\mathrm{T}_0(A, B), \mathrm{T}_0(A, B)], [\mathrm{T}_0(A, B), \mathrm{T}_0(A, B)]] \subseteq [\mathrm{Phom}(A, B), \mathrm{Phom}(A, B)]$$

and from this, we get the next result.

Lemma 356 *Suppose* $(A, +)$ *is an additive group and that* $(\mathrm{Phom}(A), +)$ *is additively solvable. Then both* $(A, +)$ *and* $\mathrm{T}_0(A, +)$ *are additively solvable.*

I characterised the commutator subgroup of $\mathrm{T}_0(A)$ in Lemma 306. Consider $\mathrm{Phom}(A, A) = \mathrm{Phom}(A)$. Suppose $\theta, \phi, \psi \in \mathrm{Phom}(A)$. We have

$$((\theta \circ (\phi + \psi)) - (\theta \circ \phi) - (\theta \circ \psi)) \in \mathrm{T}_0(A, [A, A])$$

Consequently, if $\alpha \in \mathrm{T}_0(A, [A, A])$, then

$$(\theta \circ (\phi + \alpha) - (\theta \circ \phi)) \in \mathrm{T}_0(A, [A, A])$$

I use this notation below together with Definition 86 on page 55.

Lemma 357

(i) $[\mathrm{T}_0(A), \mathrm{T}_0(A)]$ *is an ideal of* $\mathrm{Phom}(A)$.
(ii)

$$\mathcal{A}_{\mathrm{Phom}(A)} \subseteq [\mathrm{T}_0(A), \mathrm{T}_0(A)]$$

(iii)

$$\frac{\mathrm{Phom}(A)}{[\mathrm{T_0}(A), \mathrm{T_0}(A)]}$$

is either trivial or a ring with identity.

Lemma 358 *Suppose A is a free abelian group with basis S, and suppose $\omega : S \to H$ is any mapping of the set S to a group H (which may or may not be abelian). Then there is a unique phomomorphism $\Omega : A \to H$ extending ω.*

Lemma 359 *Suppose A, B and C are additive groups. Then*

$$\mathrm{Phom}(A, B \oplus C) \cong (\mathrm{Phom}(A, B) \oplus \mathrm{Phom}(A, C)) \tag{9.4}$$

Observations 360 Select a perfect group $(C, +)$ (definition—page xviii). Then, $\mathrm{Phom}(X, C) = \mathrm{T_0}(X, C)$ whence

$$\mathrm{Phom}(A \oplus B, C) \cong \mathrm{T_0}(A \oplus B, C)$$

In finite cases, this is a direct sum of $(|A|.|B| - 1)$ copies of C. Now,

$$\mathrm{Phom}(A, C) \oplus \mathrm{Phom}(B, C) = \mathrm{T_0}(A, C) \oplus \mathrm{T_0}(B, C)$$

and this is a direct sum of $(|A| + |B| - 2)$ copies of C.

Lemma 361 *Suppose $\{B_j\}_{j \in J}$ is some indexed collection of additive groups and A is any additive group. Then,*

$$\mathrm{Phom}\left(A, \prod_J B_j\right) \cong \prod_J \mathrm{Phom}\left(A, B_j\right)$$

(the isomorphism being one of additive groups).

This next result is a generalisation of Lemma 489.

Lemma 362 *Suppose A, B and C are three additive groups. Then*

$$\mathrm{Phom}\left(A, \mathrm{T_0}(B, C)\right) \cong \mathrm{T_0}\left(B, \mathrm{Phom}(A, C)\right)$$

(the isomorphism being one of additive groups).

Proof We know from Eq. (7.13)

$$\mathrm{T_0}(B, C) \cong \prod_{b \neq 0} C$$

so, from Lemma 361,

$$\text{Phom}(A, \text{T}_0(B, C)) \cong \prod_{b \neq 0} \text{Phom}(A, C) \cong \text{T}_0(B, \text{Phom}(A, C))$$

□

Corollary 363

$$\text{T}_0(B, \text{Phom}(B)) \cong Phom(B, \text{T}_0(B))$$

Lemma 364 *The set* $\{\theta \in \text{Phom}(G) : \theta[G, G] = \{0\}\}$ *is an ideal of the near-ring* Phom*(G).*

Other ideals occur generally in the phomomorphism near-ring, and there are other epimorphisms involving the homomorphism group. Suppose A is an additive group and $\theta \in \text{Phom}(A)$. I define two maps

$$\hat{\theta} \in Hom\left(\frac{A}{[A, A]}, \frac{A}{[A, A]}\right)$$
$$\hat{\theta}(a + [A, A]) = (\theta(a) + [A, A]) \quad (a \in A)$$

and

$$\theta' : \in \text{T}_0([A, A])$$
$$\theta'(b) = \theta(b) \quad (b \in [A, A])$$

(the restriction of θ to the commutator subgroup).

Lemma 365 *The mapping*

$$\Phi : \text{Phom}(A) \twoheadrightarrow \left((\text{T}_0([A, A]) \oplus Hom\left(\frac{A}{[A, A]}, \frac{A}{[A, A]}\right)\right)$$
$$\theta \mapsto \left(\theta', \hat{\theta}\right)$$

is an epimorphism of near-rings. Its kernel is the ideal of all mappings which take the commutator subgroup to zero and all elements to the commutator subgroup. This has the additive form:

$$K = \prod_{a \notin [A, A]} [A, A]$$

and it is an ideal of the near-ring $\text{T}_0(A, [A, A])$.

We now have a near-ring epimorphism:

$$\text{Phom}(A) \twoheadrightarrow \text{T}_0([A, A]) \tag{9.5}$$

9.1.1 Extending Mappings to Phomomorphisms

Suppose $(G, +)$ is an additive group and $\theta : G \to S$ is any mapping to the set S. There is a free group F_S generated by S. If we factor this out by the normal subgroup D generated by all formal relations, $\delta\theta(a, b) = \theta(b) - \theta(a + b) + \theta(a)$, we obtain an additive group $(H, +) \cong \frac{F_S}{D}$ for which the corresponding induced mapping, $\bar{\theta} : G \to H$, is a homomorphism.

In the case in which our mapping θ takes the group G to another group H, the δ-relations impose additional relations on a presentation of H, and we obtain a factor group $\frac{H}{D}$ and group homomorphism, $\bar{\theta} : G \to \frac{H}{D}$, although of course the co-domain here may well be trivial, as the next example illustrates.

Example 366 Working with the symmetric group of degree n, S_n, define $\theta : S_n \to S_n$, as $\theta(\sigma) = \sigma + (1, 2)$, where (1,2) means the usual transposition. Then, $\delta(\theta)(\sigma, (1, 2)) = (1, 2) + (1, 2) - (\sigma + (1, 2) + (1, 2)) + \sigma + (1, 2) = (1, 2)$. This makes the normal subgroup D one that contains the transposition $(1, 2)$.

9.1.2 Phomomorphism-Invariant Subgroups

Suppose $\theta \in$ Phom(G) and suppose S is a subgroup of G which is invariant under θ. In general, we cannot say whether θ acts phomomorphically on S. Suppose $\theta : G \to H$ is a zero-fixing mapping between groups. The image of θ is θG and will not usually be a group. It additively generates the group $\langle \theta G \rangle$.

Definition 367

(i) D_θ is the normal closure of $(\delta(\theta))\,(G \times G)$ in $\langle \theta G \rangle$. That is,

$$D_\theta = \langle \delta(\theta)(G \times G) \rangle_{\mathcal{N}}^{\langle \theta g \rangle}.$$

(ii) $\mathcal{K}_\theta = \{g \in G : \theta g \in D_\theta\}$ is the **pseudo-kernel** of the mapping θ.
(iii) $\frac{\langle \theta G \rangle}{D_\theta}$ is the **pseudo-image** of the mapping θ.
(iv) The **pseudo-cokernel** of the mapping θ is defined as $\frac{H}{\langle (\theta(G)) \rangle_{\mathcal{N}}^{H}}$.

Lemma 368

(i) $\langle \theta G \rangle = \theta(G) + D_\theta$.
(ii) $\mathcal{K}_\theta \lhd G$.
(iii) $\frac{G}{\mathcal{K}_\theta} \simeq \frac{\langle \theta G \rangle}{D_\theta}$.

It may be worth noting that whenever H is a subset of a group G and $[H, H]$ is the set of all commutators obtained using elements of H, then $[\langle H\rangle, \langle H\rangle] = \langle [H, H]\rangle$. The particular case that is of interest below is $[\langle \theta G\rangle, \langle \theta G\rangle] = \langle [\theta G, \theta G]\rangle$.

Lemma 369

(i) $\theta : G \to H$ *is a phomomorphism precisely when*

$$D_\theta \subseteq [H, H]$$

(ii) $\theta : G \to \langle \theta G\rangle$ *is a phomomorphism precisely when*

$$D_\theta \subseteq [\langle \theta G\rangle, \langle \theta G\rangle]$$

and when that happens, θ *will be a phomomorphism with respect to any co-domain group containing* $\theta(G)$.

(iii) If $\theta : G \to H$ *is a phomomorphism, then*

$$\{\langle \theta(G)\rangle \; - \; \{\theta(G)\}\} \subseteq [H, H]$$

(iv) If $\theta : G \to H$ *is any normalised mapping, then for all* $a, b, \in G$

- $\theta(a + b) = \theta(a) + \theta(b) - \{\delta(\theta)(a, b)\}^{\theta(b)}$
- $\theta(-a) = -(\theta(a)) + \delta(\theta)(-a, a)$
- $\theta(a + b) \equiv \theta(a) + \theta(b)$ *(modulo* D_θ*)*
- $\delta(\theta + \phi)(a, b) = (\delta(\phi)(a, b))^{-\theta(b)} + [-\theta(b), \phi(a)] + \delta(\theta)(a, b)^{\phi(a)}$

9.2 Cohomology Groups

I return to some of the ideas introduced on page 31. Some of this work first appeared in the papers [103, 104], which, so far as I know, were the first papers to discuss phomomorphisms.

Consider the direct sum of $n \geq 1$ copies of an additive group Q and all normalised functions from that group to an additive group G. The set of such mappings forms the additive group $C^n(Q, G)$. They are the **n-dimensional cochains** acting on Q with values in G (compare with Chap. 7). As is standard, I define $C^0(Q, G)$ as $(G, +)$. I use the same notation for subsets $H \subseteq G$, writing $C^n(Q, H)$. In such situations, H is to be assumed to be a non-empty subset of G containing the additive identity, and of course, we may not have additive closure. Normalisation means that $\mathrm{T}_0(Q^n, G)$ and $C^n(Q, G)$ are distinct things when $n \geq 2$. As in Definition (45), there is an operator:

$$\delta : C^n(Q, G) \to C^{(n+1)}(Q, G), \tag{9.6}$$

$$\delta(f) \mapsto \langle f \rangle$$

defined by the rule

$$\begin{aligned}\langle f \rangle (q_0, q_1, \dots, q_n) &= f(q_1, \dots, q_n) - f(q_0 + q_1, q_2, \dots, q_n) \\ +f(q_0, q_1 + q_2, \dots, q_n) &- \dots + (-1)^{n-1} f(q_0, \dots, q_{n-1} + q_n) \\ &+ (-1)^n f(q_0, \dots, q_{n-1})\end{aligned} \tag{9.7}$$

When G is abelian, this is the delta-homomorphism of group cohomology [109].

Phomomorphisms, $f : Q \to G$, are simply maps $f \in C^1(Q, G)$ for which

$$\langle f \rangle \in C^2(Q, [G, G]) \tag{9.8}$$

The reader may wish to compare the next result with Lemma 308.

Lemma 370

(i) Suppose H and K are non-empty subsets of G. Then,

$$C^n(Q, [H, K]) = [C^n(Q, H), C^n(Q, K)]$$

(ii)

$$C^n(Q, [G, G]) = [C^n(Q, G), C^n(Q, G)] \tag{9.9}$$

Proof This is proved using arguments similar to those of Lemma 306 and its relations. □

The following lemma uses Notation 94.

Lemma 371 *Let $(Q, +)$ be an additive group.*
Then, Phom(Q) =

$$\Big\{\theta \in C^1(Q, Q) \ : \ \theta \circ (\alpha + \beta) \equiv (\theta \circ \alpha) + (\theta \circ \beta)$$
$$\Big(Mod\Big[C^1(Q, Q), C^1(Q, Q)\Big]\Big)\Big\} = \mathcal{F}_{C^1(Q,Q)}$$

Proof Choose $f \in$ Phom(Q), and $\psi, \lambda, \in C^1(Q, Q)$. Then, if $q \in Q$,

$$f \circ (\psi + \lambda)(q) = f(\psi(q) + \lambda(q)) \equiv (f(\psi(q)) + f(\lambda(q)) \text{ Mod}[Q, Q]$$

since f is a phomomorphism. This means

$$(f \circ \lambda - f \circ (\lambda + \phi) + f \circ \phi) \in C^1(Q, [Q, Q]) = \Big[C^1(Q, Q), C^1(Q, Q]\Big]$$

and so, $f \in \mathcal{F}_{C^1(Q,Q)}$. Alternatively, suppose $f \in \mathcal{F}_{C^1(Q,Q)}$ and suppose q_1 and $q_2 \in Q$. If neither q_1 nor q_2 are zero, we can find q_0 in Q such that $\theta(q_0) = q_1$ and $\phi(q_0) = q_2$, for some $\theta, \phi, \in C^1(Q, Q)$. We know that $f\phi - f(\theta + \phi) + f\theta \in C^1(Q, [Q, Q])$ and that function must therefore map q_0 into $[Q, Q]$. So, $fq_2 - f(q_1 + q_2) + fq_1 \in [Q, Q]$ and therefore f is a phomomorphism. The case in which one or both of q_1 and q_2 are zero is trivial. □

I make several uses of the mapping Φ in what follows, involving different domain sets and different normal subgroups. It is always related to a canonical mapping onto an additive quotient.

Lemma 372

(i) There is a group epimorphism:

$$\Phi : C^n(Q, G) \overset{C^n}{\to} \left(Q, \frac{G}{[G, G]}\right) \tag{9.10}$$

$$\theta \mapsto \bar{\theta} \tag{9.11}$$

$$\bar{\theta}(\underline{q}) = \{\theta(\underline{q}) + [G, G]\} \tag{9.12}$$

(ii) We have the exact sequence of groups and homomorphisms:

$$C^n(Q, [G, G]) \longmapsto C^n(Q, G) \overset{\Phi}{\twoheadrightarrow} C^n\left(Q, \frac{G}{[G, G]}\right) \tag{9.13}$$

(iii) $\Phi \circ \delta = \delta \circ \Phi$

Lemma 373 *We now have a sequence of groups and phomomorphisms:*

$$\cdots \overset{\delta}{\to} C^{n-1}(Q, G) \overset{\delta}{\to} C^n(Q, G) \overset{\delta}{\to} C^{(n+1)}(Q, G) \overset{\delta}{\to} \cdots \tag{9.14}$$

Proof The square

$$\begin{array}{ccc} C^n(Q, G) & \xrightarrow{\ \delta\ } & C^{(n+1)}(Q, G) \\ \downarrow \Phi & & \downarrow \Phi \\ C^n(Q, \frac{G}{[G,G]}) & \xrightarrow{\ \delta\ } & C^{(n+1)}(Q, \frac{G}{[G,G]}) \end{array}$$

commutes. The vertical arrows are group epimorphisms. The lower horizontal map is a group homomorphism, but the upper one is usually not. That it is a *phomomorphism* follows from the commutativity of the square. □

Lemma 374 $\delta(C^{n-1}(Q, G)) \subseteq \{\theta \in C^n(Q, G) : \langle f \rangle \in C^{(n+1)}(Q, [G, G])\}$.

Lemma 375 *Suppose* $G = H \oplus K$ *(a direct sum of two subgroups).*

Then 9.14 and Lemma (225) suggest the two associated sequences:

$$\cdots \xrightarrow{\delta_1} C^{(n-1)}(Q,H) \xrightarrow{\delta_1} C^n(Q,H) \xrightarrow{\delta_1} C^{(n+1)}(Q,H) \xrightarrow{\delta_1} \cdots$$
$$\cdots \xrightarrow{\delta_2} C^{(n-1)}(Q,K) \xrightarrow{\delta_2} C^n(Q,K) \xrightarrow{\delta_2} C^{(n+1)}(Q,K) \xrightarrow{\delta_2} \cdots$$

where $\delta = \delta_1 \oplus \delta_2$.

9.2.1 Non-abelian Group Cohomology

I offer the following generalisation of the standard abelian situation (given in very many textbooks, such as [93]). In this definition, I utilise notation for subgroup closure and normal closure defined in Notation 1.

Definition 376 In what follows, $n \geq 0$ is an integer:

(i) $Z^n(Q,G) = \{f \in C^n(Q,G) : \langle f \rangle \in C^{(n+1)}(Q,[G,G])$ (**"cocycles"**).
(ii) $b^n(Q,G) = \delta(C^{(n-1)}(Q,G))$ (for $n > 1$), with $b^0(Q,G) = 0$.
(ii) $B^n(Q,G) = \langle \delta(C^{(n-1)}(Q,G)) \rangle$ (for $n \geq 0$); and $B^0(Q,G) = 0$.
(iii) $\Delta^n(Q,G) = \langle b^n(Q,G) \rangle_{\mathcal{N}}^{Z^n(Q,G)}$ (**"coboundaries"**).
(iv) $H^n(Q,G) = \frac{Z^n(Q,G)}{\Delta^n(Q,G)}$ (**"cohomology groups"**).
(v) $\mathfrak{a}^n(Q,G) = \Phi^{-1}\left(\Delta^n\left(Q, \frac{G}{[G,G]}\right)\right)$.

Lemma 377

(i) $[C^n(Q,G), C^n(Q,G)] \lhd Z^n(Q,G) \lhd C^n(Q,G)$
(ii) $Z^1(G,G)$ *is a near-ring and a normal subgroup of* $C^1(G,G)$. *It is Phom(G, G).*
(iii) $\delta(C^n(Q,G)) \subseteq Z^{(n+1)}(Q,G)$.
(iv) $Z^n(Q,G) = \Phi^{-1}\left(Z^n\left(Q, \frac{G}{[G,G]}\right)\right)$.
(v) There is the inclusion chain

$$b^n(Q,G) \subseteq B^n(Q,G) \subseteq \Delta^n(Q,G) \subseteq \mathfrak{a}^n(Q,G) \subseteq Z^n(Q,G) \subseteq C^n(Q,G) \tag{9.15}$$

Observations 378

(i) When G is abelian, $\{H^n(Q,G)\}_{n=0}^{\infty}$ are the standard cohomology groups [92].
(ii) Although $C^n(Q,G) \xrightarrow{\delta} C^{(n+1)}(Q,G)$ is a phomomorphism, there is no reason why $C^n(Q,G) \xrightarrow{\delta} B^{(n+1)}(Q,G)$ should be one. I collect a number of pertinent observations on the behaviour of δ in Chap. 10.

(iii) Lemma 372 generalises. Suppose $H \lhd G$. We have the canonical epimorphism, $\phi : G \to \frac{G}{H}$ which in turn induces the epimorphism $\Phi : C^n(Q, G) \to C^n(Q, \frac{G}{H})$ and once again the square

$$\begin{array}{ccc} C^n(Q, G) & \xrightarrow{\delta} & C^{(n+1)}(Q, G) \\ \downarrow \Phi & & \downarrow \Phi \\ C^n(Q, \frac{G}{H}) & \xrightarrow{\delta} & C^{(n+1)}(Q, \frac{G}{H}) \end{array}$$

commutes, although now neither of the horizontal maps are necessarily homomorphisms.

Lemma 379

(i) The set $b^n(Q, G)$ is not always additively closed (examples of this follow).
(ii) $B^n(Q, G)$ is a subgroup of $\{b^n(Q, G) + [C^n(Q, G), C^n(Q, G)]\}$
(iii) $\{B^n(Q, G) - b^n(Q, G)\} \subseteq [C^n(Q, G), C^n(Q, G)]$

Lemma 380

(i) There is an exact sequence of groups

$$K^n(Q, G) \mapsto H^n(Q, G) \overset{\Phi_H}{\twoheadrightarrow} H^n\left(Q, \frac{G}{[G, G]}\right) \tag{9.16}$$

and

$$K^n(Q, G) = \frac{\mathfrak{a}^n(Q, G)}{\Delta^n(Q, G)} \tag{9.17}$$

(ii)

$$H^n\left(Q, \frac{G}{[G, G]}\right) \cong \frac{Z^n(Q, G)}{\mathfrak{a}^n(Q, G)} \tag{9.18}$$

Proof

(i) I define

$$\Phi_H\left(z + \Delta^n(Q, G)\right) = \left(\Phi(z) + \Delta^n\left(Q, \frac{G}{[G, G]}\right)\right) \tag{9.19}$$

Then,

$$K^n(Q,G) = \left\{ z + \Delta^n(Q,G) \ : \ \Phi(z) \ \in \Delta^n\left(Q, \frac{G}{[G,G]}\right)\right\} \tag{9.20}$$
$$= \frac{\mathfrak{a}^n(Q,G)}{\Delta^n(Q,G)}$$

□

I might remark that (ii) above offers a construction of the well-known cohomology groups from this generalised set-up.

Lemma 381

(i) Suppose that α *is a homomorphism of the additive group* G*. There is an induced group homomorphism:* $A : C^n(Q,G) \to C^n(Q,G)$*,* $(\theta \mapsto \alpha \circ \theta)$*. Thus, we have actions:*

$$\big(hom(G,G) \times C^n(Q,G)\big) \mapsto C^n(Q,G)$$

(ii) $\delta(\alpha \circ \theta) = \alpha(\delta(\theta))$.

(iii) There are associated a sub-actions:

- $(hom(G,G) \times b^n(Q,G)) \mapsto b^n(Q,G)$
- $(hom(G,G) \times Z^n(Q,G)) \mapsto Z^n(Q,G)$

(iv) There is a similar action: $(\mathrm{Phom}(G) \times C^n(Q,G)) \mapsto C^n(Q,G)$ *which makes* $C^n(Q,G)$ *into a left* Phom(G) *module.*

Lemma 382 *Suppose that* σ *is any mapping from* Q *to* Q*. Then there is an induced mapping*

$$B : C^n(Q,G) \to C^n(Q,G) \quad (\theta \mapsto \theta^\sigma)$$

in which

$$\theta^\sigma(q_1, q_2, \ldots, q_n) = \theta(\sigma(q_1), \sigma(q_2), \ldots \sigma(q_n))$$

9.2.2 Cochain Symmetries

We have an action $C^1(G,G) \times C^n(Q,G) \mapsto C^n(Q,G)$

$$(\alpha, \theta) \mapsto (\alpha \cdot \theta)$$
$$\alpha \cdot \theta(\underline{\mathbf{q}}) = \alpha(\theta(\underline{\mathbf{q}}))$$

Remark 383 Assuming $\theta, \phi, \in C^n(Q, G)$ and $\alpha, \beta \in C^1(G, G)$

(i) $\alpha \cdot 0 = 0$
(ii) $(\alpha + \beta) \cdot \theta = (\alpha \cdot \theta + \beta \cdot \theta)$
(iii) $(\alpha \cdot \beta) \cdot \theta = \alpha \cdot (\beta \cdot \theta)$
(iv) Whenever $\alpha \in Z^1(G, G)$,

$$\alpha \cdot (\theta + \phi) \equiv (\alpha \cdot \theta) + (\alpha \cdot \phi) \text{ Mod } [C^n(Q, G), C^n(Q, G)]$$

Lemma 384

(i) $C^n(Q, G)$ is a left $Z^1(G, G)$ module under the above action.
(ii) The action implies $Z^1(G, G) \times Z^n(Q, G) \mapsto Z^n(Q, G)$.

Extensions
Suppose that we have an additive homomorphism:

$$\Theta : (Q, +) \longrightarrow \text{Phom}(G, G)$$

with kernel K and in which the subgroup $\Theta(Q)$ is a sub-near-ring. We have the action of Q on G given by

$$q \cdot g = \Theta(q)(g)$$

This action turns G into a left $\frac{Q}{K}$ module (see (1.34) and Definition 394).

Chapter 10
Specific Examples

I shall now look at how the generalisations offered in the previous chapter differ from the conventional abelian situation and give some particular examples. I continue to use the symbol $\mathbb{Z}_n$ for the cyclic group C_n.

Notation 385 In this section, Q_8 is the quaternion group of order 8 (see (5.2) on page 190). I use two other notations for the elements of this group; the reader may wish to refer to the dictionary (5.3). I shall start by using *multiplicative* notation—involving the elements

$$\{\mp 1, \mp i, \mp j, \mp k\}$$

and go on to use the more appropriate additive notation. Readers are warned that the element represented multiplicatively as -1 will later be represented in Table 10.2 as $+1$ (where $+1$ is represented by 0).

The normalised mappings in $C^1(\mathbb{Z}_3, Q_8)$ are defined by their actions on the non-zero elements of $\mathbb{Z}_3$, which I might write as 1 and 2. Each can be sent independently to each of the 8 elements of the quaternion group, so there are 64 such mappings and $C^1(\mathbb{Z}_3, Q_8)$ has additive structure: $Q_8 \oplus Q_8$. The operator δ sends an element $\theta \in C^1(\mathbb{Z}_3, Q_8)$ to $\delta(\theta) \in C^2(\mathbb{Z}_3, Q_8)$, defined as $\delta(\theta)(a, b) = \theta(b) - \theta(a+b) + \theta(a)$.

Form Equations
Suppose we assign $\theta(i) = Q_i \in Q_8,\ i = 1, 2$. One obtains these *form equations*:

$$\delta(\theta)(1, 1) = Q_1 - Q_2 + Q_1.$$
$$\delta(\theta)(1, 2) = Q_2 + Q_1.$$
$$\delta(\theta)(2, 1) = Q_1 + Q_2.$$
$$\delta(\theta)(2, 2) = Q_2 - Q_1 + Q_2.$$

R. Lockhart, *The Theory of Near-Rings*, Lecture Notes in Mathematics 2295,
https://doi.org/10.1007/978-3-030-81755-8_10

Table 10.1 Addition of two functions

Argument	(1,1)	(1,2)	(2,1)	(2,2)
$\delta(\theta)$	$-j$	$-k$	k	$-i$
$\delta(\phi)$	$-i$	$-j$	j	$-k$
$\delta(\theta)+\delta(\phi)$	$-k$	$-i$	$-i$	$-j$

Table 10.2 The quaternion group of order 8

+	0	1	2	3	4	5	6	7
0	0	1	2	3	4	5	6	7
1	1	0	3	2	6	7	4	5
2	2	3	1	0	5	6	7	4
3	3	2	0	1	7	4	5	6
4	4	6	7	5	1	2	0	3
5	5	7	4	6	3	1	2	0
6	6	4	5	7	0	3	1	2
7	7	5	6	4	2	0	3	1

If θ is the map $1 \mapsto i, 2 \mapsto j$ and if ϕ is the map $1 \mapsto k, 2 \mapsto i$, then Table 10.1 gives the mapping $\delta(\theta)+\delta(\phi)$. Additive closure of $b^2(\mathbb{Z}_3, Q_8)$ would require $\delta(\theta)+\delta(\phi)$ to satisfy the form equations. We would seek elements $Q_1, Q_2 \in Q_8$ such that $Q_1 + Q_2 = -i$, $Q_2 + Q_1 = -i$, $Q_1 - Q_2 + Q_1 = -k$ and $Q_2 - Q_1 + Q_2 = -j$. It is impossible to find such elements in Q_8, so $b^2(\mathbb{Z}_3, Q_8) \neq B^2(\mathbb{Z}_3, Q_8)$.

If $f \in C^2(\mathbb{Z}_3, Q_8)$, then $\delta(f)(a, b, c) = f(b, c) - f(a + b, c) + f(a, b + c) - f(a, b)$. Thus, using the definition of $\theta \in C^1(\mathbb{Z}_3, Q_8)$ that I gave above, we obtain that $\delta^2(\theta)(1, 1, 1) = -1 \in [Q_8, Q_8]$, and so in this case, $\delta^2 \neq 0$.

Lemma 386 *The phomomorphism*

$$\delta^2 : C^1(\mathbb{Z}_3, Q_8) \to C^3(\mathbb{Z}_3, Q_8)$$

is non-trivial.

10.1 The Group $B^2(\mathbb{Z}_3, Q_8)$

One can express the Cayley table for the *additive* group Q_8 as shown in Table 10.2 (see also (5.3) on page 191).

Notation 387 Members of $C^1(\mathbb{Z}_3, Q_8)$ are essentially pairs of elements of Q_8, and the above table identifies a natural octal (base 8) ordering for them: 00, 01, ... 77. Representing the cyclic group of order 3 as the set $\{0, 1, 2\}$, we have that function

Table 10.3 $b^2(\mathbb{Z}_3, \mathrm{Q}_8)$

0	73	146	219	292	365	438	511
513	584	659	730	806	879	948	1021
1027	1098	1168	1241	1335	1404	1445	1518
1538	1611	1681	1752	1845	1918	1959	2028
2054	2124	2205	2263	2336	2427	2481	2538
2567	2637	2718	2772	2866	2920	2979	3065
3076	3150	3231	3285	3361	3450	3504	3563
3589	3663	3740	3798	3891	3945	4002	4088

xy is the normalised function θ which takes 1 to x and 2 to y. The form equations then tell us that

$$\delta(\theta)(1,1) = x - y + x = A$$

$$\delta(\theta)(1,2) = y + x = B$$

$$\delta(\theta)(2,1) = x + y = C$$

$$\delta(\theta)(2,2) = y - x + y = D$$

Members of $C^2(\mathbb{Z}_3, \mathrm{Q}_8)$ may be represented as four-digit octal numbers. The above example is the number ABCD. We can then convert these octal numbers to denary form and list those that appear in the set $b^2(\mathbb{Z}_3, \mathrm{Q}_8)$.

Table 10.3 is the list obtained by direct computation, based on the above table.

To further illustrate, the function $\phi \in C^1(\mathbb{Z}_3, \mathrm{Q}_8)$, which takes 1 to 4 and 2 to 5, is acted on by the δ-operator to produce

$$\delta(\phi)(1,1) = 4 - 5 + 4 = 7$$

$$\delta(\phi)(1,2) = 5 + 4 = 3$$

$$\delta(\phi)(2,1) = 4 + 5 = 2$$

$$\delta(\phi)(2,2) = 5 - 4 + 5 = 6$$

The corresponding member of $C^2(\mathbb{Z}_3, \mathrm{Q}_8)$ has octal number 7326. This is the function identified in denary as 3231 in the table above. The set $b^2(\mathbb{Z}_3, \mathrm{Q}_8)$ has precisely the 64 elements given in the table, which means that the δ-function acts injectively on the 64 elements of $C^1(\mathbb{Z}_3, \mathrm{Q}_8)$. Exactly four members of the set $b^2(\mathbb{Z}_3, \mathrm{Q}_8)$ are found to be commutators. In octal notation, these are the functions: $\mathcal{F} = \{(0000), (1110), (1001), (0111)\}$; and they correspond to the functions numbered 0, 73, 513 and 584 in the table. The commutator subgroup of $C^2(\mathbb{Z}_3, \mathrm{Q}_8)$ is structurally a direct sum of four copies of the cyclic group of order 2 and has order 16. In our octal notation, it is the set of elements of four-digit octal numbers each of whose digits is 0 or 1. $\mathcal{F}$ is itself a subgroup of the commutator subgroup and is isomorphic to the Klein group. We already know that $B^2(\mathbb{Z}_3, Q_8)$ is larger than $b^2(\mathbb{Z}_3, Q_8)$. The question becomes whether or not it generates all the commutators in the group $C^2(\mathbb{Z}_3, Q_8)$; but since that is equivalent to asking whether it is additively closed (see Lemma 379), we already know that it does not.

- The set $\mathcal{F}$ is in $b^2(\mathbb{Z}_3, Q_8)$.
- Function 806 has octal form 6441, function 1335 has octal form 7642, added, they give the function with octal form 2013, and when that is added to function 1538, which has octal form 2003, we obtain the commutator 1011, which therefore lies within $B^2(\mathbb{Z}_3, Q_8)$.
- Adding the commutator 1011 to each element of $\mathcal{F}$ generates the additional commutators $\{(1011), (0101), (0010), (1100)\}$.
- One now deduces that $B^2(\mathbb{Z}_3, Q_8)$ contains the set $\{(0000), (0010), (0101), (0111), (1001), (1011), (1100), (1110)\}$. This is additively generated by the elements $\{(0010), (0101), (1100)\}$ and is a direct product of three cyclic groups of order 2.
- Function 3798 has octal form 6237; added to 1100, this gives 4337, which does not occur in our table. Adding that function to function 3740, which does and which has octal form 4327, we obtain the function 1101. This is a new commutator.
- We conclude that $B^2(\mathbb{Z}_3, Q_8) = b^2(\mathbb{Z}_3, Q_8) + \left[C^2(\mathbb{Z}_3, Q_8), C^2(\mathbb{Z}_3, Q_8)\right]$.
- So, in this case, $\Delta^2(Z_3, Q_8) = B^2(\mathbb{Z}_3, Q_8)$.

$C^2(\mathbb{Z}_3, Q_8)$ is additively a direct product of four copies of Q_8, and abelianised, it is an elementary abelian group which is the direct product of eight copies of the cyclic group of order 2. Determining $B^2(\mathbb{Z}_3, Q_8)$ involves deciding how many of the functions given in Table 10.3 are distinct, modulo the commutator subgroup.

I determined the set $b^3(\mathbb{Z}_3, Q_8)$ by direct computation. These functions are represented as eight-digit octal numbers, which are then converted into denary. We found

$$b^3(\mathbb{Z}_3, Q_8) \bigcap \left[C^3(\mathbb{Z}_3, Q_8), C^3(\mathbb{Z}_3, Q_8)\right]$$

to consist of the following (the top line identifies the function, in denary, the bottom lists the number of occurrences—that is, the number of members of $C^3(\mathbb{Z}_3, Q_8)$ mapping to that particular function under the action of δ) (Table 10.4). This means that a total of 256 members of $C^2(\mathbb{Z}_3, Q_8)$ are mapped to the commutator subgroup of $C^3(\mathbb{Z}_3, Q_8)$ by δ. This comprises the group $Z^2(\mathbb{Z}_3, Q_8)$ and includes

Table 10.4 Commutator members of $b^3(\mathbb{Z}_3, Q_8)$

0	4104	262656	266760	2130505	2134593	2392137	2396225
40	24	24	40	40	24	24	40

the entirety of its subgroup $B^2(\mathbb{Z}_3, \mathrm{Q}_8)$. Analysis of Table 10.3 indicates that such representatives of the independent cosets in

$$\frac{C^2(\mathbb{Z}_3, \mathrm{Q}_8)}{\left[C^2(\mathbb{Z}_3, \mathrm{Q}_8), C^2(\mathbb{Z}_3, \mathrm{Q}_8)\right]}$$

as occur in $B^2(\mathbb{Z}_3, \mathrm{Q}_8)$ are identified by the denary functions:

$$\{146, 1027, 1404, 2336\}$$

which are, in octal form

$$\{(2220), (3002), (4752), (0444)\} \tag{10.1}$$

From this, one concludes that

$$B^2(\mathbb{Z}_3, \mathrm{Q}_8) = \Delta^2(Z_3, \mathrm{Q}_8) = Z^2(\mathbb{Z}_3, \mathrm{Q}_8)$$

and that $H^2(\mathbb{Z}_3, \mathrm{Q}_8)$ is trivial.

Lemma 388 *Neither*

$$C^1(\mathbb{Z}_3, \mathrm{Q}_8) \xrightarrow{\delta} B^2(\mathbb{Z}_3, \mathrm{Q}_8)$$

nor

$$C^1(\mathbb{Z}_3, \mathrm{Q}_8) \xrightarrow{\delta} Z^2(\mathbb{Z}_3, \mathrm{Q}_8)$$

is a phomomorphism.

Proof The two co-domain groups are the same here, so we have only to prove one assertion. We need to determine

$$\left[B^2(\mathbb{Z}_3, \mathrm{Q}_8), B^2(\mathbb{Z}_3, \mathrm{Q}_8)\right]$$

and need only to consider commutators involving the four functions in (10.1).

In octal form, the commutator subgroup is generated by the functions:

$$\{(0, 0, 0, 1), (0, 1, 1, 0), (1, 0, 0, 0)\}$$

It is the elementary abelian 2-group of order 8. Using the functions θ and ϕ defined in Table 10.1, we obtain that $\theta + \phi$ is the map, $1 \mapsto -j$, $2 \mapsto -k$, and hence get Table 10.5.

Table 10.5 Some delta functions

Argument	(1,1)	(1,2)	(2,1)	(2,2)
$\delta(\theta+\phi)$	k	$-i$	i	j
$\delta(\theta)+\delta(\phi)$	$-k$	$-i$	$-i$	$-j$
$\delta(\theta+\phi)-(\delta(\theta)+\delta(\phi))$	-1	$+1$	-1	-1

This says, in octal notation, that

$$\delta(\theta+\phi)-\delta(\phi)-\delta(\theta)=(1,0,1,1)$$

This commutator of $C^2(\mathbb{Z}_3, \mathrm{Q}_8)$ is not a commutator of $B^2(\mathbb{Z}_3, \mathrm{Q}_8)$ (its denary form is 577, and it is not even a member of $b^2(\mathbb{Z}_3, \mathrm{Q}_8)$—see Table 10.3). □

10.2 Phom (Q_8)

I first look at the structure of $\mathrm{Phom}(\mathrm{Q}_8, \mathrm{Q}_8) = \mathrm{Phom}(\mathrm{Q}_8)$. It follows from Lemma 309 that $\mathrm{Phom}(\mathrm{Q}_8)$ has order 2^{11}, being an extension of $C^1(\mathrm{Q}_8, \mathbb{Z}_2)$ by the endomorphism ring of the Klein group. The near-ring contains the commutator subgroup of $C^1(\mathrm{Q}_8, \mathrm{Q}_8)$, which is a direct sum of seven copies of the cyclic group of order 2. The commutator subgroup of Q_8 is its centre; it is fixed by phomomorphisms, so we have that

$$\left[\mathrm{Phom}(\mathrm{Q}_8), \mathrm{Phom}(\mathrm{Q}_8)\right] : [\mathrm{Q}_8, \mathrm{Q}_8] \to \{0\}$$

Using the notation of Table 10.2, we can define two members of $C^1(\mathrm{Q}_8, \mathrm{Q}_8)$ with the recipes

$$\theta(y) = \begin{cases} 2 & \text{if } y \neq 0 \\ 0 & \text{otherwise} \end{cases} \qquad \phi(y) = \begin{cases} 4 & \text{if } y \neq 0 \\ 0 & \text{otherwise} \end{cases}$$

Then, for $y \neq 0$:

$$[\theta, \phi](y) = -\theta(y) - \phi(y) + \theta(y) + \phi(y) = 3+6+2+4 = 1$$

The relation holds when y is 1, the non-zero element of $[\mathrm{Q}_8, \mathrm{Q}_8]$, and we conclude that the commutator subgroup of $\mathrm{Phom}(\mathrm{Q}_8, \mathrm{Q}_8)$ lies *properly* inside $\left[C^1(\mathrm{Q}_8, \mathrm{Q}_8), C^1(\mathrm{Q}_8, \mathrm{Q}_8)\right]$, that is,

$$\left[\mathrm{Phom}(\mathrm{Q}_8, \mathrm{Q}_8),\ \mathrm{Phom}(\mathrm{Q}_8, \mathrm{Q}_8)\right] \subsetneq \left[C^1(\mathrm{Q}_8, \mathrm{Q}_8), C^1(\mathrm{Q}_8, \mathrm{Q}_8)\right]$$

Using the octal representation developed in (387) (page 356), $[\theta, \phi]$ can be identified as 1111111. The commutator subgroup of $C^1(\mathrm{Q}_8, \mathrm{Q}_8)$ will involve all functions

Table 10.6 Two automorphisms of Q_8

Function	0	1	2	3	4	5	6	7
θ	0	1	2	3	5	6	7	4
ϕ	0	1	5	7	2	4	3	6
$[\theta, \phi]$	0	0	1	1	1	0	1	0

Table 10.7 Two phomomorphisms of Q_8

Function	0	1	2	3	4	5	6	7
θ	0	1	2	2	2	0	2	0
ϕ	0	1	0	0	5	5	5	5
$[\theta, \phi]$	0	0	0	0	1	0	1	0

with octal representation $a_1a_2a_3a_4a_5a_6a_7$, where $a_j \in \{0, 1\}$; and we have just seen that the commutator subgroup of Phom(Q_8, Q_8) must involve commutators for which $a_1 = 0$. The commutator subgroup of Q_8 is $\{0, 1\}$, and its cosets are

$$\{0, 1\}, \{2, 3\}, \{4, 6\}, \{5, 7\}$$

Phomomorphisms must permute these cosets. If θ and ϕ are phomomorphisms, then the phomomorphism $[\theta, \phi]$ maps all members of a coset to the same thing. This implies that member of the commutator subgroup of Phom(Q_8) has the octal form 0AABCBC where A and B are 0 or 1 and gives its maximum size as 8 elements.

Table 10.6 shows two functions which are automorphisms of (Q_8, +).

Generally, one may specify endomorphisms of (Q_8, +) by extrapolating linearly from their effect on two independent elements, such as the i and j, defined in (385) (these being the elements 2 and 4 in Table 10.2). Group automorphisms are specified with the additional restriction that the image of i should be in a different coset of the commutator subgroup from the image of j and that neither i nor j may map to the commutator subgroup itself. This gives six possible images for i and further four possible images for j, which means the automorphism group has order 24.[1] From Table 10.6, the function 0111010 will be a member of the commutator subgroup of Phom(Q_8). If we select α, an automorphism of Q_8, then $[\theta, \phi]\alpha = [\theta \circ \alpha, \phi \circ \alpha]$ will be another member of $\left[\text{Phom}(Q_8), \text{Phom}(Q_8)\right]$, which permits me to claim 0110101 and 0001111 as further members of $\left[\text{Phom}(Q_8), \text{Phom}(Q_8)\right]$.

Table 10.7 shows two phomomorphisms of Q_8.

Interestingly, any endomorphism of Q_8 with a non-trivial kernel would map the whole group into $\{0, 1\}$, which perhaps indicates the extra information content of phomomorphisms. Table 10.7 shows 0001010 lies in the commutator subgroup, from which it follows that 0110000 and 0000101 do too. We have now determined that that commutator subgroup contains the seven elements

$$\{0AABCBC : A, B, C \in \{0, 1\} \text{ and } A.B.C = 0\}$$

[1] It is in fact isomorphic to S_4.

Table 10.8 Two more automorphisms

Function	0	1	2	3	4	5	6	7
θ	0	1	2	3	5	6	7	4
ϕ	0	1	5	7	4	3	6	2
$[\theta, \phi]$	0	0	1	1	1	1	1	1

Table 10.9 Equivalence classes of phomomorphism

Function	2	4
θ_1	2	4
θ_2	2	5
θ_3	4	5
θ_4	4	2
θ_5	5	2
θ_6	5	4
θ_7	2	2
θ_8	2	1
θ_9	1	5
θ_{10}	4	4
θ_{11}	4	1
θ_{12}	1	4
θ_{13}	1	2
θ_{14}	5	1
θ_{15}	5	5
θ_{16}	1	1

Table 10.8 gives a further pair of automorphisms.
which permit us to add 0111111 to our list of commutators.

Lemma 389

(i) Phom(Q_8) *has order* 2^{11}.
(ii) It contains the commutator subgroup of $C^1(Q_8, Q_8)$ *which means the* 2^7 *functions* $a_1a_2a_3a_4a_5a_6a_7$, *where* $a_j \in \{0, 1\}$.
(iii) And its own commutator subgroup is the subgroup of eight functions of the form $0AABCBC$ *where* $A, B, C \in \{0, 1\}$.

I am interested in phomomorphisms that do not map the whole of Q_8 to its commutator subgroup; the non-bijective *endomorphisms* must do this. Phomomorphisms permute the cosets of the commutator subgroup. One may define an equivalence relation on them based on the notation of Lemma (309) in which phomomorphisms are deemed equivalent provided they are in the pre-image (fibre) of the same element of Hom $\left(\frac{Q}{[Q,Q]}, \frac{Q}{[Q,Q]}\right)$ under the action of Φ. This permits us to distinguish these equivalence classes of phomomorphisms, which we identify by giving the values of a representative on the elements 2 and 4 (i and j) (Table 10.9).

Functions $\theta_1 - \theta_6$ are the equivalence classes of all of the automorphisms of Q_8. $\theta_7 - \theta_{15}$ are the non-bijective phomomorphisms, with θ_{16} the class of the trivial phomomorphisms. These mappings represent the 2^4 elements of

$$\frac{\mathrm{Phom}(Q_8)}{\left[C^1(Q_8, Q_8), C^1(Q_8, Q_8)\right]}$$

One may choose a phomomorphism, ϕ, in the equivalence class θ_{11} such that ϕ maps

$$2 \mapsto 4$$
$$3 \mapsto 4$$
$$4 \mapsto 1$$

and one can choose another phomomorphism, ψ, from the class of θ_{13} such that

$$2 \mapsto 1$$
$$3 \mapsto 0$$
$$4 \mapsto 2$$

So, $((\theta \circ (\phi + \psi)) - (\theta \circ \psi) - (\theta \circ \phi))$ maps $2 \mapsto (\theta(6) - \theta(1) - \theta(4))$ and maps $3 \mapsto \theta(4) - \theta(4) = 0$. Since we can choose a phomomorphism θ such that $\theta(6) - \theta(1) - \theta(4) = 1$, the expression $((\theta \circ (\phi + \psi)) - (\theta \circ \psi) - (\theta \circ \phi))$ cannot always represent a member of $\left[\mathrm{Phom}(Q_8), \mathrm{Phom}(Q_8)\right]$ (see Lemma 389). This gives us

Lemma 390 *Phom*(Q_8, Q_8) $=$ *Phom*(Q_8) *is not an F-near-ring (see Definition 75).*

10.3 Phom (S_3)

The phenomenon described in Lemma 390 appears to be the norm, and it is easy to find other examples of it. One involves the symmetric group: $S_3 = D_3$.

$$S_3 = \langle a, b : 3a = 2b = b + a + b + a = 0\rangle$$

(see (7.121)).

This has commutator subgroup $\langle a\rangle$, which is isomorphic to the cyclic group of order 3. Phomomorphisms of S_3 will fix this subgroup. Mappings in

[Phom(S_3), Phom(S_3)] must annihilate it. Define three phomomorphisms of S_3 as follows:

$$\theta(x) = x$$

$$\phi(x) = x + x$$

$$\psi(x) = \begin{cases} a & \text{if } x = 2a \\ 0 & \text{otherwise} \end{cases}$$

All three mappings are in Phom(S_3, S_3). Then,

$$(\psi \circ ((\theta + \phi) - (\psi \circ \phi) - (\psi \circ \theta)))\,(a) = \psi(a + 2a) - \psi(2a) - \psi(a) = 0 + a + 0 = a \neq 0$$

so Phom(S_3, S_3) is not an F-near-ring.

10.4 Phom (D_4)

D_4 has the presentation given by (5.4). Its phomomorphism group has the same order as the phomomorphism group of Q_8, 2^{11}. I shall determine [Phom(D_4), Phom(D_4)] and thence show that

$$\text{Phom}(D_4) \ncong \text{Phom}(Q_8)$$

Extending the ideas reported in Sect. 2.1.5, one can utilise the presentation to determine the phomomorphisms, with construction conditions involving congruences rather than equalities. We know that $[D_4, D_4] = \langle 2a \rangle$, and it will be seen that the conditions impose no restrictions on the images of either a or b under a phomomorphism—permitting us 8×8 choices. The remaining image values can be obtained directly by linearity (these involving the elements $\{3a, 2a, (a + b), (2a + b), (3a + b)\}$) except that in all cases we may add on an additional 2a term at will. This process accounts for $8 \times 8 \times 2 \times 2 \times 2 \times 2 \times 2 = 2^{11}$ phomomorphisms. To illustrate, we define $\theta(a) = b, \theta(b) = 3a$. The three defining relations we are working with give us

$$4\theta(a) = 4b = 0$$

$$2\theta(b) = 2a \equiv 0 \text{ (modulo } [D_4, D_4])$$

$$2(\theta(a) + \theta(b)) = 2(b + 3a) = 0$$

so extending θ linearly will result in a phomomorphism (and as I have observed, any image choice for a and b works). Extending linearly gives us a phomomorphism

Table 10.10 Thirty-two distinct phomomorphisms of D_4

Image	Linear extension	Extension plus addition
$\theta(2a)$	0	$2a$
$\theta(3a)$	b	$(2a+b)$
$\theta(a+b)$	$(3a+b)$	$(a+b)$
$\theta(2a+b)$	a	$3a$
$\theta(3a+b)$	$(3a+b)$	$(a+b)$

Table 10.11 The basic phomomorphisms of D_4

x	$\theta_1(x)$	$\theta_2(x)$	$\theta_3(x)$	$\theta_4(x)$	$\theta_5(x)$	$\theta_6(x)$	$\theta_7(x)$	$\theta_8(x)$
a	a	a	a	b	b	b	0	0
2a	0	0	0	0	0	0	0	0
3a	a	a	a	b	b	b	0	0
b	b	a	0	a	b	0	a	b
$(a+b)$	$(a+b)$	0	a	$(a+b)$	0	b	a	b
$(2a+b)$	b	a	0	a	b	0	a	b
$(3a+b)$	$(a+b)$	0	a	$(a+b)$	0	b	a	b

Table 10.12 Commutators of phomomorphisms

	$[\theta_1,\theta_2]$	$[\theta_1,\theta_3]$	$[\theta_1,\theta_4]$	$[\theta_1,\theta_5]$	$[\theta_1,\theta_6]$	$[\theta_1,\theta_7]$	$[\theta_1,\theta_8]$	$[\theta_2,\theta_6]$
a	0	0	$2a$	$2a$	$2a$	0	0	$2a$
$2a$	0	0	0	0	0	0	0	0
$3a$	0	0	$2a$	$2a$	$2a$	0	0	$2a$
b	$2a$	0	$2a$	0	0	$2a$	0	0
$(a+b)$	0	$2a$	0	0	$2a$	$2a$	$2a$	0
$(2a+b)$	$2a$	0	$2a$	0	0	$2a$	0	0
$(3a+b)$	0	$2a$	0	0	$2a$	$2a$	0	$2a$

shown as the second column in the table below; the third column indicates the other permitted value (obtained simply by adding on 2a).

Table 10.10 represents 32 distinct phomomorphisms and corresponds to the basic phomomorphism θ_4 given in Table 10.11.

Table 10.11 lists the eight basic phomomorphisms of D_4 that are not kernel elements in the exact sequence (9.1). Further phomomorphisms of this sort are obtained by adding terms 2a to the images of non-zero elements in any way at all; but since such terms are in the centre of the additive group, they are irrelevant with respect to investigating commutators of phomomorphisms.

These basic phomomorphisms can be used to generate the key mappings in $[\mathrm{Phom}(D_4), \mathrm{Phom}(D_4)]$—that is, the ones which are not in $T_0(D_4, [D_4, D_4])$. There are 56 commutators to check; the distinct ones are reported in Table 10.12.

Commutators of phomomorphisms are sums of these basic commutators. If I represent mappings as tuples with the form

$$(\theta(a), \theta(2a), \theta(3a), \theta(b), \theta(a+b), \theta(2a+b), \theta(3a+b))$$

then these commutators have the form

$$(A, 0, A, B, C, B, D)$$

where each letter is a member of the set $\{0, 2a\}$. There are $2 \times 2 \times 2 \times 2 = 16$ such mappings and so, additively,

$$[\mathrm{Phom}(\mathrm{D}_4), \mathrm{Phom}(\mathrm{D}_4)] \cong \mathrm{C}_2 \oplus \mathrm{C}_2 \oplus \mathrm{C}_2 \oplus \mathrm{C}_2$$

Lemma 391

(i) The near-ring $[\mathrm{Phom}(\mathrm{D}_4), \mathrm{Phom}(\mathrm{D}_4)]$ *is additively isomorphic to the elementary abelian group of order 16.*

(ii)

$$\left[\mathrm{Phom}(\mathrm{Q}_8), \mathrm{Phom}(\mathrm{Q}_8)\right] \ncong [\mathrm{Phom}(\mathrm{D}_4), \mathrm{Phom}(\mathrm{D}_4)]$$

(iii) Both commutator subgroups are sub-near-rings in which the product is trivial (Definitions 6).

Observations 392 One wonders what group symmetries might guarantee that two groups host the same near-rings or that they have the same phomomorphism groups. There are various possibilities, for instance, isoclinism, invented by Philip Hall and used in the theory of p-groups, or any variations on that idea. Now, Q_8 and D_4 are very similar groups. They are both 2-groups; they are *isoclinic* (see [14]); and when each is factored by its commutator subgroup, one obtains isomorphic abelian groups. The near-rings they host are quite distinct (Lemmata 207 and 209), as are their phomomorphism groups.

10.5 Phom ($\mathbb{Z}_2$, [A_4, A_4])

A_4 is the alternating group of degree four. Its commutator subgroup is a direct sum of two copies of the cyclic group of order 2, $\mathbb{Z}_2$. Phomomorphisms from the $\mathbb{Z}_2$ to A_4 must map the additive generator to an element with order 2 and consequently map into $[A_4, A_4] \cong (\mathbb{Z}_2 \oplus \mathbb{Z}_2)$.

Lemma 393

(i) $Z^1(\mathbb{Z}_2, A_4) = \mathit{Phom}(\mathbb{Z}_2, A_4) \cong \mathbb{Z}_2 \oplus \mathbb{Z}_2$.
(ii) $\mathit{Phom}(\mathbb{Z}_2, [A_4, A_4]) = \mathit{Hom}(\mathbb{Z}_2, \mathbb{Z}_2 \oplus \mathbb{Z}_2) \cong \mathbb{Z}_2 \oplus \mathbb{Z}_2$.
(iii) $\mathit{Phom}(\mathbb{Z}_2, [A_4, A_4]) \ncong [\mathit{Phom}(\mathbb{Z}_2, A_4), \mathit{Phom}(\mathbb{Z}_2, A_4)]$. *(Contrast this with (9.9).)*

Part IV
Some Traditional Constructions

Chapter 11
Modules

All near-rings in this chapter are right-distributive and zero-symmetric.

Although my chief interest is with finite structures, in this chapter,

Near-rings occurring should *not* automatically be assumed to be finite.

11.1 Introduction

The natural response to investigating near-rings is to attempt direct generalisations of important ideas in ring theory; and this is the motivation for the topics forming the next three chapters, which largely report the work of other mathematicians.

Much interesting work has been done on this, and it is a pity that it is not better known. These generalisations often split into rather more or less technical constructions.

Sometimes this complexity isolates individual constructs which are blurred by the more abelian features of standard ring theory. One might hope that this, in itself, might permit a more complete understanding of these ideas.

Sometimes this complexity is augmented by the aridity of mathematical expression—the standard lexicon of mathematical terms being over-exploited or used in multifarious and conflicting ways, diminishing intelligibility without properly reflecting difficulties that really are present—a recurrent metastasising problem for those attempting generalisations. I suspect that any postgraduate

R. Lockhart, *The Theory of Near-Rings*, Lecture Notes in Mathematics 2295,
https://doi.org/10.1007/978-3-030-81755-8_11

student of this, and related areas, will know what I mean, unless I am alone in spending long hours floundering over inconsistently expressed concepts.[1]

In the next chapters, I skim over some of the area where generalisations of this sort have been attempted, beginning with the most obvious of structures—the module.

I do need some results from Sects. 7.6 and 7.7 of Chap. 7 here, and this is particularly the case for the sections on primitive near-rings. Importantly, I need Definition 270 and the notation $\mathfrak{M}_S(\Gamma)$, although, here, I use G rather than S.

There is quite a lot about ring theory in the next two chapters, partly because it makes near-ring generalisations more accessible and partly because it is, after all, a suburb of near-rings! It is probable that the majority of texts on ring theory assume unital rings and that can have a big effect on the results that one can obtain. One has to be careful about what is being assumed when quoting from these texts, and I do hope I have committed no howlers. It is so often hard to determine what is and is not being assumed when consulting a mathematical text as an incidental reference.

11.2 Basics

In this chapter, $(R, +, \cdot)$ is a zero-symmetric, right-distributive near-ring, and $(M, +)$ is an additive group. Phom$(M, +)$ is the standard (right) near-ring of zero-fixing phomomorphisms (page 31). Recall from page 51 that $(R, +, \cdot)^{opp} = (R, +, :)$. The reader may wish to have a look at Sect. 1.32, page 65, and Definition 20, before continuing.

Definition 394

(i) A **left R-module** is a near-ring homomorphism

$$\theta(R, +, \cdot) \longrightarrow \text{Phom}(M, +).$$

(i) A **right R-module** is a near-ring homomorphism

$$\phi(R, +, \cdot)^{opp} \longrightarrow \text{Phom}(M, +).$$

On the face of it, my terminology here commits the same category error I complained of on page 17. Perhaps it would be better to speak of left and right **near-ring modules**, but in this particular work, identifying the near-ring is of some importance, so I leave things as they are. Left R-modules can be represented with

[1] Regrettably, my own notation for annihilators is a case in point—what may, perhaps, work on the printed page is wholly inappropriate for the rough pencil-and-paper calculations that potential readers would have to undertake—mea culpa! Incidentally, I am *not* advocating a Bourbaki-styled police service! .

the near-ring R acting directly on the group. Thus, rm means $\theta(r)(m)$. On the other hand,

$$\phi(r_2) \circ \phi(r_1) \ = \ \phi(r_2 : r_1) \ = \ \phi(r_1 \cdot r_2)$$

and this kind of right module action would give us

$$(r_1 \cdot r_2)\, m \ = \ (r_2(r_1 m))$$

As is standard, in this situation, we write R-action on the *right* of the group—so, mr means $\phi(r)(m)$; and we then have that $m(r_2 \cdot r_1) = m(r_1 : r_2) = (\phi(r_1) \circ (\phi(r_2))\,(m) = (mr_2)r_1$—which is *right* module action. Phom$(M, +)$ is a right-distributive near-ring (see Definition (47)). Consequently, $\phi(R)$ is a *bi-distributive* near-ring in the right-module case (see Definition (87)) and a sub-near-ring of Phom$(M, +)$. If $(R, + \cdot)$ has a multiplicative identity, $\phi(R)$ will be a ring (Corollary 89). I should prefer to represent module action without recourse to an action symbol rm rather than $r \cdot m$ (see page 8), because I always explicitly write in a symbol for near-ring products; but this policy is sometimes just confusing (I would have to write $r0$ below, for $r \cdot 0$, for instance) and I deviate from it, citing supposed clarity.

Observations 395 Left R-module action has the following properties.

(i) $r \cdot 0 = 0$ for all $r \in R$.
(ii) $(r_1 + r_2) \cdot m = r_1 \cdot m + r_2 \cdot m$ for all $r_1, r_2, \in R$ and all $m \in M$.
(iii) $(r_1 \cdot r_2) \cdot m = r_1 \cdot (r_2 \cdot m)$ for all $r_1, r_2, \in R$ and all $m \in M$.
(iv) For all $r \in R$ and all $m_1, m_2 \in M$

$$r \cdot (m_1 + m_2) \equiv (r_1 \cdot m) + (r_2 \cdot m) \text{ (modulo ([M,M]))} \tag{11.1}$$

Observations 396
Right R-module action has the following properties.

(i) $0 \cdot r = 0$ for all $r \in R$.
(ii) For all $m \in M$ and all $r_1, r_2 \in R$

$$m \cdot (r_1 + r_2) \ = \ (m \cdot r_1) + (m \cdot r_2) \tag{11.2}$$

(iii) $m \cdot (r_1 \cdot r_2) = (m \cdot r_1) \cdot r_2$ for all $m \in M$ and all $r_1, r_2 \in R$.
1. For all $r \in R$ and all $m_1, m_2 \in M$

$$(m_1 + m_2) \cdot r \equiv (m_1 \cdot r) + (m_2 \cdot r) \text{ (modulo ([M,M]))} \tag{11.3}$$

(iv) $m \cdot [rs, kl] = 0$ for all $r, s, l \in R$ and for all $m \in M$. This follows from the properties of bi-distributive near-rings (Definition (87)).

Definition 397 A (right) near-ring is said to be **right representable** provided it has a non-trivial right module.

Right representable near-rings must have proper distributor ideals (Sect. 1.25). Clearly, not all near-rings are right representable.

Remark 398 Laxton's left modules, [98], replaced (11.1) with the requirement that the set S of distributive generators should distribute over the module—that is,

$$s \cdot (m_1 + m_2) = ((s \cdot m_1) + (s \cdot m_2)) \ \ for\ s \in S,\ and\ m_1, m_2 \in M \qquad (11.4)$$

I believe my generalisation enhances, rather than dilutes, the force of his results.

(i) Left modules are always left near-ring groups. Right modules do not quite satisfy the required distributivity properties ((11.3) is only a congruence, whereas (1.16) is an equality).
(ii) The additive commutator subgroup of left modules is a left near-ring group.
(iii) When $(\Gamma, +)$ is abelian, left modules are modules in the same sense as the term is used in ring theory, although the near-ring itself might not be a ring.
(iv) N is the left near-ring group ${}_N N$. Its left ideals are precisely its left mideals.
(v) A near-ring acting faithfully upon a left module $(\Gamma, +)$ is a sub-near-ring of $\mathfrak{Ph}_S(\Gamma)$ (Corollary 28).
(vi) Right modules actually have stronger left-distributivity properties than right near-ring groups (see 11.2).
(vii) In general near-rings, R itself will be neither a left nor a right R-module.
(viii) When R is an *F-near-ring*, it will be a left R-module in the usual way; and when R is a left-distributive, d.g. near-ring, it is a right R-module (the mapping ϕ defined as $\phi(r)x \to x \cdot r$).
(ix) Any group $(G, +)$ is an additive left $\mathbb{Z}$-module, where $\mathbb{Z}$ represents the ring of integers and the obvious action $m \cdot g = g + g + \ldots + g$ (m times). Groups are also additive right $\mathbb{Z}$-modules.
(x) These definitions of right and left modules go over directly to the ring case because all additive groups there are abelian and the congruences become equalities.

Henceforth, when the near-ring is unital, all near-ring groups will be unitary.

In point of fact, many of the near-ring groups I deal with here are often automatically unitary so this condition is less of a restriction than one might think (see Lemma 30).

The next result relies on Lemma 24.

Lemma 399 *Suppose ${}_N\Gamma$ is a left module and Ω is a mideal contained in Γ. Then the near-ring group $\left(\frac{\Gamma}{\Omega}\right)$ is actually a left module.*

Proof The structure is a left near-ring group. We need to consider Observations 395, (11.1).

What is in question is the relation

$$n \cdot (\{\gamma_1 + \Omega\} + \{\gamma_2 + \Omega\}) \;=\; \{(n \cdot \gamma_1 + n \cdot \gamma_2) + \delta + \Omega\}$$

We need it to be the case that $\{\delta + \Omega\} \in \left[\frac{\Gamma}{\Omega}, \frac{\Gamma}{\Omega}\right]$. Since Γ is a left module, we can say $\delta \in [\Gamma, \Gamma]$. □

Change of Near-Rings

If M is a left S-module and $\phi : R \mapsto S$ is a near-ring phomomorphism (one which fixes the identity, if there is one), then the definition

$$(r \cdot m) \;=\; (\phi(r)) \cdot m \tag{11.5}$$

defines a left R-module action on M, just as it does in ring theory.

11.2.1 Sub-structures

I refer back to the definitions of sub-near-ring groups and mideals given in Definition 22. If we are dealing with modules, I shall refer to sub-near-ring groups as **submodules**.

On that basis, observe that a submodule might not actually be a module in its own right and that a normal subgroup satisfying (1.21) will certainly be a submodule.

The obvious modification apply to *right* R-modules, where one might define (right) submodules and mideals.

In standard ring theory, rings are modules over themselves. Additive subgroups are always additively normal, and ideals have definitions which are left and right symmetric. None of this is true for near-rings so we can expect further layers of complexity in attempted generalisations, as has already been seen in my discussion of ideals and sideals (page 16).

Bookkeeping Lemmata

I now revisit the results on modules over rings given on page 96.

Theorem 400 *Let $_RM$ be a left near-ring group and $_RN$ be a left near-ring group inside it.*

(i) If K is any mideal inside M, then

$$\frac{(N + K)}{K} \cong \frac{N}{(N \cap K)} \tag{11.6}$$

This is an isomorphism of near-ring groups.

(ii) If $K \subseteq L$ are both mideals inside M, then $\frac{L}{K}$ is a mideal inside the near-ring group $\frac{L}{K}$, and

$$\frac{(\frac{M}{K})}{(\frac{L}{K})} \cong \frac{M}{L} \tag{11.7}$$

(another isomorphism of near-ring groups).

(iii) If K is a mideal in M and L is a sub-near-ring group, then

$$(L \cap (K + N)) = (K + (L \cap N)) \tag{11.8}$$

Technical Lemmata

This lemma appears in [11] and was, I think, originally due to Wielandt. I have always found it to be quite brilliant and utterly baffling.

Lemma 401 *Suppose M is a left near-ring module ${}_N M$ and $A, BandC$ are mideals.*

Then

$$\Gamma = \frac{((A + C) \cap (B + C))}{((A \cap B) + C)} \tag{11.9}$$

is itself a left near-ring module, and for each $n \in N$, the mapping

$$\Gamma \mapsto \Gamma \quad \gamma \mapsto n \cdot \gamma \tag{11.10}$$

is an additive group endomorphism.

Proof Lemma 4 gives us that the quotient group is abelian. The results on page 18 do the rest—notice that the group in question is abelian so pseudo-homomorphic action is the same as homomorphic action. □

Corollary 402 *If M, in the previous lemma, is simply a left near-ring group, then Γ is one too. It is abelian, and the mapping (11.10) is still an additive group endomorphism.*

Proof The tricky part is showing that (11.10) actually is an endomorphism and I simply reproduce Betsch's proof.

Write $E = ((A \cap B) + C)$ and $H = ((A + C) \cap (B + C))$.

Take $n \in N$ and $h_1, h_2 \in H$. We can find $\alpha \in A, \beta \in B$ such that

$$h_1 \equiv \alpha \text{ (modulo } E)$$

$$h_2 \equiv \beta \text{ (modulo } E)$$

Now,

$$n \cdot (\alpha + \beta) \equiv (n \cdot \beta) \equiv (n \cdot \alpha + n \cdot \beta) \ (modulo\ A)$$

and, similarly,

$$n \cdot (\alpha + \beta) \equiv (n \cdot \alpha + n \cdot \beta) \ (modulo\ B)$$

So, both congruences are true modulo $(A \cup B)$ and, indeed, modulo E. □

Corollary 403 *With the notation and assumptions made in either Lemma 401 or Corollary 402,*

$$\left(\frac{N}{\begin{pmatrix} \Gamma \\ 0 \end{pmatrix}_{\mathcal{L}}} \right) \tag{11.11}$$

is a ring.

Proof First, note the left annihilator of Γ is a (two-sided) ideal. The quotient near-ring acts faithfully upon it, and Γ is abelian. Apply Lemma 27. □

The next result requires the idea of a distributive lattice (page 95). The lattice of left ideals described here has "meet" as intersection and "join" as simple addition.

Lemma 404 *Suppose $(N, +, \cdot)$ is a unital near-ring for which no non-zero epimorphic image is a ring. Then, the lattice of left ideals is distributive.*

Proof Take left ideals $A, BandC$ and form Γ as in (11.9). If $\Gamma \neq \{0\}$, its left annihilator is a proper ideal of $(N, +, \cdot)$ because the identity is not involved. The corresponding quotient is a ring from Corollary 403, so must be trivial. This means

$$((A + C) \cap (B + C)) = ((A \cap B) + C)$$

and the lattice of left ideals is distributive. □

Lemma 405 *Suppose ${}_N\Gamma$ is a near-ring group which is monogenic and $\Gamma = \{N \cdot \gamma\}$. Suppose the near-ring acts faithfully on Γ.*

Suppose $BandC$ are left ideals of N and both

$$\left(B + \begin{pmatrix} \gamma \\ 0 \end{pmatrix}_{\mathcal{L}}\right) = \left(C + \begin{pmatrix} \gamma \\ 0 \end{pmatrix}_{\mathcal{L}}\right) = N$$

$$and\ (B \cap C) \subseteq \begin{pmatrix} \gamma \\ 0 \end{pmatrix}_{\mathcal{L}}$$

Then, N is a ring.

Proof The left near-ring group

$$\Omega = \frac{((B + \binom{\gamma}{0}_{\mathcal{L}}) \bigcap (C + \binom{\gamma}{0}_{\mathcal{L}}))}{((B \cap C) + \binom{\gamma}{0}_{\mathcal{L}})} \cong \frac{N}{\binom{\gamma}{0}_{\mathcal{L}}}$$

Ω is N-isomorphic to Γ, by Lemma 51. It is abelian from Lemma 4. From Lemma 401, N induces endomorphisms on Ω and so Γ. From Lemma 27, N must be a ring. □

11.3 Primitivity: Laxton's Paper

The basic reference for this section is the paper by Laxton [99]. This was a hugely influential paper, but it restricted itself to d.g. near-rings, possibly because it was thought they might form the bedrock for a generalisation of homological algebra [42, 119]. In other words, Laxton was not, at that time, interested in general near-rings per se.

However, his results were immediately extended to more general near-rings by work that partially superseded them. I suspect their importance has become largely unrecognised.

This definition is a *modified* form of one given by Laxton [99] and not entirely equivalent to his work, being somewhat more general in its conception of a module. The extensions that these days form the basis for ideas of primitivity are covered in Sect. 11.4.

Definition 406 Suppose M is a left near-ring module, ${}_N M$, and $N \cdot M \neq \{0\}$.

(i) The module is held to be **irreducible** provided it has no proper subgroups which are near-ring groups under the action of N.
(ii) **G** is defined as the set of all near-ring group homomorphisms of M to itself.

$$G = \{\theta \in \text{Hom}(M, M) : \theta(n \cdot m) = n \cdot \theta(m) \; \forall \, n \in N, \forall \, m \in M\} \tag{11.12}$$

(iii) It is often convenient to show elements of G acting on the *right*. When I do this, I tend not to utilise Greek letters, preferring to talk about $g \in G$. The above definition is then

$$G = \{g \in \text{Hom}(M, M) : (n \cdot m) \cdot g = n \cdot (m \cdot g) \; \forall \, n \in N, \forall \, m \in M\} \tag{11.13}$$

G is a multiplicative semi-group, contained in $T_0(M)$.

The condition $N \cdot M \neq \{0\}$ implies there is some $\gamma \in M^*$ such that $\{N \cdot \gamma\}$ is a non-trivial, near-ring group, contained in M. This is a monogenic near-ring group.

If the original module were irreducible, it would mean that that module itself is monogenic and that means Laxton's *primitive modules* (see later) are 2-primitive, in the sense of Betsch (Definition 413), although Betsch's definition is more general.

Lemma 407 *When M is an irreducible left module and $N \cdot M \neq \{0\}$, M will be monogenic.*

Laxton gives this version of **Schur's lemma** ([30]). To make sense of it, you need Definition 26 and the idea of a regular automorphism (page 3).

Lemma 408 *Suppose M is a faithful, irreducible, left near-ring module. Then, the non-trivial members of G are regular group automorphisms.*

Proof Suppose $\theta \in G^*$. The kernel of θ and the image of θ are both subgroups of M and near-ring groups. Consequently, the kernel is $\{0\}$, and the image is M, and so θ is an automorphism. Now, suppose $r \in N and m \in M$,

$$\theta\left(\left(\theta^{-1}(r \cdot m)\right) - \left(r \cdot \theta^{-1}(m)\right)\right) = 0$$

Hence, the group automorphism θ^{-1} is a NR-linear mapping. Define

$$\text{Fixed}(\theta) = \{m \in M \, : \, \theta(m) = m\} \tag{11.14}$$

$\text{Fixed}(\theta)$ is an additive subgroup of M and $\theta(r \cdot m) = r \cdot \theta(m)$, so

$$N \cdot \text{Fixed}(\theta) \subseteq \text{Fixed}(\theta)$$

If θ is non-trivial and not the identity, this means it will be regular. □

Lemma 409 *The following conditions are equivalent for a near-ring N acting faithfully upon a left, irreducible module M.*

(i) *$(M, +)$ is abelian.*
(ii) *$(N, +)$ is abelian.*
(iii) *$(N, +, \cdot)$ is a ring.*

Proof

(i) Suppose $(M, +)$ is abelian. Then, for any $m \in M$,

$$[N, N] \cdot m = \{0\}$$

By faithfulness, this means $(N, +)$ is abelian.

(ii) When N actually is abelian, since M is monogenic, it must be the case that it is also abelian.

(iii) Suppose both $(N, +)$ and $(M, +)$ are abelian, and suppose $r_1, r_2, r_3 \in N$ and $m \in M$. Then,

$$(r_1 \cdot (r_2 + r_s)) \cdot m = r_1 \cdot (r_2 \cdot m + r_3 \cdot m) = (r_1 \cdot r_2 \cdot m + r_1 \cdot r_3 \cdot m)$$

And,

$$(r_1 \cdot r_2 + r_1 \cdot r_3) \cdot m = (r_1 \cdot r_2 \cdot m + r_1 \cdot r_3 \cdot m)$$

This applies to all $m \in M$ consequently,

$$(r_1 \cdot (r_2 + r_3)) = (r_1 \cdot r_2 + r_1 \cdot r_3)$$

□

The upshot of this is that the theory of faithful irreducible modules dissolves into the theory of faithful irreducible ring modules when any of the conditions of Lemma 409 apply. Sect. 11.4 describes a generalisation of these ideas, involving the notion of *2-primitive near-rings*. One aspect of this is that transformations near-rings of the form $T_0(\Gamma)$ are 2-primitive. When the additive group $(\Gamma, +)$ has order in excess of 2, these are also non-rings. Taking $(\Gamma, +)$ to be abelian gives many examples of zero-symmetric, unital, abelian, 2-primitive non-rings (the smallest such example has order 9 and is $T_0(C_3)$).

Lemma 410 *Suppose $(N, +, \cdot)$ is a finite near-ring and M is a faithful irreducible left module. Then, M is finite.*

Proof Taking some non-zero $m \in M$, we know that $N \cdot m$ $\{r \cdot m : r \in R\}$ is a near-ring group and non-zero. It must be the whole of M, so M must be finite. □

One of the glories of the generation of mathematicians that included Laxton and Betsch was the determination of the finite simple groups. A number of theorems posted the way, and an early one was Thompson's result that a finite group having a regular automorphism of prime order must be nilpotent [132, 139]. This is used in the next result.

Lemma 411 *Suppose N is a finite non-ring and finite and acts faithfully on the irreducible module M. Then, $G = \{0, Id\}$.*

Proof From Thompson's result, if G contains non-identity automorphisms, then $(M, +)$ is nilpotent. Its commutator subgroup $[M, M]$ is a proper subgroup and non-trivial, because $(M, +)$ cannot be abelian. This commutator subgroup is a near-ring group, which contradicts irreducibility. The conclusion is that $G = \{0, Id\}$. □

The proof of this last lemma suggests that when M is a non-trivial irreducible left module upon which the finite near-ring N acts faithfully, the commutator subgroup of M is either trivial, in which case N is a ring, or equal to whole of M, in which case M is additively perfect. Laxton's original formulation of these results worked with d.g. near-rings. My condition (iv) of Observations 395 relates to elements of the near-ring inducing phomomorphisms on the module.

In Laxton's d.g. version, the equivalent of this condition is the requirement that the distributively generating set induces endomorphisms. The conclusion, in his set-up, is that M should be **fully invariantly simple**, rather than merely perfect, and so my more general definition leads to an appreciably weaker result. Incidentally, I don't think much is known about fixed-point-free automorphism groups acting on infinite groups although there is interesting work on some of the finite automorphism groups that can occur in that context.

An Embedding

Suppose N is a non-ring and it has a faithful irreducible module M. If $n_1, n_2 \in N$ and $n_1 \neq n_2$, then there must be some $\gamma \in \Gamma$ such that $(n_1 - n_2) \cdot \gamma \neq 0$, by faithfulness. This means that we have an injective embedding of N into the near-ring $\mathfrak{M}_G(\Gamma)$ (defined, page 295) which is an isomorphism of N with a sub-near-ring of $\mathfrak{M}_G(\Gamma)$ $n \mapsto (\theta\gamma \mapsto (n \cdot \gamma))$.

Laxton restricted his work to d.g. near-rings and described a near-ring having a faithful irreducible module as **primitive** [99]. These days, a more complicated and more general definition of primitivity is standard, and I give it in the next section.

11.4 Primitive Near-Rings

Arguably, the theory of rings starts with Jacobson and Chevalley's work on the **density theorem**, [66, 76], and with the Artin-Wedderburn theorem on primitive rings [7]. These are aspects of the "modern algebra" that crystallised from lectures of Emmy Noether and Emil Artin, at Göttingen, given in the 1920s.

I cannot resist quoting the Wedderburn-Artin theorem [66, 71]. It's a "gold-standard" result motivating a lot of work in near-rings.

Theorem 412 (Wedderburn-Artin) *For left Artinian rings, these conditions are equivalent.*

(i) The ring is simple.
(ii) The ring is left primitive.
(iii) For some fixed positive integer $n \in \mathbb{N}$, R is isomorphic to the full ring of square matrices over some division ring D. Moreover, the size of the matrices and the division ring are unique.

There were attempts to generalise these fundamental ideas to near-rings, dating from the early 1960s, although I believe Wielandt had some results from as much as 25 years earlier, results, largely unknown, outside Germany.

The first to publish accessibly was Laxton, [98], who was working on his own and specifically concerned with d.g. near-rings. Betsch worked more generally. He was active at about the same time and published a little later [11]. Meldrum cites

Betsch, Laxton and Ramakotaiah in this context, and I know that Michael Holcombe produced a thesis on primitive near-rings in 1970. The most authoritative account of the history of these ideas is probably that of Betsch, in [11].

My story is based heavily on the two papers of Betsch, [10, 11], and on Laxton's pioneering paper, [99]. Beware! [11] describes $a \cdot (b + c) = (a \cdot b) + (a \cdot c)$ as characterising *right* near-rings (see page 63).

I have deliberately used a lot of the same notation as Betsch adopts to reflect my extensive borrowings more accurately. The following definition is fairly standard, [10, 11, 119]. I believe it was introduced by Gerhard Betsch in his doctoral dissertation at Tubingen, in 1963.

Definition 413 Suppose M is a left near-ring group acted on by the near-ring, N, and further suppose it is monogenic (see (1.19)).

(i) The near-ring group is of **type 0** provided it has no proper non-trivial mideals.
(ii) It is of **type 1** if it is already of type 0, and for each $m \in M$, either $N \cdot m = \{0\}$ or $N \cdot m = N$.
(iii) It is of **type 2** if it has no proper non-trivial subgroups which are near-ring groups under the action of N.

Of course, type 2 near-ring groups are also type 1 near-ring groups. Type 2 near-ring groups are very similar to Laxton's irreducible near-ring modules but not quite the same (because his modules had additional left-distributivity conditions, which I have generalised to (11.1)). This next definition comes straight from [11].

Definition 414 Suppose N is a near-ring and M a left near-ring group such that $N \cdot M \neq \{0\}$. Suppose N acts faithfully on M and M is of type ν, where $\nu \in \{0, 1, 2\}$. Then, N is said to be a **ν-primitive near-ring**.

If $(N, +, \cdot)$ is a ring and it acts faithfully on the type 0 near-ring group Γ, then it is easy to see that it is a primitive ring (using monogenicity and arguments similar to those of Lemma 409). I exclude primitive rings from the discussion here on the basis that they are well covered elsewhere.

Lemma 415 *Suppose N is a unital near-ring and the left near-ring group M is unitary.*

(i) If M is a type 1 near-ring group, it will also be a type 2 one.
(ii) So, type 1 is equivalent to type 2 for unital rings and unitary near-ring groups.
(iii) When N is a ring, 2-primitivity coincides with the usual notion of primitivity.

I now need some ideas and notation that appeared in Chap. 7, particularly, Sects. 7.2.4 and 7.7 of that chapter. Notation 269 is vital, and you should note that I usually apply NR-linear mappings on the *right*.

Lemma 416 *Suppose G is a group of automorphisms acting fixed-point-freely on the non-trivial additive group $(\Gamma, +)$. Any dense sub-near-ring N of $\mathfrak{M}_G(\Gamma)$ is 2-primitive on $(\Gamma, +)$.*

Proof Take $\gamma_a, \gamma_b \in \Gamma$ with $\gamma_a \neq 0$. We can find some $m \in \mathfrak{M}_G(\Gamma)$ such that $m \cdot \gamma_a = \gamma_b$, by Lemma 281. We know $\gamma_a = \gamma_j \cdot g$ for some orbit representative γ_j and some $g \in G$, and by Lemma 285, we know that there is a $n \in N$ such that $n \cdot \gamma_1 = m \cdot \gamma_1$. This means that $n \cdot \gamma_a = \gamma_b$ and that there are actually no proper, non-trivial *subsets* of Γ which are closed under left action of N and that N is 2-primitive. □

Corollary 417 *When the automorphism group G has but a finite number of orbits on* Γ, *the dense sub-near-ring is automatically unital, and its type 2 near-ring group is unitary.*

Proof This is simply an application of Lemma 279. □

Definition 418 Suppose Γ is a near-ring group acted on by the near-ring N. I define

$$\Theta_0 = \{\gamma \in \Gamma \,:\, N \cdot \gamma = \{0\}\} \tag{11.15}$$

$$\Theta_1 = \{\gamma \in \Gamma \,:\, N \cdot \gamma = \Gamma\} \tag{11.16}$$

The first of these two sets is non-empty. The second is non-empty if and only if the near-ring group is monogenic. The two sets have empty intersection unless Γ is trivial. If Γ is of type 1, then $\Theta_1 = \{\Gamma - \Theta_0\}$. If N has an identity element and is of type 2, then $\Theta_1 = \Gamma^*$. If $G = \mathrm{Aut}_n(\Gamma)$ represents the near-ring automorphisms of ${}_N\Gamma$, then

(i) $\Theta_0 \cdot G = \Theta_0$
(ii) $\Theta_1 \cdot G = \Theta_1$.

Now comes Betsch's version of Schur's lemma.

Lemma 419 *Suppose* Γ *is a left near-ring group of type 0. Suppose* $C = \mathrm{Hom}_N(\Gamma, \Gamma)$ *(a semi-group of NR-linear endomorphisms) and* $G = \mathrm{Aut}_N(\Gamma)$ *(the NR-linear automorphisms).*

(i) If $\gamma_1, \gamma_2 \in \Theta_1$ *and* $\gamma_1 = \gamma 2 \cdot c$ *for some* $c \in C$, *then* $c \in G$.
(ii) The action of G on Θ_1 *is regular, that is to say fixed-point-free.*
(iii) If $g \in G$ *and* $g \neq Id$, *then the fixed points of g—that is, the* additive subgroup

$$\mathrm{Fixed}(g) = \{\gamma \in \Gamma \,:\, \gamma \cdot g = \gamma\} \tag{11.17}$$

is contained in $\{\Gamma - \Theta_1\}$.
(iv) These statements are equivalent;

(i)

$$C = \left\{G \bigcup \{0\}\right\} \tag{11.18}$$

(the zero represents the zero mapping).

(ii) *There is no bijective NR-endomorphism mapping Γ to one of its proper non-trivial subgroups.*

(v) *If Γ is of type 1 or finite, then (11.18) applies.*
(vi) *If Γ is of type 2 and N has an identity element, then (11.18) applies, and G is a regular group of automorphisms (i.e. it acts fixed-point-free).*

Proof

(i) $\Gamma = N \cdot \gamma_1 = (N \cdot \gamma_2 \cdot g) \subset \Gamma \cdot g$, so g is surjective. The kernel of g is a mideal, and it is not the whole of Γ so g is an automorphism.
(ii) Take some $\gamma \in \Theta_1$, and suppose $\gamma = \gamma \cdot g$. Monogenicity means that any element $\omega \in \Gamma$ can be written $\omega = n \cdot \gamma$ for some n.

$$\omega \cdot g = n \cdot \gamma \cdot g = n \cdot \gamma = \omega$$

This means that $g = Id$ (the identity mapping).
(iii) Fixed(g) is obviously an additive subgroup. If $\gamma \in (\text{Fixed}(g) \cap \Theta_1)$, then, because $(n \cdot \gamma) \cdot g = (n \cdot (\gamma \cdot g)) = (n \cdot \gamma)$, it would be the identity mapping, Id.
(iv) A bijective mapping of the sort described must have a trivial kernel, since we are dealing with a type 0 near-ring group. If (ii) applies, then its image should be the group Γ, and so the mapping is an automorphism.
(v) In each case, there will be no bijective NR-endomorphism mapping Γ to one of its proper non-trivial subgroups, so (iv) applies.
(vi) Type 2 implies type 1, so (11.18) is immediate. Fixed-point-freeness comes from (ii).

□

Lemma 420 *Suppose $(N, +, \cdot)$ is a unital near-ring; Γ is a unitary, type 0 module; and $\gamma_1, \gamma_2 \in \Theta_1$.*
If

$$\begin{pmatrix} \gamma_1 \\ 0 \end{pmatrix}_{\mathcal{L}} = \begin{pmatrix} \gamma_2 \\ 0 \end{pmatrix}_{\mathcal{L}}$$

Then γ_1 and γ_2 are in the same orbit of G.

Proof There is a mapping $\theta : \Gamma \mapsto \Gamma$,

$$\theta(n \cdot \gamma_1) = (n \cdot \gamma_2) \tag{11.19}$$

The annihilator condition makes it injective, and it is obviously surjective. It is a near-ring group automorphism, and (writing the action on the right)

$$(\gamma_1)\theta = (1 \cdot \gamma_1)\theta = (1 \cdot \gamma_2) = \gamma_2$$

so γ_1 and γ_2 are in the same orbit.

□

11.4.1 I(N), A(N)andE(N)

The near-ring N acts on each of the above groups in the obvious way. Near-ring subgroups of these groups are, respectively, subgroups fixed under inner automorphisms, automorphisms and endomorphisms. These are, respectively, the normal, characteristic and fully invariant additive subgroups of $(N, +)$.

Lemma 421

(i) The near-ring $\mathrm{E}(N)$ *acts 2-primitively on* $(N, +)$ *if, and only if, the group* $(N, +)$ *is fully invariantly simple.*
(ii) The near-ring $\mathrm{A}(N)$ *acts 2-primitively on* $(N, +)$ *if, and only if, the group* $(N, +)$ *is characteristically simple.*
(iii) The near-ring $\mathrm{E}(N)$ *acts 2-primitively on* $(N, +)$ *if, and only if, the group* $(N, +)$ *is simple.*

11.4.2 The Density Theorem

The results in this subsection are due to a number of mathematicians including Wielandt, Laxton, Betsch and others. I have contributed almost nothing new to this account.

Throughout this subsection, $(N, +, \cdot)$ is a non-ring acting faithfully on the type 0 near-ring group $(\Gamma, +)$.

The proof of the next lemma is an inductive one. The symbol $\mathcal{S}(t)$ represents an inductive statement at level t. The notation is due to Betsch.

Lemma 422 *Suppose* $\{\gamma_1, \gamma_2, \gamma_3, \ldots, \gamma_s\}$ *is a subset of* Θ_1 *with the added property that no two of these elements are in the same orbit of* $G = \mathrm{Aut}_N(\Gamma)$.
Then

$$\mathcal{S}(t) = \bigcap_{j=1}^{t} \left(\begin{pmatrix}\gamma_j\\0\end{pmatrix}_{\mathcal{L}}\right) \not\subset \begin{pmatrix}\gamma_k\\0\end{pmatrix}_{\mathcal{L}} \text{ for all } t < k \leq s \tag{11.20}$$

Proof $\begin{pmatrix}\gamma_1\\0\end{pmatrix}_{\mathcal{L}}$ is a maximal left ideal by Lemma 51 (vi). Appealing to Lemma 420, none of the annihilators of the orbit representatives can be equal. This gives a basis for the induction by proving $\mathcal{S}(1)$. Suppose now that all statements up to $\mathcal{S}(t)$ are true, where $(t+1) < k \leq s$ and $s \geq 3$. Notice that the intersection in the statement cannot be trivial, at least this far. The hope is to establish $\mathcal{S}(t+1)$. Define

$$\mathcal{B} = \bigcap_{j=1}^{t} \left(\begin{pmatrix} \gamma_j \\ 0 \end{pmatrix}_{\mathcal{L}} \right) \not\subset \begin{pmatrix} \gamma_k \\ 0 \end{pmatrix}_{\mathcal{L}}$$

$$\mathcal{C} = \begin{pmatrix} \gamma_{t+1} \\ 0 \end{pmatrix}_{\mathcal{L}} \not\subset \begin{pmatrix} \gamma_k \\ 0 \end{pmatrix}_{\mathcal{L}}$$

We know the annihilator of γ_k is a maximal left ideal, from Lemma 51. Thus,

$$\left(\mathcal{B} + \begin{pmatrix} \gamma_k \\ 0 \end{pmatrix}_{\mathcal{L}} \right) = \left(\mathcal{C} + \begin{pmatrix} \gamma_k \\ 0 \end{pmatrix}_{\mathcal{L}} \right) = N$$

Appealing to Lemma 405 and noting that N is a non-ring, we obtain

$$(\mathcal{B} \cap \mathcal{C}) \not\subset \begin{pmatrix} \gamma_k \\ 0 \end{pmatrix}_{\mathcal{L}}$$

which means

$$\mathcal{S}(t+1) = \bigcap_{j=1}^{t+1} \left(\begin{pmatrix} \gamma_j \\ 0 \end{pmatrix}_{\mathcal{L}} \right) \not\subset \begin{pmatrix} \gamma_k \\ 0 \end{pmatrix}_{\mathcal{L}}$$

□

Lemma 423 (The Density Theorem) *Suppose* $\{\gamma_1, \gamma_2, \gamma_3, \ldots, \gamma_s\}$ *is a subset of* Θ_1 *with the added property that no two of these elements are in the same orbit of* $G = \mathrm{Aut}_N(\Gamma)$. *Now take any set* $\{\omega_1, \omega_2, \ldots, \omega_s\}$ *contained in* Γ. *Then, there will be some* $n \in N$ *for which* $n \cdot \gamma_j = \omega_j$, *for* $j = 1, 2, 3, \ldots, s$.

Of course, there will be elements of $\mathfrak{M}_G(\Gamma)$ for which this assertion holds (boxed note on page 300); what the lemma is saying is that one of them is in N.

Proof $\gamma_1 \in \Theta_1$ means we can find $n \in N$ for which $n \cdot \gamma_1 = \omega_1$. Another inductive argument is used. So, suppose we have some $n_t \in N$ such that

$$\left(n_t \cdot \gamma_j\right) = \left(\omega_j\right) \textit{ for } 1 \leq j \leq t \leq (s-1)$$

From Lemma 422,

$$\mathcal{K} = \bigcap_{j=1}^{t} \left(\begin{pmatrix} \gamma_j \\ 0 \end{pmatrix}_{\mathcal{L}} \right) \not\subset \begin{pmatrix} \gamma_{(t+1)} \\ 0 \end{pmatrix}_{\mathcal{L}}$$

This means that the non-trivial mideal $\{\mathcal{K} \cdot \gamma_{(t+1)}\}$ is actually Γ and that means that we can find some $k \in \mathcal{K}$ such that

$$(k \cdot \gamma_{(t+1)}) = (\omega_{(t+1)} - (n_t \cdot \gamma_{(t+1)}))$$

Then, put $n_{(t+1)} = (k + n_t)$, and get

$$n_{(t+1)} \cdot \gamma_j = \omega_j, for\ j = 1, 2, \ldots, (t+1)$$

□

11.4.3 Unital, 2-Primitive Non-rings

The near-ring $(N, +, \cdot)$ will be unital, 2-primitive and a non-ring in this subsection, and it acts upon the type 2 near-ring group, Γ. From Lemma 419, we already know that $G = \text{Aut}_N(\Gamma)$ is fixed-point-free.

Lemma 424 *$(N, +, \cdot)$ is a dense sub-near-ring of $\mathfrak{M}_G(\Gamma)$.*

Proof This result follows from Lemma 278 and from Lemma 423. □

Density
This last lemma says 2-primitive non-rings are dense sub-near-rings of some $\mathfrak{M}_G(\Gamma)$. Lemma 416 says that dense sub-near-rings of $\mathfrak{M}_G(\Gamma)$ are 2-primitive.

Section 7.7 gives an example of an *uncountable* dense sub-near-ring of some $\mathfrak{M}_G(\Gamma)$ which is not unital, and Lemma 417 says that dense 2-primitive near-rings in which the automorphism group G has only a finite number of orbits are always unital. In finite cases, density is the same thing as equality (page 235). So, finite, 2-primitive near-rings are transformation near-rings of the form $\mathfrak{M}_G(\Gamma)$. Laxton's primitive near-rings are a proper subset of the class of 2-primitive near-rings, as is my generalisation of them. These near-rings embed in some $\mathfrak{Ph}_G(\Gamma)$ (see page 296). If they are finite, this will mean

$$\mathfrak{M}_G(\Gamma) = \mathfrak{Ph}_G(\Gamma) \tag{11.21}$$

In the event that G is trivial, this becomes $T_0(\Gamma) = \text{Phom}(\Gamma)$, which, from Lemma 309, means that $(\Gamma, +)$ is additively perfect. From Lemma 411, we can say that Laxton's primitive near-rings, when finite, are transformation near-rings on perfect groups. In that case, they have the form

$$(N, +, \cdot) = \overbrace{(\Gamma \oplus \Gamma \oplus \ldots \Gamma)}^{(|N^*| \text{ times})} \tag{11.22}$$

Simple, and 2-Primitive, Near-Rings
If N is 2-primitive and acts on the faithful near-ring group $(\Gamma, +)$, and if G has but a finite number of orbits, then from Lemma 298, the near-ring is simple (defined on page 13). If $(N, +, \cdot)$ is a simple near-ring and it contains a non-trivial subgroup M which is a left sideal, and which does not itself contain any such proper subgroup, then for any $m \in M^*$, we must have that either $N \cdot m = \{0\}$ or $N \cdot m = M$. When N possesses a multiplicative (left) identity, the former cannot occur. The annihilator of M itself is an ideal of N. Faithfulness corresponds to that annihilator being trivial. When none of the elements of M are annihilated by N itself, M is a near-ring group of type 2, and so N is a 2-primitive near-ring. One way to make this happen is simply to insist that the near-ring has a left identity and a proper non-trivial left sideal which is a subgroup of $(N, +)$.

11.4.4 Left Ideal Cores

The reader may wish to refer back to Sect. 1.19. Suppose $(N, +, \cdot)$ is unital and zero-symmetric and there is a left ideal J which is actually maximal *as a left sideal*. There is a well-defined action of the near-ring $\frac{N}{\mathcal{C}(J)}$ upon the additive group $\frac{N}{J}$ as

$$\{m + \mathcal{C}(J)\} \cdot \{n + J\} = \{(m \cdot n) + J\}$$

This makes $\frac{N}{J}$ into a type 2 near-ring group, and it makes $\frac{N}{\mathcal{C}(J)}$ into a 2-primitive near-ring with maximal left ideal $\frac{J}{\mathcal{C}(J)}$ with trivial core and which is maximal as a left sideal. Now suppose that N is a zero-symmetric, unital near-ring acting faithfully on the type 2 near-ring group Γ. Using Lemma 51, and taking $\gamma \in \Gamma^*$, we can say that the left ideal

$$J = \begin{pmatrix} \gamma \\ 0 \end{pmatrix}_{\mathcal{L}}$$

is maximal and that, as near-ring groups,

$$\Gamma \cong \frac{N}{J}$$

which means no proper left sideal properly contains J. The core of the left ideal J is a two-sided ideal, and the above near-ring group correspondence argues that it annihilates Γ. It is therefore trivial.

Lemma 425 *The unital, zero-symmetric near-ring $(N, +, \cdot)$ is 2-primitive if and only if it possesses a maximal left ideal J which is also maximal as a left sideal and which has trivial core.*

11.4.5 A Famous Result of Fröhlich

Earlier (page 258), I referred to Fröhlich's famous result that the near-ring of inner automorphisms of a finite simple group, I(Δ), is additively a direct sum of one fewer copies of (Δ, +) than its order [41].[2] The near-ring I(Δ) acts on (Δ, +) as a faithful, irreducible left module, in the sense of Definition 406. The theorem follows.

11.4.6 Near-Fields

Suppose $(N, +, \cdot)$ is a near-field. Of course, we have the near-ring group ${}_N N$ and this is a type 2 near-ring group upon which N acts faithfully. So, N is a 2-primitive non-ring, hosted by an elementary abelian p-group (see Lemma 145). We become interested in $G = \mathrm{Aut}_N(N)$. This is a fixed-point-free group of automorphisms acting on an abelian near-ring group (so, potentially, non-trivial). A non-trivial fixed-point-free automorphism should map 1 to something other than itself. Acting these automorphisms on the right, we might say $(1)g = m$. Then, since g is a near-ring group homomorphism, we have $(n = n \cdot 1)g = (n \cdot m)$, so elements of G are fixed by their actions on the multiplicative identity and have the form $(n)g = (n \cdot m)$, which is fixed-point-free from the properties of near-fields. It is then easy to see that $G \cong (N^* \cdot)$, where each mapping is identified with the element to which it sends 1. It is equally obvious that G has but a single orbit on N^*. In the seven finite non-Dickson near-fields, the fixed-point-free automorphism group G is one of the groups listed in Table 3.3. Some of these groups are highly non-abelian.

Changing focus, suppose $(N, +)$ is a finite additive group and G is a multiplicative group of fixed-point-free automorphisms of $(N, +)$. Suppose $|G| = |N^*|$. Pick some arbitrary, fixed element of N^*, and call it e. Taking any $x \in N^*$, we can say there is a unique element $g \in G$ such that $(e)g = x$. Re-label that element g as ϕ_x. Then the set $\{\phi_x\}_{x \in N^*}$ is identical to G. Now select $x, y \in N^*$ and define

$$(x \cdot y) = (x)\phi_y \tag{11.23}$$

and define a more general zero-symmetric product on N extending this in the obvious way.

Lemma 426 *$(N, +, \cdot)$, as defined in (11.23), is a near-field.*

Proof The only tricky thing to prove is multiplicative associativity. This is the condition

$$(\phi_y \circ \phi_z) = \phi_{(y)\phi_z} \tag{11.24}$$

[2] I suspect that Fröhlich's result was so striking that it ignited near-ring research in the United Kingdom and elsewhere.

which is, of course, a *coupling condition*. It is established by checking that both mappings send e to the same thing. □

It is pleasant that the fact G has but a single orbit on N^* emerges from the cardinality equality. The determination of the finite near-fields is equivalent to the determination of the multiplicative groups of fixed-point-free automorphisms of elementary abelian p-groups which have order 1 less than the order of the p-group. Finite fields correspond to the multiplicative group being cyclic. Compare this construction with 2.7.1.

11.4.7 $\mathfrak{M}_S(\Gamma)$

Surviving readers will now, perhaps, realise that the inordinate amount of work on $\mathfrak{M}_S(\Gamma)$ in Chap. 7 was actually about the structure of 2-primitive near-rings. Observations 301, on page 312, tells us that when the fixed-point-free automorphism group has a finite number of orbits, the faithful type-2 near-ring group is unique up to isomorphism.

11.4.8 Alternative Generalisations

The reader will appreciate that further approximations to irreducibility in ring modules are possible in near-rings. One might imagine an irreducibility criterion for left modules in which one would demand the existence of no proper non-trivial sub-near-ring groups that were actually *modules* in their own right. That is $\{0\} \neq (\Omega, +) \subsetneq (\Gamma, +)$ such that

$$n \cdot (\omega_1 + \omega_2) \equiv ((n \cdot \omega_1) + (n \cdot \omega_2)) \ (modulo\ [\Omega, \Omega]) \qquad (11.25)$$

is forbidden. I include the idea because it does illustrate the distinction between my formulation and Laxton's. His distributive generators would, perforce, distribute over sub-structures. In this form of irreducibility, modules could, potentially, have proper non-trivial near-ring subgroups or even proper, non-trivial mideals, so the generalisation takes us beyond the generalisations described in Sect. 11.4 and may perhaps exceed what one could do usefully.

It seems likely that much further work could be done on these ideas and one wonders what could be done with respect to *right* near-ring groups, too. Perhaps the lack of progress in this direction does not in itself indicate impossibility. People have discussed this, I believe, and I remember Alan Machin doing work on it, with ideas rather different from my own right modules [107].

11.5 Module-Like Structures

There has been much work done on near-ring groups and module-like structures. Some of this is reported in [119, 128] and [23], but much is only available in research papers. I thought I might use this section to indicate a few particularly nice results. The next lemma requires information given in Sect. 2.5.6. Two operations on ideals are involved: sums, which produce new ideals, and intersections which do the same. These are, respectively, the supremum and infinum operations.

Lemma 427 *Ideals of near-rings form complete, modular lattices, but these lattices are not necessarily distributive.*

This next lemma is due to Stuart Scott, compare it with Lemma 17.

Lemma 428 *Suppose N is a near-ring and ${}_N\Gamma$ is left near-ring group. Suppose A and B are mideals of Γ. Then, for all $n \in N$, $\alpha \in A$ and $\beta \in B$,*

$$n \cdot (\alpha + \beta) \equiv (n \cdot \alpha + n \cdot \beta) \ (modulo\ (A \cap B)) \tag{11.26}$$

Proof

$$n \cdot (\alpha + \beta) - (n \cdot \beta) - (n \cdot \alpha) \in A$$

and,

$$n \cdot (\alpha + \beta) - (n \cdot \beta) - (n \cdot \alpha) = n \cdot (\alpha + \beta) - (n \cdot \alpha) + (n \cdot \alpha) - (n \cdot \beta) - (n \cdot \alpha) \in B$$

□

Definition 429 A near-ring is **completely reducible** if it is a direct sum of simple ideals. Ideals are **completely reducible** provided they are completely reducible when regarded as near-rings. A near-ring is **decomposable** provided it is a direct sum of proper non-trivial ideals. Otherwise, it is **indecomposable**.

These definitions, from Pilz [128], indicate that completely reducible near-rings are decomposable but that the reverse is probably untrue. One can extend these ideas to near-ring groups and near-ring modules in fairly obvious ways, working with mideals rather than ideals.

Pilz attributes this next result to Roth and Beidleman [128]. It is really a generalisation of standard ring theory. It is important to distinguish between *sums* of ideals and *direct sums* of ideals in what follows (see page xiv). Simple ideals are defined on page 13. The lemma concerns completely reducible near-rings.

Lemma 430 *These conditions are equivalent for any near-ring, $(N, +, \cdot)$:*

(i) Each ideal is the sum of a set of simple ideals.
(ii) The near-ring itself is a sum of a set of simple ideals.
(iii) The near-ring itself is a direct sum of a set of simple ideals.

(iv) *Every ideal of the near-ring is a direct summand of it, in the sense that each ideal J has a partner ideal K such that $N = (J \oplus K)$.*

(v) *If I is an ideal, then both I itself and the quotient ring $\frac{N}{I}$ are completely reducible.*

(vi) *The near-ring is a sum of minimal ideals.*

Proof

$(ii) \Rightarrow (iii)$ The hypothesis is that elements of N are finite sums of elements from these simple ideals, which I might express as

$$N = \Sigma_{\alpha \in A} I_\alpha$$

Define

$$\mathcal{A} = \left\{ B \subset A : \Sigma_B I_\beta = \oplus_B I_\beta \right\} \neq \phi$$

$\mathcal{A}$ is partially ordered by inclusion. Given any chain of subsets of $\mathcal{A}$, the union of the chain will still be in $\mathcal{A}$. So, every chain in $\mathcal{A}$ has an upper bound in $\mathcal{A}$, that is, $\mathcal{A}$ is an **inductive set** [26]. Zorn's lemma now says that we may pick a maximal element of $\mathcal{A}$, which I shall call C. I write,

$$M = \Sigma_{\gamma \in C} I_\gamma$$

If $\alpha \in A$, simplicity tells us that $(I_\alpha \cap M)$ is either trivial or I_α.
In the first case, $\{\{\alpha\} \cup C\}$ would be a subset of $\mathcal{A}$ that was larger than M.
The conclusion is that $M = N = \oplus_A I_\alpha$.

$(iii) \Rightarrow (iv)$ Select an ideal I and consider the class of ideals with trivial intersection with I. One can apply Zorn's lemma to this class, obtaining a maximal element, an ideal J.
Define $N_1 = (I \oplus J)$. If this is not equal to N, it must be the case that one of the simple ideals in the direct sum does not lie in N_1; call that K. Simplicity ensures $(K \cap N_1) = \{0\}$, and this implies $(K \cap I) = \{0\}$. But by maximality of J we would have $K \subset J$, which means $(K \cap N_1) = K$. The conclusion is that $N_1 = N$.

$(iv) \Rightarrow (i)$ Select any ideal I, and define J as the sum of all the simple ideals contained in I. Suppose $J \subsetneq I$. Since J is a direct summand of N, I write $N = (J \oplus K)$, for some ideal K of N. Simple ideals of J are simple ideals of N; and

$$I = (J \oplus (I \cap K))$$

The plan is to show that $(I \cap K)$ contains a simple, non-trivial ideal of N, which would contradict the definition of J. Notice that $(I \cap K) \neq \{0\}$. There are two subcases:

(a) $(I \cap K)$ is finitely generated. A variation of Lemma 140 relating to ideals rather than left ideals gives us a maximal ideal $W \subset (I \cap K)$. As above, we then have $(I \cap K) = (W \oplus V)$, and V is that simple non-trivial ideal both of $(I \cap K)$ and of N.

(b) $(I \cap K)$ is not finitely generated. $(I \cap K)$ does certainly contain finitely generated ideals which are non-trivial; select one such F. This ideal contains a maximal ideal $M \subset F$. Then, direct complements of M in F are non-trivial simple ideals of $(I \cap K)$.

$(iii) \Rightarrow (v)$ We know $(iii) \Rightarrow (iv) \Rightarrow (i)$. Each ideal I of N is the direct sum of simple ideals. We know $N = (I \oplus K)$ and that K itself is a direct sum of simple ideals.

$(vi) \Rightarrow (iv)$ Virtually the same argument as $(iii) \Rightarrow (iv)$ applies.

□

11.6 $\mathcal{A}$-Matrices

All groups in this section will be finite.

The $\mathcal{A}$-matrix notation can be useful in describing near-ring groups. I make extensive use of the transpose operation in what follows A^t.

11.6.1 Left Near-Ring Groups

Suppose $(M, +)$ is a left near-ring group. Suppose $M = \{m_0, m_1, \ldots, m_n\}$. The left action of R on M means that any $r \in M$ can be associated with a $(n+1) \times (n+1)$, $(0-1)$, column stochastic matrix $\mathcal{M}^l(r)$ which is defined by insisting that the $(i-j)$th entry of the matrix will be 1 precisely when $r \cdot m_j = m_i$ and 0 otherwise. As I have noted elsewhere, this matrix notation respects near-ring multiplication. The matrix corresponding to the element $r_1 \cdot r_2$, where $r_1, r_2 \in R$, is simply the product of the two matrices corresponding to these elements

$$\mathcal{M}^l(r) = \mathcal{M}^l(r_1) \cdot \mathcal{M}^l(r_2) \tag{11.27}$$

and, more than that, if we represent elements of M as column vectors (so the element m_j would be the $(n+1) \times 1$ column stochastic matrix with entry corresponding to row j (counting from zero) set to 1), then the product $(r \cdot m_j)$ results in a column vector obtained by left multiplying the column vector representing m_j by the matrix $\mathcal{M}^l(r)$; and, of course, this is simply the jth column of the matrix $\mathcal{M}^l(r)$.

11.6.2 Right Near-Ring Groups

For right near-ring groups one forms the associated matrices in almost the same way. One looks at products $m_i \cdot r = m_j$ and uses them to define the ith row of the matrix as having all entries set to 0 except for the jth which is set to 1. This produces a matrix $\mathcal{M}^r(r)$ associated with r. The matrix is a square, $(0 - 1)$ row stochastic matrix. Again, multiplication is respected

$$\mathcal{M}^r(r) = \mathcal{M}^r(r_1) \cdot \mathcal{M}^r(r_2) \tag{11.28}$$

and the product $(m_j \cdot r)$ is the $j^{th\cdot}$ row of $\mathcal{M}^r(r)$; or, equivalently, it is the result of multiplying the row vector obtained by transposing the vector previously identified as representing m_j, operating on the *right* by the matrix $\mathcal{M}^r(r)$, and then transposing again (to get a column vector).

I illustrate this using the group $C_5 = \{0, 1, 2, 3, 4\}$. Consider a transformation expressed as the tuple $(0, 3, 1, 3, 2)$ (Notation 253).

If this represents the action of some element $r \in M$ on the *left*, the corresponding column stochastic matrix would be

$$\begin{bmatrix} 1\,0\,0\,0\,0 \\ 0\,0\,1\,0\,0 \\ 0\,0\,0\,0\,1 \\ 0\,1\,0\,1\,0 \\ 0\,0\,0\,0\,0 \end{bmatrix}. \tag{11.29}$$

Alternatively, if this represents the action of some element $r \in M$ on the *right*, the corresponding row stochastic matrix is

$$\begin{bmatrix} 1\,0\,0\,0\,0 \\ 0\,0\,0\,1\,0 \\ 0\,1\,0\,0\,0 \\ 0\,0\,0\,1\,0 \\ 0\,0\,1\,0\,0 \end{bmatrix}. \tag{11.30}$$

(the transpose of the previous matrix). The $\mathcal{A}$-matrix notation ensures that identity mappings result in multiplicative identity matrices.

11.6.3 Opposites

I use the notation of page 51. Suppose M is some sort of near-ring group (either right or left!). There is a collection of stochastic matrices indexed by elements of the near-ring $(N, +, \cdot)$, and there is an *opposite near-ring* $(N, +, \cdot)^{opp}$, which is

$(N, +, :)$. The near-ring $(N, +, :)$ is left-distributive if and only if $(N, +, \cdot)$ was right-distributive. There remains an obvious action ($\maltese$) of this opposite near-ring on M—it is exactly the same action as occurred before; so, in the case of right near-ring groups, $(m \maltese n) = (m \cdot n)$. However, its properties are different in some respects. In the case of right near-ring groups, for example,

$$(m \maltese (n_1 : n_2)) = ((m \maltese n_2) \maltese n_1) \tag{11.31}$$

and while previous definitions applied to a near-ring $(N, +, \cdot)$ which was implicitly right-distributive, the opposite near-ring is left-distributive. I refer to the "near-ring groups" obtained by replacing the original near-ring with its opposite as **contrary structures** or **contrary near-ring groups**. So, a contrary left near-ring group is the system obtained by going from ${}_N M$ to ${}_{N^{opp}} M$. If it was the case that the near-ring acts homomorphically upon its near-ring group, then so does its opposite. The $\mathcal{A}$-matrices for a contrary near-ring group are the same as for the original, but opposite multiplication applies

$$\mathcal{M}^r(r_1 : r_2) = \mathcal{M}^r(r_2) \cdot \mathcal{M}^r(r_1) \tag{11.32}$$

It seems simpler to interpret these opposite actions as actions of the original near-ring but on the other side of the near-ring group. This would involve writing (in the case of what was originally a right near-ring group),

$$\left(n \overset{\times}{\times} m\right) = (m \maltese n) = (m \cdot n) \tag{11.33}$$

11.6.4 Bimodals

Motivated by bimodules in ring theory, I wish to look at left near-ring groups, H, which are also right near-ring groups, and for which the identity

$$((r \cdot h) \cdot s) = (r \cdot (h \cdot s)) \; \textit{for all } r, s, \in R, \textit{ and all } h \in H \tag{11.34}$$

I shall call this the **bimodal condition**. If A represents the left-action matrix of r, and B represents the right-action matrix of s, and $\underline{h}$ is the column vector representing h, the condition is

$$\left((A \cdot \underline{h})^t \cdot B\right)^t = \left(A \cdot (\underline{h}^t \cdot B)^t\right)$$

That is,

$$\left(B^t \cdot (A \cdot \underline{h})\right) = \left(A \cdot (B^t \cdot \underline{h})\right)$$

This is to apply to all vectors $\underline{h}$ so the condition is

$$\left(B^t \cdot A\right) = \left(A \cdot B^t\right) \tag{11.35}$$

Of course, the condition is equivalent to

$$\left(A^t \cdot B\right) = \left(B \cdot A^t\right) \tag{11.36}$$

Lemma 431 *Suppose H is a left and right near-ring group and the actions satisfy the bimodal condition. Then, for all, r, s ∈ R,*

$$\left(\mathcal{M}^l(r) \cdot \mathcal{M}^r(s)^t\right) = \left(\mathcal{M}^r(s)^t \cdot \mathcal{M}^l(r)\right) \tag{11.37}$$

The condition is both necessary and sufficient.

The fact that H is a near-ring group itself poses additional restrictions which are not handled in the above result. That is to say, it is not the case that *any* matrices satisfying the condition will represent near-ring groups (for the definition, see page 16). This really relates to how one adds these stochastic matrices. An illustration is given by (7.56). The addition depends on the addition in the group being acted on (in the case above, $(H, +)$). With *left* near-ring groups under action by R, we know $(r_1 + r_2) \cdot h = ((r_1 \cdot h) + (r_2 \cdot h))$, and the matrix conditions become

$$\mathcal{M}^l(r_1 + r_2) = \left(\mathcal{M}^l(r_1) +_{\textit{not standard}} \mathcal{M}^l(r_2)\right) \tag{11.38}$$

You should note that the addition of matrices on the right-hand side of this identity is emphatically *not* the standard addition of matrices (see (7.56), again!).

Bimodal Near-Ring Groups
One can consider more general bimodal conditions of the form ${}_RM_N$, where R and N may actually be distinct near-rings (so, a *left* near-ring group, a *right* N-group—in old terminology (see page 17)). In these circumstances, I describe M as a **bimodal near-ring group**. Notice that the corresponding $\mathcal{A}$-matrices will be the same size (one set row, one set column, stochastic). In the event that M is both a left and a right near-ring group under these different actions, I will of course call it a **bimodal**. Now, the $\mathcal{A}$-matrices $\mathcal{M}^l(r)$ are defined for elements of R, and the $\mathcal{A}$-matrices $\mathcal{M}^r(n)$ are defined for elements of N. Lemma 431 still applies, though, even in this situation of different near-rings.

Lemma 432 *Suppose M is a bimodal near-ring group ${}_RM_S$. Then, using vector notation, if $r \in R, s \in Sandm \in M$,*

$$\begin{aligned}(r \cdot m \cdot s) &= \left(\left(\left(\mathcal{M}^r(s)\right)^t \cdot \left(\mathcal{M}^l(r)\right)\right) \cdot (\underline{m})\right) \\ &= \left(\left(\left(\mathcal{M}^l(r)\right) \cdot \left(\mathcal{M}^r(s)\right)^t \cdot\right) \cdot (\underline{m})\right)\end{aligned}$$

This is simply a left multiplication by a column stochastic $(0-1)$ *matrix.*

11.6.5 Kronecker Products

Given two matrices, A and B, the Kronecker product is a matrix $(A \otimes B)$ comprising "blocks" of the form $a_{i,j} \cdot B$—which is a block of elements with the same size as B and whose general term is $(a_{i,j} \cdot b_{j,k})$. Actually, this is the *right* Kronecker product. The *left* Kronecker product would be defined similarly, but we would be talking about blocks of the form $A \cdot b_{i,j}$. Kronecker products crop up all over the place, particularly in representation theory [30]. They are described more formally in [95], and I obtained this "left/right" terminology from that book because many authors talk simply of Kronecker products and seem to mean just the right one.

In this book, my Kronecker products will be right Kronecker products.

The Kronecker product of two row stochastic $(0-1)$ matrices is row stochastic and $(0-1)$ (similarly for column stochastic matrices). An associated operation applying to matrices is provided by the **Vec** operator. This simply stacks the columns of a matrix, starting from the left, turning a $(m \times n)$ matrix into a $(m \cdot n \times 1)$ one; thus,

$$Vec\left(\begin{bmatrix} \alpha & \beta \\ \gamma & \delta \end{bmatrix}\right) = \begin{bmatrix} \alpha \\ \gamma \\ \beta \\ \delta \end{bmatrix} \tag{11.39}$$

Properties of the Kronecker Product

I list the properties of the Kronecker product that I need. They are generally well-known, and they are given in [95]. They apply equally to right and to left Kronecker products. The reader will be able to see that I misrepresent a much more profound set of results since I only wish to apply them to my immediate needs. In particular, simply assume all of these matrices are square, of the same size, $(0-1)$, and that they are all either row stochastic or all column stochastic, with no mixing. Vectors are column vectors.

(i)

$$((A \otimes B) \otimes C) = (A \otimes (B \otimes C)) \tag{11.40}$$

(ii)

$$(A \otimes B)^t = \left(A^t \otimes B^t\right) \tag{11.41}$$

(iii)

$$((A \otimes B) \cdot (C \otimes D)) = ((A \cdot C) \otimes (B \cdot D)) \tag{11.42}$$

(iv)

$$Vec(A \cdot B \cdot C) = \left(\left(C^t \otimes A\right) \cdot Vec(B)\right) \tag{11.43}$$

From this, Lemma 432 may be written

$$(r \cdot m \cdot s) = \left(\left(\underline{m}^t \otimes (\mathcal{M}^r(s))^t\right) \cdot Vec\left(\mathcal{M}^l(r)\right)\right) = \tag{11.44}$$
$$\left(\left(\underline{m} \otimes (\mathcal{M}^r(s))\right)^t \cdot Vec\left(\mathcal{M}^l(r)\right)\right)$$

11.6.6 Applications to Bimodal Near-Ring Groups

I continue to restrict to finite cases.

Suppose we do have a bimodal near-ring group ${}_RM_S$.

The $\mathcal{A}$-matrix notation associates each element of R with a square $(0-1)$ column stochastic matrix and each element of S with a square $(0-1)$ row stochastic matrix. The matrices have as many rows as there are elements in the group $(M, +)$. Irritatingly, I shall assume that this group has m elements (previously this was $(m+1)$ for reasons connected to numbering). The actions of R and S on M provide us with column, and row, stochastic $(0-1)$ matrices. I use the notation on page 394. For each $r \in R$ and each $s \in S$, I obtain two new matrices,

$$\mathcal{M}^l(s \otimes r) = \left((\mathcal{M}^r(s))^t \otimes \left(\mathcal{M}^l(r)\right)\right) \tag{11.45}$$

which is column stochastic, $(0-1)$ and square and has m^2 rows or columns and

$$\mathcal{M}^r(r \oslash s) = \left(\mathcal{M}^l(r)^t \otimes (\mathcal{M}^r(s))\right) \tag{11.46}$$

which is row stochastic, $(0-1)$ and square and has m^2 rows or columns. Note that the symbols $(r \otimes s)$ and $(r \oslash s)$ presently have notational significance only. No new product has been defined between the near-rings R and S. These matrices can be regarded as $\mathcal{A}$-matrices: the first representing left actions and the second right actions, on the additive group

$$(M_m, +) = \left(\overbrace{(M \oplus M \oplus \ldots \oplus M)}^{\text{(m times)}}\right) \tag{11.47}$$

which is simply a direct sum of m copies of $(M, +)$.

Lemma 433 *Suppose $r_1, r_2 \in Rands_1, s_2 \in S$, then*

$$\left(\left(\mathcal{M}^r(s_1)^t \otimes \mathcal{M}^l(r_1)\right) \cdot \left(\mathcal{M}^l(s_2)^t \otimes \mathcal{M}^r(r_2)\right)\right) =$$
$$\left((\mathcal{M}^r(s_1)^t \cdot (\mathcal{M}^r(s_2)^t)) \otimes \left(\mathcal{M}^l(r_1) \cdot \mathcal{M}^l(r_2)\right)\right) =$$
$$\left((\mathcal{M}^r(s_2 \cdot s_1)^t) \otimes \left(\mathcal{M}^l(r_1 \cdot r_2)\right)\right)$$

and

$$\left(\mathcal{M}^l(r_1)^t \otimes \mathcal{M}^r(s_1)\right) \cdot \left(\mathcal{M}^l(r_2)^t \otimes \mathcal{M}^r(s_2)\right) =$$
$$\left(\mathcal{M}^l(r_2 \cdot r_1) \otimes \mathcal{M}^r(s_1 \cdot s_2)\right)$$

Consequently,

$$\left(\mathcal{M}^l(s_1 \otimes r_1) \cdot \mathcal{M}^l(s_2 \otimes r_2)\right) = \mathcal{M}^l((s_2 \cdot s_1) \otimes (r_1 \cdot r_2)) \tag{11.48}$$
$$(\mathcal{M}^r(r_1 \oslash s_1) \cdot \mathcal{M}^r(r_2 \oslash s_2)) = \mathcal{M}^r((r_2 \cdot r_1) \oslash (s_1 \cdot s_2)) \tag{11.49}$$

Each of the matrices $\mathcal{M}^l(r \otimes s)$ can be identified with a unique mapping in $T_0(M_m)$, although several might map to the same mapping. Each of the matrices $\mathcal{M}^r(r \oslash s)$ acts, on the *right*, on the group $(M_m, +)$.

Lemma 434

$$\left(\mathcal{M}^l(s_2 \otimes r_2) \cdot \mathcal{M}^r(r_1 \oslash s_1)^t\right) = \left(\mathcal{M}^r(r_1 \oslash s_1)^t \cdot \mathcal{M}^l(s_2 \otimes r_2)\right)$$

Proof

$$\left(\mathcal{M}^l(s_2 \otimes r_2) \cdot \mathcal{M}^r(r_1 \oslash s_1)^t\right) =$$
$$\left(\mathcal{M}^r(s_2)^t \otimes \mathcal{M}^l(r_2)\right) \cdot \left(\mathcal{M}^l(r_1)^t \otimes \mathcal{M}^r(s_1)\right)^t =$$
$$\left(\left(\mathcal{M}^r(s_2)^t \cdot \mathcal{M}^l(r_1)\right) \otimes \left(\mathcal{M}^l(r_2) \cdot \mathcal{M}^r(s_1)^t\right)\right) =$$
$$\left(\left(\mathcal{M}^l(r_1) \cdot \mathcal{M}^r(s_2)^t\right) \otimes \left(\mathcal{M}^r(s_1)^t \cdot \mathcal{M}^l(r_2)\right)\right) =$$
$$\left(\mathcal{M}^l(r_1) \cdot \mathcal{M}^r(s_1)^t\right) \otimes \left(\mathcal{M}^r(s_2)^t \cdot \mathcal{M}^l(r_2)\right) =$$
$$\left(\mathcal{M}^r(r_1 \oslash s_1)^t \cdot \mathcal{M}^l(s_2 \otimes r_2)\right)$$

□

Compare this with Lemma 431.

Lemma 435

(i) The set $\mathcal{K}_1 = \{\mathcal{M}^l(s \otimes r) : r \in R, s \in S\}$ *is a semi-group under standard matrix multiplication.*
(ii) The set $\mathcal{K}_2 = \{\mathcal{M}^r(r \oslash s) : r \in R, s \in S\}$ *is a semi-group under standard matrix multiplication.*

We have mappings

$$\left(S^{opp} \times R\right) \longmapsto \mathcal{K}_1 \quad (s, r) \mapsto \mathcal{M}^l(s \otimes r) \tag{11.50}$$

$$\left(R^{opp} \times S\right) \longmapsto \mathcal{K}_2 \quad (r, s) \mapsto \mathcal{M}^r(r \oslash s) \tag{11.51}$$

These are semi-group homomorphisms. The semi-group $\mathcal{K}_1$ maps homomorphically onto a multiplicative semi-group of $T_0(M_m)$. There will be a smallest sub-near-ring of $T_0(M_m)$ containing this semi-group; I write this $\mathfrak{G}(\mathcal{K}_1)$ (see page 336). $\mathfrak{G}(\mathcal{K}_1)$ acts on M_m as a left near-ring group.

11.6.7 Multiple Near-Ring Groups

Suppose R and S are near-rings, A is a right near-ring group A_R and B is a bimodal near-ring group ${}_RB_S$. Given $r_1, r_2 \in R$, we may form the matrices

$$\mathcal{M}^l(r_1 \circledast r_2) = \left(\mathcal{M}^r(r_1)^t \times \mathcal{M}^l(r_2)\right) \tag{11.52}$$

and

$$\mathcal{M}^r(r_1 \boxtimes r_2) = \left(\mathcal{M}^l(r_1)^t \otimes \mathcal{M}^r(r_2)\right) \tag{11.53}$$

These are $(0-1)$ square matrices with $|A|.|B|$ row and columns. The first is column stochastic and the second row stochastic. They represent left and right actions on two additive groups. Note that although all matrices are square and $(0-1)$, they may be of different sizes. Using the notation of (11.47), the first additive group is

$$(B_{|A|}, +) = \left(\overbrace{(B \oplus B \oplus \ldots \oplus B)}^{(|A| \text{ times})}\right) \tag{11.54}$$

and the second is

$$(A_{|B|}, +) = \left(\overbrace{(A \oplus A \oplus \ldots \oplus A)}^{(|B| \text{ times})}\right) \tag{11.55}$$

One thinks of the elements of these groups as row vectors, tuples, $\underline{v}$.

Lemma 436 *These actions are near-ring group actions.*

We hypothesised a right action of S on the group $(B, +)$. This translates to a right action of S on $B_{|A|}$

$$\left(b_1, b_2, \ldots, b_{|A|}\right)^t \cdot s = \left(b_1 \cdot s, b_2 \cdot s, \ldots, b_{|A|} \cdot s\right)^t \tag{11.56}$$

and that action is a right near-ring group action.

Lemma 437 *Suppose* $r_1, r_2 \in R$ *and* $s \in S$, *and* $\underline{v} \in B^t_{|A|}$, *then*

$$\left(\left(\mathcal{M}^l(r_1 \circledast r_2) \cdot \underline{v}^t\right) \cdot s\right) = \left(\mathcal{M}^l(r_1 \circledast r_2) \cdot \left(\underline{v}^t \cdot s\right)\right) \tag{11.57}$$

This is bimodal.

The reader will realise that there are many structural combinations of matrices that might be explored here and perhaps guess that at least some of my motivation relates to the tensor product functor of homological algebra.

Chapter 12
Radicals

All near-rings in this chapter are right-distributive and zero-symmetric.

The radical theory of rings is a big subject, and it was inevitable that people would seek to generalise it to near-rings [64]. Of course, radical theory is not a domain of ring theory per se and, indeed, originated in non-associative settings [94]. Good sources on radicals in ring theory include [34, 50] and [84]. For near-rings, I relied on [128] and [119], and my personal grasp of these issues owes much to conversations with John Hartney.

The pioneers of radical theory in rings and algebras include Wedderburn, Köthe, Jacobson, Amitsur, Divinsky and Kurosh, although there were many people thinking about radicals in the 1940s and 1950s and interesting stuff produced by more mathematicians than I have space to list.[1]

I should own that I lack expertise in this area and am simply reporting it from a duty of completeness. My account is far less comprehensive or insightful than those of Meldrum or Pilz [119, 128]; and they, naturally, do not cover the considerable body of work on these ideas occurring since their books were published [64]. I also do not give comprehensive proofs of everything in what follows, though I do cite references when complete proofs are not given, and most of these incomplete proofs are actually almost there.

[1] For example, Nagata worked on generalised radicals in the early 1950s. I was unaware of this until I encountered it in the book by Gray [58].

R. Lockhart, *The Theory of Near-Rings*, Lecture Notes in Mathematics 2295,
https://doi.org/10.1007/978-3-030-81755-8_12

12.1 Abstract Radicals

In the mid-twentieth century, there were a number of "radicals" occurring in ring theory and the theory of algebras, and a number of mathematicians formalised the idea of a radical property. The key names in this endeavour are Kurosh and Amitsur, and the description they invented characterises **Kurosh-Amitsur radicals**. I offer a definition in terms of rings, but the idea has much wider scope.

Definition 438 One starts with some class of rings, $\mathfrak{C} \neq \phi$. This class is to have the following properties.

(i) Homomorphic images of rings in $\mathfrak{C}$ are rings in $\mathfrak{C}$.
(ii) For any ring, R, the sub-ring generated by those ideals of the ring that are themselves in $\mathfrak{C}$, which is of course an ideal, is itself in $\mathfrak{C}$. I call this sub-ring $\mathfrak{r}(R)$ or $\mathfrak{r}_{\mathfrak{C}}(R)$.
(iii) For any ring, N, $\mathfrak{r}\left(\frac{N}{\mathfrak{r}(N)}\right) = \{0\}$.

From this, some terminology emerges.

(i) $\mathfrak{r}(N)$ is, of course, called the **radical** or the **$\mathfrak{C}$-radical** of the class $\mathfrak{C}$. (I do use other symbols for particular radicals—notably, $J(R)$ for the Jacobson radical and its offspring.)
(ii) Rings in which $\mathfrak{r}(R) = R$ are called **radical rings** of the class $\mathfrak{C}$ or **C-radical rings**.
(iii) Rings in which $\mathfrak{r}(R) = \{0\}$ are called **semi-simple rings** of the class $\mathfrak{C}$ or **$\mathfrak{C}$-semi-simple rings**.

The study of radicals of this sort is sometimes called **torsion theory**. In torsion theory, it is important that ideals can be rings in their own right, and so people do tend not to insist that rings should be unital. The above definition is a sort of guideline for what radicals should look like and a model from which basic properties may be deduced.

Unfortunately, this description of the ingredients of a radical theory of rings does not work out quite so neatly with near-rings, and there are even radical-like structures in ring theory which don't quite fit into it. The definition was never intended as legislation, but it is a useful guide.

12.1.1 General Radicals

Radicals occur very widely in ring theory as this example shows. Suppose we take any non-empty class of rings and select any ring at all, that is, not necessarily in the class. Then we look at epimorphisms from our selected ring to members of the class. Each such epimorphism has a kernel, and the intersection of all possible kernels gives us an ideal. The ring modulo this ideal cannot have an epimorphism onto a ring of the original class with non-trivial kernel, so in this quotient ring, the kernel intersection is trivial. It can also be shown that this business of having trivial intersections characterises rings which are subdirect products of members of the class—a typical "radical property" in rings [2]. Given any class of rings, there will be a "smallest" radical class containing it. This defines the **lower radical** of the class. There is also an **upper radical class** associated with certain classes of rings that have additional properties. It is actually disjoint from the class which engenders it.

12.1.2 *Divinsky's Definition*

Divinsky tailored Kurosh-Amitsur radicals for rings [34, 58].

Definition 439 A **radical property** is some property of rings for which:

(i) Homomorphic images of rings R with the property have the property.
(ii) Each R ring contains an ideal $\tau(J)$ which has the property in question when regarded as a ring in its own right and which contains all ideals of the original ring that had the property.
(iii) For all rings R, $\tau\left(\frac{R}{\tau(R)}\right) = \{0\}$.

In rings, the sum of nil ideals is nil. This is proved by a simple trick involving quotients (Lemma 67 or [94]). So, nilness is a radical property. Commutative rings have **nilradicals**; non-commutative rings have **upper nilradicals**, also called **Köthe radicals** (see Sect. 12.2.1). . However, the sum of nilpotent ideals is not always a nilpotent ideal, and so nilpotence is not a radical property in this sense. One interesting feature of this definition is that it permits one to compare radical properties. It may be that we have two radical properties and any ring having the first must have the second. The first radical property may be said to be *contained* in the second, and certainly any radical of the first property is contained in one of the second.

12.1.3 *Hoehnke Radicals*

Another concept that is useful in radical theory is that of a **Hoehnke radical** [50]. This is any construction that assigns each ring R, to an ideal of it, $\mathcal{H}(R)$. One further requires that if $\theta R \mapsto S$ is any homomorphism of rings, then

$$\theta(\mathcal{H}(R)) \subseteq \mathcal{H}(\theta(R)) \tag{12.1}$$

and one requires that

$$\mathcal{H}\left(\frac{R}{\mathcal{H}(R)}\right) = 0 \tag{12.2}$$

There is a theory to Hoehnke radicals and useful additional properties that one might require. For example, **complete Hoehnke radicals** have the property that whenever J is an ideal of R with the property $\mathcal{H}(J) = J$, then it must be true that $J \subseteq \mathcal{H}(R)$. **Idempotent Hoehnke radicals** have the property that

$$\mathcal{H}(\mathcal{H}(R)) = \mathcal{H}(R) \tag{12.3}$$

It turns out that the Kurosh-Amitsur radicals correspond to the complete, idempotent, Hoehnke radicals [50]. This characterisation in terms of the radicals themselves seems more useful in near-ring theory, although some of the radical-like structures are not even ideals there! (see $J_{\frac{1}{2}}(N)$, page 409).

12.2 Rings and Modules

One tends to think of radicals as relating to rings, but it is possible to develop the theory in terms of modules and get back to rings by thinking of them as modules over themselves [7]. Ideals of rings and of near-rings are, respectively, rings and near-rings in their own right. This means it is possible and sometimes helpful to define radicals of ideals. I do that later when looking at the prime radical.

12.2.1 The Köthe Radical

The radical was invented by Köthe, following suggestions from Wedderburn. It appeared in the 1930s and was first applied to commutative rings. It is the ideal of nilpotent elements. Nilpotent elements are "roots of zero", in the sense $a^n = 0$, so Köthe called the corresponding ideals, $\mathfrak{r}(R)$, "radicals", which is where the name originally comes from. This radical is itself a nil ideal.

Köthe posed the question of whether the upper nilradical (as we would now call it) contains all the nil left ideals: an unsolved problem that is nearly 100 years old and one of the outstanding open problems in the theory of rings—*the Köthe conjecture*.

12.2.2 The Jacobson Radical for Rings

The classic radical in ring theory is the Jacobson radical; this definition is in [66] and refers to ring modules.

Definition 440 The **Jacobson radical** of a ring is defined as the intersection of the annihilators of all of its irreducible left modules. If there are no irreducible left modules, the radical is defined as the ring itself.

The Jacobson radical is a chameleon and has many characterisations, some of which are specific to particular structures. For example, many ring theory texts insist on unital rings. Such rings always have maximal left ideals—an application of Zorn's lemma. The Jacobson radical is then often defined as the intersection of the maximal left ideals [84]. In unital rings, the Jacobson radical is a proper ideal. Its closest near-ring equivalent, the $J_2(N)$ radical, described below, can have the property $J_2(N) = N$ even in unital near-rings (something reported with surprise in [50]).

The Jacobson radical, as defined, is obviously a two-sided ideal of the ring, and our definition is based on left modules. If we were to work with *right* ring modules instead, we would obtain exactly the same ideal (a theorem of ring theory). In the case of unital commutative rings, the Jacobson radical is the intersection of all the maximal ideals. The intersection of all the prime ideals is **the Baer radical**, and for this class of rings, that is identical to the nil radical, so we have

$$\textit{Nil radical} \equiv \textit{Baer radical} \subset \textit{Jacobson radical} \tag{12.4}$$

for unital, commutative rings. Another characterisation of the Jacobson radical comes directly from the previous work on modular ideals (Sect. 1.28). Given a ring, define $\mathfrak{m}$ as the class of maximal, proper, left ideals that are modular. It is

$$J(R) = \cap \{\rho : \rho \in \mathfrak{m}\} \tag{12.5}$$

Other characterisations are given in [66] and in [84]. The radical class that characterises the Jacobson radical is the class of *primitive rings*. Alternatively, it is the class of all left quasi-regular rings. Alternatively, it is the class of rings R for which the *circle operation* is a group (page 61).

Properties of the Jacobson Radical

The Jacobson radical has astonishing properties and, on its own, is the subject of a number of books. I now list some of its best-known properties, results garnered from a number of ring theory texts. Note that in (v) and in (vi), the nilpotence referred to is the stronger form (see Definition 66).

(i) It is the unique, maximal, left quasi-regular left ideal in the ring R.
(ii) One can define a *right* Jacobson radical, but it is equal to the left one.
(iii) The Jacobson radical is a two-sided ideal.
(iv) The Jacobson radical contains all the nil left or right ideals of its ring.
(v) When the ring R is *left Artinian*, the Jacobson radical, $J(R)$, is a strongly nilpotent ideal.
(vi) In a left Artinian ring, any nil left, right or two-sided ideal is also strongly nilpotent.

(vii) If A is an ideal of a ring R, then $J(A) = (A \cap J(R))$. One consequence of that is that $J(J(N)) = J(N)$.
(viii) The Jacobson radical of the ring of square matrices over a ring R is the corresponding (same size) ring of square matrices over $J(R)$.
(ix) The Jacobson radical of a direct product of a family of rings is the corresponding direct product of the individual Jacobson radicals.
(x) If a ring R is *Jacobson semi-simple*—that is, $J(R) = 0$, then so is any ideal of the ring.

Some of these results follow directly from the work of Sects. 1.28 and 1.29. Complete details can be found in most standard ring theory texts, particularly [66, 84] and [71]. I don't give the proofs here because it would take up too much space and the results are widely available elsewhere. One example, perhaps. Nilpotent elements are left quasi-regular (Lemma 107). The upper nilradical is a nil ideal, so it consists entirely of left quasi-regular elements and lies in the Jacobson radical. That is, the Jacobson radical contains all the nil ideals. [(vi)] is known as **Hopkins theorem**. And this important result should be mentioned; this version is in [15].

Lemma 441 (Nakayama's Lemma) *Suppose R is a unital ring and A is a right ideal of R and $A \subseteq J(R)$. Then, these two conditions are equivalent for every finitely generated module, M.*

A If N is a submodule of M and $(N + M \cdot A) = M$, then $N = M$.
B If $M \cdot A = M$, then $M = \{0\}$.

Nakayama's lemma has various forms and was first proved for commutative rings. Here is Atiyah's version.

Lemma 442 *Suppose R is a commutative unital ring and K is an ideal contained in $J(R)$. Suppose M is a finitely generated module; then*

$$K \cdot M = \{0\} \Rightarrow M = \{0\} \tag{12.6}$$

Nakayama's lemma is more often known as the **Krull-Azumaya theorem** nowadays. As with so many results in algebra, the rather restricted situation in which it was first proved provides a more startling result than its generalisations, at first sight. Obviously, it is attractive to see whether results of this sort obtain in near-rings and to an extent they do, but the situation becomes more complicated and the classical Jacobson radical splits into at least three cases (just as primitive rings do). Ring theory boasts many other radicals [34, 50, 51]. Some of these radicals can be at least partially extended to near-rings.

12.2.3 *Chain Conditions in Rings*

I've already sketched over some chain conditions in Sect. 2.5.7. Recall that one defines left Noetherian and left Artinian rings in terms of the behaviour of chains of left ideals. The next two results are copied directly from [7]. Ring theory definitions are in Sect. 1.8.9.

Theorem 443 (Artin-Wedderburn) *For any unital ring R, these are equivalent conditions.*

(i) R is both left Artinian and $J(R) = \{0\}$.
(ii) The module ${}_R R$ is a direct sum of simple R-modules.
(iii) R is isomorphic to a finite direct sum of complete rings of square matrices over division rings.

Theorem 444 (Hopkins) *Any left Artinian unital ring is left Noetherian.*

12.3 Near-Rings

12.3.1 *Radical Maps*

The best suggestion so far of a framework for treating near-ring radicals seems to be the following [119, 128], which I believe owes its origins to universal algebra.

Definition 445 A **radical map** is a mapping which assigns to each near-ring N and ideal $\mathfrak{r}(N)$ which is its associated radical. The mapping should have these properties.

(i)

$$\mathfrak{r}\left(\frac{N}{\mathfrak{r}(N)}\right) = \{0\} \tag{12.7}$$

(ii) If $\theta : N \mapsto W$ is a homomorphism of near-rings, then

$$\theta\left(\mathfrak{r}(N)\right) \subseteq \mathfrak{r}\left(\theta(N)\right) \tag{12.8}$$

Again, we define a near-ring as **semi-simple**, with respect to this radical mapping provided $\mathfrak{r}(N) = \{0\}$, and we define it as **radical** with respect to it, if $\mathfrak{r}(N) = N$.

This definition is a bit less restrictive than Divinsky's and considerably less restrictive than Definition 438.

12.3.2 Near-Ring Radicals

This definition requires knowledge of Definition 414.

Definition 446

(i) Suppose $(N, +, \cdot)$ is a near-ring, and suppose J is an ideal of N such that the corresponding quotient near-ring is ν-primitive; then J is called a **ν-primitive ideal** ($\nu = 0, 1, 2$).

(ii) A left ideal J is said to be ν-modular provided it is modular and the quotient $\left(\frac{N}{J}\right)$ is a type ν near-ring group under the obvious N-action

$$(m \cdot \{n + J\}) = (\{(m \cdot n) + J\}) \tag{12.9}$$

Clearly, ν-primitive ideals are exactly the annihilators of ν-primitive near-ring groups (see Lemma 105).

Lemma 447 *Suppose $(N, + \cdot)$ is a zero-symmetric near-ring. An ideal J of N is ν-primitive if and only if it has the form*

$$J = \binom{N}{K}_{\mathcal{L}} \tag{12.10}$$

where K is some modular left ideal of N.

Proof Suppose the ideal J is ν-primitive. The factor near-ring $\left(\frac{N}{J}\right)$ acts primitively upon some type ν near-ring group. We can extend this action so that N acts on that near-ring group with annihilator J. Acting N on a monogenic generator and using Lemma 105, we have that the near-ring group is isomorphic to the quotient of N by the modular left ideal which is the annihilator of the generator; call that left ideal K. K contains J. From Lemma 56,

$$J = \binom{N}{K}_{\mathcal{L}} \tag{12.11}$$

Going the other way, suppose J is an ideal for which (12.11) holds for some ν-modular left ideal K. We know $\left(\frac{N}{K}\right)$ is a type ν near-ring group, acted on by N. The annihilator of this near-ring group is

$$\{n \in N \; : \; n \cdot N \subset L\} = J$$

So, J is a ν-primitive ideal. □

Definition 448 The **ν-radical** of the near-ring $(N, +, \cdot)$ is defined as

$$J_\nu(N) = \bigcap_{\mathcal{X}} \left(\binom{\Gamma}{0}_{\mathcal{L}}\right) \tag{12.12}$$

where $\mathcal{X}$ is here defined as the class of type ν near-ring groups, Γ, of N. If there are no type ν near-ring groups, the corresponding radical is defined to be N itself.

From the definitions,

$$J_0(N) \subseteq J_1(N) \subseteq J_2(N) \tag{12.13}$$

From Lemma 415, we can say that when the near-ring has an identity,

$$J_1(N) = J_2(N) \tag{12.14}$$

When N is a ring, these three radicals collapse into the standard Jacobson radical, but there are examples of near-rings in which they are distinct. Both Pilz and Meldrum state and prove this result which I merely state, although the proof is fairly immediate.

Lemma 449 *The mapping $N \mapsto J_\nu(N)$ is a radical map for $\nu = 0, 1, 2$.*

These radicals do not fit into the classification of radical classes which can be used in ring theory. For example, it can be shown that there are zero-symmetric near-rings for which $J_0(J_0(N)) \neq J_0(N)$ and there are zero-symmetric near-rings for which $J_1(J_1(N)) \neq J_1(N)$ [50] (see Lemma 453). This lemma is directly from [128]. It comes from the results I have listed in Sects. 1.28, 1.29, and 12.3. First, let $(N, +, \cdot)$ be a near-ring and define

(i) $\mathfrak{W}_\nu$ is the class of ν-primitive ideals of N.
(ii) $\mathfrak{U}_\nu$ is the class of ν-modular left ideals of N.

Lemma 450 *The radical $J_\nu(N)$ is alternatively described as*

(i) $J_\nu(N) = \cap\{X : X \in \mathfrak{W}_\nu\}$.
(ii) $J_\nu(N) = \cap\left\{\binom{N}{X}_{\mathcal{L}} : X \in \mathfrak{U}_\nu\right\}$.
(iii) When $\nu \neq 0$, $J_\nu(N) = \cap\{X : X \in \mathfrak{U}_\nu\}$.

This last result is illustrative of two points relating to the radical theory of near-rings. First, there really is one. Second, it's much more complicated than the ring version. I say this because people have gone on to define

$$J_{\frac{1}{2}}(N) = \cap\{X : X \in \mathfrak{U}_0\} \tag{12.15}$$

This is certainly a left ideal but not usually a two-sided one—so we are already out of the comfort zone provided by Sect. 12.1.2. It turns out that

$$J_0(N) \subseteq J_{\frac{1}{2}}(N) \subseteq J_1(N) \subseteq J_2(N) \tag{12.16}$$

which explains the notation (due to Pilz). This next result is due to Ramakotaiah [50, 128].

Lemma 451 *$J_0(N)$ is the left ideal core of $J_{\frac{1}{2}}(N)$.*

Proof Lemma 105

$$\left(\left(\bigcap\{X : X \in \mathfrak{U}_0\}\right)^{\{N\}}\right)^{N}_{\mathcal{L}} = \bigcap_{\{X : X \in \mathfrak{U}_0\}} \left(\left(\frac{N}{X}\right)^{N}_{\mathcal{L}}\right) = J_0(N)$$

□

Meldrum proves that $J_{\frac{1}{2}}(N)$ (which, in his terms, is not actually a radical, being often not an ideal) is both a left quasi-regular left ideal and contains all the other left quasi-regular left ideals [119]—Theorem 5.34. From Lemma 107, this means

Lemma 452 *All nil ideals of a near-ring N are contained in $J_0(N)$.*

Pilz gives this lemma as his Theorem 5.20. My account automatically restricts to zero-symmetric near-rings so my version of this is less general than his.

Lemma 453 *Suppose $\{N_a\}_{a \in A}$ represents some family of near-rings. For $\nu \in \{0, 1, 2\}$,*

(i)

$$J_\nu\left(\prod_A N_a\right) \subseteq \prod_A (J_\nu(N_a)) \tag{12.17}$$

(ii)

$$J_\nu(\oplus_A N_a) \subseteq \oplus_A (J_\nu(N_a)) \tag{12.18}$$

and if $\nu = 2$, equality occurs here.

12.3.3 Chain Conditions in Near-Rings

It's obvious from the results I have quoted for rings that finiteness conditions, particularly chain conditions, often give sharp results in ring theory. This is true for near-rings too except that, inevitably, the number of possible chain conditions proliferates. I only find a small amount of space for this. Pilz and Meldrum both give a much more comprehensive story [119, 128]. Chain conditions defined for near-rings include:

Definition 454

A The descending chain condition on ideals.
B The descending chain condition on left ideals.

C The descending chain condition on right ideals.
D The descending chain condition on two-sided near-ring subgroups.
E The descending chain condition on left near-ring subgroups.
F The descending chain condition on right near-ring subgroups.

and

Definition 455

A The ascending chain condition on ideals.
B The ascending chain condition on left ideals.
C The ascending chain condition on right ideals.
D The ascending chain condition on two-sided near-ring subgroups.
E The ascending chain condition on left near-ring subgroups.
F The ascending chain condition on right near-ring subgroups.

Note that $(455)_A$ implies all ideals are finitely generated (see Lemma 140). Authors habitually refer to these different conditions using simple codes. For example, Pilz would use DCCI to mean 455_A. This is, of course, transparently clear to everyone except me. I find my more ponderous notation clearer. There are a great many interesting theorems involving these chain conditions and others. For example, this theorem, reported in [128] as Theorem 5.48, is by Ramakotaiah.

Theorem 456 *Suppose N is a near-ring satisfying* 454_E*. Then, these conditions are equivalent.*

(i) $J_2(N)$ *is nil.*
(ii) $J_2(N)$ *is nilpotent.*
(iii) $J_2(N)$ *is left quasi-regular.*
(iv) $J_2(N) = J_1(N) = J_{\frac{1}{2}}(N) = J_0(N)$.
(v) Every type 0 near-ring group is a type 2 near-ring group.
(vi) 0-primitive ideals are 1-primitive and they are 2-primitive.
(vii) Non-trivial prime ideals are 2-primitive.

and this result, attributed to Betsch, occurs as Corollary 4.47 in the second editions of Pilz's book (compare with Theorem 412).

Theorem 457 *Suppose N is a (zero-symmetric), non-trivial, near-ring containing a left identity and J is an ideal. Suppose N satisfies* 454_E*.*

(i) N is 1-primitive if and only if it is 2-primitive, and this occurs if and only if it is simple.
(ii) J is 1-primitive if and only if it is 2-primitive, and this occurs if and only if it is maximal.

12.3.4 The Ring Radical

The map $(N, +, \cdot) \mapsto \mathcal{A}_N$ (page 55) is a radical map defining what one might call the **ring radical**. This is also a complete Hoehnke radical, but not idempotent. The mapping

$$\mathcal{A}_N \longmapsto N \overset{\pi}{\twoheadrightarrow} \left(\frac{N}{\mathcal{A}_N}\right) \tag{12.19}$$

links rings to near-rings. Of course, for some near-rings, the image of π is trivial.

There are important classes of near-rings in which $J_2(N)$ is nil (see Theorem 456). This makes $\pi(J_2(N))$ a nil ideal of the quotient and so

$$\pi(J_2(N)) \subseteq J\left(\frac{N}{\mathcal{A}_N}\right) \tag{12.20}$$

Radicals of ideals in rings are often defined using inverse images, and perhaps it would be worth exploring the inverse images of various ring radicals using this exact sequence. I am unaware of any work relating radicals in the near-ring N to radicals in the ring $\left(\frac{N}{\mathcal{A}_N}\right)$ using this mapping.

12.3.5 The Nil Radical

The nil radical is defined as the sum of the nil ideals of a near-ring. Lemma 67 tells us that this is itself a nil ideal. From Lemma 452, we can say that the nil radical is contained in $J_0(N)$. It's easy to show that the nil radical gives us a radical map.

12.3.6 The Prime Radical

Definition 458 The prime radical of an ideal J is defined as the intersection of the prime ideals containing it, $\mathfrak{p}(J)$. The prime radical of a near-ring is defined as the prime radical of the trivial ideal. I shall call it $\mathfrak{p}(N)$.

From Lemma 37, one can say that the prime radical of an ideal is itself a semi-prime ideal. It can be shown that the prime radical of a near-ring offers a radical map [119].

Lemma 459 *Suppose N is a near-ring and $w \in N$ is an element of the prime radical of some ideal J. Then for some finite $n \in \mathbb{N}$, $w^n \in J$.*

Proof If this is not the case, one can consider the collection of ideals containing J but not containing any power of w. A Zorn's lemma argument gives us a maximal such ideal, M with that property. Suppose W and V are two ideals such that $W.V \subset$

M. One of these two must not contain a power of w. Suppose it is W. If the ideal $(W + M)$ is strictly larger than M, it contains powers of w, so $(W + M) \cdot V$ must contain such powers. Yet, $(W + V) \cdot M \subseteq M$. The conclusion is that $(W + M) = M$, that is, $W \subset M$. This makes M a prime ideal containing J, and that means $w \in M$.

□

Lemma 460

(i) *If J is an ideal in the near-ring N, then $\mathfrak{p}(J)$ contains all the nilpotent ideals of N.*
(ii) *The ideal $\mathfrak{p}(\{0\})$ is nil.*

Proof

(i) This result essentially uses Lemma 43. A nilpotent ideal K gives rise to a nilpotent ideal $\frac{(K+\mathfrak{p}(J))}{\mathfrak{p}(J)}$, and this must be trivial.
(ii) This is an application of Lemma 459.

□

12.3.7 Prime Ideals in Ring Theory

Prime ideals are central in ring theory where the **prime radical** is defined as the intersection of the prime ideals and **prime rings** are rings in which the trivial ideal is prime. **Semi-prime rings** are defined as rings in which the prime radical is trivial. These are exactly the rings with no non-trivial nilpotent ideals. There is this theorem [71]:

Theorem 461 *A ring is semi-prime if and only if it is a subdirect product of prime rings.*

The Spectrum
The collection of prime ideals of a commutative unital ring is called its **spectrum**. This collection is topologised by the **Zariski topology** and has important applications in algebraic geometry.

12.3.8 Further Work

Radical theory traditionally works with individual rings. In near-rings, the **host problem** seems more insistent, and one wonders whether one could consider classes of near-rings hosted by a given group in terms of their possession of various sorts of radical. So far as I am aware, this is a problem that has not been considered.

12.4 The Trouble with Generalisations

A lot of the early work on near-rings sought to exploit clear successes from ring theory and commutative algebra. This was most notable in representation and radical theory, with many papers on these topics in the 1960s and 1970s [128], but it also included generalisations of standard topics such as group near-rings [119], tensor products [110], category theory [69] and universal algebra.

Some of this work was remarkable, but the utility of these structures is not yet proven in all cases. I wanted to use the chapters in this section of the book to give something of the flavour of generalisations of this sort, with the next chapter, on matrices, representing the most accessible and immediate case I could think of.

The two chapters just given were partly a response to Fröhlich's foray into near-rings, although there was an equally strong German current in their construction, stemming from the work of Zassenhaus and Wielandt and initially based on permutation groups and near-fields.

Fröhlich had been motivated by a desire to give a reasonable and perhaps useful extension of what was, in those days, the nascent, abelian theory of homological algebra. His work starts with a generalisation of endomorphism rings—the d.g. near-rings, and he goes on to build the non-abelian constructs needed for this generalisation. In considering the "addition" of endomorphisms, he took on board work by Hanna Neumann ([124] and Sect. 8.5) incorporating that into his general theory. What set Fröhlich's work apart was that the generalisations he produced, including generalisations of group rings and the functors of homological algebra, were motivated by the exigencies of what he was attempting—a desiderata that not all of us can manage.

As a sort of mathematical grandson (my doctoral supervisor was his doctoral student), when I came to do my own research in near-rings, I followed the path he had originated, but I found his ideas, though interesting, were very complicated and not always in keeping with subsequent developments in homological algebra; and I invented new generalisations in an effort to match the very first papers in the subject—going back to basics. This is how I came to discover F-near-rings and phomomorphisms, and I found them sufficiently manipulable to obtain a cohomology theory that fitted in with the abelian origins of the subject (Sect. 9.2), although it perhaps lacked the sophistication of the modern theory of derived functors which was something that Fröhlich had achieved.

What stopped me from working on my own generalisation was the lack of immediate application to algebra, though I consoled myself that results like Lemma 380 are at the very least suggestive, in that they provide a direct connection with the standard abelian situation, and even Eilenberg and Maclane initially struggled to give meaning to the higher cohomology groups [37], not always being spectacularly successful!

Some of the people who read the first drafts of this book thought that I might give clearer motivations for new material and I am attempting that here. On that basis, I include all the papers by Fröhlich on near-rings in the references. They tell a coherent story and are well worth reading in their own right, although you may find, as I did, that they might appear to be a little forced if you are equipped with the fashionable modern luxury of hindsight [41–48].

Chapter 13
Matrices

The reader may wish to review the material covering cores (around page 43) and also Notation 61, before continuing.

Suppose $(N + .)$ is any general near-ring. We form standard $n \times n$ matrices over N and consider their properties.

They will be multiplicatively zero-symmetric exactly when the original near-ring was. Similarly, possession of an identity is preserved.

However, the structure obtained is not generally even a pre-near-ring. This happens exactly when the subgroup $r^{(2)}(N)$ is abelian.

Supposing that to be the case, we might attempt to determine the associativity core of this pre-near-ring—matrices of the form $\left(a_{i,j}\right)$ for which

$$\sum_{s=1}^{n}\left(a_{i,s}.\left(\sum_{t=1}^{n} b_{s,t}.\mathrm{C}_{t,j}\right)\right) = \sum_{t=1}^{n}\left(\left(\sum_{s=1}^{n} a_{i,s}.b_{s,t}\right).\mathrm{C}_{t,j}\right)$$

This is the requirement that the elements $a_{i,j}$ right distribute over elements of $r^{(2)}(N)$.

This is certainly satisfied when the matrix $\left(a_{i,j}\right)$ entirely consists of elements from the distributivity core of N, and when the original near-ring possesses either a left or a right multiplicative identity, the correspondence is exact.

That is, the associativity subgroup consists of matrices $\left(a_{i,j}\right)$, where $a_{i,j} \in \mathrm{C}_{\mathcal{D}}(N)$ for all i, j, $\in \{1, 2, \ldots n\}$.

Of course, $\mathrm{C}_{\mathcal{D}}(N)$ is additively closed and hence a sub-near-ring in that case and it is actually a ring (see Corollary 89).

So, if the original near-ring has either a left or a right identity, the matrix structure will be a near-ring exactly when the original near-ring is a ring, and then of course it is itself a ring.

This attempt to construct a theory of "matrix near-rings" dissolves into the standard ring case or structures with intractable properties.

R. Lockhart, *The Theory of Near-Rings*, Lecture Notes in Mathematics 2295,
https://doi.org/10.1007/978-3-030-81755-8_13

13.1 The Work of Meldrum and van der Walt

There have been various attempts to generalise matrix rings. This idea dates from 1984 and is due to John Meldrum and Andries van der Walt.

The idea is based on the role elementary matrices play within the general linear group over a finite field.

These elementary matrices have the form

$$E_{i,j} = \left(e_{r,s}\right) \textit{ where } e_{r,s} = (\delta_{r,i}.\delta_{j,s}) \tag{13.1}$$

(using Kronecker's delta—page xv).

They are important in the analysis of linear groups [132], and they are additive generators of the general linear group over a finite field.

Meldrum and van der Walt insist on a zero-symmetric near-ring with identity, $(N, +, \cdot)$.

They form the direct sum N^n, and, for each $r \in N$, and $1 \le i, j \le n$ define mappings:

$$f^r_{i,j} : N^n \longmapsto N^n \tag{13.2}$$

$$(r_1, r_2, \ldots, r_n) \mapsto (r'_1, r'_2, \ldots r'_n) = (0, 0, \ldots, 0, r.r_j, 0, \ldots 0) \tag{13.3}$$

$$r'_k = \delta_{i,k}(r.r_j) \tag{13.4}$$

(only the i_{th} position is non-zero, and that is $(r.r_j)$).

Their definition of a matrix near-ring $M_n(N)$ is the sub-near-ring of $T_0(R^n)$ (see Chap. 7) generated by

$$\left\{f^r_{i,j} : r \in N, 1 \le i, j \le n\right\} \tag{13.5}$$

This is obviously a zero-symmetric, right-distributive near-ring, with identity

$$I = \left(\sum_{i=1}^{n} f^1_{i,i}\right) \tag{13.6}$$

(see [12]).

There is considerably more to know about matrix near-rings [119, 120], and there are a number of papers on them listed in the Linz bibliography [101].

Chapter 14
F-Near-Rings

Near-rings occurring here are usually assumed to be zero-symmetric.

Recall that a **right** *F-near-ring* is a zero-symmetric, right-distributive near-ring satisfying:

$$m \cdot (n + k) \equiv (m \cdot n) + (m \cdot k) \text{ (modulo } ([N, N]))$$

$\forall\, m, n, k, \in N$ (Definition 75).

d.g. near-rings are F-near-rings (page 50). F-near-rings represent a generalisation of d.g. near-rings, which are a part of near-ring theory that has been extensively explored ever since it became a focus of attention in the 1950s.

F-near-rings form a subclass of the class of near-rings which have homomorphic images which are rings. They are near-rings in which the commutator subgroup is an ideal containing the distributivity core. F-near-rings hosted by abelian groups will be rings.

Standard d.g. conditions apply to *some* set of additive generators: usually, not all of them (after all, the set of all elements will not normally be distributive). In the case of F-near-rings, we can establish the F-property by verifying it for *any* set of additive generators at all.

d.g. near-rings are zero-symmetric. I have to specifically insist on this for F-near-rings. If I do not, the elements $x \cdot 0$ will be in $[N, N]$ and have the property

$$(x \cdot 0) \cdot y = (x \cdot 0) \tag{14.1}$$

(see Sect. 1.24.6), so the set

$$\{(x \cdot 0) : x \in N\} \tag{14.2}$$

R. Lockhart, *The Theory of Near-Rings*, Lecture Notes in Mathematics 2295,
https://doi.org/10.1007/978-3-030-81755-8_14

is an additive subgroup of $[N, N]$, a two-sided sideal and a constant near-ring.

The seven non-isomorphic near-rings with identity hosted by D_4 are listed in Appendix B. Some of them are F-near-rings, and the ones that are are identified there.

Although sub-near-rings of F-near-rings need not themselves be F-near-rings, so ideals are perhaps not F-near-rings in their own right, the quotient structures they supply will be F-near-rings. Ring theorists sometimes describe a property that is always inherited by ideals as a **hereditary** property. On that basis, the F-near-ring property would not be hereditary. However, the term is much more commonly used in relation to modules being projective, which is how it occurs in [18]. Direct sums and products of F-near-rings will be F-near-rings, and so will ultra-products (page 101).

Sub-near-rings of near-rings which actually *are* F-near-rings in their own right are rather intractable. One cannot easily obtain results about intersections and possible cores because, if H and K are subgroups of a group, one cannot expect that the identity

$$[H \cap K, H \cap K] \subseteq ([H, H] \cap [K, K]) \tag{14.3}$$

would often supply us with an equality. In other words, we cannot expect the intersection of sub-near-rings which are themselves F-near-rings to furnish us with further F-near-rings. I don't think that situation is entirely hopeless, and I hope someone might look at the prospects for some sort of lattice, perhaps using techniques that involve topologising.

14.1 The Pseudo-Centre

I have already observed that in near-rings the multiplicative centre is not usually a sub-near-ring.

Definition 462 Suppose $(N, +, \cdot)$ is an F-near-ring. The **pseudo-centre** is defined as

$$\mathcal{PZ}(N, +, \cdot) = \{z \in N \ : \ (z \cdot a) \equiv (a \cdot z)(modulo\ [N, N]_{(add)}) \forall a \in N\} \tag{14.4}$$

Lemma 463 *The pseudo-centre is a sub-near-ring of the F-near-ring* $(N, +, \cdot)$.

14.2 Ideals

Lemma 464 *Suppose* $(N, +, \cdot)$ *is a zero-symmetric F-near-ring and* $(I, +) \lhd (N, +)$. *I is a left ideal of N if and only if* $\forall\, f \in N, \forall\, i \in I$

$$(f \cdot i) = (i' + \delta) \text{ where } i' \in I \text{ and } \delta \in [N, N]_{(add)} \tag{14.5}$$

14.3 F-Stems

Definition 465

(1) A stems $(N + \cdot)$ for which

$$(a + b) \cdot c \equiv (a \cdot c) + (b \cdot c) \pmod{[N, N]}$$

will be called a **right F-stems**.

(ii) The opposite property

$$a \cdot (b + c) \equiv (a \cdot b) + (a \cdot c) \pmod{[N, N]}$$

characterises **left F-stems**.

(iii) **Bi-F-stems** are stems which are simultaneously both left and right F-stems.

Thus, any *right near-ring* is also a *right F-stem*; and both right F-near-rings and left F-near-rings are bi-F-stems. The distributivity property characterising right F-stems is sometimes called *pseudo-right distributivity*.

Notation 466 The set of all bi-F-stems hosted by the additive group $(N, +)$ will be denoted by **BiF**(N) (see Sect. 15.5.1).

A general defect in the theory of F-stems is that sub-stems are not necessarily themselves F-stems, and we see from Lemma 390 that near-rings of phomomorphisms might not themselves be pseudo-left-distributive—that is, they may well not be F-near-rings.

F-near-rings occur throughout this book. Examples of this are to be found on pages 48, 63, 75, 296, 363, 372 and 498 and elsewhere. I had intended to give more time to them than I found. I have been claiming for years that ideas present in the theory of d.g. near-rings often transfer easily to them. Imagine my surprise when writing this book to find that was so often the case!

There is an example of an F-near-ring which is not d.g. in Appendix B. It seems plausible that only a small percentage of F-near-rings are d.g., which would make the generalisation potentially important. I do hope other people might find them interesting and worthy of further study.

There are a number of further generalisations that may be worth investigating. I believe Tharmaratnam considered topologising d.g. near-rings,[1] and it's probable one could do something similar here, investigating topological near-rings containing dense sub-near-rings which are themselves F-near-rings [128].

[1] I last encountered him at a near-ring conference, which he told me he had crossed a minefield in the Jaffna peninsula to get to.

Part V
Product Theory

Chapter 15
Product Theory

In this chapter, I develop some of the ideas introduced in Sect. 8.7. All mappings here will be *normalised* in the sense that they shall preserve additive group identities (Definition 45). The reader is advised that one does need to pay very careful attention to the notation used in what follows and might wish to remind themselves of the concept of a *stem* (page 7).

Many of the near-rings in this chapter are left-distributive.

This chapter is about obtaining a connection between the associative and distributive aspects of the near-ring product and the splitting of an exact sequence. It mirrors work on group extensions associated with work of Otto Schreier and others [93] but applies specifically to near-ring products. The reader may find the next few pages rather dry and technical and might be motivated by the fact that they lead to Lemma 477.

15.1 General Case

Normally, $(N\ +\ \cdot)$ will be a right-distributive near-ring, together with a left ideal $R \lhd_l N$ (Definition 12) and quotient group $B = \frac{N}{R}$. Left-distributive near-fields do crop up in Sect. 15.6 and Chap. 16. Initially, no assumption of zero symmetry is made. There is the exact sequence of groups

$$R \mapsto N \stackrel{\pi}{\twoheadrightarrow} B \tag{15.1}$$

where π is the canonical epimorphism.

R. Lockhart, *The Theory of Near-Rings*, Lecture Notes in Mathematics 2295,
https://doi.org/10.1007/978-3-030-81755-8_15

Select a **lifting** or **section** of π
(that is a mapping

$$i : B \to N \tag{15.2}$$

for which $\pi \circ i = id_B$). We may regard elements of N as pairs $\{(i(b), r)\}_{b\in B, r\in R}$.
I fix new operations $\oplus$ and $\otimes$, on the set N, using the formulae

$$(i(b_1)\ ,\ r_1)\ \oplus\ (i(b_2)\ ,\ r_2)\ =\ (i \circ \pi\ (i(b_1) + i(b_2))\ ,\ r_1 + r_2) \tag{15.3}$$

$$(i(b_1)\ ,\ r_1)\ \otimes\ (i(b_2)\ ,\ r_2)\ =\ (i \circ \pi\ (r_1 \cdot i(b_2))\ ,\ r_1 \cdot r_2) \tag{15.4}$$

I refer to this continually throughout this chapter.
Notice that

$$\begin{aligned}(i(b_1), r_1) \oplus (i(b_2), r_2) &= (i \circ \pi\ (i(b_1) + i(b_2))\ , r_1 + r_2)\\ &= (i\ (b_1 + b_2)\ , r_1 + r_2)\end{aligned} \tag{15.5}$$

Lemma 467 *(N $\oplus$ $\otimes$) is a right-distributive near-ring.*

Proof (N $\oplus$) is straightforwardly a group with identity element (0, 0). (This needs mapping normalisation.) For right distributivity,

$$\begin{aligned}((i(b_1), r_1) \oplus (i(b_2), r_2)) \otimes (i(b_3), r_3) &=\\ (i\ (b_1 + b_2)\ , r_1 + r_2) \otimes (i(b_3), r_3) &=\\ (i \circ \pi\ ((r_1 + r_2) \cdot i(b_3))\ , (r_1 + r_2) \cdot r_3) &=\\ (i \circ \pi\ (r_1 \cdot i(b_3) + r_2 \cdot i(b_3))\ , (r_1 \cdot r_3 + r_2 \cdot r_3)) &=\\ (i\ (\pi(r_1 \cdot i(b_3)) + \pi(r_2 \cdot i(b_3)))\ , (r_1 \cdot r_3 + r_2 \cdot r_3)) &=\\ (i \circ \pi\ (i \circ \pi(r_1 \cdot i(b_3)) + i \circ \pi(r_2 \cdot i(b_3)))\ , (r_1 \cdot r_3 + r_2 \cdot r_3)) &=\\ (i \circ \pi(r_1 \cdot i(b_3)), r_1 \cdot r_3) \oplus (i \circ \pi(r_2 \cdot i(b_3)), r_2 \cdot r_3) &=\\ ((i(b_1), r_1) \otimes (i(b_3), r_3)) \oplus ((i(b_2), r_2) \otimes (i(b_3), r_3))&\end{aligned}$$

For associativity,

$$\begin{aligned}(i(b_1), r_1) \otimes ((i(b_2)\ ,\ r_2) \otimes (i(b_3)\ ,\ r_3)) &=\\ (i(b_1), r_1) \otimes (i \circ \pi(r_2 \cdot i(b_3))\ ,\ r_2 \cdot r_3) &=\\ (i \circ \pi\ (r_1 \cdot i \circ \pi(r_2 \cdot i(b_3)))\ ,\ r_1(r_2 \cdot r_3))&\end{aligned}$$

and

$$((i(b_1), r_1) \otimes (i(b_2), r_2)) \otimes (i(b_3), r_3) =$$
$$(i \circ \pi (r_1 \cdot i(b_2)), r_1 \cdot r_2) \otimes (i(b_3), r_3) =$$
$$(i \circ \pi (r_1 \cdot r_2 \cdot i(b_3)), (r_1 \cdot r_2) \cdot r_3)$$

The equality

$$(i \circ \pi (r_1 \cdot i \circ \pi(r_2 \cdot i(b_3))), r_1(r_2 \cdot r_3)) =$$
$$(i \circ \pi (r_1 \cdot r_2 \cdot i(b_3)), (r_1 \cdot r_2) \cdot r_3)$$

follows from the fact that R is a left ideal. □

Corollary 468 *If the original near-ring* $(N, +, \cdot)$ *is zero-symmetric, then so is* $(N, \oplus, \otimes)$.

Corollary 469 *The group* $(N, \oplus)$ *is the direct sum of* $(B, +)$ *and* $(R, +)$.

Corollary 470 *The set* $\{(0, r) : r \in R\}$ *is a left ideal of* $(N, \oplus, \otimes)$. *It is isomorphic to R as a near-ring, and I identify it with R in what follows.*

Corollary 471 *When R is a two-sided ideal, any two choices of lifting produce isomorphic near-rings.*

Proof The product defined by (15.4) becomes

$$(i(b_1), r_1) \otimes (i(b_2), r_2) = (0, r_1 \cdot r_2)$$

when R is a right ideal. □

15.2 The Stem $(N, +, \star)$

In general, the multiplicative structure of the near-ring $(N, \oplus, \otimes)$ depends on the choice of lifting i in (15.2). That is, the symbol $\otimes$ is dependent on the chosen lifting i.

With that in mind, suppose that we have another lifting of π in (15.1)

$$j : B \to N \quad (\pi \circ j = \mathrm{Id}_B) \tag{15.6}$$

(see Definition 342 on page 333).

Of course, for all $b_1, b_2 \in B$

$$j(b_i) = (i(b_i), r_i) \quad (i = 1, 2)$$

(here r_1 and r_2 are elements of R).

This new lifting permits us to define two operations, this time, on the set B—with the prescription

$$\left(b_1 \left(\frac{\oplus}{\oplus}\right) b_2\right) = \pi \left(j(b_1) \oplus j(b_2)\right) \tag{15.7}$$

$$(b_1 \star b_2) = \pi \left(j(b_1) \otimes j(b_2)\right) \tag{15.8}$$

Equation (15.7) brings us nothing new; it is the usual addition on the quotient group, and I shall not employ the special symbol $\left(\frac{\oplus}{\oplus}\right)$ again.

Equation (15.8) is different. It uses (15.4) and so involves our original fixed lifting i of (15.2). This new multiplicative product will not necessarily have the usual algebraic properties of near-ring products and may not even be zero-symmetric. This product ($\star$) is useful in characterising the near-rings and general binary products hosted by the group $(B, +)$.

Definition 472 $(B, +, \star)$ is the **stem defined using the lifting j, relative to i.**

Notice that the policy of normalising mappings does at least ensure that for all $b \in B$, we have that $0 \star b = 0$. If the original near-ring $(N, +, \cdot)$ was zero-symmetric, then $(B, +, \star)$ will actually be a *bi-zero-symmetric stem.*

We can write the product ($\star$) more directly in terms of the original near-ring operations as

$$(b_1 \star b_2) = \pi \left(r_1 \cdot i(b_2)\right) \tag{15.9}$$

In the case in which j and i are identical, the resulting product has $r_1 = 0$, and so $b_1 \star b_2 = 0, \ \forall \, b_1, b_2 \in B$. That is, we have the trivial near-ring (Definition 6). Equation (15.9) represents a right-distributive product precisely when, for all $b_1, b_2, b_3 \in B$, we have (all congruences here relate to the left ideal R which is the kernel of π in (15.1))

$$r' \cdot i(b_3) \equiv r_1 \cdot i(b_3) + r_2 \cdot i(b_3) \tag{15.10}$$

where

$$j(b_1 + b_2) = \left(i(b_1 + b_2) \, , \, r'\right)$$

Similarly, the condition for (15.9) to be associative is

$$r_1 \cdot i(\pi(r_2 \cdot i(b_3))) \equiv r'' \cdot i(b_3) \tag{15.11}$$

where

$$j \circ \pi(r_1 \cdot i(b_2)) = \left(i \circ \pi(r_1 \cdot i(b_2)) \, , \, r''\right)$$

15.3 Additive and Multiplicative Splitting

I utilise the notation just developed in what follows. In particular, the definition relies on the near-ring $(N, \oplus, \otimes)$ associated with the fixed lifting i.

Recall from Lemma 50 on page 33 that

$$\binom{N}{R}_{\mathcal{L}} = \{x \in N \ : \ x \cdot N \subseteq R\}$$

is an ideal of $(N, +, \cdot)$.

Definition 473 Suppose

$$j : B \to N$$

is a lifting defining the associated stem $(B, +, \star)$.

(i) j **splits additively** in case

$$j(b_1 + b_2) \equiv (j(b_1) \oplus j(b_2))$$

(ii) j **splits multiplicatively** in case

$$j(b_1 \star b_2) \equiv (j(b_1) \otimes j(b_2))$$

Here $b_1, b_2 \in B$ and congruences are relative to the ideal $\binom{N}{R}_{\mathcal{L}}$.

Lemma 474

(a) The product $(\star)$ hosted by the additive group $(B, +)$ is right-distributive if and only if the lifting j splits additively.

(b) The product $(\star)$ hosted by the additive group $(B, +)$ is associative if and only if the lifting j splits multiplicatively.

Proof

(a) Using the notation already developed, and (15.10), the condition for right distributivity is

$$r' \cdot i(b_3) \equiv r_1 \cdot i(b_3) + r_2 \cdot i(b_3) \quad \text{(modulo } R\text{)}$$

If we write

$$r' = r_1 + r_2 + q$$

this is precisely the condition that $q \in \binom{N}{R}_{\mathcal{L}}$, and that is the condition that j splits additively.

(b) From (15.11), the condition for associativity is

$$r_1 \cdot i(\pi(r_2 \cdot i(b_3))) \equiv r'' \cdot i(b_3) \quad (\text{modulo } R)$$

We can write

$$i(\pi(r_2 \cdot i(b_3))) = r_2 \cdot i(b_3) + \hat{r}$$

for some $\hat{r} \in R$. R is a left ideal of $(N, +, \cdot)$ so, $r_1 \cdot i(\pi(r_2 \cdot i(b_3))) =$

$$r_1 \cdot \left(r_2 \cdot i(b_3) + \hat{r}\right) \equiv r_1 \cdot r_2 \cdot i(b_3) \text{ (modulo } R)$$

Associativity requires

$$r_1 \cdot r_2 \equiv r'' \left(\text{modulo} \binom{N}{R}_{\mathcal{L}}\right)$$

and that is the condition that j should split multiplicatively.

□

15.4 A Particular Situation

I develop the situation just described, using the same notation, but restricting my attention to $(N, +, \cdot)$ being a right-distributive near-ring with multiplicative idempotent $a = a^2$. Now, B is defined as the group $N \cdot a = \{n \cdot a \ : \ n \in N\}$. R is the annihilator left ideal

$$\binom{a}{0}_{\mathcal{L}} = \{x \in N \ : \ x \cdot a = 0\}$$

Both B and R are sub-near-rings of $(N, +, \cdot)$, and $(N, +)$ is additively a semi-direct sum

$$\left(\binom{a}{0}_{\mathcal{L}} \rtimes N \cdot a\right) = (R \rtimes B)$$

(see Notation 2). The lifting i of (15.2) can be thought of as if it were the identity mapping on $B = N \cdot a$. If this be done, we can regard $(B, +)$ as a subgroup of $(N, +)$ or as $\frac{N}{R}$, though these are logically distinct things; and when i is fixed in this way, so is the corresponding product $(\otimes)$ of (15.4).

Equation (15.3) can now be written

$$(n_1 \cdot a\ ,\ r_1) \oplus (n_2 \cdot a\ ,\ r_2) = ((n_1 + n_2) \cdot a\ ,\ r_1 + r_2) \qquad (15.12)$$

while Eq. (15.4) becomes

$$(n_1 \cdot a\ ,\ r_1) \otimes (n_2 \cdot a\ ,\ r_2) = (r_1 \cdot n_2 \cdot a\ ,\ r_1 \cdot r_2) \qquad (15.13)$$

It is simpler to write $(n_j + r_j)$ for (n_j, r_j) in what follows, and I shall do this.

These two binary operations on the set N are connected to the original binary operations of the near-ring by the relations

$$(n_1 \cdot a + r_1) \oplus (n_2 \cdot a + r_2) = (n_1 \cdot a + r_1) + (n_2 \cdot a + r_2) + [n_2 \cdot a, r_1]^{r_2} \qquad (15.14)$$

and

$$(n_1 \cdot a + r_1) \otimes (n_2 \cdot a + r_2) = (r_1 \cdot n_2 \cdot a + r_1 \cdot r_2) \qquad (15.15)$$

Notation 475 When $(N,\ +,\ \cdot)$ is a near-ring with idempotent a, and the lifting i prescribed, the associated near-ring $(N,\ \oplus,\ \otimes)$ will be denoted by $\mathbf{S}(N, a)$.

Again we choose some lifting

$$j : B = N \cdot a \longrightarrow N = \left(\begin{pmatrix} a \\ 0 \end{pmatrix}_{\mathcal{L}} \rtimes N \cdot a\right)$$

and obtain the stem product $(N,\ +,\ \star)$.

Lemma 476 *In this situation,*

$$\begin{pmatrix} N \\ R \end{pmatrix}_{\mathcal{L}} \subseteq R$$

Proof If $x \cdot a \in R$, then $0 = (x \cdot a) \cdot a = x \cdot a^2 = x \cdot a$ □

Suppose that $(N,\ +,\ \cdot)$ is not zero-symmetric. We can choose a to be the zero element of the near-ring (see Lemma 80). Now, R is $\begin{pmatrix} 0 \\ 0 \end{pmatrix}_{\mathcal{L}}$, and $(N,\ \oplus,\ \otimes) = \mathbf{S}(N, 0)$ will be denoted by $\mathbf{SN}$.

15.4.1 The Stemhome

One final particularisation takes us to the transformation near-ring $\mathrm{T}(B)$ (see Chap. 7 and, in particular, Sect. 8.7 on page 332). The idempotent element a is now the mapping which takes any element x to zero ($x \mapsto 0$). In this special case, I call the near-ring $\mathbf{ST}(B)$ the **stemhome near-ring** of the additive group $(B, +)$ (see Definition 8.7.1).

Elements of this near-ring have the form (b, θ) where $b \in B$ and $\theta \in \mathrm{T}_0(B)$. The relations (8.29) and (8.30), on page 332, detail the additive and multiplicative operations on $\mathbf{ST}(B)$. There is the exact sequence of groups

$$R \longmapsto \mathbf{ST}(B) \overset{\pi}{\twoheadrightarrow} B \tag{15.16}$$

B is now $\{\mathrm{T}(B) \cdot 0\}$ and is essentially the set of constant mappings (identifying an element of B with the mapping that takes all elements of B to that element). The left ideal R is now just the near-ring $\mathrm{T}_0(B)$ of normalised mappings. I am interested in the possible stem products that we may construct relating to the various choices of lifting in (15.16).

Writing

$$j(b) = (b, \theta_b) \tag{15.17}$$

Equation (15.9) becomes

$$b_1 \star b_2 = \theta_{b_1}(b_2) \tag{15.18}$$

The ideal

$$\binom{N}{R}_{\mathcal{L}} = \binom{\mathbf{ST}(B)}{R}_{\mathcal{L}} = \{\theta \in \mathrm{T}_0(B) \ : \ \theta \circ \mathrm{T}(B) \subseteq \mathrm{T}_0(B)\}$$

is trivial, and the lifting conditions of Definition 473 therefore relate to *equalities* rather than *congruences*. This gives me the main result of this chapter.

Lemma 477

(i) There is a one-to-one correspondence between the distinct liftings

$$j : B \to \mathbf{ST}(B)$$

and the distinct bi-zero-symmetric products which may be defined on the additive group $(B, +)$.

(ii) Right-distributive products correspond to liftings which split additively in the sense

$$(j \circ \pi\ (j(a) \oplus j(b))) \ = \ (j(a) \oplus j(b))$$

(iii) Associative products correspond to liftings which split multiplicatively in the sense

$$(j \circ \pi\ (j(a) \otimes j(b))) \ = \ (j(a) \otimes j(b))$$

(iv) Near-ring products correspond to liftings which split both ways.

(v) The set of distinct bi-zero-symmetric products hosted by an additive group $(B, +)$ is in one-to-one correspondence with the group of zero-fixing mappings from $(B, +)$ to the near-ring of zero-fixing mappings from $(B, +)$ to itself

$$\mathrm{T}_0\,(B, \mathrm{T}_0(B)) \tag{15.19}$$

and this is the Wurzel group, $\mathbf{W}(B)$ (see Definition 343).

Proof

(i) Given a bi-zero-symmetric product (*) on $(B, +)$ and $b_1, b_2 \in B$, we have an associated mapping $\theta_{b_1} \in \mathrm{T}_0(B)$ defined by

$$\theta_{b_1}(b_2) \ = \ (b_1 * b_2)$$

Our stem-lifting, j, is constructed using the formula

$$j(b_1) \ = \ \left(b_1, \theta_{b_1}\right)$$

It is apparent that

$$(b_1 \star b_2) \ = \ (b_1 * b_2)$$

(v) This follows directly from looking at the ways in which j may be defined.

□

The distributivity core of a pre-near-ring is defined on page 43. Here, I extend the definition slightly, applying it to the more general class of bi-zero-symmetric stems. Appealing to (15.18), we have that $b_1 \in \mathrm{C}_{\mathcal{D}}(N) \Leftrightarrow \theta_{b_1} \in \mathrm{End}(N, +)$ and so,

Corollary 478 $\mathrm{C}_{\mathcal{D}}(N) \ = \ \{b \in B \ : \ \theta_b \in \mathit{End}(B, +)\}$.

Observations 479 The group addition on $\mathbf{W}(B)$ translates to an addition $(+_p)$ of stem products. Two mappings,

$$k, j \in \mathbf{W}(B)$$

may be thought of as representing (bi-zero-symmetric) stems

$$(B,\ +,\ \cdot) \text{ and } (B,\ +,\ :)$$

The sum

$$(j+k) : B \to \mathrm{T}_0(B,\, B)$$

corresponds to the stem product $(\cdot +_p :)$ defined by

$$\left(b_1\ \cdot +_p :\ b_2\right) \ = \ ((b_1 \cdot b_2) + (b_1 : b_2)) \tag{15.20}$$

This is an additive group essentially consisting of a direct product of copies of $(B,\ +)$. (See Sects. 8.8 and 482.)

15.4.2 Associative and Distributive Elements

Given a stem $(B,\ +,\ \cdot)$, we are interested in

(i) Elements of the set

$$\mathbf{D} \ = \ \{b_3\ :\ (b_1 + b_2) \cdot b_3 = (b_1 \cdot b_3) + (b_2 \cdot b_3)\ \forall\, b_1, b_2\ \in B\ \} \tag{15.21}$$

(ii) Elements of the set

$$\mathbf{A} \ = \ \{b_3\ :\ (b_1 \cdot b_2) \cdot b_3 = (b_1 \cdot (b_2 \cdot b_3)\ \forall\, b_1, b_2,\ \in\ B\} \tag{15.22}$$

(compare with Definition 69).

Notation 480 Suppose $b \in B$ and $b \neq 0$, then

$$\mathrm{T}_{\{0,b\}}(B) \ = \ \{\theta \in \mathrm{T}_0(B)\ :\ \theta(b) = 0\}$$

$\mathrm{T}_{\{0,b\}}(B)$ is a left ideal of the near-ring $\mathrm{T}_0(B)$ and, of course,

$$\bigcap_{b \in B} \mathrm{T}_{\{0,b\}}(B) \ = \ \{0\}$$

If $b_i \in B, i \in \{1, 2, 3\}$ and $j(b_i) = (b_i, \theta_i)$, we write

$$j(b_1 + b_2) \ = \ \left((b_1 + b_2)\ ,\ \theta'\right)$$

$$j(\theta_1(b_2)) \ = \ \left(\theta(b_2), \hat{\theta}\right)$$

Corollary 481

(i) $b_3 \in \mathbf{D}$ *if and only if*

$$\theta' \equiv (\theta_1 + \theta_2) \text{ (} modulo \text{ } \mathrm{T}_{\{0,b_3\}}(B)) \tag{15.23}$$

(ii) $b_3 \in \mathbf{A}$ *if and only if*

$$\hat{\theta} \equiv (\theta_1 \circ \theta_2) \text{ (} modulo \text{ } \mathrm{T}_{\{0,b_3\}}(B)) \tag{15.24}$$

Suppose $(B, +, \cdot)$ is a near-field (page 16). The distributivity core $\mathcal{D} = \mathrm{C}_{\mathcal{D}}(B)$ will then be a field (see page 43 and Chap. 3). Then, $(\mathcal{D}, +) < (B, +)$ and mappings $\theta \in \mathrm{T}_0(B)$ project to mappings $\bar{\theta} \in \mathrm{T}_0(\mathcal{D})$ via the obvious formula

$$\bar{\theta}(x) = \begin{cases} 0 & \text{if } x \notin \mathcal{D} \\ \theta(x) & \text{otherwise} \end{cases}$$

We can replace the exact sequence

$$\mathrm{T}_0(B) \longmapsto \mathbf{ST}(B) \xrightarrow{\pi} B$$

by the exact sequence

$$\mathrm{T}_0(\mathcal{D}) \longmapsto \mathbf{ST}(\mathcal{D}) \xrightarrow{\pi} \mathcal{D}$$

and the original lifting $j : B \to \mathbf{ST}(B)$, with the modified lifting $\bar{j} : \mathcal{D} \to \mathbf{ST}(\mathcal{D})$, where

$$\bar{j}(d) = \left(b, \bar{\theta}_b\right)$$

Of course, the problem with this is the implicit assumption that $\bar{\theta}_b : B \to \mathcal{D}$. This is resolved by (15.17), because, if $d_1, d_2 \in \mathcal{D}$, then from (15.18) and the supposed multiplicative closure of $\mathcal{D}$, we must have $\theta_{d_1}(d_2) \in \mathcal{D}$. It will be seen that the product on $\mathcal{D}$ corresponds to the modified lifting $\bar{j}$ and that we may write down the usual splitting conditions for distributivity and associativity, obtaining an alternative formulation of Corollary 481. I shall return to this in Sect. 15.6.

15.4.3 Opposites

The lifting (15.17) gives rise to the stem product $(\star)$ of (15.18). It may have occurred to the reader that we could write

$$b_1 \overset{\star}{\star} b_2 = \theta_{b_2}(b_1)$$

If that is done, then, of course,

$$b_1 \overset{\star}{\star} b_2 = b_2 \star b_1 \tag{15.25}$$

and we have the *opposite* stem (page 51).

15.4.4 Coupling

I now return to the ideas developed in Sect. 2.2 on page 77 and use some of the notation developed there, particularly, Eq. (2.5). We start with a lifting

$$j : N \to \mathbf{ST}(N)$$

and write

$$j(n) = (n, \phi_n)$$

Associativity in the related stem product corresponds to the condition

$$\phi_{\phi_{n_1}(n_2)} = \phi_{n_1} \circ \phi_{n_2}$$

This is simply (2.8) applied to the *constant* stem $(N, +, \cdot)$ (defined page 50) and, as such, is a *coupling condition*. The mapping (2.5) can be associated with an obvious lifting which in turn defines the *left* p.n.r. $(N, +, :)$. In that case, (2.6) defines a stem product from *any* pre-existing stem $(N, +, \cdot)$ by

$$m * n = (n : m) \cdot n \tag{15.26}$$

- When $(N, +, \cdot)$ is bi-zero-symmetric, so is $(N, +, *)$.
- When $(N, +, \cdot)$ is a *right* p.n.r., so is $(N, +, *)$.

Observations 482 Equation (15.26) suggests investigating possible products hosted by the Wurzel group. Such a product should take two general stems and deliver a third. The Wurzel already has a clear group structure which can be thought of as involving the binary operation $(+_p)$ described in Observations 479. This new product ought to fit in with that. This leads me to reject (15.26) for something slightly different.

Notation 483 Suppose $(N, +, \cdot)$ and $(N, +, :)$ are stems. $(N, +, \cdot \times_p :)$ is the stem with product

$$\left(a \cdot \times_p : b\right) = (a \cdot (a : b)) \tag{15.27}$$

Supposing that the liftings

$$j : a \mapsto (a, \theta_a)$$
$$k : a \mapsto (a, \phi_a)$$

are the liftings associated with these stem products $(\cdot)$ and $(:)$, respectively, then, of course,

$$\left(a \cdot \times_p : b\right) = \theta_a\left(\phi_a(b)\right)$$

and so the lifting associated with the stem $(N, +, \cdot\times_p :)$ will be

$$a \mapsto (a, (\theta_a \circ \phi_a)) \tag{15.28}$$

The next result involves some abuse of notation and requires Definition 6.

Lemma 484

$$\left(\mathbf{W}(B), +_p, \times_p\right)$$

is a right-distributive near-ring in which the trivial near-ring forms the additive zero.

We continue to identify $\mathbf{W}(B)$ with the collection of bi-zero-symmetric stems that may be defined upon the additive group $(B, +)$ although these things are logically distinct. The next result uses the notation just developed and the notation from Definition 69 and Notation 11.

Lemma 485

$$\mathrm{C}_{\mathcal{D}}(\mathbf{W}(B)) = \mathbf{D}_l(B, +)$$

15.4.5 Further Symmetries

Lemma 486 *Suppose that $(N, +, \cdot)$ is a zero-symmetric near-ring and $(B, +)$ is an additive group. Suppose there is an exact sequence of additive groups*

$$K \mapsto N \overset{\phi}{\twoheadrightarrow} B$$

and that K is a left ideal of the near-ring. Then there exists a unique near-ring homomorphism $\Theta : N \to \mathbf{ST}(B)$ such that the following diagram commutes.

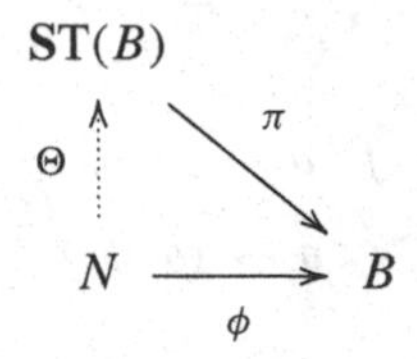

Proof We take a lifting $i : B \to N$, so that $\phi \circ i = Id_B$. We define Θ by prescribing that

$$\Theta(m) = (\phi(m), \theta_m)$$

defining

$$\theta_m(b) = \phi(m \cdot i(b)) \tag{15.29}$$

The fact that K is a left ideal makes this definition independent of the choice of lifting. The assertion is that Θ is a homomorphism of near-rings and we need to recall the operations of (8.29) and (8.30) for that. The additive part is essentially the assertion that

$$\theta_{m+n} = (\theta_m + \theta_n)$$

which is obvious. The multiplicative part requires

$$(\phi(m \cdot n), \theta_{m \cdot n}) = (\theta_m(\phi(n)), \theta_m \circ \theta_n)$$

Now, $i \circ \phi(n) = n + k$ for some $k \in K$ which means

$$\theta_m(\phi(n)) = \phi(m \cdot i \circ \phi(n)) = \phi(m \cdot n)$$

(again using the fact K is a left ideal). To show

$$\theta_{m \cdot n} = \theta_m \circ \theta_n$$

we must show

$$\phi(m \cdot n \cdot i(b)) = \phi(m \cdot i \circ \phi(n \cdot i(b)))$$

which, once more, follows from the left ideal property.

As to uniqueness, suppose Ψ were another near-ring homomorphism from N to $\mathbf{ST}(B)$ for which $\phi = \pi \circ \Psi$. We can write $\Psi(m) = (\phi(m), \psi_m)$, for some $\psi_m \in \mathrm{T}_0(N)$. For some m, and some $b \in B$, it must be that $\theta_m(b) \neq \psi_m(b)$. We can write $\phi(n) = b$ for some suitable $n \in N$. Then,

$$\Theta(m \cdot n) = \Theta(m) \otimes \Theta(n) = (\phi(m), \theta_m) \otimes (\phi(n), \theta_n) = (\theta_m(b), \theta_m \circ \theta_n)$$

and, similarly,

$$\Psi(m \cdot n) = (\psi_m(b), \psi_m \circ \psi_n)$$

But these relations would require $\theta_m(b) = \psi_m(b)$. □

Observations 487 Given the above set-up, we can define a lifting

$$j_i : B \to \mathbf{ST}(B)$$

by the formula

$$j_i(b) = \Theta \circ i(b) = \left(b, \theta_{i(b)}\right)$$

This corresponds to a stem product hosted by $(B, +)$ which has the form

$$(b_1 *_i b_2) = \phi(i(b_1) \cdot i(b_2))$$

Now, if $m, n \in N$ and

$$i \circ \phi(m) = m + k_1$$
$$i \circ \phi(n) = n + k_2$$

(where $k_1, k_2 \in K$), then, using the left ideal property,

$$\begin{aligned}\phi(m) *_i \phi(n) &= \phi(i(\phi(m)) \cdot i(\phi(n))) \\ &= \phi\left((m + k_1) \cdot (n + k_2)\right) \\ &= \phi(m \cdot n) + \phi(k_1 \cdot n)\end{aligned}$$

So, ϕ is a near-ring homomorphism from the near-ring $(N, +, \cdot)$ onto the stem $(B, +, *_i)$ exactly when K is an ideal, and in that case, $(B, +, *_i)$ is the corresponding quotient near-ring.

Suppose

$$j : B \to N$$

is another lifting of ϕ. Equation (15.9) gives the stem product on B associated with the lifting j (though the mapping there given as π is here the mapping ϕ).

That product here has the form

$$(b_1 \star b_2) = \phi\left(k_1 \cdot i(b_2)\right)$$

where $j(b_t) = i(b_t) + k_t,\ t \in \mathbb{N},\ k_t \in K$. As before, we have that $j_j = (\Theta \circ j)$ is a lifting of the mapping π in Lemma 486 and, of course,

$$j_j(b) = \Theta \circ j(b) = \big(b, \theta_{j(b)}\big)$$

This lifting of π also defines a stem product on B, the formula is

$$\big(b_1 \star_j b_2\big) = \theta_{j(b_1)}(b_2) = \phi\left(j(b_1) \cdot i(b_2)\right) = \phi\left((i(b_1) + k_1) \cdot i(b_2)\right)$$

and so,

$$\big(b_1 \star_j b_2\big) = (b_1 \star_i b_2) + (b_1 \star b_2)$$

15.4.6 *Isomorphisms*

As always, stem products on $(B,\ +)$ are to be identified with liftings

$$j : B \longrightarrow \mathbf{ST}(B)$$

In this subsection, we shall suppose that the above lifting corresponds to the stem $(B,\ +,\ \star)$. I shall write $j(b) = (b, \theta_b)$.[1] Let δ be some automorphism of the additive group $(B,\ +)$, and recall the definition of the functor **ST** given as Definition 8.7.1, on page 333. The following square is commutative.

$$\begin{array}{ccc} B & \xrightarrow{\ \delta\ } & B \\ {\scriptstyle j}\big\downarrow & \copyright & \big\downarrow{\scriptstyle j^{\delta}} \\ \mathbf{ST}(B) & \xrightarrow[\mathbf{ST}(B)(\delta)]{} & \mathbf{ST}(B) \end{array}$$

Here, j^{δ} is the lifting

$$j^{\delta}(b) = \left(b, \delta \circ \theta_{\delta^{-1}(b)} \circ \delta^{-1}\right)$$

I shall call j^{δ} the **lifting derived from j by** δ and note that j is similarly the lifting derived from j^{δ} and δ^{-1} and that derivation represents an equivalence relation on the class of stem products.

[1] θ_b just tags the mapping in this case; the notation is unrelated to (15.29).

Lemma 488

(i) Two stem products hosted by an additive group $(B, +)$ are isomorphic if and only if their liftings are derived from each other.
(ii) The isomorphism class of a particular stem product hosted by an additive group $(B, +)$ has a natural group structure isomorphic to a homomorphic image of the group of all automorphisms of the group.

Proof

(i) We have the lifting j and stem $(B, +, \star)$ mentioned above. Suppose that the lifting $j^{\delta}(b)$ corresponds to the stem $(B, +, \diamond)$. We have

$$(\delta(b_1) \diamond \delta(b_2)) = \left(\delta \circ \theta_{(b_1)}(b_2)\right) = \delta(b_1 \star b_2)$$

Consequently,

$$(B, +, \star) \overset{\delta}{\cong} (B, +, \diamond)$$

On the other hand, if $(B, +, \star)$ and $(B, +, \diamond)$ are two stems corresponding to the liftings j and k, they are isomorphic under an isomorphism δ. We could write

$$k(b) = (b, \psi_b)$$

so that

$$\psi_{\delta b_1}(\delta(b_2)) = \delta\left(\theta_{b_1}(b_2)\right)$$

which requires

$$\psi_b = \delta \circ \theta_{\delta^{-1}(b)} \circ \delta^{-1}$$

that is,

$$k = j^{\delta}$$

(ii) In principle, it may be possible for two products derived from each other to be identical. Fixing a lifting, j, we define the class of stem products or liftings that might be derived from j, using automorphisms of $(B, +)$ as $\mathcal{J}$. Returning to our original commutative square, we see that two liftings in $\mathcal{J}$ will have the form

$$\left(\delta^{-1} \circ j \circ \mathbf{ST}\right)(\delta)$$

and

$$\left(\omega^{-1} \circ j \circ \mathbf{ST}\right)(\omega)$$

The obvious product of j^{ω} with j^{δ} is $j^{(\omega \circ \delta)}$ and under that definition $\mathcal{J}$ becomes a homomorphic image of Aut$(B, +)$. □

15.4.7 Isotopies

I use the ideas of isotopy and allotropy given on page 484. Thus, two stems $(B, +, *)$ and $(B, +, \cdot)$ are held to be isotopic provided there are automorphisms of the additive group, α, $\beta and \gamma$, such that

$$\gamma\left(\alpha(b_1) * \beta(b_2)\right) = (b_1 \cdot b_2)$$

We shall assume that $(\star)$ corresponds to the lifting j and write

$$j(b) = (b, \theta_b)$$

and we define a new lifting i as

$$i(b) = \left(b, \gamma \circ \theta_{\alpha(b)} \circ \beta\right) \tag{15.30}$$

This lifting defines a stem product $(\star)$ as

$$(b_1 \star b_2) = \left(\gamma \circ \theta_{\alpha(b_1)} \circ \beta\right)(b_2) = \gamma\left(\alpha(b_1) * \beta(b_2)\right) = (b_1 \cdot b_2)$$

I should wish to relate this version of isotopy to symmetries of the stemhome near-ring. However, the presence of the $\theta_{\alpha(b)}$ term in (15.30) is a particularisation which makes this difficult. According to Lemma 510, we may satisfy our search for stem products by restricting our view of isotopy to products of the form

$$a \wedge b = \theta\left(a \cdot \phi(b)\right)$$

where θ and ϕ are group automorphisms. In this particular case, that comes down to setting $\alpha = Id$ (the identity mapping) in the relations of this subsection. From (15.30), we obtain the related lifting

$$k(b) = (b, \gamma \circ \theta_b \circ \beta)$$

and this commutative diagram

$$\begin{array}{ccc} B & \xrightarrow{\text{Id}} & B \\ j \downarrow & © & \downarrow k \\ \mathbf{ST}(B) & \xrightarrow[\kappa]{} & \mathbf{ST}(B) \end{array}$$

where

$$\kappa\,(b, \theta) \;=\; (b, \gamma \circ \theta \circ \beta)$$

The mapping κ encapsulates the associated symmetry in the stemhome. It represents an additive group automorphism, though not an automorphism of the associated stemhome near-ring. All allotropies of the sort considered do arise in this way.

15.5 Restricted Co-domains

Stems correspond to liftings

$$j : B \to \mathbf{ST}(B)$$

but it may be convenient to consider liftings to subgroups of $\mathbf{ST}(B)$. One obvious possibility is to consider liftings to the group $(B \oplus \text{Phom}(B))$. The relations (8.29) and (8.30), on page 332, apply, and it will be seen that $(B \oplus \text{Phom}(B))$ is a sub-near-ring of $\mathbf{ST}(B)$. Stems associated with liftings of this sort are already left F-stems. We have this restricted version of Lemma 477.

Lemma 489

(i) The set of distinct bi-zero-symmetric left F-stems hosted by an additive group $(B, +)$ is in one-to-one correspondence with the group of zero-fixing mappings from $(B, +)$ to the near-ring of phomomorphisms of $(B, +)$

$$\text{T}_0\,(B, \text{Phom}(B)) \tag{15.31}$$

(ii) The set of distinct bi-zero-symmetric right F-stems hosted by an additive group $(B, +)$ is in one-to-one correspondence with the group of phomomorphisms from the additive group $(B, +)$ to the transformation near-ring of $\text{T}_0(B)$

$$\text{Phom}((B, \text{T}_0(B)) \tag{15.32}$$

(iii)

$$T_0\,(B, \mathrm{Phom}(B)) \cong \mathrm{Phom}((B, T_0(B)) \tag{15.33}$$

(as additive groups).

Proof We only need to prove (iii). Both (15.31) and (15.32) are subgroups of the Wurzel group, $\mathbf{W}(B)$, which was defined on page 334. From symmetry considerations, they must be isomorphic. Alternatively, use the corollary to Lemma 362. □

15.5.1 Bi-F-Stems

If we turn to a consideration of Bi-F-stems (page 421), these are certainly obtained when the lifting is of the form $j \in \mathrm{Phom}(B, \mathrm{Phom}(B, B))$; but, of itself, this condition is too severe. The bi-F-stems produced by such liftings are left-pseudo-distributive (page 421) but exhibit the stronger right-pseudo-distributivity property

$$(b_1 + b_2)\cdot c = (b_1 \cdot c) + (b_2 \cdot c) + \delta(c), \quad (\delta \in [\mathrm{Phom}(B, B), \mathrm{Phom}(B, B)])$$

In more general bi-F-stems, one merely requires $\delta \in [T_0(B, B), T_0(B, B)]$. (Lemma 306 may make that observation clearer.) In the next lemma, the symbol $\equiv$ is intended to indicate sets of the same cardinality.

Lemma 490

$$\mathbf{BiF}(B) \equiv \left(T_0\,(B, \mathrm{Phom}(B)) \bigcap \mathrm{Phom}\,(B, T_0(B))\right)$$

$\mathbf{BiF}(B)$ is to be thought of as a subgroup of the Wurzel group with the group $\mathrm{Phom}(B, \mathrm{Phom}(B, B))$ lying inside it. The addition on $\mathbf{BiF}(B)$ is given by (15.20). It is not generally true that $\mathbf{BiF}(B)$ is a sub-near-ring of $\left(\mathbf{W}(B), +_p, \times_p\right)$—the group is not usually multiplicatively closed.

15.6 Abelian Groups

One can investigate other important subclasses of products by moving to sub-near-rings of $\mathbf{ST}(B)$. For instance, rings require that $(B, +)$ should be abelian and we require us to look at

Definition 491

$$\mathbf{RT}(B) = \{(b, \theta) : \theta \in Hom(B, B)\}$$

which I call the **Ringhome** ring.

As is obvious, the Ringhome ring is actually a ring. Liftings to the Ringhome define stems that are already left-distributive—that is, they are *left* p.n.r., and they are hosted by abelian groups (see Corollary 478). The modifications to be made to Lemma 477 are obvious. Chapter 16 gives more attention to this in the special context in which near-rings are hosted by finite elementary abelian p-groups.

Observations 492 Irritatingly, our products are left-distributive now, and expressions of the form

$$\overbrace{(x + x + \ldots . + x)}^{(m \text{ times})}$$

should be written xm (see (1.5)). This can be surprisingly confusing. In Chap. 16, when considering near-fields, I look at expressions $x \cdot \pi$, which means $\overbrace{(x + x + \ldots . + x)}^{(\pi \text{ times})}$, and ask for the unique near-field element a for which $x \cdot \pi = a \cdot x$. The whole point is that when π is not central, it can happen that $a \neq \pi$.

15.7 General Groups

There is a matrix representation of products in finite non-abelian cases, very similar to Definition 242. This uses some of the ideas explored in Chap. 16. Following Sect. 16.1, our representation involves matrices whose first columns specify the element involved. We enumerate the group as $N = \{n_0 = 0, n_1, n_2, \ldots, n_m\}$ and represent individual elements (n_k) as column vectors $\left(\theta_i, 0 = \delta_{i,k}\right)$. The transformations $\theta \in \mathrm{T}_0(N)$ are represented by $(m \times m)$ matrices in which, if

$$\theta(g_i) = g_k$$

then

$$\theta_{i,j} = \delta_{j,k}$$

and whose top row has the form $(1, 0, 0, \ldots, 0)$. We then produce $(m+1) \times (m+1)$ matrices $\left(\theta_{i,j}\right)$ by additionally setting $\theta_{0,j} = 0$ for all $j \in \{0, 1, \ldots, m\}$. The reader will have no difficulty in formulating this generalisation.

Coda

I believe this treatment of product theory is original to me. It was first published in [105]. I have come to appreciate that at least some of these ideas, though none of the formalism, occur in combinatorial form in early papers on the host problem for near-rings dating back to the 1960s, particularly the papers of J. R. Clay, as reported in [128].

Chapter 16
Product Theory on Finite Elementary Abelian Groups

16.1 Finite Elementary Abelian p-Groups

Most of the near-rings occurring in this chapter are left-distributive.

I use the Gothic "Q", "F", "E", "C" and "N" symbols in this chapter: $\mathfrak{Q}$, $\mathfrak{F}$, $\mathfrak{E}$, and $\mathfrak{N}$; and S_2 is always a Sylow 2-subgroup (as in page 105).

I restrict our considerations to situations in which $(B, +)$ is a finite elementary abelian p-group, and I insist near-rings must be zero-symmetric. The group will have r-independent additive generators $\{b_j\}_{j=1}^{r}$. Additive endomorphisms of the group are fixed by $r \times r$ matrices. Writing

$$\theta(b_k) = \sum_i \theta_{i,k} b_i$$

$$b = \sum_k \theta_{k,0} b_k$$

$$\theta(b) = \sum_i \left(\sum_k \theta_{i,k}\theta_{k,0} \right) \cdot b_i$$

one gets the standard

$$\theta(b) = \begin{bmatrix} \theta_{1,1} & \theta_{1,2} \cdots & & \theta_{1,r} \\ \theta_{2,1} & \theta_{2,2} \cdots & \vdots & \theta_{2,r} \\ \vdots & \ddots & \vdots & \ddots \vdots \\ \theta_{r,1} \cdots & \cdots & \cdots\cdots & \theta_{r,r} \end{bmatrix} \cdot \begin{bmatrix} \theta_{1,0} \\ \theta_{2,0} \\ \vdots \\ \theta_{r,0} \end{bmatrix}$$

R. Lockhart, *The Theory of Near-Rings*, Lecture Notes in Mathematics 2295,
https://doi.org/10.1007/978-3-030-81755-8_16

We may now represent typical elements $(b, \theta) \in \mathbf{RT}(B)$ as $(r+1) \times (r+1)$ matrices

$$(b, \theta) = \begin{bmatrix} 0 & 0 & 0 \ldots & 0 \\ \theta_{1,0} & \theta_{1,1} & \theta_{1,2} \cdots & \theta_{1,r} \\ \theta_{2,0} & \theta_{2,1} & \theta_{2,2} \cdots \vdots & \theta_{2,r} \\ \vdots & \ddots & \vdots \ \ddots & \vdots \\ \theta_{r,0} & \theta_{r,1} \cdots & \cdots \cdots \cdots & \theta_{r,r} \end{bmatrix} \tag{16.1}$$

16.1.1 Matrix Representations

The set of all such matrices forms a matrix ring $\mathbf{MT}(B)$ under the standard matrix addition and multiplication in the Galois field of order p. This ring is, of course, isomorphic to the ringhome ring.

Notation 493

(i) A particular lifting $j \to \mathbf{RT}(B)$ defines a set of matrices in $\mathbf{MT}(B)$, which I shall write as $\mathbf{m}_j\mathbf{T}(B)$.
(ii) The set of matrices $\mathbf{m}_j\mathbf{T}(B)$ in turn generates a matrix ring inside $\mathbf{MT}(B)$, which I denote by $\mathbf{M}_j\mathbf{T}(B)$.

What we have is an association between left p.n.r. on finite elementary abelian p-groups and rings. The analysis can be extended to more general abelian groups, but the matrix operations then require more complicated modular reductions.

Lemma 477 can now be recast in terms of matrix operations. Additive splitting corresponds to additive closure in $\mathbf{m}_j\mathbf{T}(B)$, whereas multiplicative splitting is multiplicative closure. $\mathbf{MT}(B)$ offers a measure as to how far a given lifting is from representing a ring. Rings correspond to the equality: $\mathbf{M}_j\mathbf{T}(B) = \mathbf{m}_j\mathbf{T}(B)$.

Non-standard Matrix Operations
The operations on B can be thought of as matrix operations. The rule is that one forms the appropriate sum or product of matrices in $\mathbf{m}_j\mathbf{T}(B)$ and thereby obtains a matrix in $\mathbf{M}_j\mathbf{T}(B)$. We then replace that matrix with the unique member of $\mathbf{m}_j\mathbf{T}(B)$ which has the same first column. Opposite structures (Sect. 15.4.3) are straightforward transposes of the matrices in $\mathbf{m}_j\mathbf{T}(B)$.

16.1.2 The Cayley-Hamilton Theorem

The Cayley-Hamilton theorem applies to square matrices over commutative rings. It says that they satisfy their own characteristic equations. The theorem appears to be restricted to standard matrix operations. In multiplicatively associative stems, the matrix products are the same, but the addition involves this business of supplanting a sum by the unique element with the same first column.

Imagining the Cayley-Hamilton theorem applying simply to the first column of the matrices would extend the result to our matrices; but the uniqueness of first columns means that the matrix obtained by substituting into the characteristic equation, performing the standard matrix multiplications (associative stem products) and this modified matrix addition should indeed be the zero matrix. In this way, one obtains a generalisation of the Cayley-Hamilton theorem to the matrices used in this representation of near-rings.

Lemma 494 *For associative stems, the matrices obtained satisfy their own characteristic equations using the modified, non-standard matrix addition described above.*

16.1.3 Restricted Transposes

It occurs to me that one could perhaps restrict the transpose operation to the interior matrices and obtain stems of interest. If this is applied to the near-field given in Sect. 16.4, the resulting structure is not associative. Although, of course, the interior matrices do form a multiplicatively closed subgroup, the full matrices do not; and we have a p.n.r. but not a near-ring. This idea reappears on page 464.

16.1.4 Notational Laxity

Remark 495 In what follows in this chapter, I am sometimes lax about the distinction between an element of a p.n.r., its stem-lifted matrix representation and even the interior of the associated matrix. In the work on near-fields, for instance, I will sometimes blithely switch between 3×3 and 2×2 matrices, overloading notation, because we only really need to track the internal matrices. For example, on page 467, $\mathcal{P}$ is contextually regarded either as a 3×3 or as a 2×2 matrix.

I hope the reader will forgive this imprecision, which arises in an attempt to reduce notational complexity, perhaps at the expense of immediate intelligibility on first reading.

16.1.5 Identities

Suppose we do have a near-ring and that there is a left multiplicative identity, (a, θ). Basic linear algebra tells us that the interior matrix θ will have to be a square, identity, matrix.

It is certainly true that we could re-label the additive generators of our group in such a way as to force a to be the column vector

$$\begin{bmatrix} 0 \\ 1 \\ 0 \\ 0 \\ .. \\ .. \\ 0 \end{bmatrix}$$

In the two-dimensional case, the element (a, θ) would be

$$\begin{bmatrix} 0\,0\,0 \\ 1\,1\,0 \\ 0\,0\,1 \end{bmatrix}$$

and under this re-labelling, the interior matrices would still form a multiplicatively closed semi-group. One can relate this to basic linear algebra over the Galois field with order p. I do this for the two-dimensional case first. We wish to re-order the field elements via a linear transformation which changes the vector associated with the first column of the matrix whose interior is the identity matrix to the column matrix

$$\begin{bmatrix} 0 \\ 1 \\ 0 \end{bmatrix}$$

Change of Basis
This linear transformation notionally converts a basis $\{x_1, x_2\}$ to a basis $\{y_1, y_2\}$. One might write

$$(\lambda \cdot x) = \begin{bmatrix} \lambda_{1,1} \, \lambda_{1,2} \\ \lambda_{2,1} \, \lambda_{2,2} \end{bmatrix} \cdot \begin{bmatrix} x_1 \\ x_2 \end{bmatrix} = \begin{bmatrix} y_1 \\ y_2 \end{bmatrix} = y$$

This re-writing may be shown schematically as

$$(a, \theta) \longmapsto \left(\lambda \cdot a, \left(\lambda \cdot \theta \cdot \lambda^{-1}\right)\right) \tag{16.2}$$

Clearly, when θ is scalar—in particular, the identity matrix, re-writing of this sort leaves it fixed. This is a derived lifting in the sense of Lemma 488, and the two near-rings are isomorphic. The whole thing works for higher dimensions too. Consequently, if we have a near-ring with a left multiplicative identity on our elementary abelian group, we can re-order the additive group generators in such a way that the stem-lifting representing the left identity has the form

$$\begin{bmatrix} 0 & 0 & \dots & \dots & \dots & 0 \\ 1 & 1 & 0 & \dots & \dots & 0 \\ 0 & 0 & 1 & 0 & \dots & 0 \\ 0 & 0 & 0 & 1 & \dots & 0 \\ \dots & \dots & \dots & \dots & \dots & \dots \\ 0 & 0 & 0 & 0 & \dots & 1 \end{bmatrix} \tag{16.3}$$

(where the dots indicate sequences of zero elements with exactly one, diagonal, 1). In the event that the near-ring has a (two-sided) identity, it can be represented by just this element. More, in such situations, the matrices of the near-ring element, in the stem representation, have the form (illustrating just the two-dimensional case, though all dimensions apply)

$$\begin{bmatrix} 0 & 0 & 0 \\ x & x & z \\ y & y & w \end{bmatrix} \tag{16.4}$$

One of our matrices will then have the form

$$\mathcal{X} = \begin{bmatrix} 0 & 0 & 0 \\ 0 & 0 & A \\ 1 & 1 & B \end{bmatrix} \tag{16.5}$$

This symbol repeatedly occurs in this chapter, always with the meaning given here.

The symbol $\mathcal{X}$ is important for the rest of the chapter.

From this, we can deduce that every last column of our matrix set occurs as a first column too, that is,

$$\begin{bmatrix} 0 & 0 & 0 \\ a & a & c \\ b & b & d \end{bmatrix} \Rightarrow \begin{bmatrix} 0 & 0 & 0 \\ c & c & (Aa + Bc) \\ d & d & (Ab + Bd) \end{bmatrix} \tag{16.6}$$

As I say, this result applies in all dimensions.

Lemma 496 *When we are dealing with near-rings with identity and we order generators in the way described, columns of any of our matrices are unique, with no other matrix having the same first or second column.*

This result reappears in Lemma 500. The matrix representation makes it easier to see connections between what are, on the face of it, wildly different stem structures. This analysis also replaces the standard calculations associated with verifying associativity by matrix multiplications aiming to verify closure, which seems a little simpler and is certainly more easily programmable. Bi-distributivity similarly corresponds to additive closure of the matrices. The restrictions on the form of the matrices obtained above are used throughout the rest of this chapter and are illustrated in the next subsection.

16.1.6 The Klein Group

I illustrate these ideas by considering left near-rings hosted by

$$(V, +) = (C_2 \oplus C_2) \tag{16.7}$$

We have column vectors

$$\underline{a} = \begin{bmatrix} 1 \\ 0 \end{bmatrix}, \ \underline{b} = \begin{bmatrix} 0 \\ 1 \end{bmatrix}, \ \underline{c} = \begin{bmatrix} 1 \\ 1 \end{bmatrix} \tag{16.8}$$

associated with 2 × 2 (0–1) (Boolean) matrices

$$\begin{bmatrix} x & z \\ y & w \end{bmatrix} \tag{16.9}$$

A particular stem-lifting results in 3 × 3 matrices of the form

$$j(a, b) = \begin{bmatrix} 0 & 0 & 0 \\ a & x & z \\ b & y & w \end{bmatrix} \tag{16.10}$$

one for each element of V, with zero corresponding to the 3 × 3 zero matrix: multiplicative closure of these matrices corresponding to left near-rings. There are 16 possible Boolean matrices, and since the interior matrix for the zero element is already fixed, $16^3 = 2^{12}$ distinct stem products that may be defined.

A Ring
These matrices represent a ring hosted by the Klein group.

$$\left\{ \begin{bmatrix} 0&0&0 \\ 0&0&0 \\ 0&0&0 \end{bmatrix}, \begin{bmatrix} 0&0&0 \\ 1&1&1 \\ 0&0&0 \end{bmatrix}, \begin{bmatrix} 0&0&0 \\ 0&0&0 \\ 1&1&1 \end{bmatrix}, \begin{bmatrix} 0&0&0 \\ 1&1&1 \\ 1&1&1 \end{bmatrix} \right\} \tag{16.11}$$

The ring has two right identities and no left identity.

Near-Rings

I restrict myself to allocating interior matrices to the vectors $\{\underline{a}, \underline{b}, \underline{c}\}$ in such a way that their interior matrices are multiplicatively closed. The first two, obvious, cases involve all the interior matrices being zero matrices, the trivial case ($x \cdot y = 0$), and cases in which the non-zero interior matrices are equal to the identity—the constant case ($x \cdot y = y$). The invertible Boolean matrices form the set $\mathcal{A}$ =

$$\left\{ \begin{bmatrix} 1 & 0 \\ 0 & 1 \end{bmatrix}, \begin{bmatrix} 0 & 1 \\ 1 & 0 \end{bmatrix}, \begin{bmatrix} 0 & 1 \\ 1 & 1 \end{bmatrix}, \begin{bmatrix} 1 & 0 \\ 1 & 1 \end{bmatrix}, \begin{bmatrix} 1 & 1 \\ 1 & 0 \end{bmatrix}, \begin{bmatrix} 1 & 1 \\ 0 & 1 \end{bmatrix} \right\} \tag{16.12}$$

Tagged as $\{\mathcal{A}_j\}_{j=1,2,\ldots 6}$. The matrices have multiplicative orders $\{1, 2, 3, 2, 3, 2\}$.

Near-Rings with Identity

I now make the further restriction of only considering near-rings which have a multiplicative identity element, and I apply the ideas of Sect. 16.1.5, so $\mathcal{A}_1$ always occurs and the form of our matrices is given by (16.4). Our near-rings will additionally be zero-symmetric. My methods are in some sense canonical, and, of course, the symmetries of the Klein group mean that any particular near-ring is isomorphic to, possibly, three others. However, any near-ring occurring should simply be isomorphic to the ones reported, using the result given in Sect. 15.4.6. If either of the matrices $\{\mathcal{A}_3, \mathcal{A}_5\}$ were to occur as an interior matrix of the lifting, the other two non-zero terms would have to be powers of that matrix (in the near-ring case). The inverse of $\mathcal{A}_3$ is $\mathcal{A}_5$, so $\{\mathcal{A}_3, \mathcal{A}_5, \mathcal{A}_1\}$ is the only cyclic group with order 3 made from these Boolean matrices. From this, the field with order 4 is represented by the matrices

$$\left\{ \begin{bmatrix} 0 & 0 & 0 \\ 0 & 0 & 0 \\ 0 & 0 & 0 \end{bmatrix}, \begin{bmatrix} 0 & 0 & 0 \\ 1 & 1 & 0 \\ 0 & 0 & 1 \end{bmatrix}, \begin{bmatrix} 0 & 0 & 0 \\ 0 & 0 & 1 \\ 1 & 1 & 1 \end{bmatrix}, \begin{bmatrix} 0 & 0 & 0 \\ 1 & 1 & 1 \\ 1 & 1 & 0 \end{bmatrix} \right\} \tag{16.13}$$

There cannot be a near-ring with identity containing elements from

$$\{\mathcal{A}_2, \mathcal{A}_4, \mathcal{A}_6\}$$

(perhaps with repetitions). Such a near-ring would represent a finite field with several elements with multiplicative order 2, although another way to see this is to try various possibilities and observe that multiplicative closure does not occur. So, remaining, non-isomorphic, near-rings will contain some elements with interior matrices having determinant 0. The non-invertible 2×2 Boolean matrices form the set

$$\mathcal{B} = \left\{ \begin{bmatrix} 0 & 0 \\ 0 & 0 \end{bmatrix}, \begin{bmatrix} 1 & 1 \\ 1 & 1 \end{bmatrix}, \begin{bmatrix} 1 & 0 \\ 0 & 0 \end{bmatrix}, \begin{bmatrix} 0 & 1 \\ 0 & 0 \end{bmatrix}, \begin{bmatrix} 0 & 0 \\ 0 & 1 \end{bmatrix}, \right.$$
$$\left. \begin{bmatrix} 0 & 0 \\ 1 & 0 \end{bmatrix}, \begin{bmatrix} 1 & 1 \\ 0 & 0 \end{bmatrix}, \begin{bmatrix} 0 & 1 \\ 0 & 1 \end{bmatrix}, \begin{bmatrix} 0 & 0 \\ 1 & 1 \end{bmatrix} \begin{bmatrix} 1 & 0 \\ 1 & 0 \end{bmatrix}, \right\} \tag{16.14}$$

Of these, only

$$\mathcal{B}_6 = \begin{bmatrix} 0\,0\,0 \\ 0\,0\,0 \\ 1\,1\,0 \end{bmatrix} \quad or \quad \mathcal{B}_9 = \begin{bmatrix} 0\,0\,0 \\ 0\,0\,0 \\ 1\,1\,1 \end{bmatrix}$$

and

$$\mathcal{B}_{10} = \begin{bmatrix} 0\,0\,0 \\ 1\,1\,0 \\ 1\,1\,0 \end{bmatrix} \quad or \quad \mathcal{B}_2 = \begin{bmatrix} 0\,0\,0 \\ 1\,1\,1 \\ 1\,1\,1 \end{bmatrix}$$

are possibilities (applying the column repetition idea). Clearly, $\{\mathcal{B}_2, \mathcal{B}_6\}$ cannot coexist. $\{\mathcal{B}_2, \mathcal{B}_9\}$ and $\{\mathcal{B}_{10}, \mathcal{B}_6\}$ are possible near-rings. I obtain

$$\left\{ \begin{bmatrix} 0\,0\,0 \\ 0\,0\,0 \\ 0\,0\,0 \end{bmatrix}, \begin{bmatrix} 0\,0\,0 \\ 1\,1\,0 \\ 0\,0\,1 \end{bmatrix}, \begin{bmatrix} 0\,0\,0 \\ 0\,0\,0 \\ 1\,1\,1 \end{bmatrix}, \begin{bmatrix} 0\,0\,0 \\ 1\,1\,1 \\ 1\,1\,1 \end{bmatrix} \right\} \tag{16.15}$$

and

$$\left\{ \begin{bmatrix} 0\,0\,0 \\ 0\,0\,0 \\ 0\,0\,0 \end{bmatrix}, \begin{bmatrix} 0\,0\,0 \\ 1\,1\,0 \\ 0\,0\,1 \end{bmatrix}, \begin{bmatrix} 0\,0\,0 \\ 0\,0\,0 \\ 1\,1\,0 \end{bmatrix}, \begin{bmatrix} 0\,0\,0 \\ 1\,1\,0 \\ 1\,1\,0 \end{bmatrix} \right\} \tag{16.16}$$

These are two isomorphic near-rings where (in the notation of 16.2)

$$\lambda = \begin{bmatrix} 1\,1 \\ 0\,1 \end{bmatrix}$$

The matrices in this representation are not additively closed. That is,

$$\mathcal{A}_1 \neq (\mathcal{B}_2 + \mathcal{B}_6)$$

so, the near-ring is not bi-distributive. This is a non-ring and represents a smallest zero-symmetric near-ring with identity which is not actually a ring. There are several of them, all the same size; and the next largest such near-ring will be one of several hosted by D_4. I consider co-structures to the near-ring (16.15) (see page 70) and use the elements associated with the matrices $\mathcal{A}_1$ and $\mathcal{B}_9$ as generators. One obtains a *right*-distributive near-ring by this procedure. Representing that as left-distributive gives the near-ring

$$\left\{ \begin{bmatrix} 0\,0\,0 \\ 0\,0\,0 \\ 0\,0\,0 \end{bmatrix}, \begin{bmatrix} 0\,0\,0 \\ 1\,1\,0 \\ 0\,0\,1 \end{bmatrix}, \begin{bmatrix} 0\,0\,0 \\ 0\,0\,0 \\ 1\,1\,1 \end{bmatrix}, \begin{bmatrix} 0\,0\,0 \\ 1\,1\,0 \\ 1\,1\,0 \end{bmatrix} \right\} \tag{16.17}$$

which is $\{\mathcal{B}_9, \mathcal{B}_{10}\}$. This is a proper near-ring and not isomorphic to (16.15). This leaves us to consider near-rings in which only one non-identity element has interior determinant 0. The possibilities are

$$\{\{\mathcal{A}_2, \mathcal{B}_2\}, \{\mathcal{A}_2, \mathcal{B}_{10}\}, \{\mathcal{A}_4, \mathcal{B}_6\}, \{\mathcal{A}_4, \mathcal{B}_9\}\}$$

All but the first of these violate closure even of the interior matrices. I obtain the near-ring

$$\left\{\begin{bmatrix} 0&0&0 \\ 0&0&0 \\ 0&0&0 \end{bmatrix}, \begin{bmatrix} 0&0&0 \\ 1&1&0 \\ 0&0&1 \end{bmatrix}, \begin{bmatrix} 0&0&0 \\ 0&0&1 \\ 1&1&0 \end{bmatrix}, \begin{bmatrix} 0&0&0 \\ 1&1&1 \\ 1&1&1 \end{bmatrix}\right\} \tag{16.18}$$

The matrices shown are additively closed so this structure is a bi-distributive near-ring with identity—that is, it is a (in this case, commutative) ring. J.R. Clay determined all the near-rings hosted by the Klein group using computers.[1] His results are reported in [128].[2]

16.1.7 *Subfields*

The idea behind product theory can be applied to sub-structures. For instance, the distributivity core of the near-field (page 43) is a subfield of the near-field, and we could arrange our matrix representation in such a way that the first additive generators in our basis belong to this core. We can then recognise the sub-block of our matrices relating to actions on the core, though of course these do not preserve the core itself. If this be done, product theory applies, and these sub-blocks will be additively closed within the matrix structures we produce. In the case that will concern me here, we are only dealing with two generators, one of which is from the core, and this observation is of limited help; but it is likely to be of use if more generators were considered.

16.2 Finite Near-Fields

The reader may wish to refer to Chap. 3; to Sect. 15.6, Notation 160; and to Sect. 15.4.2 since I use some of the information and notation given in these places.

---

[1] I suspect this was the first paper to utilise computers in near-ring research.

[2] The reader will not make the mistake I did. Pilz's book concerns *right* near-rings; the matrix representation shown here refers to *left* near-rings. If one forgets that, one can while away many happy hours negotiating with recalcitrant matrices.

The results in Sect. 16.1.5 apply too. **We are dealing with *left* near-fields here** and require the idea of a quadratic residue and a limited amount of linear algebra, group representation theory and field theory. All finite near-fields are hosted by elementary abelian p-groups (Lemma 145). Their non-zero elements form a group under multiplication so the corresponding non-zero interior matrices $\{(\theta_{i,j}) : 1 \leq i, j, \leq r\}$ must constitute a group of multiplicatively invertible matrices. I find it helpful to represent this schematically in the form

$$(a, \theta) \cdot (b, \phi) = ((\theta(b), (\theta \circ \phi)) \tag{16.19}$$

The matrix representation detailed earlier is entirely consistent, the vector b above corresponding to a *column* vector, acted on the *left* by the interior matrix of the element a—that is what $\theta(b)$ means, in this instance. The *opposite* (Definition 79) structure has the form

$$(a, \theta) : (b, \phi) = ((\phi(a), (\phi \circ \theta)) \tag{16.20}$$

and now the matrix representation represents the vector a as a *row* vector, acted on the *right* by the interior matrix of the element b.

Lemma 497 *Every finite near-field with order p^n is associated with a representation of a group with order $(p^n - 1)$ over the Galois field with order p.*

(i) The degree of the representation is n.
(ii) The representation is irreducible and faithful.
(iii) The representation is fixed-point-free.
(iv) In this representation, the matrices act regularly (defined in Notation 3) on the elements of the associated two-dimensional space.

Proof The representation is unveiled by exploiting Sect. 16.1. Our near-field corresponds to a lifting, and the representation comes by associating elements of N^* with their corresponding interior matrices. The representation essentially maps the vector given by the first column in the stem representation to its own interior matrix (forgetting the first row, of course). That the representation is faithful and irreducible comes from standard near-field properties. More, the group representation cannot have fixed points in the associated vector space (it would be equivalent to requiring $a \cdot b = a$ for some non-integral b of the near-field). So the associated matrices will have eigenvalues that do not include the number 1. □

A theorem of Wielandt says that a finite group having a faithful irreducible and fixed-point-free representation must be a Frobenius complement; but we already know this in the near-field case, of course (Chap. 3).

16.3 Near-Fields Hosted by $(C_p \oplus C_p)$

For almost the whole of the rest of this chapter, I shall look at proper, left near-fields hosted by additive groups of the form

$$(N, +) = (C_p \oplus C_p) \tag{16.21}$$

with p prime. It is known that this includes all possible finite non-Dickson near-fields, but I do not have a simple proof of this (Sect. 3.7), and that is a serious deficiency in my account. $\mathcal{D}$ (Notation 160) must be the Galois field with order p, and $\mathcal{C}$ (Notation 162) must be equal to $\mathcal{D}$. $(\mathcal{D}, +)$ is additively generated by a primitive root π. $(\mathcal{D}^*, \cdot)$ is multiplicatively generated by it. I am interested in the conjugates of π, and, of course, π is central, in Dickson cases and in four of the seven non-Dickson ones (page 163).

In what follows, I use Notation 162, and Φ represents the Euler phi-function. Suppose

$$(x \cdot \pi) = (a \cdot x)$$

(where $a \in N^*$ is the unique element that fits and, obviously, a conjugate of π). Perhaps it happens that

$$(y \cdot \pi) = (a \cdot y)$$

In that case,

$$((x - y) \cdot \pi) = (a \cdot (x - y))$$

(recall our near-fields are *left-distributive* and that $\pi \in \mathcal{D}^*$). We conclude that there will be an additive subgroup $(\mathcal{C}_a, +)$, where

$$\mathcal{C}_a^* = \{x \in N^* : x \cdot \pi = a \cdot x\} \tag{16.22}$$

When π is central, we have that $a = \pi$ and $\mathcal{C}_a = N$.

16.3.1 Non-central Cases (the Fantastic Four)

In non-central (which would have to be non-Dickson) cases, the subgroup $(\mathcal{C}_a, +)$ must have (additive) order p; for otherwise, $-1 \in \mathcal{C}_a \Rightarrow a = -1 \Rightarrow a = \pi \Rightarrow$ *π is central.* The four finite near-fields for which this is true comprise the fantastic four (page 164). The number of (additive) subgroups with order p is calculated by the **Gaussian binomial coefficient**. In this case, this means that there are then exactly $(p+1)$ possible values for a. Each possible value has the same multiplicative order as π, that is, $(p-1)$; and the set of them constitute the distinct possible conjugates of π.

The normaliser of $\langle\pi\rangle_{(mult)}$ is related to its distinct conjugates. (I introduced the symbol $\mathfrak{N}^*$ for this subgroup on page 120 and will use that symbol here.)

Some conjugates of π may remain within this subgroup itself, having the effect of reducing the number of conjugates of the *subgroup* to a factor of $(p+1)$ and ensuring that the subgroup normaliser is actually larger than the subgroup itself.

To illustrate, suppose there are m distinct conjugates of π which are elements of $\langle\pi\rangle_{(mult)}$. The subgroup will then have $\frac{(p-1)}{m}$ distinct conjugates, and its normaliser has that index in N^*.

Conjugates of π have the form

$$\left(x^{-1}\cdot\pi\cdot x\right) = ((\overbrace{x^{-1}+x^{-1}+\ldots+x^{-1}}^{(\pi\text{ times})})\cdot x) = \left(\left(x^{-1}\pi\right)\cdot x\right) \quad (16.23)$$

We know that (again using (3.24))

$$\mathcal{C}^* = \mathcal{C}^*_\pi = \{x\in N^* : x\cdot\pi = \pi\cdot x\} = \mathcal{D}^*$$

and we must have additive subgroups $\{\mathrm{C}_{a_j}\}_{j=0}^{p}$, (where $a_0 = \pi$), each with order p, and such that

$$\left(\mathrm{C}_{a_i}\bigcap\mathrm{C}_{a_j}\right) = \{0\}\ \textit{when}\ i\neq j$$

(see Observations 492). The collection $\{a_j\}_{j=0}^{p}$ is the distinct conjugates of π. Additively, for any particular $j>0$, we have

$$(N,\ +) = \left(\mathrm{C}_\pi\oplus\mathrm{C}_{a_j}\right) \quad (16.24)$$

and the disjoint sets $\{C^*_{a_j}\}_{j=0}^{p}$ form a partition of N^*. We have

$$(x\cdot\pi) = (a_j\cdot x)\Rightarrow\left(x\cdot\pi\cdot x^{-1}\cdot\pi^{-1}\right) = (a_j\cdot\pi^{-1}) \in [N^*,N^*]_{(mult)}$$

implying that $|[N^*,N^*]_{(mult)}| \geq (p+1)$. More, each a_j is a member of the multiplicative coset $\{[N^*,N^*]\cdot\pi\}$. Also, we know that for all i, j,

$$(a_j\cdot\pi^{-1})\cdot(a_k\cdot\pi^{-1})^{-1} = (a_j\cdot a_k^{-1}) \in [N^*,N^*]_{(mult)}$$

We are dealing with a subgroup of the general linear group here. Having a determinant equal to 1 is a necessary, but perhaps insufficient, condition for an element to be in the commutator subgroup.

Lemma 498 *Suppose* $(N, +, \cdot)$ *is a finite near-field in which* $\mathcal{D}^* \supsetneq \mathcal{Z}^*$. *Then the order of the commutator subgroup* $[N^*, N^*]_{(mult)}$ *exceeds the prime characteristic of the near-field.*

This lemma is, of course, also observationally true. Among the non-Dickson near-fields of Table 3.3, only case B involves a commutator subgroup smaller than the prime characteristic, and that near-field is not a member of the fantastic four.

The lemma implies that $\mathcal{L} = N$ (Notation 3.11) using Lagrange's theorem on the additive groups, and that fits in with observations in Sect. 3.7.

One wonders about that commutator subgroup. We already know that it has even order precisely when the Sylow 2-subgroup S_2 is quaternion (Lemma 167) and if that is not the case, that N^* is a 3-group in which the multiplicative commutator is cyclic and has odd order.

If $x \cdot \pi = a_j \cdot x$, then $(x \cdot \pi^s) \cdot \pi = (x \cdot \pi \cdot \pi^s) = (a_j \cdot x \cdot \pi^s)$. We conclude that if $x \in C^*_{a_j}$, then

$$C^*_{a_j} = \{(x \cdot \pi^s) : 1 \le s \le (p-1)\} = \{\langle x \rangle_{(add)}\}^* \tag{16.25}$$

However, $x \cdot \pi^s = a_j^s \cdot x$; therefore,

$$C^*_{a_j} = \{(a_j^s \cdot x) : x \in C^*_{a_j}, 1 \le s \le (p-1)\} \tag{16.26}$$

We know

$$\begin{aligned}(x \cdot \pi = a \cdot x) \Rightarrow (b \cdot x) \cdot \pi &= ((b \cdot a) \cdot x \cdot) \cdot x^{-1} \cdot b^{-1} \cdot (b \cdot x) \\ &= (b \cdot a \cdot b^{-1}) \cdot (b \cdot x)\end{aligned}$$

so it follows

$$\{b \cdot C^*_a\} = C^*_{(b \cdot a \cdot b^{-1})}$$

The set

$$\mathcal{S} = \{a_0 = \pi, a_1, a_2, \ldots, a_p\} \tag{16.27}$$

which constitutes the distinct conjugates of π is a normal subset of N^*. No more than $\Phi(p-1)$ of these elements can lie in $\mathcal{D}^*$, and they each have the same determinant as does π. It turns out that only two conjugates of π can possibly lie in $\mathcal{D}$ (Lemma 505). Observationally, only one near-field in the fantastic four has the property that $\pi \notin [N^*, N^*]_{(mult)}$. This is case F of Table 3.3 where $\pi^7 \in [N^*, N^*]_{(mult)}$.

16.4 The Near-Field with Order 9

I have already determined this particular near-field, which is the smallest proper near-field. Its multiplicative group of non-zero elements is isomorphic to Q_8 (see page 113), and its matrix representation looks like this (which illustrates the notational laxity I mentioned in Remark 495).

$$1 = \begin{bmatrix} 1 & 0 \\ 0 & 1 \end{bmatrix} \quad -1 = \begin{bmatrix} -1 & 0 \\ 0 & -1 \end{bmatrix} \quad i = \begin{bmatrix} 0 & -1 \\ 1 & 0 \end{bmatrix}$$

$$-i = \begin{bmatrix} 0 & 1 \\ -1 & 0 \end{bmatrix} \quad j = \begin{bmatrix} 1 & 1 \\ 1 & -1 \end{bmatrix} \quad -j = \begin{bmatrix} -1 & -1 \\ -1 & 1 \end{bmatrix}$$

$$k = \begin{bmatrix} -1 & 1 \\ 1 & 1 \end{bmatrix} \qquad -k = \begin{bmatrix} 1 & -1 \\ -1 & -1 \end{bmatrix}$$

The integers are to be interpreted as in C_3, so -1 is identical to 2. Each of these matrices has determinant 1. We have a non-abelian subgroup of order 8 of the two-dimensional special linear group over the Galois field with order 3 (which has order 24). The two central elements are, inevitably, scalar matrices. Notice that there are elements with determinant 1 which are not in the commutator subgroup of the group. The eigenvalues of these first two matrices are 1 and -1, respectively. The remaining six matrices have eigenvalues which are the two square roots of 2, neither of which occur in the prime subfield C_3. If we regard these matrices as constituting a representation of the group Q_8, it will be seen that that representation is faithful, irreducible (else each matrix would have eigenvalues in C_3) and fixed-point-free (see [144]).

Interestingly, although this near-field does have a normal subgroup with index 2 in N^*, the matrices do all have determinant 1 (see page 462). More, the set of matrices given is preserved under transposition, and each matrix is fixed by any of its rows or columns (i.e. no row or column occurs more than once).

The representation just obtained reminds me, irresistibly, and perhaps wholly irrelevantly, of the standard representation of the Hamiltonian quaternions by the **Pauli spin matrices**—matrices over the ring of Gaussian integers. These are (perhaps confusing the reader with my double use of the symbol i)[3]

$$i = \begin{bmatrix} i & 0 \\ 0 & -i \end{bmatrix}, \; j = \begin{bmatrix} 0 & 1 \\ -1 & 0 \end{bmatrix}, \; k = \begin{bmatrix} 0 & i \\ i & 0 \end{bmatrix}$$

This covers the $p = 3$ case; I now suppose $p > 3$.

[3] I felt slightly ashamed of my laziness here, but was heartened to find it replicated in [36].

16.5 Finite Division Rings

I now look at finite division rings hosted by

$$(N, +) = (C_p \oplus C_p) \ (p > 2)$$

I define

$$\mathcal{X} = \begin{bmatrix} 0 & 0 & 0 \\ 0 & 0 & A \\ 1 & 1 & B \end{bmatrix} \tag{16.28}$$

which was first defined in (16.5). Notice $A \neq 0$. Division rings are bi-distributive, so our representing matrices are additively closed. This means that elements of N^* have the form

$$\mathcal{Y} = \begin{bmatrix} a & bA \\ b & (a + bB) \end{bmatrix} \tag{16.29}$$

and that the binary quadratic form

$$\mathfrak{F}_{A,B} = x^2 + xyB - Ay^2 \tag{16.30}$$

is non-zero so long as x and y are not simultaneously zero. The matrices that have this form are both additively and multiplicatively closed—they represent bi-distributive near-rings whose non-zero elements form a group—**division rings**. But, matrices of the form (16.29) commute with one another—so all division rings occurring must be commutative fields, a special (and reassuring) case of Wedderburn's theorem. Any choice of A and B meeting these conditions represents a field hosted by $(N, +)$.

If we attempt to make all the matrices symmetric, we are faced with the additional conditions

$$b = A^2 b$$

$$(a + Bb) = (a + ABb)$$

applying for all $a, b \in C_p^*$. These conditions translate into the two possibilities

$$A = 1 \textit{ and } \mathfrak{F}_{1,B} = (x^2 + Bxy + y^2) \textit{ is positive definite,}$$

giving

$$\mathcal{Y} = \begin{bmatrix} a & b \\ b & (a + Bb) \end{bmatrix}$$

Note that $p \equiv 1$ (modulo 4) would mean that $B \neq 0$ in this case.

The other possibility is

$$A = -1 \text{ and } B = 0 \text{ and } \mathfrak{F}_{-1,0} = (x^2 - y^2) \text{ is positive definite,}$$

which is clearly impossible.

16.6 Near-Fields Hosted by $(C_p \oplus C_p)$ with $p > 3$

The multiplicative group has order $(p^2 - 1)$, and we should like to isolate non-abelian groups of (2×2) matrices with this order, having the property that each first column is unique and all possible columns occur as both first and second columns of the matrices (Lemma 496). Viewed this way, the near-field determination problem is a problem in the representation of finite groups. It seems likely that Lemma 71 might be useful here too. Our matrices are non-singular, so we have a sequence of groups and homomorphisms

$$\mathcal{G} \longmapsto N^* \longrightarrow C_{(p-1)} \tag{16.31}$$

provided using the determinant function. $\mathcal{G}$ is the set of matrices which have determinant 1 and must contain the commutator subgroup. The group $SL(2, p)$ has order $p.(p^2 - 1)$, and the order of $\mathcal{G}$ should divide that. If N^* does not have a proper normal subgroup with index some divisor of $(p - 1)$, the representation given must involve matrices exclusively within $SL(2, p)$. Although there is such a normal subgroup in the example given in Sect. 16.4, the matrices occurring still have determinants equal to 1.

There is some possibility for confusion here. Some of the near-fields that are discussed do have representations in $SL(2, p)$, for example, case A in Table 3.3 is isomorphic to $SL(2, 3)$. The characteristic of that near-field is five, not three, and the matrices which concern us using product theory are 2×2 matrices from the Galois field with order 5. These may or may not have determinants equal to 1. Discovering this is non-trivial.

The additive group $\langle e \rangle$ is the prime subfield of our near-field. When the near-field is not a field, its elements form the *distributivity core* (page 43), which, I shall continue to denote by $\mathcal{D}$ (Notation 160).

In my near-field chapter I observed that the multiplicative centre of the near-field lies in this core. Situations in which that centre is strictly smaller than it are known, though rare (the "fantastic four"), and, in this case at least, correspond to central elements not being additively closed (as distributive elements are) (but see Lemma 161). The next result comes from Sect. 16.1.5.

Lemma 499 *Members of* $\mathbf{m}_j\mathbf{T}(B)$ *have the form*

$$\begin{bmatrix} 0 & 0 & 0 \\ x & x & z \\ y & y & s \end{bmatrix} \tag{16.32}$$

with $xs \not\equiv yz \pmod p$.

Near-field products correspond to particular two-dimensional representations of groups of order (p^2-1) (the multiplicative groups of the possible near-fields). Using Lemma 488, it will be seen that isomorphic near-fields correspond to equivalent representations. In these representations, the (p^2-1) first columns correspond to all pairs

$$\begin{bmatrix} x \\ y \end{bmatrix}$$

with $x, y \in N^*$.

The Matrix $\mathcal{X}$

One wonders whether $\mathcal{X}$ from (16.5) may be similar to a matrix of the form

$$\begin{bmatrix} 0 & D \\ 1 & 0 \end{bmatrix}$$

where, of course, $D = A$—that is, an **anti-diagonal matrix** (defined—page xviii).

Can one write?

$$\begin{bmatrix} 0 & A \\ 1 & \mathcal{B} \end{bmatrix} \cdot \begin{bmatrix} x & z \\ y & w \end{bmatrix} = \begin{bmatrix} x & z \\ y & w \end{bmatrix} \cdot \begin{bmatrix} 0 & A \\ 1 & 0 \end{bmatrix}$$

It seems to me that simple algebraic manipulations reveal this is impossible, so, potentially, $\mathcal{X}$ exists in one of two forms, depending on whether $\mathcal{B}$ is or is not zero.

Notice that

$$\begin{bmatrix} 0 & A \\ 1 & 0 \end{bmatrix} \cdot \begin{bmatrix} 0 & A \\ 1 & 0 \end{bmatrix} = \begin{bmatrix} A & 0 \\ 0 & A \end{bmatrix} \tag{16.33}$$

a diagonal matrix. Moreover, if

$$\mathcal{Y} = \begin{bmatrix} 0 & b \\ a & c \end{bmatrix} \in N^* \tag{16.34}$$

and $\mathcal{B} = 0$, then

$$(\mathcal{X}.\mathcal{Y}) = \begin{bmatrix} Aa & Ac \\ 0 & b \end{bmatrix} \in N^* \Rightarrow c = 0 \tag{16.35}$$

Furthermore,

$$\mathcal{X}.\begin{bmatrix} a & c \\ b & d \end{bmatrix} = \begin{bmatrix} Ab & Ad \\ a & c \end{bmatrix} \tag{16.36}$$

which means that each first row occurs as a second row and each second row occurs as a first row.

Note (16.35) and (16.36) are consequences of $\mathcal{B} = 0$ and not claimed generally.

One wonders whether each *possible* first row occurs. If two distinct matrices, perhaps associated with the column vectors $\underline{a}$, $\underline{A}$, occur and have the same first rows (and, of course, distinct second rows), then defining

$$\mathfrak{Q}^* = \left\{ \begin{bmatrix} 0 & c \\ b & d \end{bmatrix} : b \neq 0 \right\} \tag{16.37}$$

we can say that when $x \in N^*$,

$$\left(\underline{A} \cdot \underline{x} - \underline{a} \cdot \underline{x}\right) \in \mathfrak{Q}^*$$

(this uses the star notation of Notation 7). The additive subgroup $\mathfrak{Q}$ has order p, so we should be able to find $x, y \in N^*$, with $x \neq y$, for which

$$\left(\underline{A} \cdot \underline{x} - \underline{a} \cdot \underline{x}\right) = \left(\underline{A} \cdot \underline{y} - \underline{a} \cdot \underline{y}\right)$$

But, then

$$\left(\underline{A} \cdot (x - y)\right) = \left(\underline{A} \cdot \underline{x} - \underline{A} \cdot \underline{y}\right) = \left(\underline{a} \cdot \underline{x} - \underline{a} \cdot \underline{y}\right) = \left(\underline{a} \cdot (x - y)\right)$$

And this implies $A = a$.

Lemma 500 *Suppose that in (16.5) we have that $B = 0$. Then, the matrices of our near-field representation have unique rows and unique columns. Every possible row vector and column vector must occur.*

Other Symmetries

When $B = 0$, the uniqueness of rows, particularly the top row, suggests other symmetries (see page 449). If we have a matrix representation which one might represent as

$$\left(\underline{a}, A\right)$$

we know, from Lemma 500, that the top row of the matrix A is a unique vector of the form $\underline{b}$. This means that

$$\left(\underline{b}, A^t\right)$$

(where A^t is the transpose of the (interior) matrix) represents another near-field. In the case of possible finite division rings, with $B = 0$, we already know that the two structures related in this way are isomorphic (Sect. 16.5). If we had a finite division ring whose first rows were unique to each element (whether or not $B = 0$), we have:

$$\mathcal{X}^t = \begin{bmatrix} 0 & 1 \\ A & B \end{bmatrix}$$

and additive closure means that the new structure must contain the matrix

$$\begin{bmatrix} 0 & A^{-1} \\ 1 & BA^{-1} \end{bmatrix}$$

and so general matrices will have the form

$$\begin{bmatrix} a & \frac{b}{A} \\ b & (a + \frac{bB}{A}) \end{bmatrix}$$

This would be a finite field, isomorphic to the original one.

Symmetric Matrices
One might suspect that the matrices in the representation should be symmetric, but the product of two symmetric matrices is only again symmetric if they commute, so this cannot be the case when N^* is non-abelian. As one would expect, the matrices in Sect. 16.4 are not all symmetric; but the matrices in (16.13) certainly are.

The Form of the Matrices
We are now concerned with invertible matrices of the form

$$\begin{bmatrix} a & c \\ b & d \end{bmatrix}$$

The characteristic equation is

$$x^2 - x \cdot (a + d) + (ad - bc) = 0 \qquad (16.38)$$

and the corresponding inverse matrix is

$$\frac{1}{D} \cdot \begin{bmatrix} d & -c \\ -b & a \end{bmatrix}$$

(where $D = (ad - bc) \neq 0$). These matrices may not have 1 as an eigenvalue, so

$$1 - (a + d) + (ad - bc) \not\equiv 0 \ (modulo\ p\ )$$

– unless $a = d = 1$ and $b = c = 0$. If we impose the further restriction that our matrices should be members of SL(2, p), which is the case in Sect. 16.4, then A, in (16.5), is equal to -1, and $(a + d) \neq 2$. We do know that matrices of the form

$$\begin{bmatrix} 0 & y \\ x & z \end{bmatrix} \tag{16.39}$$

occur in our representation ($x \neq 0$, $y \neq 0$). Their inverses are

$$\begin{bmatrix} -x^{-1}zy^{-1} & x^{-1} \\ y^{-1} & 0 \end{bmatrix} \tag{16.40}$$

The Distributivity Core

I now use notation developed in Notation 160 and Notation 3. Choosing a primitive root π modulo p (see (3.39), we admit the matrix

$$\mathcal{P} = \begin{bmatrix} 0 & 0 & 0 \\ \pi & \pi & E \\ 0 & 0 & \pi^r \end{bmatrix} \tag{16.41}$$

as the matrix in our representation corresponding to π, which, in this left-distributive situation, should be regarded as $1 \cdot \pi \; = \overbrace{(1 + 1 + \ldots. + 1)}^{(\pi \text{ times})}$ (see (3.17). Of course, there are typically several different values one might choose for π, but in what follows, I simply suppose one has been selected and fixed. Here $1 \leq r \leq (p - 1)$ and $E \in \{0, 1, 2, \ldots (p - 1)\}$. This matrix has eigenvalues π and π^r. It's easy to spot the vector

$$e = \begin{bmatrix} 1 \\ 0 \end{bmatrix}$$

(the group identity) as the eigenvector corresponding to π. For the eigenvalue π^r, we need

$$(1 \cdot \pi) \cdot x = x\pi^r$$

and $\{1, x\}$ is a basis for the two-dimensional space (the eigenvectors are independent). Choosing

$$x = \begin{bmatrix} 0 \\ 1 \end{bmatrix}$$

as the other eigenvector forces $E = 0$ and makes the matrix corresponding to π to be

$$\mathcal{P} = \begin{bmatrix} 0 & 0 & 0 \\ \pi & \pi & 0 \\ 0 & 0 & \pi^r \end{bmatrix}$$

Each of the powers of π up to the $(p-1)$th represents different elements of N^*; consequently, π^r must have multiplicative order $(p-1)$ (by Lemma 496), and so π^r is another primitive root modulo p. From Lemma 188, we can say that the interior matrix commutes with all the others when the near-field is Dickson—so $r = 1$ in that case and, again, $E = 0$.

The symbol $\mathcal{P}$ is important for the rest of the chapter.

Summarising, our matrix has the form

$$\mathcal{P} = \begin{bmatrix} 0 & 0 & 0 \\ \pi & \pi & 0 \\ 0 & 0 & \pi^r \end{bmatrix} \tag{16.42}$$

The case $\mathcal{D}^* = \mathcal{Z}^*$ is r=1, and this includes all Dickson cases and three of the non-Dickson ones (see Sect. 3.7). π^r is always a primitive root. Now,

$$\left(\mathcal{X}^{-1} \cdot \mathcal{P} \cdot \mathcal{X}\right) = \begin{bmatrix} \pi^r & \mathcal{B}(\pi^r - \pi) \\ 0 & \pi \end{bmatrix}$$

From this,

Lemma 501

(i) The case $\mathcal{D}^* \supsetneq \mathcal{Z}^*$ corresponds to $\pi \neq \pi^r$ and only occurs in four non-Dickson near-fields.
(ii) When we do have $\pi \neq \pi^r$, it must be that

$$\mathcal{X} = \begin{bmatrix} 0 & A \\ 1 & 0 \end{bmatrix}$$

and $\pi^{r^2} = \pi$.
That is to say,

$$r^2 \equiv 1 \ (modulo\ (p-1)) \tag{16.43}$$

(iii) Generally,

$$x = \begin{bmatrix} a & c \\ b & d \end{bmatrix} \Rightarrow (\mathcal{P})^x = \begin{bmatrix} \frac{(ad\pi - cb\pi^r)}{D} & \frac{(cd\pi - cd\pi^r)}{D} \\ \frac{(ab\pi^r - ab\pi)}{D} & \frac{(ad\pi^r - bc\pi)}{D} \end{bmatrix} \tag{16.44}$$

where $D = (ad - bc)$.

Squaring the matrix $\mathcal{X}$ above gives us the scalar matrix

$$\mathcal{X}^2 = \begin{bmatrix} A & 0 \\ 0 & A \end{bmatrix}$$

and the deduction that $A = \pi^s = \pi^{rs}$.

If N^* were a $\mathcal{Z}$-group (and we know, but have not proved, it is not), the element

$$b^s = \begin{bmatrix} \pi^s & 0 \\ 0 & \pi^{rs} \end{bmatrix} \tag{16.45}$$

would stand revealed as the multiplicative centre of N^* (see (3.29)).

My interest in this centre is based on Lemma 167. If it does not intersect with $[N^*, N^*]$, we have a direct proof that S_2 is quaternionic (which, as I say, is known already but not yet proved here).

Any matrix with the form

$$\begin{bmatrix} 0 & c \\ a & d \end{bmatrix}$$

must have $d = 0$, and if we write $c = \pi^t$ above, then $a = \pi^{(rt-s)}$.

16.7 The Fantastic Four

In this section, I look at the four non-Dickson near-fields in which $\mathcal{D}^*$ is non-central. The reader may assume that $\pi \neq \pi^r$ in what follows. These are the four cases of the fantastic four (page 164). The multiplicative centre of the general linear group is simply the group of scalar matrices, and it's fairly easy to see (by direct calculation) that the centre consists of scalar matrices in our situation too. Interior matrices in $\mathcal{D}^*$ have the form (overloading our notation in the way we have already described in Remark 495)

$$\mathcal{P}^k = \begin{bmatrix} \pi^k & 0 \\ 0 & \pi^{kr} \end{bmatrix} \tag{16.46}$$

$k = 1, 2, 3, \ldots, (p-1)$. These are the $(p-1)$ matrices of $\mathcal{D}^*$. There is some s such that $A = \pi^s = \pi^{rs}$ (see (16.33)).

First Possibility for s

A possibility is $s = \frac{p-1}{2}$, which would give $A = -1$. If that holds, the simple multiplication with $\mathcal{X}$ gives that

$$\begin{bmatrix} a & c \\ b & d \end{bmatrix} \in N^* \Longrightarrow \left\{ \begin{bmatrix} c & -a \\ d & -b \end{bmatrix}, \begin{bmatrix} -a & -c \\ -b & -d \end{bmatrix}, \begin{bmatrix} -c & a \\ -d & b \end{bmatrix}, \right\} \subset N^*$$

and also

$$\left\{ \begin{bmatrix} -b & -d \\ a & c \end{bmatrix}, \begin{bmatrix} b & d \\ -a & -c \end{bmatrix}, \begin{bmatrix} d & -b \\ -c & a \end{bmatrix} \right\} \subset N^*$$

Negations are there too so,

$$\begin{bmatrix} -d & b \\ c & -a \end{bmatrix} \in N^*$$

If

$$\begin{bmatrix} a & c \\ b & d \end{bmatrix} \in \mathrm{SL}(2, p)$$

its multiple inverse is present too and so,

$$\left\{ \begin{bmatrix} d & -c \\ -b & a \end{bmatrix}, \begin{bmatrix} -c & -d \\ a & b \end{bmatrix}, \begin{bmatrix} -b & a \\ -d & c \end{bmatrix}, \begin{bmatrix} a & b \\ c & d \end{bmatrix}, \right\} \subset N^*$$

plus the negations

$$\left\{ \begin{bmatrix} -d & c \\ b & -a \end{bmatrix}, \begin{bmatrix} c & d \\ -a & -b \end{bmatrix}, \begin{bmatrix} b & -a \\ d & -c \end{bmatrix}, \begin{bmatrix} -a & -b \\ -c & -d \end{bmatrix}, \right\} \subset N^*$$

Each of these 16 matrices is a member of the subgroup of N^* generated by

$$\left\{ \begin{bmatrix} 0 & -1 \\ 1 & 0 \end{bmatrix}, \begin{bmatrix} a & c \\ b & d \end{bmatrix} \right\} \tag{16.47}$$

We cannot immediately claim they are all distinct; but it is clear that in this situation the interior rows of our matrices have the same uniqueness property as the interior columns (Lemma 496).

Second Possibility for s
Another possibility is that $s = (p-1)$, which means $A = 1$. However,

$$\begin{bmatrix} 0 & 1 \\ 1 & 0 \end{bmatrix}^2 = \begin{bmatrix} 1 & 0 \\ 0 & 1 \end{bmatrix}$$

which would mean that although $\mathcal{X} \neq -1$, we have $|\mathcal{X}| = 2$. This is impossible, from Lemma 146.

Third Possibility for s
Other values for s are possible, but only one of the fantastic four exhibits this. I discuss it below.

16.7.1 The Condition $\pi^s = \pi^{rs}$

The condition $\pi^s = \pi^{rs}$ is interesting. It is expressed alternatively as

$$((r-1) \cdot s) \equiv 0 \,(modulo\,(p-1)\,) \tag{16.48}$$

and (16.43) is a special case of it (with $s = (p+1)$). There are primes in which the only s values fitting the condition are $s = \frac{(p-1)}{2}$ and $s = (p-1)$. The non-Dickson near-field primes $p = 5, 7, 11$, of fantastic four cases A, C and E, are in that category (page 164). But, there are primes in which other values for s may occur. The fantastic four non-Dickson prime $p = 29$ is one of these, where the value pairs $\{s = 14, r = 13\}$, and $\{s = 16, r = 15\}$ work; and these r values fit (16.43) too. s-values fitting the condition reveal elements $\mathcal{P}^s$ which are central. For near-fields in which $\mathcal{D}^*$ is not central, these other values of s reveal elements of $\mathcal{Z}^*$ lying outside C_2, since the element $\mathcal{P}^s$ is one of these, as are its powers.

Lemma 502 *In the non-Dickson near-fields given by cases A, C and E of Table 3.3,*

$$\mathcal{Z}^* = C_2$$

More,

$$\mathcal{X} = \begin{bmatrix} 0 & -1 \\ 1 & 0 \end{bmatrix} \tag{16.49}$$

which is an element with multiplicative order 4. And,

$$\mathcal{P} = \begin{bmatrix} \pi & 0 \\ 0 & \pi^{-1} \end{bmatrix}$$

Moreover, $\mathcal{D}^* \subset SL(2, p)$.

The representation for $\mathcal{X}$ matches the representation for i in Sect. 16.4. If one could establish the corresponding matrix for j occurs too, one would have a proof that S_2 is quaternion for these three cases.

The fantastic four near-fields do have non-trivial inner automorphism groups (page 46).

$$\begin{bmatrix} \pi^{-1} & 0 \\ 0 & \pi \end{bmatrix} \cdot \begin{bmatrix} 0 & -1 \\ 1 & 0 \end{bmatrix} \cdot \begin{bmatrix} \pi & 0 \\ 0 & \pi^{-1} \end{bmatrix} \cdot \begin{bmatrix} 0 & 1 \\ -1 & 0 \end{bmatrix} = \begin{bmatrix} \pi^{-2} & 0 \\ 0 & \pi^2 \end{bmatrix}$$

which is an element with order $\frac{(p-1)}{2}$ in the commutator subgroup.

When $(p-1) \equiv 0$ (*modulo* 4), this shows that the commutator subgroup has even order, and so S_2 is quaternion. The congruence holds for cases A and F of the fantastic four.

Case F
In case F, $p = 29$, and

$$\mathcal{Z}^* = C_{14} = C_{\frac{(p+1)}{2}} \subsetneq \mathcal{D}^* = C_{29}$$

In this case, $(r, (p-1)) = 1$.

There are four possible values for r fitting (16.43), and these are $\{1, 13, 15, 27\}$.

We may reject $r = 1$. The possible s values fitting (16.48) are as follows:

(i) $r = 13, s \in \{7, 14, 21, 28\}$.
(ii) $r = 15, s \in \{2, 4, 5, 6, 10, \ldots, 28\}$.
(iii) $r = 27, s \in \{14, 28\}$.

Case (iii) represents a possible near-field with centre C_2. Case (i) represents one with centre C_4. Neither apply to case F so case (ii) must be correct.

This gives

Lemma 503 *In the non-Dickson near-field which is case F of Table 3.3,*

$$\mathcal{P} = \begin{bmatrix} \pi = 3 & 0 \\ 0 & \pi^{15} = 26 \end{bmatrix} \tag{16.50}$$

and

$$\mathcal{Z}^* = \left\{ \mathcal{P}^{2t} \right\}_{t=1}^{14} \tag{16.51}$$

Elements of $\mathcal{D}^*$ have the form (16.46). If we require a centraliser of such an element to be outside $\mathcal{D}^*$, we can deduce that

$$\pi^k = \pi^{rk}$$

which actually forces $\mathcal{P}^k \in \mathcal{Z}^*$.

Lemma 504 *Non-central elements of $\mathcal{D}^*$ have centralisers that are equal to $\mathcal{D}^*$.*

16.7.2 The Subgroup $\mathcal{M}^*$

Consider subgroups of the form

$$\mathcal{M}^* = \left(\mathcal{D}^* \bigcap (\mathcal{D}^*)^x\right) \tag{16.52}$$

We have

$$\begin{bmatrix} 0 & c \\ b & 0 \end{bmatrix} \cdot \begin{bmatrix} \pi & 0 \\ 0 & \pi^r \end{bmatrix} \cdot \begin{bmatrix} 0 & 1/b \\ 1/c & 0 \end{bmatrix} = \begin{bmatrix} \pi^r & 0 \\ 0 & \pi \end{bmatrix}$$

So, $\mathcal{M}^*$ is actually equal to $\mathcal{D}^*$ (and strictly bigger than $\mathcal{Z}^*$) for some values of x. In fact, it is easy to see that the relation above identifies the only distinct conjugate of π inside $\mathcal{D}^*$.

Lemma 505 *The element $\mathcal{P}^{\mathcal{X}}$ represents the only non-identical conjugate of π that could possibly lie in $\mathcal{D}^*$. There are other conjugates, but these lie outside $\mathcal{D}^*$ (see page 459).*

We have that

$$\mathcal{P}^{\mathcal{X}} \in \left(\mathcal{D}^* \bigcap (\mathcal{D}^*)^{\mathcal{X}}\right) = \mathcal{D}^* \tag{16.53}$$

It's also straightforward to establish that (see page 120 (3.26), for notation)

$$\mathfrak{N}^* = \left\langle \left\{ \mathcal{D}^* \cup \left\{ \begin{bmatrix} 0 & c \\ a & 0 \end{bmatrix} \in N^* \right\} \right\} \right\rangle_{mult} \tag{16.54}$$

This is a multiplicative group with order $2(p-1)$ and index $\frac{(p+1)}{2}$. It consists entirely of the diagonal and anti-diagonal matrices (page xviii) that lie in N^*. Suppose $x \notin \mathfrak{N}^*$. Then

$$\left(\mathcal{D}^* \bigcap (\mathcal{D}^*)^x\right) = \mathcal{Z}^*$$

If $x, y \in N^*$ are in the same coset of $\mathfrak{N}^*$, so $x = n \cdot y$, $(n \in \mathfrak{N}^*)$, we shall have

$$\left((\mathcal{D}^*)^x \bigcap (\mathcal{D}^*)^y\right) = (\mathcal{D}^*)^x = (\mathcal{D}^*)^y$$

On the other hand, if they are in distinct cosets of $\mathfrak{N}^*$,

$$a \in \left((\mathcal{D}^*)^x \bigcap (\mathcal{D}^*)^y\right) \Rightarrow$$
$$(a)^{x^{-1}} \in \left((\mathcal{D}^*) \bigcap (\mathcal{D}^*)^{yx^{-1}}\right) \Rightarrow$$
$$(a)^{x^{-1}} \in \mathcal{Z}^* \Rightarrow x \in \mathcal{Z}^*$$

Lemma 506 *Suppose $x, y \in N^*$. The intersection*

$$\left((\mathcal{D}^*)^x \bigcap (\mathcal{D}^*)^y\right)$$

is equal to $(\mathcal{D}^*)^x$ exactly when $x = n \cdot y$, for some $n \in \mathfrak{N}^*$. In other circumstances, this intersection is equal to $\mathcal{Z}^*$.

Consider

$$\mathcal{E}^* = \bigcup_{x \in N^*} (\mathcal{D}^*)^x \tag{16.55}$$

We can say

$$\begin{aligned} |\mathcal{E}^*| &= \left(\frac{(p+1)}{2}((p-1) - |\mathcal{Z}^*|) + |\mathcal{Z}^*|\right) \\ &= \left(\frac{(p^2-1)}{2} - \frac{(p-1)}{2}|\mathcal{Z}^*|\right) \end{aligned} \tag{16.56}$$

Moreover, from Lemma 502, in the near-fields of cases A, C and E,

$$\mathcal{E}^* \subset \mathrm{SL}(2, p) \tag{16.57}$$

16.7.3 Determinants for the Fantastic Four

The fantastic four are cases A, C, E and F, of Table 3.3. Commutators will have determinants equal to 1, and we have the exact sequence (16.31). Case E represents a perfect group, so it will itself be a subgroup of SL(2, 11). In cases A, C and F, the index of the commutator subgroup is prime (3, 2 and 7, respectively). So the subgroup $\mathcal{G}$ of elements with determinant 1 is either the commutator subgroup or the whole group in each case; but, we know already that in case F there are matrices not in SL(2, 29) (page 471), so $\mathcal{G}$ is the commutator subgroup, then. In case A, the element

$$\begin{bmatrix} 1 & 0 \\ 1 & 1 \end{bmatrix}.$$

is in SL(2, 5) and has order 5, so it is not in SL(2, 3). Similarly, the order 6 element

$$\begin{bmatrix} 1 & -1 \\ 1 & 0 \end{bmatrix}.$$

is in SL(2, 5), but its square

$$\begin{bmatrix} 0 & -1 \\ 1 & -1 \end{bmatrix}.$$

is not in SL(2, 3) (Lemma 502). Therefore, we have an order 5 and an order 3 element both of which are not in SL(2, 3), and, consequently, for this case, $\mathcal{G}$ is the commutator subgroup. Last, for case C, according to me (and GAP), the binary octahedral group is not a subgroup of SL(2, 7), so, again $\mathcal{G}$ is the commutator subgroup.

16.8 Eigenvalues

I now turn to considering the eigenvalues of the matrices that appear in our representation. Suppose $\lambda = \pi^s$ is an eigenvalue for its matrix and that matrix is associated with the (column) vector v. Then there is some y such that

$$v \cdot y = y\lambda \tag{16.58}$$

(recall the notation of (1.5)). This means that

$$(y \cdot e\lambda \cdot y^{-1}) = v \tag{16.59}$$

and so v is an element of the set which we have defined as $\mathcal{E}^*$, which now stands revealed as the set of elements having eigenvalues in the Galois field with order p. The complement of $\mathcal{E}^*$ represents the elements whose eigenvalues are not in C_p (their eigenvalues are in a quadratic extension of it). This set includes elements whose order is coprime with $(p-1)$ and possibly some other elements. The set of these elements is

$$\mathfrak{E} = \left(\mathcal{E}^*\right)^c \tag{16.60}$$

The set $\mathfrak{C}$ has cardinality $\left(\frac{(p^2-1)}{2} + \frac{(p-1)}{2}|\mathcal{Z}^*|\right)$. Recall that each element of N^* lies in exactly one of the additive subgroups $\{C^*_{a_j}\}$ More,

$$\begin{aligned} y \in C^*_{a_j} &\Rightarrow (y \cdot \pi) = (a_j \cdot y) \\ &\Rightarrow (y \cdot \pi^s) = ((a_j)^s \cdot y) \end{aligned}$$

This means that $y \in C^*_{a_j}$ is an eigenvector for each of the "matrices" $\{a^s_j\}^{(p+1)}_{s=0}$, having associated eigenvalues $\{\pi^s\}^{(p+1)}_{s=0}$. Suppose it happens that for some $i \neq j$, we have that $(a_i)^s = (a_j)^t$. We select $x_r \in C^*_{a_r}$,where $r \in \{i, j\}$. Then,

$$(x_k \cdot \pi) = (a_k \cdot x) \Rightarrow (x_k \cdot \pi^v) = (a^v_k \cdot x)$$

The eigenvalue condition (16.58) is

$$v \cdot y = y\lambda = y\pi^s$$

Assuming $y \in C^*_{a_j}$, we can say $v = (a_j)^s$. So, it happens that the only elements that may have eigenvalues in the Galois field with order p are those that lie in the set

$$\mathfrak{C} = \{(a_j)^s \ : 0 \leq j \leq p,\ 1 \leq s \leq (p-1)\} \tag{16.61}$$

This set has cardinality no more than $p(p-1)$, which implies at least $(p-1)$ elements of N^* have eigenvalues that are not in the prime field. With respect to determinants,

$$\pi \in \mathrm{SL}(2, p) \iff \pi = \begin{bmatrix} \pi & 0 \\ 0 & \pi^{-1} \end{bmatrix} \tag{16.62}$$

(a further example of the overloading we admitted in Remark 495). If one replaces the first column of such (3×3) matrices with a column of zeros and appends the obvious new row to each of them, so, for instance,

$$\begin{bmatrix} 0 & 0 & 0 \\ a & a & c \\ b & b & d \end{bmatrix} \longrightarrow \begin{bmatrix} 0 & a & c \\ 0 & a & c \\ 0 & b & d \end{bmatrix} \tag{16.63}$$

we obtain a corresponding right near-field (see 16.20). This is a distinct process from the simple transposition of 3×3 matrices that switches us from a left to a right near-field. I now return to more general matrices.

Conjugacy in GL(2, *p*)

Algebraists have worked out the conjugacy classes for elements of GL(2, p). There are four distinct groupings of conjugacy classes based on the kinds of eigenvalues that can occur. The matrices just described belong to the final grouping, which involves $p.\frac{(p-1)}{2}$ distinct classes, each containing $(p^2 - p)$ elements, and is usually referred to as the *elliptic* conjugacy classes. The classes that involve repeated eigenvalues within the Galois field are the *scalar* class (of diagonalisable matrices) (comprising $(p - 1)$ classes each of size 1) and the *parabolic* class (of non-diagonalisable matrices) ($(p - 1)$ classes each of size $(p^2 - 1)$). The final class, called the *hyperbolic* class, does not concern us, but involves diagonalisable matrices with distinct eigenvalues and involves $(\frac{(p-1).(p-2)}{2})$ classes each of size $(p^2 + p)$ elements.[4]

A simple calculation of the characteristic equation shows that each of the matrices $\{\mathcal{P}^r.\mathcal{M}\}_{r=1,2,\ldots,(p-1)}$ has distinct eigenvalues (which are powers of π times the eigenvalues of $\mathcal{M}$). We know that our representation involves $((p^2 - 1) - (p - 1)) = p.(p - 1)$ elliptic matrices.

16.9 Apologia

When I planned this book, I was under the impression that I could use product theory to determine the finite non-Dickson near-fields. It should be possible to do this, and I imagined I had the bones of a proof but must have been mistaken since I found I was unable to flesh out that work. I do hope others might take up the challenge. It would also be interesting to link these ideas to traditional group representation theory, which seems their natural home—another of my original intentions for this book.

[4] I lack a succinct reference for this particular piece of folklore but it can be easily located by web searching. The reader may wish to note that the work actually applies to $GL(2, p^r)$ and one obtains the correct sizes by replacing each occurrence of p, above, with p^r.

Appendix A
Isotopy

All sets are finite in this appendix.

A.1 Introduction

I became interested in the ideas in this appendix in connection with my work on near-fields and the more general host problem for near-rings. One sees various sorts of transformation which turn one sort of algebra into another (Sects. 2.2, 2.1.3 and elsewhere), and it is enticing to imagine that, if the process is reversible, the algebras are in some sense "the same" even when they clearly are not isomorphic. I don't attempt to define this idea of sameness more formally, feeling that the way to that may very well be via formal recursion theory; but I do present a rather simple view of this sameness, a view which I feel to have a certain coherence, particularly in finite situations. What it amounts to is the crazy idea that asymmetric structures require a more general view of identity than isomorphism. Some of this material has previously been published [106], but much of it is quite new.

I start with isotopy—a direct generalisation of the algebraic notion of isomorphism and one which was discussed by a number of mathematicians in the mid-twentieth century, proving useful in various theories of finite geometries. My account of that is based on the book by Kurosh [91] and has a few small additions of my own.

A **groupoid** is a set in which the product of any two elements of the set is unambiguously defined as a specified member of the set. Bourbaki uses the term *magma* for these structures; but I shall adopt the term originally used by the mathematicians who first discussed isotopy.

Some modern authors use the term "groupoid" to refer to the sets with *partial* binary operations that arise naturally in category theory; my usage is perhaps more old-fashioned, although it does conform to that of Howie [70].

R. Lockhart, *The Theory of Near-Rings*, Lecture Notes in Mathematics 2295,
https://doi.org/10.1007/978-3-030-81755-8

Table A.1 Cayley table for two products hosted by a three-element set

·	A	B	C
A	A	A	A
B	B	C	B
C	C	B	C

Table A.2 Conjoint product to Table A.1

·	A	B	C
A	A	B	C
B	A	C	B
C	A	B	C

Groupoids may be specified by a *Cayley table* such as Table A.1. Table A.1 defines two groupoids; therefore, we would also need to be told, for example, that it is the product $A \cdot B = A$, that is here intended. The *opposite* product, in which $A \cdot B = B$, is inherent in this tabular representation. If we adopt the convention that products from Cayley tables are to be read so that the element in the ith row and jth column represents the product of the element heading the ith row with the element heading the jth column, then the alternative product given by the table would itself have an associated Cayley table in which the inner matrix of product terms was the transpose of the above one. Its table is shown as Table A.2.

Definition 507 I shall say that the two distinct products implied directly by a Cayley table are **conjoint** to each other, sometimes referring to a product and its **conjoint**. **Abelian groupoids** are simply self-conjoint groupoids.

Conjoint groups are isomorphic under the bijection that takes elements to their inverses.

A.1.1 Additive Conjoints and Near-Rings

Suppose $(N, +, \cdot)$ is a near-ring. We have the isomorphic opposite additive group $(N, \overset{+}{+})$, and, because

$$\left(\left(a \overset{+}{+} b\right) \cdot c\right) = (b+a) \cdot c = ((b \cdot c) + (a \cdot c)) = \left((a \cdot c) \overset{+}{+} (b \cdot c)\right) \quad (A.1)$$

we also have the **additively opposite near-ring** $(N, \overset{+}{+}, \cdot)$. The groups $(N, +)$ and $(N, \overset{+}{+})$ are isomorphic under the mapping

$$\theta(a) = (-a)$$

but since, in general,

$$-(a \cdot b) = \theta(a \cdot b) \neq (\theta(a) \cdot \theta(b)) = -(a \cdot (-b)) \tag{A.2}$$

the mapping θ will not generally be an isomorphism of near-rings. I give a specific example of an additively opposite near-ring hosted by D_4 in Appendix B.2 on page 500.

A.1.2 Multiplicative Conjoints and Opposites

Algebraists tend to conjoin only the product operation, when discussing rings (see Definition 79). Doing this, one obtains the **opposite ring**. In that situation, isomorphism between conjoints involves the existence of an **anti-isomorphism** in the original ring.

Definition 508 An anti-isomorphism is a bijective group automorphism satisfying the multiplicative relation

$$\theta(x \cdot y) = \theta(y) \cdot \theta(x)$$

The ring

$$\begin{bmatrix} \mathcal{Z} & \mathcal{Q} \\ 0 & \mathcal{Q} \end{bmatrix}$$

has no such anti-isomorphism and, consequently, is not isomorphic to its own conjoint. Two groupoids hosted by a set $(G, \cdot) and (G, *)$ are held to be isomorphic when there exists a bijection $\theta : G \mapsto G$ such that $\forall\, a, b, \in G$

$$a * b = \theta^{-1}(\theta(a) \cdot \theta(b))$$

and mathematicians are normally happy to classify their structures up to such symmetries. However, the existence of non-isomorphic conjoints might suggest that other information-preserving transformations could be worth a look. What "went wrong" in the ring example above was that the transposition that was needed would take us into another set of matrices. One might expect a wider lack of symmetries in such asymmetric structures as near-rings.

The *host problem* for near-rings is the problem of determining the distinct near-ring products that can be supported by a given additive group. Conventionally, one decides to calculate either left- or right-distributive near-rings and seeks the number of distinct isomorphism classes. It is well-known, for example, that the dihedral group of order 8 supports seven isomorphism classes of near-rings with identity

[78]. It can be seen that the group supports seven isomorphism classes of right near-rings and seven of left.

Suppose $(N, +, \cdot)$ is a right-distributive near-ring, i.e. $((x+y)\cdot z = x\cdot y+x\cdot z)$, and suppose that $(N, +, *)$ is a left-distributive near-ring. The two are isomorphic precisely when there is an automorphism θ of the additive group, such that $\theta(a*b) = \theta(a)\cdot\theta(b)$. This implies that $\theta((a+b)*c) = \theta(a+b)\cdot\theta(c) = (\theta(a)+\theta(b))\cdot\theta(c) = ((\theta(a)\cdot\theta(c)) + (\theta(b)\cdot\theta(c))$, and, θ being bijective, we conclude from this that the two near-rings are both *bi-distributive*—a situation which does apply in the ring case, where it imposes only a vacuous restriction. We see that bi-distributivity is needed for left and right near-rings to be isomorphic and that, for example, the number of *non-isomorphic* near-rings definable on the dihedral group of order 8 is actually 14—although these 14 involve 7 pairs with essentially the same information content, but opposite distributivity (see Corollary 89).

Of course, for rings with anti-isomorphisms, left- and right-distributive products are isomorphic (both being bi-distributive); so one does not always simply double the host number to obtain the number of distinct isomorphism classes.

A.1.3 Isotopy

Given a particular groupoid product $(\cdot)$ together with full knowledge of the symmetries of a set G, one can construct additional products by selecting bijections θ, ϕ and ψ of G and setting

$$a * b = \psi\,(\theta(a)\cdot\phi(b)) \tag{A.3}$$

This is the notion of **isotopy** that Kurosh offers in his textbook, [91], and which I mentioned in Sect. 5.2.1. Isomorphism relates to the special subcase in which $\psi = \theta^{-1}$ and $\theta = \phi$.

Isotopic products can look very different, but each can be obtained from the other, and in some sense, they might be thought to have the same information content, albeit much changed. In particular, although isotopes of groups are not usually themselves groups, it is known (and proved here) that non-isomorphic groups cannot be isotopic. This fact is probably the reason that early researchers applied isotopy only to the *product* operation, when discussing rings—something that I shall do myself in Lemma 514.

It seems that if one could formalise a notion for the information content of a finite static algebraic structure such as a groupoid that one would expect isotopic or conjoint groupoids to have the same such content.

Some of my interest in isotopy relates to the discovery of the near-ring products that may be defined upon a given group. It therefore becomes important to reduce the number of symmetries one must consider, which motivates the next definition.

Definition 509 Two groupoids, $(G, \wedge)$ and $(G, \cdot)$, are **allotropic** provided there are bijections θ, ϕ of the host set such that $\forall\, a, b, \in G$

$$a \wedge b = \theta\,(a \cdot \phi(b)) \tag{A.4}$$

The point of this idea is to observe that one can get anything from allotropy that one gets from isotopy, up to isomorphism. I shall state and prove that, but first mention that either of the alternative definitions

$$a \wedge b = \theta\,(\phi(a) \cdot b) \tag{A.5}$$

or

$$a \wedge b = (\theta(a) \cdot \phi(b)) \tag{A.6}$$

can be used instead. Allotropy involves *two* rather than *three* bijections.[1] As such, it stands midway between isotopy and isomorphism. The missing bijection is reinstated as part of the isomorphism in the next result.

Lemma 510 *All isotopes of a groupoid $(G, \cdot)$ are isomorphic to allotropes of it.*

Proof Suppose $a * b = \gamma\,(\alpha(a) \cdot \beta(b))$. We seek bijections θ, ϕ, ψ such that the groupoid $(G, \wedge)$ defined by

$$a \wedge b \;=\; \theta\,(a \cdot \phi(b))$$

satisfies

$$a \wedge b \;=\; \psi^{-1}\,(\psi(a) * \psi(b)) \tag{A.7}$$

The prescription

$$\psi = \alpha^{-1},\; \phi = \beta\alpha^{-1},\; \text{ and } \theta = \alpha\gamma$$

does the trick. This proves the result using Definition (A.4) and, of course, a symmetrical proof may be given if (A.5) is adopted. For Definition (A.6), notice our search is for bijections θ, ϕ, ψ such that the groupoid $(G, \wedge)$ defined by

$$a \wedge b \;=\; \theta(a) \cdot \phi(b)$$

satisfies (A.7). In this case, the prescription is

$$\psi = \gamma \text{ and } \theta = \alpha\psi \text{ and } \phi = \beta\psi$$

□

[1] It would be more correct to say that one of the three bijections is the identity.

Table A.3
$x * y = \theta\,(\theta(x) \cdot \theta(y))$

*	A	B	C
A	C	A	A
B	B	B	B
C	A	C	C

A.1.4 Isomorphs Which Are Not Allotropes

It is possible to find isotopes which are not actually allotropes. In fact, there are isotopes which cannot be reached, in one move, by *any* of the three definitions of allotropy (although they obviously could be reached by multiple move compositions of these definitions).

Take the groupoid given by Table A.1 and the bijection θ which simply swaps A and B. Define an isotopic (and isomorphic) groupoid by the prescription

$$a * b \;=\; \theta\,(\theta(a)\;\cdot\;\theta(b)) \tag{A.8}$$

This has Cayley table given by Table A.3. We seek to obtain Table A.3 by applying one of the three definitions of allotropy to the groupoid defined by Table A.1.

Method (A.4) One requires bijections θ and ϕ such that $x * y = \theta\,(x \cdot \phi(y))$. We must have both that $C = A * A = \theta\,(A \cdot \phi(A)) = \theta(A)$ and that $A = A * C = \theta\,(A \cdot \phi(C)) = \theta(A)$. Thus, there is no consistent definition for θ.

Method (A.5) One requires bijections θ and ϕ such that $x * y = \theta\,(\phi(x) \cdot y)$. We know that $C = A * A = \theta\,(\phi(A) \cdot A) = \theta(\phi(A))$. We know that $A = A * C = \theta\,(\phi(A) \cdot C) = \theta(\phi(A))$. We conclude that there is no consistent definition for these mappings.

Method (A.6) One requires bijections θ and ϕ such that $x * y = \theta(x) \cdot \phi(y)$. Now, $C = A * A = \theta(A) \cdot \phi(A)$; and $A = A * B = \theta(A) \cdot \phi(B)$. So, whichever element $\theta(A)$ is, it must contain both A and C in the row it defines. However, Table (A.1) shows that to be quite impossible.

Notice that since $\theta = \theta^{-1}$, (A.8) is an example of two isomorphic groupoids that are not allotropic.

A.1.5 Basic Results in Isotopy

I now mention some more standard concepts.

Definition 511 A **quasi-group** is a groupoid in which, for all elements a, b, the equations

$$a \cdot x \;=\; b, \quad y \cdot a \;=\; b \tag{A.9}$$

are uniquely solvable for x and y. A **loop** is a quasi-group with unit element (some authors call that a **neutral element**).

Lemma 512

- *Every groupoid isotopic to a quasi-group is itself a quasi-group.*
- *Every quasi-group is isotopic to a loop.*
- *Any groupoid with identity which is isotopic to a semi-group is isomorphic to it too.*

Proof

- The first assertion is trivial.
- Select any element e in the quasi-group $(G, \cdot)$. I shall construct an allotrope of $(G, \cdot)$ which is a loop. Consider the two conditions

$$\theta(b \cdot e) = b \qquad (e \cdot \phi(b)) = \theta^{-1}(b)$$

  Quasi-group's structure ensures that these conditions define bijections θ and ϕ on G. The allotrope defined by $a * b = \theta(a \cdot \phi(b))$ is a loop with identity e.
- Suppose $(G, \cdot)$ is a groupoid with identity e and $(G, *)$ is a semi-group such that $a * b = \theta(a \cdot \phi(b))$. The associativity of (*) implies

$$a \cdot \phi\theta(b \cdot \phi(c)) = \theta(a \cdot \phi(b)) \cdot \phi(c) \tag{A.10}$$

  Writing $a = \phi(c) = e$ gives us this basic relation, true for all elements b.

$$\phi\theta(b) = \theta\phi(b) \tag{A.11}$$

  From (A.10),

$$a \cdot \phi\theta(b) = \theta(a \cdot \phi(b)) \tag{A.12}$$

  From (A.10), we obtain this relation, which is true for all values of a and b

$$\phi\theta(b \cdot a) = (\theta\phi(b)) \cdot a \tag{A.13}$$

  Finally, use Eqs. (A.12), (A.11) and (A.13) to get

$$\begin{aligned} \theta\phi(a * b) &= \theta\phi\theta(a \cdot \phi(b)) \\ &\overset{(A.12)}{=} \theta\phi(a \cdot \phi\theta(b)) \\ &\overset{(A.11)}{=} \phi\theta(a \cdot \phi\theta(b)) \\ &\overset{(A.13)}{=} \theta\phi(a) \cdot \phi\theta(b) \\ &\overset{(A.11)}{=} \theta\phi(a) \cdot \theta\phi(b) \end{aligned}$$

  which demonstrates that $\theta\phi$ is an isomorphism.

□

Table A.4 Loops not isotopic to any group

*	A	B	C	D	E
A	A	B	C	D	E
B	B	A	D	E	C
C	C	E	A	B	D
D	D	C	E	A	B
E	E	D	B	C	A

From this result, it follows:

- Loops isotopic to groups are themselves groups.
- No two non-isomorphic groups are isotopic.

The matrix representation of products in Sect. A.2 makes Lemma 512 even more transparent.

Lemma 513 *Isotopic groupoids have isotopic conjoints.*

The proof of this proposition is straightforward.

Table A.4 represents two loops of order 5. Neither is cyclic, so neither is isotopic to a group. The conjoint loops are isomorphic under the bijection which simply swaps the elements E and D. Recall that in isotopic near-rings we apply transformations to the product operation alone and the group operation is left unaffected (compare with Definition 79). These transformation operations are often assumed to themselves be group automorphisms, as is the case in this next lemma.

Lemma 514 *Isotopic near-rings with identity are isomorphic. That is, if* $(N, +, *)$ *is a near-ring with identity that is isotopic to the near-ring with identity* $(N, +, \cdot)$*, under the isotopy*

$$a * b = \psi\left(\theta(a) \cdot \phi(b)\right) \tag{A.14}$$

(all mappings being group automorphisms), then the two near-rings are isomorphic.

Proof Taking the isotopy defined above, associativity for $*$ implies

$$\theta\psi\left(\theta(a) \cdot \phi(b)\right) \cdot \phi(c) = \theta(a) \cdot \phi\psi\left(\theta(b) \cdot \phi(c)\right)$$

From this relation, put $\phi(c) = 1$ to obtain

$$\theta\psi\left(\theta(a) \cdot \phi(b)\right) = \theta(a) \cdot \phi\psi\theta(b) \tag{A.15}$$

Writing $\theta(a) = 1$ in that relation gives us this relation, applying to any b

$$\theta\psi\phi(b) = \phi\psi\theta(b) \tag{A.16}$$

and writing $\theta(a) = 1$ in our original associativity condition gives us

$$\theta\psi\phi(b) \cdot \phi(c) = \phi\psi\,(\theta(b) \cdot \phi(c)) \tag{A.17}$$

Finally, use Eqs. (A.15), (A.16) and (A.17) to get

$$\begin{aligned}\theta\psi\phi(a) \cdot \theta\psi\phi(b) &\overset{(A.16)}{=} \theta\psi\phi(a) \cdot \phi\psi\theta(b) \\ &\overset{(A.15)}{=} \theta\psi\,(\theta\psi\phi(a) \cdot \phi(b)) \\ &\overset{(A.17)}{=} \theta\psi\,(\phi\psi\,(\theta(a) \cdot \phi(b))) \\ &= \theta\psi\phi\,(a * b)\end{aligned}$$

which demonstrates that $\theta\psi\phi$ is an isomorphism and completes the proof. □

Observations 515 The proof just given relies on both near-rings possessing two-sided multiplicative identities and requires the product operation $(*)$ to be associative, but no use was made of the fact that the mappings were group automorphisms. Of course, we cannot rely on (A.14) giving us a pre-near-ring product unless both θ and ψ are automorphisms (ϕ merely having to be bijective and zero-fixing to preserve zero symmetry), and for near-ring isomorphism, we would require that $(\theta \circ \psi \circ \phi)$ were a group automorphism.

Situations occur in which one might wish to permit a more generous view of isotopy in near-rings (see page 226, e.g.) and to permit the mappings associated with the product isotopy to be bijections rather than having to be group automorphisms.

A.1.6 Alleles

I shall say that two groupoids on a given set are *alleles* of each other if each can be obtained from the other by isotopy, by conjoining or by some sequence of these procedures. One might offer that "static information content" could be related directly to alleles although perhaps there are other symmetries that should be explored too; co-structures come to mind. An alternative approach to the problem of information content is to seek some canonical structure which is uniquely present in each isotopy class. I explore this next.

A.2 $\mathcal{A}$-Matrices

In this section, I discuss groupoids hosted by sets with cardinality $n < \infty$, referring to products of *degree* n. Any product implies both a Cayley table and a conjoint product. We might list the set elements $\{a_1, a_2, \ldots a_n\}$ and associate a $n \times n$ column stochastic (0, 1) matrix m_i with each element a_i by defining the jth column of m_i to be $\left(\delta_{i,k}\right)_{i=1}^{n}$ where

$$a_i \cdot a_j = a_k$$

So, in this representation, the above relation says that the kth column of the ith matrix would have 1 in the jth row and zeros elsewhere. The groupoid in Table A.1

would have these three associated column stochastic matrices

$$m_A = \begin{bmatrix} 1\,1\,1 \\ 0\,0\,0 \\ 0\,0\,0 \end{bmatrix} m_B = \begin{bmatrix} 0\,0\,0 \\ 1\,0\,1 \\ 0\,1\,0 \end{bmatrix} m_C = \begin{bmatrix} 0\,0\,0 \\ 0\,1\,0 \\ 1\,0\,1 \end{bmatrix}$$

I call these the $\mathcal{A}$-**matrices** of the groupoid. Obviously, there is a one-to-one correspondence between n-tuples of (0, 1) column stochastic matrices and products of degree n. There are n^{n^2} such products and n^n distinct (0, 1) column stochastic matrices. Quasi-group products will show up as *doubly stochastic* matrices in this representation. There are $n!$ doubly stochastic matrices.

Matrices rely on an implicit ordering of the elements of the groupoid. Changing the ordering involves a permutation from S_n (n being the groupoid order). An individual matrix associated with a given element would then transform as

$$(m_x)' = (P)\,.\,(m_x) \cdot \left(P^t\right) \tag{A.18}$$

(compare page 268, Eqs. (7.89) and (7.90)). The tuple of matrices representing the product would include this matrix in a position determined by P.

For example, the permutation

$$P = \begin{bmatrix} 0\,0\,1 \\ 1\,0\,0 \\ 0\,1\,0 \end{bmatrix}$$

maps $A \mapsto B, B \mapsto CandC \mapsto A$ and corresponds to representing the product from Table A.1 as Table A.5. The matrix corresponding to B will be

$$(m_B)' = \begin{bmatrix} 0\,0\,1 \\ 1\,0\,0 \\ 0\,1\,0 \end{bmatrix} . \begin{bmatrix} 0\,0\,0 \\ 1\,0\,1 \\ 0\,1\,0 \end{bmatrix} . \begin{bmatrix} 0\,1\,0 \\ 0\,0\,1 \\ 1\,0\,0 \end{bmatrix}$$

$$= \begin{bmatrix} 0\,0\,1 \\ 0\,0\,0 \\ 1\,1\,0 \end{bmatrix}$$

Table A.5 Cayley table for the groupoid in Table A.1 with elements re-ordered

·	C	A	B
C	C	C	B
A	A	A	A
B	B	B	C

This matrix will be the *third* in the tuple of matrices representing the product of Table A.5. An alternative expression of the original product is as a function

$$m : \{1, 2, \ldots, n\} \times \{1, 2, \ldots, n\} \longrightarrow \{1, 2, \ldots, n\} \quad m(i, j) = k \quad (A.19)$$

The ith $\mathcal{A}$-matrix of the product is

$$(m_i)_{r,l} = \begin{cases} 1 \text{ if } a_i \cdot a_l = a_r \\ 0 \text{ if } a_i \cdot a_l \neq a_r. \end{cases}$$

In functional notation, this would be

$$(m_i)_{r,l} = \left(\delta_{r,m(i,l)}\right) \quad (A.20)$$

I should note that there is a rich theory of column stochastic matrices. They occur in the theory of stochastic processes. They have interesting structural and geometric properties as I have already mentioned on page 271.

A.2.1 Matrix Symmetries

The effect of the symmetries associated with alleles is of more immediate interest. Consider a groupoid $(G, \cdot)$. I wish to consider the effects on the $\mathcal{A}$-matrices of these transformations associated with the bijection θ

A: $a * b = (a \cdot \theta(b))$
B: $a * b = (\theta(a) \cdot b)$
C: $a * b = \theta(a \cdot b)$
D: $a * b = (b \cdot a)$

A: This process corresponds to permuting the *columns* of the original Cayley table. It permutes the columns of the individual stochastic matrices, as specified by the permutation θ, without changing their order:

$$m_i' = (m_i) \cdot \Theta \quad \text{(where } \Theta \text{ is a permutation matrix).}$$

B: This process corresponds to permuting the *rows* of the original Cayley table. It re-orders the n-tuple of stochastic matrices as specified by the permutation θ:

$$m'_i \;=\; m_{\theta(i)}$$

C: This process corresponds to re-labelling the interior entries of the Cayley table as specified by θ. It permutes the rows of the individual stochastic matrices, as specified by the permutation θ, without changing their order:

$$m'_i \;=\; \Theta \cdot (m_i) \quad \text{(where } \Theta \text{ is a permutation matrix).}$$

D: This process corresponds to transposing the original Cayley table. It has an interesting, shuffling effect on the $\mathcal{A}$-matrices, resulting in an entirely new n-tuple of matrices in which the n columns of the original matrices m_i are cannibalised to give the ith columns, of the n new matrices, in order. I use the notation

$({}_i m)$ for the ith conjoint matrix produced by this procedure.

So,

$$({}_i m)_{r,l} = (m_l)_{r,i} \tag{A.21}$$

One can express the isotopy $(a * b) = \psi\,(\theta(a) \cdot \phi(b))$ in the matrix form

$$m'_i \;=\; \Phi \cdot m_{\Theta(i)} \cdot \Psi \tag{A.22}$$

These symmetries make the equivalence of the three definitions of allotropy more transparent. One could represent products using *row* stochastic matrices, in which case all row permutations of the $\mathcal{A}$-matrices become column permutations and vice versa and the conjoint matrices of case D are formed from the *rows* of the $\mathcal{A}$-matrices. Alternatively, one could interpret Cayley tables as representing the conjoints directly. Using the column representation, $\mathcal{A}$-matrices are then permuted by columns in case A, and by rows in case C, with case B permuting the matrices themselves. The row stochastic representation dualises these symmetries in exactly the same way as before.

Each of the first three symmetries leaves the ranks of the constituent matrices unchanged. This matrix representation makes the preservation of quasi-group products under isotopy quite obvious (Proposition (512)). Isotopic products will have the same rank pattern for their constituent matrices. The pattern is closely linked to isotopy but does not characterise it. The conjoint symmetry can change rank—just consider the conjoint of the product all of whose $\mathcal{A}$-matrices are identities—the matrices change from rank n to rank 1.

A.2.2 *Listing Classes of Products*

The preservation of matrix rank means that the symmetries associated with isotopy can be used to give a systematic ordering to the classes of products hosted by a set. One may permute:

- The order of the n matrices.
- All the rows of the n matrices—as long as the same permutation is applied to each.
- All the columns of the n matrices—as long as the same permutation is applied to each.

Permutations of rows and of columns leave matrix rank unchanged, and we have seen that classes of products must exist within superclasses of all possible rank n-tuples.

Allotropy

In the case of the allotropy given by (A.5), for instance, n-tuples of matrices are to be regarded as equivalent if each may be obtained from the other by a single re-ordering and applying a single row permutation to *all* the $\mathcal{A}$-matrices. One associates each matrix m_i with exactly one of a canonical sequence of (0, 1) column stochastic matrices $\left(A_{r,l}\right)$ with these characteristics:

1. Each matrix is upper triangular: $A_{r,l} = 0 \;\; r > l$.
2. $A_{r,l} = 1 \;\Rightarrow\; \forall\, j < l \;\exists\, s$ such that $A_{s,j} = 1$.

The idea is that each matrix is a row permutation of exactly one of these matrices. I shall call them the *base matrices of degree n* or the *n* × *n base matrices*. Here are the five 3 × 3 column stochastic base matrices corresponding to this prescription

$$\begin{bmatrix} 1\,1\,1 \\ 0\,0\,0 \\ 0\,0\,0 \end{bmatrix} \begin{bmatrix} 1\,1\,0 \\ 0\,0\,1 \\ 0\,0\,0 \end{bmatrix} \begin{bmatrix} 1\,0\,1 \\ 0\,1\,0 \\ 0\,0\,0 \end{bmatrix} \begin{bmatrix} 1\,0\,0 \\ 0\,1\,1 \\ 0\,0\,0 \end{bmatrix} \begin{bmatrix} 1\,0\,0 \\ 0\,1\,0 \\ 0\,0\,1 \end{bmatrix}$$

These base matrices are multiplicatively closed. They may be generated in a systematic way which also orders them. I shall illustrate this process by listing the 4 × 4 base matrices. There are 15 4 × 4 base matrices

$$\begin{bmatrix} 1\,1\,1\,1 \\ 0\,0\,0\,0 \\ 0\,0\,0\,0 \\ 0\,0\,0\,0 \end{bmatrix} \begin{bmatrix} 1\,1\,1\,0 \\ 0\,0\,0\,1 \\ 0\,0\,0\,0 \\ 0\,0\,0\,0 \end{bmatrix} \begin{bmatrix} 1\,1\,0\,1 \\ 0\,0\,1\,0 \\ 0\,0\,0\,0 \\ 0\,0\,0\,0 \end{bmatrix} \begin{bmatrix} 1\,1\,0\,0 \\ 0\,0\,1\,1 \\ 0\,0\,0\,0 \\ 0\,0\,0\,0 \end{bmatrix} \begin{bmatrix} 1\,1\,0\,0 \\ 0\,0\,1\,0 \\ 0\,0\,0\,1 \\ 0\,0\,0\,0 \end{bmatrix}$$

$$\begin{bmatrix}1&0&1&1\\0&1&0&0\\0&0&0&0\\0&0&0&0\end{bmatrix}\begin{bmatrix}1&0&1&0\\0&1&0&1\\0&0&0&0\\0&0&0&0\end{bmatrix}\begin{bmatrix}1&0&1&0\\0&1&0&0\\0&0&0&1\\0&0&0&0\end{bmatrix}\begin{bmatrix}1&0&0&1\\0&1&1&0\\0&0&0&0\\0&0&0&0\end{bmatrix}\begin{bmatrix}1&0&0&0\\0&1&1&1\\0&0&0&0\\0&0&0&0\end{bmatrix}$$

$$\begin{bmatrix}1&0&0&0\\0&1&1&0\\0&0&0&1\\0&0&0&0\end{bmatrix}\begin{bmatrix}1&0&0&1\\0&1&0&0\\0&0&1&0\\0&0&0&0\end{bmatrix}\begin{bmatrix}1&0&0&0\\0&1&0&1\\0&0&1&0\\0&0&0&0\end{bmatrix}\begin{bmatrix}1&0&0&0\\0&1&0&0\\0&0&1&1\\0&0&0&0\end{bmatrix}\begin{bmatrix}1&0&0&0\\0&1&0&0\\0&0&1&0\\0&0&0&1\end{bmatrix}$$

These 15 base matrices are shown in their induced order. I wrote them all out to make the way in which they are generated absolutely clear. Matrices of this form are canonical representatives for the classes of (0, 1) column stochastic matrices which are formed from each other by row permutations. The recurrence associated with the construction of base matrices of degree $(n+1)$ from base matrices of degree n leads me to claim the number of base matrices of degree n is given by $B(n)$, the nth *Bell number*

$$\sum_{r=1}^{n} S(n,r) \;=\; B(n) \;=\; \frac{1}{e}\sum_{k=0}^{\infty}\frac{k^n}{n!}$$

where

$$S(n,r) = (1/r!)\sum_{i=0}(-1)^i\binom{r}{i}(r-i)^n$$

is a Stirling number of the second kind. An easy way to see that is to identify each $n \times n$ matrix with a distinct partition of a set with cardinality n. For example, the 13th matrix shown

$$\begin{bmatrix}1&0&0&0\\0&1&0&1\\0&0&1&0\\0&0&0&0\end{bmatrix}$$

corresponds to the set partition

$$\{\{1\},\{2,4\},\{3\}\}$$

One could enumerate the allotropy classes for products of degree n. To do this, onc would:

1. Generate the $B(n)$ base matrices.
2. Make systematic selections (with replacement) of n of these base matrices at a time.
3. Order the matrix selections according to their induced ordering (this corresponds to symmetry B, page 487).
4. Apply row permutations to the individual matrices in some systematic fashion, in order to generate the actual $\mathcal{A}$-matrices of the product.
5. Select a single representative from each global row permutation of the resulting list of matrices (this corresponds to symmetry C, page 487).

I should say that this procedure applies only to the definition of allotropy that we are using, but that similar procedures apply to the other cases. In this way, one could, conceivably, generate allotropy classes systematically.

Isotopy

Handling full isotopy requires a further reduction of the base matrices, one which is connected to the theory of partitions. In the allotropy case that we have been discussing (case A.5), this involves introducing the symmetry that permutes *columns* (symmetry A). Let $p_k(n)$ denote the number of distinct partitions of n into k non-zero components. That is, $p_3(4) = 1$ since $4 = 1 + 1 + 2$. Now look at the matrices of rank three that can occur as base matrices. They are the 5th, 8th, 11th, 12th, 13th and 14th matrices that appear. Since there is only one way to partition 4 into three integers, all of these have a single row with two elements and two rows with one. Choose the first such matrix appearing as the reduced base matrix—that is,

$$\begin{bmatrix} 1\,1\,0\,0 \\ 0\,0\,1\,0 \\ 0\,0\,0\,1 \\ 0\,0\,0\,0 \end{bmatrix}$$

We can generate the other five base matrices with rank three by column and row permutations of the first one. Any $\mathcal{A}$-matrix of degree n can be written in the form PMQ where P and Q are permutation matrices and M is one of our reduced base matrices. Isotopic products correspond to selections of n of such matrices.

It is possible to imagine listing out canonical representatives for each allotropy class and for each isotopy class of products. Notice that the counting process is a bit more complicated than we have suggested since in some situations different permutations can produce the same $\mathcal{A}$-matrix.

Alleles

The conjoint operation simply pairs up sets of canonical representatives and may involve *two* distinct rank classes. In the case of products of degree 3, for example, the rank 3-tuple $(1, 1, 1)$ occurs in essentially (i.e. up to isotopy) three ways, which may be expressed by the matrix triples:

First class

$$\begin{bmatrix} 1\,1\,1 \\ 0\,0\,0 \\ 0\,0\,0 \end{bmatrix} \begin{bmatrix} 1\,1\,1 \\ 0\,0\,0 \\ 0\,0\,0 \end{bmatrix} \begin{bmatrix} 1\,1\,1 \\ 0\,0\,0 \\ 0\,0\,0 \end{bmatrix}$$

Second class

$$\begin{bmatrix} 1\,1\,1 \\ 0\,0\,0 \\ 0\,0\,0 \end{bmatrix} \begin{bmatrix} 1\,1\,1 \\ 0\,0\,0 \\ 0\,0\,0 \end{bmatrix} \begin{bmatrix} 0\,0\,0 \\ 1\,1\,1 \\ 0\,0\,0 \end{bmatrix}$$

Third class

$$\begin{bmatrix} 1\,1\,1 \\ 0\,0\,0 \\ 0\,0\,0 \end{bmatrix} \begin{bmatrix} 0\,0\,0 \\ 1\,1\,1 \\ 0\,0\,0 \end{bmatrix} \begin{bmatrix} 0\,0\,0 \\ 0\,0\,0 \\ 1\,1\,1 \end{bmatrix}$$

Each corresponds to a distinct isotopy class. The first of these is self-conjoint, the conjoint of the second belongs to the class of isotopies of the rank class $(2, 2, 2)$, and the conjoint of the third belongs to the class of isotopies of the rank class $(3, 3, 3)$. Its specific isotopy class is the class of the groupoid $x \cdot y = y \; \forall \, x, y$ (which is the groupoid defined by the conjoint matrices).

I might summarise these observations by saying that they offer a sequence of canonical products, characterising isotopy classes, which pair up to characterise allele classes. It seems that one can, at least potentially, determine the exact number of allele classes associated with a set, or the exact number of isotopy classes, or the exact number of allotropy classes and that one can also determine the exact number of products within each such class.

A.2.3 *Associativity*

Using (A.20), we arrive at the following for the product of two $\mathcal{A}$-matrices:

$$\left((m_i) \cdot (m_j)\right)_{r,l} = (m_i)_{r,m(j,l)} \tag{A.23}$$

Associativity is something conjoints have in common. It is the condition

$$m\,(m\,(i, j)\,, l) = m\,(i, m\,(j, l))$$

The $\mathcal{A}$-matrix expression of this is

$$\left(m_{m(i,j)}\right)_{r,l} = (m_i)_{r,m(j,l)} \tag{A.24}$$

or, equivalently,

$$\left(m_{m(i,j)}\right) = (m_i) \cdot \left(m_j\right) \tag{A.25}$$

which is a strong closure condition.

To illustrate, the groupoid of Table A.2 has $\mathcal{A}$-matrices

$$m_A = \begin{bmatrix} 1\,0\,0 \\ 0\,1\,0 \\ 0\,0\,1 \end{bmatrix} m_B = \begin{bmatrix} 1\,0\,0 \\ 0\,0\,1 \\ 0\,1\,0 \end{bmatrix} m_C = \begin{bmatrix} 1\,0\,0 \\ 0\,1\,0 \\ 0\,0\,1 \end{bmatrix} \tag{A.26}$$

and

$$(m_B) \cdot (m_A) = m_B \neq (m_{B \cdot A = A})$$

so this left quasi-associative product is not associative.

The n column stochastic matrices of a product themselves generate an associative algebra. Multiplicative closure of these generator matrices is therefore connected to associativity in the original product. The precise connection involves a quasi-group-like property which I shall now describe.

Quasi-Associativity

Definition 516 I shall say that a specific product m is **left quasi-associative** provided that, for all possible $i, j, \in \{1, 2, \ldots, n\}$, there exists a fixed k, depending perhaps on i and j, and such that, for all l,

$$m\,(k, l) = m\,(i, m\,(j, l)) \tag{A.27}$$

There may be several candidates for k, but at least *one* of them should work for all possible l.

The condition is that $a_i \cdot (a_j \cdot a_l) = b \cdot a_l$ for some element $b = a_k$, which is why I call it *quasi-associativity*. In associative cases, one can simply take $b = (a_i \cdot a_j)$. The matrix expression of this condition is that for all, i, j, there exists k such that for any choice of l

$$(m_k)_{r,l} = (m_i)_{r,m(j,l)} \tag{A.28}$$

Definition 517 Dually, **right quasi-associativity** is the condition that, for each choice of

$$i, j, \in \{1, 2, \ldots, n\}$$

there exists at least one k, probably depending on that choice, and such that, for all l,

$$m(l, k) = m(m(l, j), i) \tag{A.29}$$

This expresses the condition that given any x_i, x_j, x_l, we can find a k, perhaps depending on i and j, but working for all l and such that

$$(x_l \cdot x_k) = ((x_l \cdot x_j) \cdot x_i)$$

in associative cases, one simply takes $x_k = (x_j \cdot x_i)$. The matrix form of right quasi-associativity is, of course,

$$(m_l)_{r,k} = (m_{m(l,j)})_{r,i} \tag{A.30}$$

l being allowed to vary and k probably depending on i and j. Table A.6 represents a non-associative product with both quasi-associative properties. The product given by Table (A.1) is not left quasi-associative. To be so, we should need an element X such that $X \cdot Y = B \cdot (A \cdot Y)$ for all choices of Y. Now,

$$B = B \cdot (A \cdot B) = X \cdot B$$

forces $X = C$, but then

$$C = C \cdot C \neq B \cdot (A \cdot C) = B$$

so no such choice can be made. It is fairly easy to see that the product actually is right quasi-associative and this would also follow as a consequence of the next lemma.

Lemma 518

1. *Products are right quasi-associative if and only if their conjoints are left quasi-associative.*
2. *A product m is left quasi-associative if and only if its corresponding $\mathcal{A}$-matrices form a multiplicatively closed set.*

Table A.6
$B = B \cdot (C \cdot B) \neq$
$(B \cdot C) \cdot B = A$

·	A	B	C
A	B	A	B
B	A	B	A
C	A	B	A

Proof

1. If we represent the $\mathcal{A}$-matrices as $\{m_i\}_{i=1}^n$ and their conjoints as $\{{}_im\}_{i=1}^n$, then from (A.21), the conjoint condition is

$$({}_im)_{r,l} = (m_l)_{r,i} \tag{A.31}$$

The conjoint of m is the product $\overline{m}(i, j) = m(j, i)$. Assuming m is left quasi-associative,

$$({}_lm)_{r,k} = (m_k)_{r,l} \overset{(A.28)}{=} (m_i)_{r,m(j,l)} \overset{(A.31)}{=} \left({}_{m(j,l)}m\right)_{r,i} = \left({}_{\overline{m}(l,\,j)}m\right)_{r,i}$$

So the conjoint form of right quasi-associativity (A.30) is valid.

2. Suppose $m_i \cdot m_j = m_k$; then, for all r, l,

$$(m_k)_{r,l} = \sum_s \delta_{r,m(i,s)}\delta_{s,m(j,l)} = \delta_{r,m(i,m(j,l))} = (m_i)_{r,m(j,l)}$$

□

Associative products automatically satisfy both quasi-associativity conditions. Their $\mathcal{A}$-matrices, and their conjoints, are multiplicatively closed.

Notation 519 I shall use the symbol

$$\mathcal{M} = \{m_A\}$$

to represent the column stochastic $\mathcal{A}$-matrices of a groupoid.

In left quasi-associative products, the association

$$\Pi : A \mapsto m_A \tag{A.32}$$

may well not be a bijection because distinct elements can map to the same matrix. Selecting a section $i : m_A \mapsto A$ of Π (i.e. $\Pi \circ i = id_{\mathcal{M}}$) allows us to define

$$A * B = i\,(\Pi(A) \cdot \Pi(B)) = i\,(m_A \cdot m_B) \tag{A.33}$$

This will only be an allotropy when Π and i are both bijective. However,

$$\begin{aligned}((A * B) * C) &= i\,(\Pi\,(A * B) \cdot \Pi(C)) = i\,(\Pi \circ i\,(\Pi(A) \cdot \Pi(B)) \cdot \Pi(C)) = \\ &i\,((\Pi(A) \cdot \Pi(B)) \cdot \Pi(C)) = i\,(\Pi(A) \cdot (\Pi(B) \cdot \Pi(C))) = \\ &i\,(\Pi(A) \cdot \Pi \circ i\,(\Pi(B) \cdot \Pi(C))) = (A * (B * C))\end{aligned}$$

so the product is associative. When the original product is actually associative, the association (A.32) can be thought of as bijective even when this is not actually the case, by regarding the matrices m_A and m_B as formally distinct simply because A and B are, and resolving the matrix product $m_A \cdot m_B$ using the relation (A.25) $m_A \cdot m_B = m_{(A \cdot B)}$. The resulting Cayley table for *matrices* is isomorphic to the original groupoid under (A.32).

Appendix B
Near-Ring Products on D_4

This appendix simply lists some near-ring products for D_4.

Finite Dihedral Groups

The general presentation for a finite dihedral group is given in (2.19). D_n has order $2n$. Its commutator subgroup is $\langle 2a \rangle$. When n is odd, the group is centreless, and the commutator subgroup is $\langle a \rangle$ and has index 2. When n is even, the group has centre $\langle \frac{n}{2}a \rangle$, and its commutator subgroup has index 4. The group exponent is n, when n is even, and $2n$, when n is odd (the group then not being I-appropriate).

When n is odd,

$$D_{2n} \cong D_n \oplus C_2 \tag{B.1}$$

B.1 Seven Non-isomorphic Left Near-Rings Hosted by D_4

The Cayley for $(D_4, +)$ is given as Table B.8 on page 501. The seven near-rings in question are cases 1, 2, 5, 6, 9, 10 and 14 of Table 5.12, in that order. The commutator subgroup of D_4 is $\langle 0, 2a \rangle$. The seven multiplication tables of these near-rings are arranged so that products of the form $x \cdot (y + 2a)$ are two columns apart from products of the form $x \cdot y$. None of these near-rings is negatively symmetric (page 9).

The reader will see that these products take values which are congruent modulo the commutator subgroup. That is, each of these near-rings is in the class $\mathcal{F}$ described on page 49. All these near-rings have identity elements, and so in each case, $r^{(2)}(D_4) = D_4$. I observed on page 196 that all these near-rings are non-simple and that we may suppose that the set

$$\{0, 2a, b, 2a + b\}$$

is a two-sided ideal.

R. Lockhart, *The Theory of Near-Rings*, Lecture Notes in Mathematics 2295,
https://doi.org/10.1007/978-3-030-81755-8

I list the tables and note the corresponding distributivity cores $C_{\mathcal{D}}(D_4)$ (see Definition 69). In each case, a is the multiplicative identity of the near-ring, and the product (A * B) should read A from the left-hand column and B from the top row. So, in the first table, $3a * (a + b)$ is $(a + b)$.

In each case, I list nilpotent and generalised idempotent elements

$$\mathcal{N} = \{x \in D_4 : x^r = 0, \text{ for some } r \in \mathbb{N}\}$$
$$\mathcal{I} = \{x \in D_4 : x^r = x, \text{ for some } r \in \mathbb{N}\}$$

We can establish that any of these near-rings is an F-near-ring, simply by verifying that for all elements $x, y \in D_4$

$$b \cdot (x + y) \equiv (b \cdot x) + (b \cdot y) \pmod{\{0, 2a\}}$$

and we can show that the near-ring is not an F-near-ring by simply offering values of x and y for which this does not hold.

It turns out that the first two near-rings are F-near-rings (the second actually being d.g.) but none of the remaining near-rings is one.

$C_{\mathcal{D}}$ (Table B.1) $= \{0, a\}$. $\mathcal{N} = \{0, 2a\}$. $\mathcal{I} = \{0, a, 3a, b, a+b, 2a+b, 3a+b\}$. This is an F-near-ring. It is not a d.g. near-ring.

$C_{\mathcal{D}}$ (Table B.2) $= \{0, a, b\}$. $\mathcal{N} = \{0, 2a\}$. $\mathcal{I} = \{0, a, 3a, b, a+b, 3a+b\}$. The distributivity core of this near-ring is $C_{\mathcal{D}}$ (Table B.2) $= \{0, a, b\}$ so this is actually a d.g. near-ring. The distributivity core is multiplicatively closed (Observations 70).

$C_{\mathcal{D}}$ (Table B.3) $= \{0, a\}$. $\mathcal{N} = \{0, 2a, b, 2a+b\}$. $\mathcal{I} = \{0, a, 3a, a+b, 3a+b\}$. In this near-ring, the product $b \cdot (a+b) = 0$ is not congruent, modulo the commutator subgroup, to the sum of the products $b \cdot a = b$ and $b \cdot b = 2a$; consequently, the near-ring is not an F-near-ring. Arguments involving the same products show than none of the other near-rings is an F-near-ring.

$C_{\mathcal{D}}$ (Table B.4) $= \{0, a\}$. $\mathcal{N} = \{0, 2a, b, 2a+b\}$. $\mathcal{I} = \{0, a, 3a, a+b, 3a+b\}$.
$C_{\mathcal{D}}$ (Table B.5) $= \{0, a\}$. $\mathcal{N} = \{0, 2a, b, 2a+b\}$. $\mathcal{I} = \{0, a, 3a, a+b, 3a+b\}$.
$C_{\mathcal{D}}$ (Table B.6) $= \{0, a\}$. $\mathcal{N} = \{0, 2a, b, 2a + b\}$. $\mathcal{I} = \{0, a, 3a, 3a + b\}$
$C_{\mathcal{D}}$ (Table B.7) $= \{0, a\}$. $\mathcal{N} = \{0, 2a\}$. $\mathcal{I} = \{0, a, 3a, b, a+b, 2a+b, 3a+b\}$.

Table B.1 Case 1

*	0	a	$2a$	$3a$	b	$(a+b)$	$(2a+b)$	$(3a+b)$
0	0	0	0	0	0	0	0	0
a	0	a	$2a$	$3a$	b	$(a+b)$	$(2a+b)$	$(3a+b)$
$2a$	0	$2a$	0	$2a$	0	0	0	0
$3a$	0	$3a$	$2a$	a	b	$(a+b)$	$(2a+b)$	$(3a+b)$
b	0	b	0	b	b	0	$(2a+b)$	0
$(a+b)$	0	$(a+b)$	$2a$	$(3a+b)$	0	$(a+b)$	0	$(3a+b)$
$(2a+b)$	0	$(2a+b)$	0	$(2a+b)$	b	0	$(2a+b)$	0
$(3a+b)$	0	$(3a+b)$	$2a$	$(a+b)$	0	$(a+b)$	0	$(3a+b)$

Table B.2 Case 2

$*$	0	a	$2a$	$3a$	b	$(a+b)$	$(2a+b)$	$(3a+b)$
0	0	0	0	0	0	0	0	0
a	0	a	$2a$	$3a$	b	$(a+b)$	$(2a+b)$	$(3a+b)$
$2a$	0	$2a$	0	$2a$	0	0	0	0
$3a$	0	$3a$	$2a$	a	b	$(a+b)$	$(2a+b)$	$(3a+b)$
b	0	b	0	b	b	0	b	0
$(a+b)$	0	$(a+b)$	$2a$	$(3a+b)$	0	$(a+b)$	$2a$	$(3a+b)$
$(2a+b)$	0	$(2a+b)$	0	$(2a+b)$	b	0	b	0
$(3a+b)$	0	$(3a+b)$	$2a$	$(a+b)$	0	$(a+b)$	$2a$	$(3a+b)$

Table B.3 Case 5

$*$	0	a	$2a$	$3a$	b	$(a+b)$	$(2a+b)$	$(3a+b)$
0	0	0	0	0	0	0	0	0
a	0	a	$2a$	$3a$	b	$(a+b)$	$(2a+b)$	$(3a+b)$
$2a$	0	$2a$	0	$2a$	0	0	0	0
$3a$	0	$3a$	$2a$	a	b	$(a+b)$	$(2a+b)$	$(3a+b)$
b	0	b	0	b	$2a$	0	0	0
$(a+b)$	0	$(a+b)$	$2a$	$(3a+b)$	$(2a+b)$	$(a+b)$	$(2a+b)$	$(3a+b)$
$(2a+b)$	0	$(2a+b)$	0	$(2a+b)$	$2a$	0	0	0
$(3a+b)$	0	$(3a+b)$	$2a$	$(a+b)$	$(2a+b)$	$(a+b)$	$(2a+b)$	$(3a+b)$

Table B.4 Case 6

$*$	0	a	$2a$	$3a$	b	$(a+b)$	$(2a+b)$	$(3a+b)$
0	0	0	0	0	0	0	0	0
a	0	a	$2a$	$3a$	b	$(a+b)$	$(2a+b)$	$(3a+b)$
$2a$	0	$2a$	0	$2a$	0	0	0	0
$3a$	0	$3a$	$2a$	a	b	$(a+b)$	$(2a+b)$	$(3a+b)$
b	0	b	0	b	0	0	0	0
$(a+b)$	0	$(a+b)$	$2a$	$(3a+b)$	b	$(a+b)$	$(2a+b)$	3a +b
$(2a+b)$	0	$(2a+b)$	0	$(2a+b)$	0	0	0	0
$(3a+b)$	0	$(3a+b)$	$2a$	$(a+b)$	b	$(a+b)$	$(2a+b)$	$(3a+b)$

Table B.5 Case 9

*	0	a	$2a$	$3a$	b	$(a+b)$	$(2a+b)$	$(3a+b)$
0	0	0	0	0	0	0	0	0
a	0	a	$2a$	$3a$	b	$(a+b)$	$(2a+b)$	$(3a+b)$
$2a$	0	$2a$	0	$2a$	0	0	0	0
$3a$	0	$3a$	$2a$	a	b	$(a+b)$	$(2a+b)$	$(3a+b)$
b	0	b	0	$(2a+b)$	0	0	0	0
$(a+b)$	0	$(a+b)$	$2a$	$(a+b)$	b	$(a+b)$	$(2a+b)$	$(3a+b)$
$(2a+b)$	0	$(2a+b)$	0	b	0	0	0	0
$(3a+b)$	0	$(3a+b)$	$2a$	$(3a+b)$	b	$(a+b)$	$(2a+b)$	$(3a+b)$

Table B.6 Case 10

*	0	a	$2a$	$3a$	b	$(a+b)$	$(2a+b)$	$(3a+b)$
0	0	0	0	0	0	0	0	0
a	0	a	$2a$	$3a$	b	$(a+b)$	$(2a+b)$	$(3a+b)$
$2a$	0	$2a$	0	$2a$	0	0	0	0
$3a$	0	$3a$	$2a$	a	b	$(a+b)$	$(2a+b)$	$(3a+b)$
b	0	b	0	$(2a+b)$	0	$2a$	0	0
$(a+b)$	0	$(a+b)$	$2a$	$(a+b)$	b	$(3a+b)$	$(2a+b)$	$(3a+b)$
$(2a+b)$	0	$(2a+b)$	0	b	0	$2a$	0	0
$(3a+b)$	0	$(3a+b)$	$2a$	$(3a+b)$	b	$(3a+b)$	$(2a+b)$	$(3a+b)$

Table B.7 Case 14

*	0	a	$2a$	$3a$	b	$(a+b)$	$(2a+b)$	$(3a+b)$
0	0	0	0	0	0	0	0	0
a	0	a	$2a$	$3a$	b	$(a+b)$	$(2a+b)$	$(3a+b)$
$2a$	0	$2a$	0	$2a$	0	0	0	0
$3a$	0	$3a$	$2a$	a	b	$(a+b)$	$(2a+b)$	$(3a+b)$
b	0	b	0	$(2a+b)$	b	0	$(2a+b)$	0
$(a+b)$	0	$(a+b)$	$2a$	$(a+b)$	0	$(a+b)$	0	$(3a+b)$
$(2a+b)$	0	$(2a+b)$	0	b	b	0	$(2a+b)$	0
$(3a+b)$	0	$(3a+b)$	$2a$	$(3a+b)$	0	$(a+b)$	0	$(3a+b)$

B.2 Additive Conjoints

I briefly return to the topic of additively opposite near-rings, first mentioned on page 478. The addition table for the group D_4 may be expressed by Table B.8.

The policy that the "product" $A \times B$ is represented by the element in the Ath row and Bth column is implicit in the symbols used to denote group elements. If we wish to express the additive conjoint (an isomorphic group), it is confusing merely to transpose the above matrix because the symbols representing group elements and heading the rows and columns would not be entirely consistent with the group operations. The conjoint group looks like Table B.9.

Table B.8 $(D_4,\ +)$

$(+)$	0	a	$2a$	$3a$	b	$(a+b)$	$(2a+b)$	$(3a+b)$
0	0	a	$2a$	$3a$	b	$(a+b)$	$(2a+b)$	$(3a+b)$
a	a	$2a$	$3a$	0	$(a+b)$	$(2a+b)$	$(3a+b)$	b
$2a$	$2a$	$3a$	0	a	$(2a+b)$	$(3a+b)$	b	$(a+b)$
$3a$	$3a$	0	a	$2a$	$(3a+b)$	b	$(a+b)$	$(2a+b)$
b	b	$(3a+b)$	$(2a+b)$	$(a+b)$	0	$3a$	$2a$	a
$(a+b)$	$(a+b)$	b	$(3a+b)$	$(2a+b)$	a	0	$3a$	$2a$
$(2a+b)$	$(2a+b)$	$(a+b)$	b	$(3a+b)$	$2a$	a	0	$3a$
$(3a+b)$	$(3a+b)$	$(2a+b)$	$(a+b)$	b	$3a$	$2a$	a	0

Table B.9 $(D_4,\ \ddagger)$

$(\ddagger)$	0	a	$2a$	$3a$	b	$(a+b)$	$(2a+b)$	$(3a+b)$
0	0	a	$2a$	$3a$	b	$(a+b)$	$(2a+b)$	$(3a+b)$
a	a	$2a$	$3a$	0	$(3a+b)$	b	$(a+b)$	$(2a+b)$
$2a$	$2a$	$3a$	0	a	$(2a+b)$	$(3a+b)$	b	$(a+b)$
$3a$	$3a$	0	a	$2a$	$(a+b)$	$(2a+b)$	$(3a+b)$	b
b	b	$(a+b)$	$(2a+b)$	$(3a+b)$	0	a	$2a$	$3a$
$(a+b)$	$(a+b)$	$(2a+b)$	$(3a+b)$	b	$3a$	0	a	$2a$
$(2a+b)$	$(2a+b)$	$(3a+b)$	b	$(a+b)$	$2a$	$3a$	0	a
$(3a+b)$	$(3a+b)$	b	$(a+b)$	$(2a+b)$	a	$2a$	$3a$	0

Table B.10 Revised $(D_4,\ \ddagger)$

$(\ddagger)$	0	$3A$	$2A$	A	$(2A+B)$	$(3A+B)$	B	$(A+B)$
0	0	$3A$	$2A$	A	$(2A+B)$	$(3A+B)$	B	$(A+B)$
$3A$	$3A$	$2A$	A	0	$(A+B)$	$(2A+B)$	$(3A+B)$	B
$2A$	$2A$	A	0	$3A$	B	$(A+B)$	$(2A+B)$	$(3A+B)$
A	A	0	$3A$	$2A$	$(3A+B)$	B	$(A+B)$	$(2A+B)$
$(2A+B)$	$(2A+B)$	$(3A+B)$	B	$(A+B)$	0	$3A$	$2A$	A
$(3A+B)$	$(3A+B)$	B	$(A+B)$	$(2A+B)$	A	0	$3A$	$2A$
B	B	$(A+B)$	$(2A+B)$	$(3A+B)$	$2A$	A	0	$3A$
$(A+B)$	$(A+B)$	$(2A+B)$	$(3A+B)$	B	$3A$	$2A$	A	0

To make this addition table consistent with respect to the row and column labels, we should re-label $a, 2aand3a$, as 3A, 2A and A, respectively, and re-label $b, a+b, 2a+band3a+b$ as 2A+B, 3A+B, B and A+B, respectively. The revised table is Table B.10 and this is better expressed as Table B.11.

With the re-labelling just described, Table B.1 now looks like Table B.12.

Tables B.11 and B.12 represent a near-ring hosted by D_4 in which the identity element is $3A$. Our tables for near-rings hosted by D_4 involve the element a being the multiplicative identity. That element is represented by $3A$ so direct comparisons are difficult and a further re-write is needed if we wish to use those tables. Again, re-labelling is needed, the regime being $\{A \mapsto 3a, 3A \mapsto a, B \mapsto b, 2A+B \mapsto$

Table B.11 $(D_4, \stackrel{\pm}{+})$

$(\stackrel{\pm}{+})$	0	A	$2A$	$3A$	B	$(A+B)$	$(2A+B)$	$(3A+B)$
0	0	A	$2A$	$3A$	B	$(A+B)$	$(2A+B)$	$(3A+B)$
A	A	$2A$	$3A$	0	$(A+B)$	$(2A+B)$	$(3A+B)$	B
$2A$	$2A$	$3A$	0	A	$(2A+B)$	$(3A+B)$	B	$(A+B)$
$3A$	$3A$	0	A	$2A$	$(3A+B)$	B	$(A+B)$	$(2A+B)$
B	B	$(3A+B)$	$(2A+B)$	$(A+B)$	0	$3A$	$2A$	A
$(A+B)$	$(A+B)$	B	$(3A+B)$	$(2A+B)$	A	0	$3A$	$2A$
$(2A+B)$	$(2A+B)$	$(A+B)$	B	$(3A+B)$	$2A$	A	0	$3A$
$(3A+B)$	$(3A+B)$	$(2A+B)$	$(A+B)$	B	$3A$	$2A$	A	0

Table B.12 Case 1—re-labelled

$*$	0	$3A$	$2A$	A	$(2A+B)$	$(3A+B)$	B	$(A+B)$
0	0	0	0	0	0	0	0	0
$3A$	0	$3A$	$2A$	A	$(2A+B)$	$(3A+B)$	B	$(A+B)$
$2A$	0	$2A$	0	$2A$	0	0	0	0
A	0	A	$2A$	$3A$	$(2A+B)$	$(3A+B)$	B	$(A+B)$
$(2A+B)$	0	$(2A+B)$	0	$(2A+B)$	$(2A+B)$	0	B	0
$(3A+B)$	0	$(3A+B)$	$2A$	A+B	0	$(3A+B)$	0	$(A+B)$
B	0	B	0	B	$(2A+B)$	0	B	0
$(A+B)$	0	$(A+B)$	$2A$	$(3A+B)$	0	$(3A+B)$	0	$(A+B)$

Table B.13 Re-labelled product for Table B.1

$*$	0	a	$2a$	$3a$	b	$(a+b)$	$(2a+b)$	$(3a+b)$
0	0	0	0	0	0	0	0	0
a	0	a	$2a$	$3a$	b	$(a+b)$	$(2a+b)$	$(3a+b)$
$2a$	0	$2a$	0	$2a$	0	0	0	0
$3a$	0	$3a$	$2a$	a	b	$(a+b)$	$(2a+b)$	$(3a+b)$
b	0	b	0	b	b	0	$(2a+b)$	0
$(a+b)$	0	$(a+b)$	$2a$	$(3a+b)$	0	$(a+b)$	0	$(3a+b)$
$(2a+b)$	0	$(2a+b)$	0	$(2a+b)$	b	0	$(2a+b)$	0
$(3a+b)$	0	$(3a+b)$	$2a$	$(a+b)$	0	$(a+b)$	0	$(3a+b)$

$2a+b, A+B \mapsto 3a+b, 3A+B \mapsto a+b\}$. The resulting table, plus re-ordering, is Table B.13.

It is, of course, identical to Table B.1. The new multiplication table can be produced from the old by swapping b with $2a+b$ and re-ordering. The mapping of D_4 that simply swaps b with $2a+b$ is an additive isomorphism. In the case of this near-ring, it is also a multiplicative isomorphism. This mapping is not an isomorphism in products Tables B.2, B.3, and B.6.

Working with Table B.6, and the swap $b \leftrightarrow 2a+b$, I obtain Table B.14.

After re-ordering, one again obtains Table B.6. It's easier to represent this new product as was done in Chap. 5, as Table B.15.

Table B.14 Re-labelled product for Table B.6

$*$	0	a	$2a$	$3a$	b	$(a+b)$	$(2a+b)$	$(3a+b)$
0	0	0	0	0	0	0	0	0
a	0	a	$2a$	$3a$	$(2a+b)$	$(a+b)$	b	$(3a+b)$
$2a$	0	$2a$	0	$2a$	0	0	0	0
$3a$	0	$3a$	$2a$	a	$(2a+b)$	$(a+b)$	b	$(3a+b)$
$(2a+b)$	0	$(2a+b)$	0	b	0	$2a$	0	0
$(a+b)$	0	$(a+b)$	$2a$	$(a+b)$	$(2a+b)$	$(3a+b)$	b	$(3a+b)$
b	0	b	0	$(2a+b)$	0	$2a$	0	0
$(3a+b)$	0	$(3a+b)$	$(2a+b)$	$(3a+b)$	$(2a+b)$	$(3a+b)$	b	$(3a+b)$

Table B.15 The new product

A	$2a$	$3a$	b	$(a+b)$	$(2a+b)$	$(3a+b)$
$b \cdot A$	0	b	$(2a+b)$	0	$(2a+b)$	0

Table B.16 Another new product

A	2a	$3a$	b	$(a+b)$	$(2a+b)$	$(3a+b)$
$b \cdot A$	0	b	0	0	2a	0

Table B.17 Transformed near-rings

Case	1	2	5	6	9	10	14
Case	1	4	8	6	9	10	14

This corresponds to row four of Table 5.10, on page 200. Applying the same ideas to Table B.3, our new product is given by Table B.16 and this corresponds to row eight of Table 5.10. Table B.17 lists the seven products given here together with the new products to which they correspond. The notation is that of Table 5.11.

The transformation process either transforms the near-ring into itself, or (in two cases) it transforms it into an isomorphic, but distinct, near-ring. Using the definition given in (A.14) on page 484, it seems that additively opposite near-rings will be isotopic to their parent. From Lemma 514, this should mean that, for near-rings with identity, they should be isomorphic.

Appendix C
Other Structures of Interest

C.1 Some Related Algebras

In this appendix, I gather the definitions of various structures that have interested algebraists in the past and which seem to be related to my subject matter. Most of these structures would not be well-known even to professional algebraists. It should be stressed that the definitions offered cannot be definitive and that individual authors sometimes adopt the same terms for rather different constructs.

I should own that some of my hyphenated names appear in the literature more commonly in unhyphenated form: semiring, semifield, quasigroup and neofield—even *nearring, nearfield,* etc. Reluctantly, I, did not hyphenate some terms which usually are not hyphenated, such as "subgroup". I did hyphenate "semi-group", though.

C.1.1 Semi-Rings

A **semi-ring** is a structure $(S, +, \cdot)$ such that:

(i) $(S, +)$ is a commutative monoid with identity element $0 \in S$.
(ii) $(S, \cdot)$ is a monoid with identity element $1 \in S$.
(iii) Multiplication is left- and right-distributive over addition.
(iv) For all $a \in R, 0 \cdot a = 0 = a \cdot 0$.

The natural number $\mathbb{N}$, under standard operations, forms a semi-ring, as do the set of ideals of a given ring. There are lots of really wonderful examples of these structures—such as the set of cardinal numbers smaller than a given infinite cardinal or the isomorphism classes in a distributive category under coproduct and product operations (the **Burnside semi-ring**). They even crop up in automata theory (I believe Samuel Eilenberg discussed them in his work in the early 1970s). **Tropical**

R. Lockhart, *The Theory of Near-Rings*, Lecture Notes in Mathematics 2295,
https://doi.org/10.1007/978-3-030-81755-8

semi-rings, which are semi-rings defined on the extended real numbers, are a very active area of current research.

It seems to me that these are intriguing constructs worthy of further study. Their properties dictate a more abelian nature than is natural to the present work, and I believe that some aspects of their theory proved to be very ring-like, module structure, in particular.

There is a matrix theory over semi-rings, and there have been a number of recent papers on the subject.

Some authors call these structures "rigs", which seems to relate to the sentence "rings *without* **n**egatives" [56]. The term started off as a joke but has now gained some credence—some of my own ideas are the other way round. There are a few recent books on semi-rings. The ones I have seen seem rather expensive, and the subject is by no means well-known. There are quite a few papers and theses on semi-rings available free of charge on the Internet.

C.1.2 Planar Ternary Rings

Definition 520 A **planar ternary ring** is a set R with distinguished elements 0, 1 $\in R$ together with a ternary operation $T : R^3 \mapsto R$ such that for $a, b, c, d, f, g \in R$:

(i) T(1,a,0) = T(a,1,0) = a.
(ii) T(a,0,b) = T(0,a,b) = b.
(iii) There is exactly one f such that T(a,b,f) = c.
(iv) If $a \neq b$ there is exactly one f such that T(f,a,c) = T(f,b,d).
(v) If $a \neq b$, there are exactly one f and exactly one g such that T(a,f,g) =c and T(b,f,g) = d.

The last condition is otiose in finite ternary rings. Dembowski prefers the term **ternary field** for these structures and says he is following the distinguished German geometer Günther Pickert in that there are no proper homomorphisms between these structures, so they are more field-like than ring-like.

One defines two binary operations from this

$$(x + b) = T(x, 1, b) \tag{C.1}$$

$$(x \wedge m) = T(x, m, 0) \tag{C.2}$$

These operations make $(R, +, 0)$ into a loop with identity element 0 (loop definition page 482). The set $R^* = \{R - \{0\}\}$ is a loop under the product ($\wedge$).

A planar ternary ring is called **linear** provided that

$$T(a, b, c) = (a \wedge b) + c \tag{C.3}$$

and a linear planar ternary ring in which $(R, +)$ is a group is called a **Cartesian group**. Finally, a Cartesian group in which the right distributivity law

$$((x + y) \wedge z) = ((x \wedge z) + (y \wedge z))$$

is called a **quasi-field** (see below).

C.1.3 Alternative Division Rings

An **alternative ring** is an abelian group $(A, +)$ together with a left- and right-distributive multiplication which satisfies the two laws

$$\begin{aligned} ((x \cdot x) \cdot y) &= (x \cdot (x \cdot y)) \\ (x \cdot (y \cdot y)) &= ((x \cdot y) \cdot y) \end{aligned}$$

An **alternative division ring** is an alternative ring with two-sided identity in which the non-zero elements have two-sided inverses (i.e. a has a corresponding a^{-1} which acts as an inverse on each side). Apparently, the choice of name was made by Zorn, and its reason is that the group A_3 acts trivially on the *associator* (see (1.41)). A famous theorem attributed to Artin and Zorn states that a finite alternative division ring is a commutative field (see [62]). Faulkner reports the "Bruck-Kleinfeld" theorem that an alternative division algebra is either associative or an "octonion algebra" over its own centre [39].

C.1.4 Semi-Fields

A **semi-field** $(S, +, \cdot)$ is an abelian group $(S, +, 0)$ such that, for all $a, b, c \in S$,

(i) $(a \cdot (b + c)) = ((a \cdot b) + (a \cdot c))$.
(ii) $((a + b) \cdot c) = ((a \cdot c) + (b \cdot c))$.
(iii) There is a multiplicative identity 1.
(iv) There are unique solutions x, y, to the equations $(x \cdot a) = b$ and $(a \cdot y) = b$.

Weibel says this idea represents a class of linear planar ternary rings which complements near-fields and that there are semi-fields for every prime power p^n, provided $n \geq 3$ and $n \neq 8$, [142].

C.1.5 Neo-fields

A **neo-field** is a structure that is in all respects the same as a field except for having weaker additive properties, the addition now being that of a loop (see Definition 511). Formally, a **right neo-field** $(N, +, \cdot)$ with order n is a set N of n symbols together with two binary operations such that

(i) $(N, +)$ is a loop with identity element 0.
(ii) $(N^*, \cdot)$ is a group (where, as usual, $N^* = \{N - \{0\}\}$).
(iii) $((a + b) \cdot c) = ((a \cdot b) + (a \cdot c))$ for all $a, b, c \in N$.

A **neo-field** is simply a left and right neo-field. There are neo-fields of all finite orders [123]. One can find out more about neo-fields in the article [85].

C.1.6 Quasi-Fields

A right **quasi-field** is a structure $(Q, +, \cdot)$ for which

(i) $(Q, +)$ is a group.
(ii) $(Q^*, \cdot)$ is a loop.
(iii) For all, $a, b, c \in Q$, $(a + b) \cdot c = ((a \cdot c) + (b \cdot c))$.
(iv) For $a, b, c \in Q$ with $a \neq b$, there is a unique x such that $(a \cdot x) = ((b \cdot x) + c)$.

There is a conflicting use of this term in the theory of semi-rings.

Quasi-fields obeying both distributivity laws occur in projective geometry where they are called **semi-fields**. Quasi-fields have also been called **Veblen-Wedderburn rings** or **Veblen-Wedderburn systems**. It follows from the axioms that $(Q, +)$ is abelian. Confusingly, quasi-fields are often called abelian provided that $(Q^*, \cdot)$ is abelian.

A semi-field is simply a quasi-field satisfying both distributivity laws. The kernel of the quasi-field Q is the set K of all elements c for which, for all $a, b, \in Q$,

$$(a \cdot (b \cdot c)) = ((a \cdot b) \cdot c) \tag{C.4}$$

$$(c \cdot (a + b)) = ((c \cdot a) + (c \cdot b)) \tag{C.5}$$

It can be shown $(K, +, \cdot)$ is a division ring and that Q is a right vector space over it. This means any finite quasi-field has order a power of a prime. Near-fields are precisely the quasi-fields in which the multiplication is associative.

C.1.7 Quasi-Groups and Loops

Quasi-groups are systems with binary operations for which the corresponding Cayley table is a **Latin square**. That is, no row or column of the table involves the repetition of an element. Loops are simply quasi-groups which possess identity elements. I define these structures more formally on page 482.

C.1.8 Near-Domains

A **near-domain** is a sort of additively non-associative near-field. Specifically, Pilz [128] defines a near-domain as a set $(N, +, \cdot)$, such that

(i) $(N, +)$ is a loop with identity element 0.
(ii) For all $a, b, \in N, (a+b) = 0 \Leftrightarrow (b+a) = 0$.
(iii) $(N^*, \cdot)$ is a group.
(iv) For all a, $a \cdot 0 = 0 \cdot a = 0$.
(v) For all $a, b, c \in N, (a+b) \cdot c = ((a \cdot b) + (a \cdot c))$.
(vi) For all $a, b \in N$, there exists $D = D(a, b) \in N^*$, such that $(a + (b + c)) = ((a+b) + D \cdot c)$.

Finite near-domains are known to be near-fields; but, so far as I am aware, no near-domain which is not a near-field is known (see page 159).[1]

C.1.9 L-R Systems

Slick teachers sometimes define groups as semi-groups with left identities and such that all elements have left inverses. If one slips up, requiring left identities and *right* inverses, say, one obtains structures that are a little different from groups: **L-R systems**. Henry Mann investigated **L-R systems** in the 1940s [115]. L-R systems are direct products of groups with **idempotent L-R systems** (which are essentially trivial).

C.1.10 Other Structures and Structural Coding

There is a bewildering variety of algebraic structures with a sea of conflicting terminology and definitions. Gouvêa mentions **pseudorings**, for example (no

[1] I was fortunate in the reviewers of this book. One of them, kindly, referred me to [130], in which it is shown that infinite near-domains that are not near-fields do exist.

hyphen)—which he describes simply as rings without identity [56]. Others use the term "rngs" for such things. I purloined that term for a quite different structure (page 63), making a poor distinction between them by using a hyphen for my ones: *pseudo-ring* (poor, because word processor line breaking ruins it, but since I only ever discuss my version, there is no problem).

It seems to me that it might be useful to develop a straightforward code for these structures, rather like an ISBN number or a library of congress number. Such a code could hardly encompass every possible structure but could certainly record basic algebraic properties; and one usually finds that when one develops something like this, the resulting grammar itself suggests further structures.

The reader might like to consider it and would need to be aware of Internet possibilities, iterations and extensions, universal algebra and subject classification issues—so each code might map to several terms already in use. One would also be well-advised to consider Neil Sloane's wonderful online encyclopaedia of number sequences in this context.

The code would not necessarily have to be immediately intelligible to the human reader, of course. One could hyper-link to web pages giving detailed explanations, references and open problems; and one could expand this to include ambiguous definitions ("partial order" comes to mind!)—even triangulating on particular books and eccentric terminology. An endeavour of this sort might best be constructed, bottom up and organically, relying on a community of individuals rather than a dirigiste authority of the Bourbaki sort.

C.2 Some Geometry

C.2.1 Projective Spaces

Projective spaces are ubiquitous in modern pure mathematics and occur in various guises in group theory, geometry, algebraic geometry and just about everywhere else. There are a great many references for these structures and many books and research articles.

Artin [4] is a good place to start, for general projective spaces. Suzuki [139] is particularly clear on them, as he is on so much else. Suzuki's definition of $(n-1)$-**dimensional projective space** starts with a vector space with dimension n over an arbitrary field. The space $\mathcal{P}$ is simply defined to be the set of all one-dimensional subspaces of the original vector space. An arbitrary subset of $\mathcal{P}$ is deemed to be a projective subspace of it, provided that there is a subspace of the original vector space for which that subset is the collection of all one-dimensional subspaces. The projective subspace is held to have dimension one less than that of the vector space subspace. Elements of $\mathcal{P}$ are called **points** and are zero-dimensional (projective) subspaces. One-dimensional projective subspaces are termed **lines**; two-dimensional **planes**; and so on.

Perspective in Art
Projective spaces have a direct connection with the theory of perspective in art—each collection of parallel lines defining a single point at infinity (the "vanishing point") and the collection of such points defining a line at infinity (the "horizon", the "vanishing line").

Perspective in art is more complicated than I suggest, and there are various versions of it—terraced perspective, vertical perspective and the late renaissance idea of **focussed perspective**, which is the one best known to most of us.

Topologically, one first meets projective spaces in the theory of two-dimensional surfaces, where they have a key classification role. The **real projective plane** is represented as a rectangle in the Euclidean plane in which opposite edges are identified with opposing orientations—so diagonally opposing vertices are regarded as identical. The resulting topological structure has no boundary, is non-orientable and has *Euler characteristic* 1. It is a compact, non-orientable, two-dimensional manifold. The book by Armstrong [3] is a good introduction to these ideas.

An alternative, topological view is that of a sphere in which a disc has been removed and replaced by a Möbius band in such a way that the edge of the area removed is identified with the boundary of the band. The fact that the real projective plane contains the Möbius band ensures its non-orientability and that it is not representable in three-dimensional space. The Klein bottle is similarly not representable in three space and for the same reason.[2]

Embeddability of projective structures in higher-dimensional structures is of recurrent interest. In the case of finite projective planes, we wish to embed in three-dimensional projective space, and whether this can be done or not relates directly to the geometry of the plane, which in turn relates to the algebraic properties of its associated planar ternary rings q.v. The real projective plane is actually representable in four-dimensional space. One can recognise more straightforwardly geometric realisations of the real projective plane, including:

1 The collection of distinct lines through the origin in Euclidean three-dimensional space $\mathbb{R}^3$.
2 The unit sphere with antipodal points identified.[3]
3 The equivalence classes of $\{\mathbb{R}^3 - (0, 0, 0)\}$ under the relation

$$(x, y, z)\ (a, b, c) \Leftrightarrow (x, y, x) = (\lambda \cdot a, \lambda \cdot b, \lambda \cdot c)$$

This is the projective plane in Euclidean space, but generally one may associate a **n-dimensional projective space** with any $(n + 1)$-dimensional vector space over any field F by simply applying the same ideas to the higher-dimensional vector spaces. When $n = 1$, we have **projective lines**; when $n = 2$, we have **projective planes**.

---

[2] Something which has not prevented people from producing and selling "Klein bottles".

[3] Opposite ends of the Earth!—planet surface only!

The complex projective line is better known as the **Riemann sphere**. The simplest manifestation of this construction uses the third realisation above and applies to finite fields with order q and to more general algebras, such as infinite fields, division rings and the quaternions (see Sect. 3.3.1).

I shall describe the idea using finite fields, concerning myself mainly with projective planes. Charles Weibel has written an excellent survey article on these planes [142], and I steal from that article here. It is freely available on the Internet. I also pillage Marshall Hall's book [62]. The reader would probably find [63] useful too, if that can be obtained.

C.2.2 *Finite Projective Planes*

To define a projective plane over a finite field, start with a three-dimensional vector space over that field.

Points are simply "lines" of the space which pass through the origin. Equivalently, this is the set of all one-dimensional subspaces of the space [138]. "Lines" are defined as collections of the projective points which are themselves projective points of some two-dimensional subspace of the original space; but a more immediate formulation is that lines are collections of projective points (x, y, z) from (C.2.1) for which there are fixed vectors $(a, b, c) \neq (0, 0, 0)$ for which

$$(a \cdot x) + (b \cdot y) + (c \cdot z) = 0$$

Since this would imply $(\lambda \cdot a \cdot x) + (\lambda \cdot b \cdot y) + (\lambda \cdot c \cdot z) = 0$, it seems that lines are very similar to points and this is very much the case in projective planes—so that some authors represent points are row vectors and lines as column vectors and apply matrix multiplication.

This duality between lines and points means that one can interchange the terms, obtaining a "principal of duality" applying to statements about projective planes and associating with each projective plane its formal "dual plane"—something quite generally true in all projective planes. Again, **projective n-space** would be the set of one-dimensional subspaces of a (n+1)-dimensional space. Its points are the "lines" through the origin, but our current interest is in finite projective planes. The definition of points involves equivalence classes. Class representatives give rise to *homogeneous coordinates* which are equivalence class representatives. One recognises classes with the forms

$(x, y, 1)$, where $x, y, \in F$, there are q^2 of these.
$(x, 1, 0)$, where $x \in F$, a further q points.
$(1, 0, 0)$, a single point.

Such finite projective planes have $(q^2 + q + 1)$ points. Arising from the finite field F, they are called **field planes**. Not all finite projective planes do arise in this way [133].

Applying the above definition to the field with two elements gives the **Fano plane**, which obviously has exactly seven points and seven lines. General finite projective planes have $(n^2 + n + 1)$ points $(n \geq 2)$ [62]. In all known cases, n is a power of a prime. It is called the **order** of the projective plane. The associated symmetry group is a quotient group of the group of all linear transformations. It is called the **projective linear group**. It is the quotient of the general linear group by its centre, which is the group of scalar matrices.

Subplanes are defined as one would expect. Field planes are straightforward; the subplanes correspond to subfields. For other planes, a theorem of Bruck says that a plane with order n has a subplane of order m only if either $n = m^2$ or $n \geq (m^2+m)$. Knowing about subplanes would be equivalent to knowing about possible orders. One may define finite projective planes axiomatically [62].

Definition 521 A **finite projective plane** is a set of **points** of which certain distinguished subsets constitute **lines**, with these properties:

1 Any two points are contained in exactly one line.
2 Any two lines contain exactly one common point.
3 There are four points, no three of which lie in a common line.

There is a standard body of mathematics relating to finite projective planes though much is still unknown.

Several different ways to classify them have been suggested. Dembowski [32] reports a classification based on work by Lenz and Barlotti which lists 53 possible "Lenz-Barlotti types for collineation groups" and Weibel references that [142]. It all seems rather opaque.

One basic result is that a plane of order n has exactly $(n + 1)$ lines and each line has exactly $(n + 1)$ points.

Coordinatisation and Algebras

Axiom (3) above can be used to coordinatise a finite projective plane. Weibel [142] attributes the original idea to von Staudt and Hilbert. Once the plane has been coordinatised, a natural ternary algebra arises. Blumenthal [16] gives a particularly clear account of this. One can view this process either in terms of simply labelling existing structures or in terms of constructing identical abstract models of them. Hall offers this version of the idea [62]. The reader may find a diagram helpful in making sense of it. One chooses a set of four points $\{0, 1, X, Y\}$ sometimes called the **coordinatising quadrilateral**. The first two points are identified as the origin (0,0) and (1,1), and the other two define a line L_∞. There is a unique point which is the intersection of the lines defined by the first two and the last two points. This is given the coordinate (1). We then give distinct coordinates (b, b) to the other points of the line between 0 and 1. If P is a point not on L_∞ and so far without a coordinate, then there is a unique line XP, and that intersects the line between 0 and 1 at some point (b,b). Similarly, YP intersects that line at (a,a); we give P the coordinates (a,b). Notice that there is a point (1,m) defined. It is essentially built from the line joining 1 and Y. Then, there is a line joining (0,0) and (1,m), and that line intersects L_∞ at a point M which is given the coordinate (m). The only point not yet given a coordinate is Y, and that is given the coordinate (∞).

Once a coordinatisation is established, a ternary operation is defined. Any line not through the point Y ≡ (∞) intersects L_∞ at some point (m) and intersects the line through O and Y at some point (0,b). If (x,y) is a point of this line, we write

$$y = (x \cdot m \circ b) \tag{C.6}$$

This is our ternary operation. Notice that the symbol ∞ does not figure in it. The structure defined in this way is a planar ternary ring (see Sect. C.1.2). From it, we can define two binary operations

$$(x + b) = (x \cdot 1 \circ b) \tag{C.7}$$

$$(x \wedge m) = (x \cdot m \circ 0) \tag{C.8}$$

Weibel [142] says that near-fields are planar ternary rings for which

$$T(a, b, c) = (a \wedge b) + c \tag{C.9}$$

One can reverse the process—going from a given planar ternary ring to a projective plane. It can happen that distinct quadrilaterals in a given projective plane give distinct and non-isomorphic planar ternary rings. That is, non-isomorphic planar ternary rings can produce isomorphic projective planes. There is a notion of **isotopism** (see page 477) in planar ternary rings, and isotopic planar ternary rings do produce isomorphic projective planes (see [142]).

Desargues and Pappus

Two results from the classical geometry of Euclidean spaces figure in the theory of projective planes:

(i) The theorem of Pappus. This says that when you have two distinct lines, each containing three points not on the other, and you join the points of one line to the points of the other, the intersections of these joining lines comprise a set of points which all lie on a single line. The theorem is usually associated with Pappus of Alexandria, who is routinely described as the last great ancient Greek geometer and who lived in Alexandria round about 300 A.D.

(ii) The theorem of Desargues. This says that when you have two triangles for which the lines through the pairs of vertices meet at a point ("perspective from a point"), then the extended lines formed from the triangle sides intersect on a line ("perspective from a line"). You need to be careful about pathological cases in which sides are parallel; indeed, it was the need to handle such situations that gave rise to projective geometry. Desargues' theorem originated in the seventeenth century, so one might object to the term "classical".

I believe it was Hilbert who first observed that there were projective planes for which these theorems do not hold. Whether they do or not is related to the kinds of symmetries that may occur. In finite projective planes, Desargues theorem implies the theorem of Pappus. Known proofs of that are algebraic in nature,[4] and people have looked for more geometric proofs. There are infinite Desarguesian projective planes which are not Pappian.[5] Every Pappian projective plane is Desarguesian. These theorems do hold in three-dimensional projective space but can fail in dimension 2.

Room and Kirkpatrick give three examples of projective planes with order 9 for which these theorems fail [133]. Interestingly, these examples utilise the smallest possible proper near-field (see Sect. 3.3.1). Projective planes for which the general theorem of Desargues fails are called **non-Desarguesian**. Marshall Hall states that projective geometries with dimension higher than 2 are infallibly Desarguesian and that Desarguesian projective planes are precisely the planes that may be embedded in a higher-dimensional projective space [63], by which he means three-dimensional space and all higher dimensions [4]. There are weaker geometric versions of these theorems which occur in some projective planes, for instance, the "little Desargues condition", mentioned below.

Clearly, a projective plane for which these theorems fail cannot be embedded in three-dimensional projective space. How to tell whether or not a given projective plane is or is not *Desarguesian*? It turns out that in Desarguesian planes, the associated planar ternary ring is a specific associative division ring—so in finite cases it will be a field.

Hall [62] gives theorems stating:

1 A projective plane is a "translation plane" with respect to L_∞ if and only if the planar ternary ring is a **Veblen-Wedderburn** system.
2 A projective plane is a "Moufang plane" if and only if the planar ternary ring is linear and an alternative division ring.

Faulkner says:

1 A projective plane is coordinatised by a field if and only if it satisfies the Pappus condition.
2 A projective plane is coordinatised by an associative division ring if and only if it satisfies the Desargues condition.
3 A projective plane is coordinatised by an alternative division ring if and only if it satisfies the "little Desargues condition".

The essential message is that geometric properties of the projective plane correspond to algebraic properties of the corresponding planar ternary rings.

[4] In fact they depend on Wedderburn's theorem that finite division rings are fields [66].

[5] The terminology is self-explanatory! (and like all self-explanatory things, requires this further explanation).

Appendix D
Semi-Linear Mappings

A **semi-linear mapping** is a transformation on a finite field K which has the form

$$x \longmapsto ((\theta(x) \cdot a) + b) \tag{D.1}$$

where $a \neq 0$ and θ is a field automorphism (sometimes called a θ-semi-linear mapping). The semi-linear mappings form a group under composition which is fairly standardly defined as $\Gamma(K)$. When the order of K is p^α, this is sometimes written as $\Gamma(p^\alpha)$. The group is both *doubly transitive* (Notation 3) and solvable [75, 132]. Its order is $\alpha p^\alpha(p^\alpha - 1)$. Restricting to $\theta = Id$ gives the **one-dimensional affine group** over K, $AGL_1(K)$ [35, 125].

This group has a venerable history dating back to Galois. It is 2-transitive, and its subgroup of **translations** (fix a=1) constitutes a normal abelian subgroup of it—it is a semi-direct product of that group by an abelian group—a situation entirely typical of affine structures. The one-dimensional affine group has order $|K| \cdot (|K| - 1)$.

I had thought that it was this group that caused Dickson, researching transitive permutation groups, to generalise the idea of a field, inventing near-fields. That seems plausible, but one of the reviewers of this book asked for some substantiation for this belief, and I have to confess that on reflection that is not to hand.

One may define m-dimensional affine groups by considering the vector space K^m and acting the general linear group $GL(m, K)$ upon it, by linear transformations—the resulting semi-direct product being the group $AGL_m(K)$ (see page 160).

An Alternative Definition

Artin uses the term for a related idea, mappings on finite dimensional vector spaces, V,W, over the same field K with the form $\lambda : V \mapsto W$ and such that

(i) $\lambda(a + b) = \lambda(a) + \lambda(b)$.
(ii) $\lambda(k \cdot a) = \theta(k) \cdot \lambda(a)$.

R. Lockhart, *The Theory of Near-Rings*, Lecture Notes in Mathematics 2295,
https://doi.org/10.1007/978-3-030-81755-8

where θ is some fixed field automorphism [4]. Artin relates this sort of semi-linear mapping to collineations in projective space (the "fundamental theorem of projective geometry"—see also [134]).

Appendix E
Zsigmondy's Theorem

Zsigmondy's theorem (sometimes spelt "Zsigmonde's theorem") was established at the end of the nineteenth century and is often useful in algebra. It does not seem to occur in the basic number theory texts with which I am familiar so I give it here.

The theorem says that if $a, b, n \in \mathbb{N}$, with $(a, b) = 1$ and $n \geq 1$, then there will be a prime divisor (called a "Zsigmondy prime" or sometimes a "primitive prime divisor") of $(a^n - b^n)$ which does not divide any $(a^k - b^k)$, with $1 \leq k \leq (n - 1)$, except in these obvious specific cases

(i) $n = 1$, $(a - b) = 1$, since then, trivially, $(a^n - b^n)$ has no prime divisors.
(ii) $a = 2, b = 1, n = 6$; because $(a^6 - b^6) = 63 = ((a^2 - b^2)^2.(a^3 - b^3))$.
(iii) $n = 2$ and $(a + b) = 2^r$ for some $r \in \mathbb{N}$. This because all odd prime factors of $(a^2 - b^2) = (a + b).(a - b)$ must then divide $(a - b)$.

Zsigmondy primes can never be equal to 2. I believe that when a is a power of a prime and b is 1, which is what concerns us in near-field applications, a Zsigmondy prime must have the form $r = (an + 1)$. More, n is the order of a modulo p. This version of the theorem is a subcase, which I believe may be known as **Bang's theorem**.

Zsigmondy's theorem is proved in [74], which also gives the corollary:

Corollary 522 *When p is prime and $n > 1$, there will be a prime $q > n$ such that q divides $(p^n - 1)$, unless $p = 3$ and $n = 2$.*

R. Lockhart, *The Theory of Near-Rings*, Lecture Notes in Mathematics 2295,
https://doi.org/10.1007/978-3-030-81755-8

Afterword

That I managed to maintain an interest in near-rings despite a varied career that took me rather far from them is in no small part due to the inexhaustible kindness of Robert Laxton, of the University of Nottingham, and of John Meldrum, of the University of Edinburgh. The bewildered tolerance and bewildering generosity of the near-ring group of mathematicians,[1] led by Günter Pilz, and centred on Johannes Kepler Universität, Linz, was another hugely supportive factor—I would turn up at their conferences once in 10 years, producing something weird. This is how the work on cohomology, product theory and isotopy started; and I would not have even got to those conferences without the support of my sisters. I might add that the fact that Professor Pilz, who has been synonymous with advances in near-rings for more than 40 years, has written a foreword to this book is an unexpected honour on my part. The spelling of his first name actually does differ between current usage and what appears in his book [128]—this is not a typing error!

I am well aware that I am not abreast of the vast amount of technical work done by algebraists in this subject and my account is eccentric, to say the least. It may be that I have not given sufficient credit to the many mathematicians working in this field and I hope they will forgive me. In general, if someone claims to have preceded me in a result, they can be regarded as correct. The mistakes, on the other hand, are wholly original to me and unlikely to have been produced without my direct input. I deserve full credit for them (perhaps there is some sort of award?—one imagines the "Wolfgang Pauli" prize, the "Albert Ross" pendant, etc.).

Near-rings resemble Sauron's "Ruling Ring"—algebraists donning them tend to vanish into a grey half-life, which is a pity, my precious, because the subject contains interesting work from clever people. I think that the fashionable lack of attention to near-rings is more to do with the debilitating effects of centralised accountancy and its concomitant overweening bureaucratisation of research than intrinsic worth.

[1] Definition 20.

R. Lockhart, *The Theory of Near-Rings*, Lecture Notes in Mathematics 2295,
https://doi.org/10.1007/978-3-030-81755-8

I guessed this book might take me 5 years: it took me 10. I got rather bogged down in near-fields and the account I had in mind for Parts III and IV was correspondingly reduced; the reader will be horrified to hear that I intended these sections to be at least twice as big! Throughout that time, I had the vague feeling that marriage and my wife Susan were preventing me from working seriously. The opposite is true: the book would have been started, but not finished, without Suzie.[2] Lemma 188 was assembled whilst baby-sitting her granddaughter, Sofia, and the "Fantastic Four" were christened following a suggestion from Sofia's brother, Magnus (his first idea being "Tim"). Finally, $\mathcal{A}$-Matrices were named at a 1990s near-ring conference (as I thought, temporarily!) after my daughter, $\mathcal{A}$nnie.

When you are young, you imagine that mathematical ability is the sole criterion for mathematical achievement. When you are older, you realise how much it depends on other things, including emotional balance and pure chance; and I suspect that is true even for young children and their traditional, almost institutionalised distaste for mathematics. I only really started to imagine I might finish this book after help from Dr Molodynski and his team at the Ridgeway centre in Didcot.

There's a saying in Nietzsche to the effect that people who remain pupils serve their teachers badly. I wish some of my own excellent teachers, or, indeed, my parents, could have seen this book, though they might well have concluded I needed further coaching. More than that, I wish Sofia and Magnus could have the same chance of excellent teachers that I had, unlikely though that is, in these progressive days.

My own, diminishing belief in what one might describe as the official channels of research mathematics was invigorated by my experience of Springer Verlag. They received a large, unsolicited book in a relatively obscure area and written by an unknown, elderly mathematician. Much of the book was new and the notation variable and eccentric (I have visual and coordination problems and am probably too used to working alone). They responded with support and generosity, sending the book out to three painstaking reviewers who were clearly of a stature that might preclude their spending time on this stuff. The reviewers responded with kindness and real scholarship. I regret the presentational problems I caused them and can only thank them, and Dr. Remi Lodh, Ms. Narmadha Nedounsejiane, and their colleagues at Springer, for their help. Truly, research as it is meant to be conducted.

[2] Our wedding reception was on the site of Cayley's study in Cambridge.

Bibliography

1. E. Aichinger, G. Pilz, A survey on polynomials and polynomial functions, in *Third International Algebra Conference*, ed. by Y. Fong, L.S. Shiao, E. Zelmanov, Tainan, Taiwan (Kluwer, Doordrecht, 2003), pp. 1–16
2. F.W. Anderson, K.R. Fuller, *Rings and Categories of Modules*, 2nd edn. (Springer, Berlin, 1992)
3. M.A. Armstrong, *Basic Topology* (Springer, Berlin, 1983)
4. E. Artin, *Geometric Algebra* (Dover, New York, 2016)
5. M.F. Atiyah, I.G. MacDonald, *Introduction to Commutative Algebra* (Westview Press, 2016)
6. J. Barwise (ed.), *Handbook of Mathematical Logic* (North Holland, Amsterdam, 1978)
7. J.A. Beachy, in *Introductory Lectures on Rings and Modules*. London Mathematical Society Student Texts, vol. 47 (1999)
8. J.L. Bell, A.B. Slomson, *Models and Ultraproducts—an Introduction* (Dover, New York, 1997)
9. J.L. Bell, M. Machover, *A Course in Mathematical Logic* (North Holland, Amsterdam, 1977)
10. G. Betsch, in *Some Structure Theorems on 2-Primitive Near-Rings*. Colloquia Mathematica Societatis János Bolyai, vol. 6. Rings, Modules and Radicals (Keszthely, 1971)
11. G. Betsch, Primitive near-rings. Math Z. **130**, 351–361 (1973)
12. G. Betsch, G. Pilz, H. Wefelscheid, Near-rings and near-fields, in *Proceedings of the 1989 Oberwolfach Conference*, (5–11, November, 1989)
13. Y. Fong, E. Bell, W.-F. Ke, G. Mason, G. Pilz, *Near-Rings and Near-Fields* (Kluwer, Dordecht, 1993)
14. F.R. Beyl, J. Tappe, in *Group Extensions, Representations, and the Schur Multiplicator*. Lecture Notes in Mathematics, vol. 958 (Springer, Berlin, 1982)
15. P.E. Bland, *Rings and Their Modules* (De Gruyter, Berlin, 2011)
16. L.M. Blumenthal, *A Modern View of Geometry* (Freeman, New York, 1961)
17. P.J. Cameron, in *Permutation Groups*. London Mathematical Society Student Texts, vol. 45 (1999)
18. H. Cartan, S. Eilenberg, *Homological Algebra* (Princeton University Press, Princeton, 1956)
19. R. Casse, *Projective Geometry, an Introduction* (Oxford, 20006)
20. A.J. Chandy, Rings generated by the inner automorphisms of nonabelian groups. Proc. Am. Math. Soc. **30**(1) (1971)
21. A.W. Chatters, C.R. Hajarnavis, *Rings with Chain Conditions* (Pitman, London, 1980)
22. J.R. Clay, The near rings on a finite cyclic group. Am. Math. Mon. **71**(1) (1964)
23. J.R. Clay, *Near Rings: Geneses and Applications* (Oxford Science Publications, Oxford, 1992)

R. Lockhart, *The Theory of Near-Rings*, Lecture Notes in Mathematics 2295,
https://doi.org/10.1007/978-3-030-81755-8

24. P.M. Cohn, *Universal Algebra* (Reidel, Dordrecht, 1981)
25. P.M. Cohn, in *Skew Fields*, in Encyclopedia of Mathematics and Its Applications, vol. 57 (Cambridge, 1995)
26. P.M. Cohn, *An Introduction to Ring Theory* (Springer, Berlin, 2000)
27. P.M. Cohn, *Basic Algebra* (Springer, Berlin, 2002)
28. P.M. Cohn, *Further Algebra and Applications* (Springer, Berlin, 2003)
29. J.H. Conway, N.J.A. Sloane, *Sphere Packings, Lattices and Groups* (Springer, Berlin, 1999)
30. Curtis, C.W., I. Reiner, *Representation Theory of Finite Groups and Associative Algebras* (Wiley Interscience, Chichester, 1962)
31. B.A. Davey, H.A. Priestley, *Introduction to Lattices and Order* (Cambridge, 1990)
32. P. Dembowski, in *Finite Geometries*. Classics in Mathematics (Springer, Berlin, 1991)
33. L.E. Dickson, Definitions of a group and a field by independent postulates. Trans. Am. Math. Soc. **6**, 198–204 (1905)
34. N.J. Divinsky, *Rings and Radicals* (Allen and Unwin, 1965)
35. J.D. Dixon, B. Mortimer, *Permutation Groups*. Graduate Texts in Mathematics (Springer, Berlin, 1991)
36. B.A. Dubrovin, A.T. Fomenko, S.P. Novikov, *Modern Geometry—Methods and Applications, (Part 1)* (Springer, Berlin, 1984)
37. S. Eilenberg, S. Maclane, *Eilenberg-Maclane: Collected Works* (Academic Press, London, 1986)
38. A. Eddington, *Fundamental Theory* (Cambridge, 1946)
39. J.R. Faulkner, in *The Role of Nonassociative Algebra in Projective Geometry*. Graduate Studies in Mathematics, vol. 159 (American Mathematical Society, Providence, 2014)
40. C.C. Ferrero, G. Ferrero, *Near-Rings—Some Developments Linked to Semigroups and Groups* (Kluwer, Dordecht, 2002)
41. A. Fröhlich, The near-ring generated by the inner automorphisms of a finite simple group. J. Lond. Math. Soc. **33** (1958)
42. A. Fröhlich, Distributively generated near-rings: I. Ideal Theory. Proc. Lond. Math. Soc. (3) **8** (1958)
43. A. Fröhlich, Distributively generated near-rings: II. Representation Theory. Proc. Lond. Math. Soc. (3) **8** (1958)
44. A. Fröhlich, On groups over a d.g. near-ring, I. Sum constructions and free R-groups. Q. J. Math. (Oxford) (2) **11** (1960)
45. A. Fröhlich, On groups over a d.g. near-ring, II. Categories and functors. Q. J. Math. (Oxford) (2) **11** (1960)
46. A. Fröhlich, Non-abelian homological algebra. I. Derived functors and satellites. Proc. Lond. Math. Soc. **11** (1961)
47. A. Fröhlich, Non-abelian homological algebra. II. Varieties. Proc. Lond. Math. Soc. **12** (1962)
48. A. Fröhlich, Non-abelian homological algebra. III. The functors EXT and TOR. Proc. Lond. Math. Soc. **111** (1962)
49. O. Ganyushkin, V. Warfield, *Classical Finite Transformation Semi-Groups* (Springer, Berlin, 2009)
50. B.J. Gardner, H. Wiegandt, *Radical Theory of Rings* (Marcel Dekker, New York, 2004)
51. B.J. Gardner, L. Shaoxue, R. Wiegandt, *Rings and Radicals* (Longman, London, 1994)
52. S. Givant, P. Halmos, *Introduction to Boolean Algebras* (Springer, Berlin, 2009)
53. R. Goldblatt, *Topoi—the Categorial Analysis of Logic* (Dover, New York, 2006)
54. O.K.R. Goodearl, R.B. Mazorchuk, in *An Introduction to Noncommutative Noetherian Rings*. London Mathematical Society Student Texts, vol. 61 (2004)
55. D. Gorenstein, *Finite Groups* (AMS Chelsea Publishing, New York, 1980)
56. F.Q. Gouvêa, *A Guide to Groups, Rings, and Fields* (MAA, 2012)
57. G. Grätzer, *Lattice Theory* (Dover, New York, 2009)
58. M. Gray, *A Radical approach to Modern Algebra* (Addison-Wesley, Reading, 1970)
59. T. Grundhöfer, H. Zassenhaus, A criterion for infinite non-Dickson near-fields of dimension two. Results Math (1989)

60. T. Grundhöfer, C. Hering, Finite near-fields classified again. J. Group Theory **20**, 829–839 (2017)
61. K.W. Gruenberg, in *Cohomological Topics in Group Theory*. Lecture Notes in Mathematics, vol. 143 (Springer, Berlin, 1970)
62. M. Hall, *The Theory of Groups*, 2nd edn. (AMS Chelsea Publishing, 1999)
63. M. Hall, *Projective Planes and Related Topics* (California Institute of Technology, Pasadena, 1954)
64. J.F.T. Hartney, *Radicals and antiradicals in near-rings*, Ph.D. Thesis, University of Nottingham University (1979)
65. H.E. Heatherly, One-sided ideals in near-rings of transformations. J. Aust. Math. Soc. **13** (1972)
66. I.N. Herstein, *Noncommutative Rings* (The Mathematical Association of America, 1968)
67. P.J. Hilton, U. Stammbach, *A Course in Homological Algebra* (Springer, Berlin, 1970)
68. J. Hocking, G. Young, *Topology* (Dover, New York, 1961)
69. W.M.L. Holcombe, Categorial representations of endomorphism near-rings. J. Lond. Math. Soc. (2) **16** (1977)
70. J.M. Howie, *Fundamentals of Semi-Group Theory* (Oxford Science Publications, Oxford, 1995)
71. T.W. Hungerford, *Algebra* (Springer, Berlin, 2003)
72. I.M. Isaacs, in *Finite Group Theory*. Graduate Studies in Mathematics, vol. 92 (American Mathematical Society, Providence, 2008)
73. B. Huppert, *Endliche Gruppen I* (Springer, Berlin, 1979)
74. B. Huppert, *Finite Groups II* (Springer, Berlin, 1982)
75. B. Huppert, N. Blackburn, *Finite Groups III* (Springer, Berlin, 1982)
76. N. Jacobson, in *Structure of Rings*. American Mathematical Society Colloquium Publications, vol. 37 (1956)
77. J.P. Jans, *Rings and Homology* (Holt, Rinehart & Winston, 1964)
78. M.J. Johnson, Near-ring with identities on dihedral groups. Proc. Ed. Math. Soc. **2**(18), (1973)
79. M.J. Johnson, Right ideals and right submodules of transformation near-rings. J. Algebra **24** (1973)
80. M.J. Johnson, Maximal right ideals of transformation near-rings. J. Aust. Math. Soc. **19** (1975)
81. D.L. Johnson, The group of formal power series under substitution. J. Aust. Math. Soc. **45**(3) (1988)
82. K.D Joshi, *Introduction to General Topology* (Wiley, London, 1993)
83. I. Kaplansky, *Commutative Rings* (University of Chicago Press, Chicago, 1974)
84. G. Karpilovsky, *The Jacobson Radical of Classical Rings* (Longman, London, 1991)
85. D. Keedwell, Construction, properties and applications of finite neofields. Comment. Math. Univ. Carolin. **41**(2), 283–297 (2000) [I found this article on the web]
86. D. Keedwell, J. Dénes, *Latin Squares and Their Applications*, 2nd edn. (North Holland, Amsterdam, 2015)
87. J.L. Kelley, *General Topology* (Ishi Press International, 2008)
88. B Klopsch, N. Nikolov, C. Voll, *Lectures on Profinite Topics in Group Theory* (Cambridge, 2011) [Edited by Dan Segal]
89. J. Krimmel, *Conditions on Near-Rings with Identity and the Near-Rings with Identity on Some Metacyclic Groups*. University of Arizona, Ph.D. Thesis (1970)
90. P.A. Krylov, A.V. Mikhalev, A.A. Tuganbaev, *Endomorphism Rings of Abelian Groups* (Kluwer, Dordecht, 2003)
91. A.G. Kurosh, *Lectures on General Algebra* (Chelsea Publishing Company, 1963)
92. A. G. Kurosh, *Theory of Groups*, vol. 1 (Chelsea Publishing Company, 1955)
93. A.G. Kurosh, *Theory of Groups*, vol. 2 (Chelsea Publishing Company, 1955)
94. T.Y. Lam, *A First Course in Noncommutative Rings* (Springer, Berlin, 2001)
95. P. Lancaster, M. Tismenetsky, *The Theory of Matrices*, 2nd edn. (Academic Press, New York, 1985)

96. S. Lang, *Algebra* (Revised Third Edition) (Springer, Berlin, 2002)
97. H. Lausch, W. Nöbauer, *Algebra of Polynomials* (North Holland, Amsterdam, 1973)
98. R.R. Laxton, Ph.D. Thesis, King's College, University of London (1961)
99. R.R. Laxton, Primitive distributively generated near-rings. Mathematika **8** (1961)
100. R.R. Laxton, R. Lockhart, The near-rings hosted by a class of groups. Proc. Ed. Math. Soc. **23**, 69–86 (1980)
101. The Linz bibliography of near-ring papers. www.algebra.uni-linz.ac.at/Nearrings/nrbibl.html
102. R. Lockhart, *The Near-Rings on a Class of Groups*. Ph.D. Thesis, Nottingham University (1977)
103. R. Lockhart, in *The Addition of Endomorphisms and Default Properties*. Tagungsbericht 16/1980: Fastringe und Fastkörper. Mathematisches Forschungsinstitut Oberwolfach
104. R. Lockhart, in *A Note on Non-Abelian Homological Algebra and Endomorphism Near-Rings*. Proc. Royal Soc. Edinburgh, Section A, vol. 92, Issue 1–2 (1982)
105. R. Lockhart, in *Products on Groups*. Contributions to General Algebra, vol. 8. (Hölder-Pichler-Tempsky, Wien, 1992)
106. R. Lockhart, in *Products on Products on Groups*. Near-Rings, Near-Fields and K-Loops, Hamburg (1995) (Kluwer Academic Publishers, Dordrecht, 1997)
107. A.W. Machin, *On a Class of Near-Rings*. Doctoral Dissertation, University of Nottingham (1971)
108. S. Maclane, *Categories for the Working Mathematician*, 2nd edn. (Springer, Berlin, 1998)
109. S. Maclane, *Homology* (Springer, Berlin, 1995)
110. S.J. Mahmood, M.F. Mansouri, in *Tensor Products of Near-Ring Modules*. Near-Rings, Near-Fields and K-Loops, Hamburg (1995) (Kluwer Academic Publishers, Dordrecht, 1997)
111. J.J. Malone, A non-abelian 2-group whose endomorphisms generate a ring and other examples of E-groups. Proc. Edinb. Math. Soc. **23** (1978)
112. J.J. Malone, C.G. Lyons, Endomorphism near-rings. Proc. Edinb. Math. Soc. **17** (1970)
113. J.J. Malone, C.G. Lyons, Finite dihedral groups and d.g. Near-Rings 1. Comp. Math. **26** (1973)
114. J.J. Malone, C.G. Lyons, Finite dihedral groups and d.g. near-rings 11. Comp. Math. **24**(3) (1972)
115. H.B. Mann, On certain systems that are almost groups. Bull. Am. Math. Soc. **50** (1944)
116. C.R.F. Maunder, *Algebraic Topology* (Cambridge University Press, Cambridge, 1970)
117. C.J. Maxson, Dickson near-rings. J. Algebra **14**, 152–169 (1970)
118. B.R. McDonald, *Finite Rings with Identity* (Marcel Dekker, New York, 1974)
119. J. Meldrum, in *Near-Rings and Their Links with Groups*. Research Notes in Mathematics, vol. 134 (Pitman, London, 1985)
120. J. Meldrum, in *Matrix Near-Rings*. Contributions to General Algebra, vol. 8 (Hölder-Pichler-Tempsky, Wien, 1992)
121. G. Michler, in *Theory of Finite Simple Groups*. New Mathematical Monographs, vol. 8 (Cambridge, 2006)
122. C.P. Milies, K.S. Sudarshan, *An Introduction to Group Rings* (Kluwer, Dordecht, 2002)
123. G. Mullen, C. Mummert, in *Finite Fields and Applications*. Student Mathematical Library, vol. 41 (American Mathematical Society, Providence, 2000)
124. H. Neumann, *Varieties of Groups*, Springer reprint of 1967 edition (2012)
125. P. Neumann, G. Stoy, E. Thompson, *Groups and Geometry* (Oxford, 1993)
126. D.S. Passman, *Permutation Groups* (Dover, New York, 2012)
127. D.S. Passman, *A Course in Ring Theory* (AMS Chelsea, 2004)
128. G. Pilz, *Near-Rings: The Theory and Its Application*, 2nd edn. (Elsevier, Amsterdam, 1983)
129. G. Pilz, F. Weber, W. Mueller, J.R. Schaefer, *Statistical Methods to Support Difficult Diagnoses*. Research Square (2020). https://www.researchsquare.com/article/rs-14046/v1
130. E. Rips, Y. Segev, K. Tent, A sharply 2-transitive group without a non-trivial, abelian, normal subgroup. J. Eur. Math. Soc. **19** (2017)
131. B. Henschenmacher, *Non-Associative Algebras and Quantum Physics*
132. D.J.S. Robinson, *A Course in the Theory of Groups*, 2nd edn. (Springer, Berlin, 1995)

133. T.G. Room, P.B. Kirkpatrick, *Miniquaternion Geometry* (Cambridge University Press, Cambridge, 1971)
134. J.J. Rotman, *An Introduction to the Theory of Groups*, 4th edn. (Springer, Berlin, 1999)
135. L.H. Rowen, *Ring Theory*, 2nd edn. (Academic Press, London, 1991)
136. R. Schafer, *An Introduction to Nonassociative Algebras* (Dover, New York, 1995)
137. W.R. Scott, *Group Theory* (Dover, New York, 1987)
138. K. Smith, L. Kahanpää, P. Kekäläinen, W. Traves, *An Invitation to Algebraic Geometry* (Springer, Berlin, 2000)
139. M. Suzuki, *Group Theory 1 & 2* (Springer, Berlin, 1982, 1986)
140. O. Veblen, J.H.M. Wedderburn, Non-Desarguesian and non-Pascalian geometries. Trans. Am. Math. Soc. **8**, 379–388 (1907)
141. H. Wähling, *Theorie der Fastkörper* (Thales Verlag Essen, 1987)
142. C. Weibel, in *Survey of Non-Desarguesian Planes*. Notices of the AMS, vol. 54(10) (2007)
143. H. Wielandt, *Finite Permutation Groups* (Academic Press, London, 1964)
144. J.A. Wolf, *Spaces of Constant Curvature* (AMS Chelsea, 2000)
145. H. Zassenhaus, *The Theory of Groups* (Chelsea, 1949)

Index

R. Lockhart, *The Theory of Near-Rings*, Lecture Notes in Mathematics 2295,
https://doi.org/10.1007/978-3-030-81755-8